AF580875

BIRKHÄUSER

Operator Theory: Advances and Applications

Vol. 189

New Developments in Pseudo-Differential Operators

ISAAC Group in Pseudo-Differential Operators (IGPDO),
Middle East Technical University,
Ankara,Turkey, August 2007

Luigi Rodino
M.W. Wong
Editors

Birkhäuser
Basel · Boston · Berlin

Editors:

Luigi Rodino
Dipartimento di Matematica
Università di Torino
Via Carlo Alberto, 10
10123 Torino
Italy
e-mail: luigi.rodino@unito.it

M.W. Wong
Department of Mathematics and Statistics
York University
4700 Keele Street
Toronto, Ontario M3J 1P3
Canada
e-mail: mwwong@mathstat.yorku.ca

2000 Mathematical Subject Classification: Primary 22A10, 32A40, 32A45, 35A17, 35A22, 35B05, 35B40, 35B60, 35J70, 35K05, 35K65, 35L05, 35L40, 35S05, 35S15, 35S30, 43A77, 46F15, 47B10, 47B35, 47B37, 47G10, 47G30, 47L15, 58J35, 58J40, 58J50, 65R10, 92A55, 94A12; Secondary 22C05, 30E25, 35G05, 35H10, 35J05, 42B10, 42B35, 47A10, 47A53, 47F05, 58J20, 92C55, 94A12

Library of Congress Control Number: 2008936167

Bibliographic information published by Die Deutsche Bibliothek.
Die Deutsche Bibliothek lists this publication in the Deutsche Nationalbibliografie; detailed bibliographic data is available in the Internet at http://dnb.ddb.de

ISBN 978-3-7643-8968-0 Birkhäuser Verlag AG, Basel - Boston - Berlin

Basel · Boston · Berlin
P.O. Box 133, CH-4010 Basel, Switzerland
Part of Springer Science+Business Media
Printed on acid-free paper produced from chlorine-free pulp. TCF∞
Printed in Germany
ISBN 978-3-7643-8968-0
9 8 7 6 5 4 3 2 1

e-ISBN 978-3-7643-8969-7
www.birkhauser.ch

Contents

Preface vii

M. de Gosson
Phase-Space Weyl Calculus and Global Hypoellipticity of a Class of Degenerate Elliptic Partial Differential Operators 1

T. Gramchev, S. Pilipović and L. Rodino
Classes of Degenerate Elliptic Operators in Gelfand–Shilov Spaces 15

A. Dasgupta and M.W. Wong
Weyl Transforms and the Heat Equation for the Sub-Laplacian on the Heisenberg Group 33

C. Iwasaki
Construction of the Fundamental Solution and Curvature of Manifolds with Boundary 43

J. Abed and B.-W. Schulze
Operators with Corner-Degenerate Symbols 67

A. Dasgupta
Ellipticity of Fredholm Pseudo-Differential Operators on $L^p(\mathbb{R}^n)$ 107

M. Oberguggenberger
Hyperbolic Systems with Discontinuous Coefficients: Generalized Wavefront Sets 117

C. Garetto
Generalized Fourier Integral Operators on Spaces of Colombeau Type 137

M. Ruzhansky
On Local and Global Regularity of Fourier Integral Operators 185

J. Johnsen
Type 1,1-Operators Defined by Vanishing Frequency Modulation 201

G. Garello and A. Morando
Regularity for Quasi-Elliptic Pseudo-Differential Operators with Symbols in Hölder Classes 247

C. Bouzar and A. Dali
Multi-Anisotropic Gevrey Regularity of Hypoelliptic Operators 265

Q. Guo, S. Molahajloo and M.W. Wong
Modified Stockwell Transforms and Time-Frequency Analysis 275

Y. Liu
Localization Operators for Two-Dimensional Stockwell Transforms ... 287

S. Molahajloo and M.W. Wong
Pseudo-Differential Operators on $\mathbb{S}^1$ 297

M. Ruzhansky and V. Turunen
On Pseudo-Differential Operators on Group SU(2) 307

A. Mohammed and M.W. Wong
Sampling and Pseudo-Differential Operators 323

Operator Theory:
Advances and Applications, Vol. 189, vii–viii

Preface

On the occasion of the Sixth Congress of the International Society for Analysis, its Applications and Computation (ISAAC), the ISAAC Group in Pseudo-Differential Operators (IGPDO) met again at the Middle East Technical University in Ankara, Turkey on August 13–18, 2007. Papers related to the talks given at the special session on pseudo-differential operators and invited papers from experts in the field are peer-reviewed for inclusion in this volume.

This volume consists of seventeen papers on pseudo-differential operators and related topics. Results on twisted Laplacians and their extensions, and the sub-Laplacian on the Heisenberg group are arranged together in the first three papers. Global analysis on manifolds with boundaries, pseudo-differential analysis on manifolds with singularities, and the ellipticity and Fredholmness of pseudo-differential operators are presented in the next three papers. Following these are three papers on hyperbolic equations and systems, Fourier integral operators and Colombeau algebras. Then there are three papers on exotic pseudo-differential operators and regularity results on quasi-elliptic operators and hypoelliptic operators. Two papers on the Stockwell transforms in time-frequency analysis are then given. The last three papers are devoted to pseudo-differential operators on the unit circle, the torus, SU(2) and related results on sampling.

This volume should be useful to graduate students and researchers in mathematical analysis with applications in sciences and engineering. It is a valuable complement to the volumes "Advances in Pseudo-Differential Operators", "Pseudo-Differential Operators and Related Topics" and "Modern Trends in Pseudo-Differential Operators" published in the same series edited by Professor Israel Gohberg in, respectively, 2004, 2006 and 2007. It should also be read in conjunction with the volume "Pseudo-Differential Operators: Partial Differential Equations and Time-Frequency Analysis", Editors: Luigi Rodino, Bert-Wolfgang Schulze and M.W. Wong, Fields Institute Communications Series 57, American Mathematical Society, 2007 and the volume "Pseudo-Differential Operators: Quantization and Signals", Editors: Luigi Rodino and M.W. Wong, Fondazione C.I.M.E., Firenze, Lecture Notes in Mathematics 1949, Springer, 2008.

In an era of explosive advances in quantitative sciences and information technologies, it is envisaged that mathematical analysis in general and pseudo-differential operators in particular will be developed hand in glove with applications and computation in the physical, biological and medical sciences. This theme will play an important role in the forthcoming volumes on pseudo-differential operators originating from IGPDO.

The Editors

Operator Theory:
Advances and Applications, Vol. 189, 1–14

Phase-Space Weyl Calculus and Global Hypoellipticity of a Class of Degenerate Elliptic Partial Differential Operators

Maurice de Gosson

Abstract. In a recent series of papers M.W. Wong has studied a degenerate elliptic partial differential operator related to the Heisenberg group. It turns out that Wong's example is best understood when replaced in the context of the phase-space Weyl calculus we have developed in previous work; this approach highlights the relationship of Wong's constructions with the quantum mechanics of charged particles in a uniform magnetic field. Using Shubin's classes of pseudodifferential symbols we prove global hypoellipticity results for arbitrary phase-space operators arising from elliptic operators on configuration space.

Mathematics Subject Classification (2000). Primary 47F30; Secondary 35B65, 46F05.

Keywords. Degenerate elliptic operators, hypoellipticity, phase space Weyl calculus, Shubin symbols.

Introduction

In a recent series of interesting papers [25, 26, 27] M.W. Wong (also see Dasgupta and Wong [5]) discusses various properties of the partial differential operator

$$W = -\frac{1}{2}(Z\overline{Z} + \overline{Z}Z) \tag{1}$$

where Z and $\overline{Z}$ are the vector fields on $\mathbb{R}^2$ defined by

$$Z = \frac{\partial}{\partial z} + \frac{1}{2}z \quad , \quad \overline{Z} = \frac{\partial}{\partial \overline{z}} + \frac{1}{2}\overline{z}. \tag{2}$$

Writing $z = x + iy$ the operator W has the explicit form

$$W = -\Delta - i\left(x\frac{\partial}{\partial y} - y\frac{\partial}{\partial x}\right) + \frac{1}{4}(x^2 + y^2) \tag{3}$$

where Δ is the usual Laplace operator in the x, y variables. It turns out that Wong's results are only the "tip of an iceberg" because the operator (3) can be viewed as the phase-space version of the Hermite operator $-\Delta + x^2$ obtained by using a quantization procedure we have introduced in previous work. To understand this, let us make the two following independent observations:

- The operator W has a well-known meaning in physics. Consider in fact an electron with mass m and charge e placed in a strong uniform magnetic field $\mathbf{B}$ directed along the z axis: $\mathbf{B} = (0, 0, B_z)$. The Hamiltonian operator is, in a particular choice of gauge (see Section 2),
$$H = -\frac{\hbar^2}{2m}\Delta + i\hbar\omega_L\left(y\frac{\partial}{\partial x} - x\frac{\partial}{\partial y}\right) + \frac{m\omega_L^2}{2}(x^2 + y^2) \tag{4}$$
where ω_L is the "Larmor frequency". This operator reduces to Wong's operator (3) if x and y are swapped and units are chosen so that $\hbar = 1$, $m = 1/2$, and $\omega_L = 1$; the spectrum of the operator (4) is well known; it consists of the sequence "energy levels" $E_N = (2N+1)\hbar\omega_L$ (see Subsection 2.1); the spectrum of W is thus given by the sequence of numbers, $E_N = 2N + 1$ for $N = 0, 1, \ldots$, a fact which Wong rediscovers in [25] using complicated calculations involving the Wigner formalism and special function theory.
- On the other hand, a straightforward calculation shows that we can rewrite the operator W more compactly as
$$W = \left(-i\frac{\partial}{\partial x} - \frac{1}{2}y\right)^2 + \left(-i\frac{\partial}{\partial y} + \frac{1}{2}x\right)^2; \tag{5}$$
setting $p = -y$ this operator becomes
$$W = \left(-i\frac{\partial}{\partial x} + \frac{1}{2}p\right)^2 + \left(i\frac{\partial}{\partial p} + \frac{1}{2}x\right)^2 \tag{6}$$
which makes apparent that W is obtained from the harmonic oscillator Hamiltonian $H(x,p) = \frac{1}{2}(p^2 + x^2)$ using the "quantization rules"
$$x \longrightarrow X = i\frac{\partial}{\partial p} + \frac{1}{2}x \ , \ p \longrightarrow P = -i\frac{\partial}{\partial x} + \frac{1}{2}p \tag{7}$$
we have studied and exploited in [10, 11] in connection with the phase space Schrödinger equation of Torres-Vega and Frederick [20, 21]; notice that X and P satisfy the same commutation relation $[X, P] = i$ as satisfied by the operators x and $-i\partial/\partial x$.

This paper consists of two parts. In the first part (Sections 1–2) we will focus on the "phase space Weyl calculus" aspect of Wong's operator and its generalizations; we also briefly review our previous definitions and results from [10, 11]. In the second part of this paper (Section 3) we show that our phase-space Weyl calculus together with Shubin's pseudodifferential calculus (Shubin [19], Chapter IV) allows us not only to recover (in a trivial way) the global hypoellipticity of Wong's operator W, but to prove that every operator

$$\widetilde{A} = a\left(i\frac{\partial}{\partial p} + \frac{1}{2}x, -i\frac{\partial}{\partial x} + \frac{1}{2}p\right)$$

obtained by replacing formally (x,p) by $(i\frac{\partial}{\partial p} + \frac{1}{2}x, -i\frac{\partial}{\partial x} + \frac{1}{2}p)$ in any positive-definite quadratic form $a = a(x,p)$ is globally hypoelliptic (we will in fact prove a $2n$-dimensional statement, allowing the symbol a to be defined on $\mathbb{R}^{2n}$).

Notation

We denote by $\mathcal{S}(\mathbb{R}^n)$ the Schwartz space of all smooth complex-valued functions on $\mathbb{R}^n$ which decrease, together with their derivatives, faster than the inverse than any polynomial when $|x| \to \infty$. The dual $\mathcal{S}'(\mathbb{R}^n)$ of $\mathcal{S}(\mathbb{R}^n)$ is the space of tempered distributions on $\mathbb{R}^n$.

Operators acting on functions (or distributions) defined on $\mathbb{R}^n$ will be denoted by capital letters $A, B, C, \ldots$ while operators acting on functions (or distributions) defined on $\mathbb{R}^n$ defined on the symplectic space $(\mathbb{R}^{2n}, \sigma)$ will denoted by covering capital letters with a tilde: $\widetilde{A}, \widetilde{B}, \widetilde{C}, \ldots$ Functions on $\mathbb{R}^n$ will usually be denoted by lower-case Greek letters $\psi, \phi, \ldots$ while functions on $\mathbb{R}^{2n}$ are denoted by upper-case Greek letters $\Psi, \Phi, \ldots$ We will use standard multi-index notation: if $\alpha = (\alpha_1, \ldots, \alpha_n)$ is a sequence of non-negative integers we write $x^\alpha = x_1^{\alpha_1} \cdots x_n^{\alpha_n}$ if $x = (x_1, \ldots, x_n)$, and $D_x^\alpha = (-i)^{|\alpha|}\partial_{x_1}^{\alpha_1} \cdots \partial_{x_n}^{\alpha_n}$ with $|\alpha| = \alpha_1 + \cdots + \alpha_n$.

We will denote by σ the standard symplectic form on the vector space $\mathbb{R}^n \times \mathbb{R}^n \equiv \mathbb{R}^{2n}$:

$$\sigma(z, z') = p \cdot x' - p' \cdot x$$

for $z = (x,p)$, $z' = (x',p')$. The symplectic group of $(\mathbb{R}^{2n}, \sigma)$ will be denoted by $\mathrm{Sp}(n)$: it is the group of all linear automorphisms s of $\mathbb{R}^{2n}$ such that $\sigma(sz, sz') = \sigma(z, z')$ for all $z, z' \in \mathbb{R}^{2n}$.

1. Phase-space Weyl calculus

For proofs and a detailed exposition we refer to de Gosson [11] (the phase space calculus was introduced in de Gosson [9, 10] following a suggestion in the paper by Grossmann *et al.* [14]).

1.1. Definitions

Let $A : \mathcal{S}(\mathbb{R}^n) \longrightarrow \mathcal{S}'(\mathbb{R}^n)$ be a continuous linear operator. In view of Schwartz's kernel theorem there exists a distribution $K \in \mathcal{S}'(\mathbb{R}^n \times \mathbb{R}^n)$ such that

$$Af(x) = \int_{\mathbb{R}^n} K(x,y)f(y)dy$$

(the integral being interpreted as a partial distribution bracket). The Weyl symbol of A is by definition (see, *e.g.*, [11], Theorem 6.12, or Wong's book [24]) the tempered distribution a given by

$$a(x,p) = \int_{\mathbb{R}^n} e^{-ip\cdot y} K(x + \tfrac{1}{2}y, x - \tfrac{1}{2}y) dy.$$

Defining the twisted Weyl symbol a_σ as being the symplectic Fourier transform of a, that is

$$a_\sigma(z) = \left(\frac{1}{2\pi}\right)^n \int_{\mathbb{R}^{2n}} e^{-i\sigma(z,z')} a(z) dz$$

the operator A is given by the Bochner integral

$$A = \left(\frac{1}{2\pi}\right)^n \int_{\mathbb{R}^{2n}} a_\sigma(z_0) T(z_0) dz_0; \tag{8}$$

here $T(z)$ is the Heisenberg–Weyl operator defined by

$$T(z_0)\psi(x) = e^{i(p_0\cdot x - \frac{1}{2}p_0\cdot x_0)} \psi(x - x_0) \tag{9}$$

if $z_0 = (x_0, p_0)$. The operator A can be (at least formally) shown to act on $\psi \in \mathcal{S}(\mathbb{R}^n)$ by the formula

$$A\psi(x) = \left(\frac{1}{2\pi}\right)^n \int_{\mathbb{R}^n \times \mathbb{R}^n} e^{ip\cdot(x-y)} a\left(\frac{1}{2}(x+y), p\right) \psi(y) dp dy.$$

We now associate to A an operator $\widetilde{A} : \mathcal{S}(\mathbb{R}^{2n}) \longrightarrow \mathcal{S}'(\mathbb{R}^{2n})$ by the formula

$$\widetilde{A} = \left(\frac{1}{2\pi}\right)^n \int_{\mathbb{R}^{2n}} a_\sigma(z_0) \widetilde{T}(z_0) dz_0 \tag{10}$$

where $\widetilde{T}(z_0)$ acts on $\mathcal{S}'(\mathbb{R}^{2n})$ via

$$\widetilde{T}(z_0)\Psi(z) = e^{-\frac{i}{2}\sigma(z,z_0)} \Psi(z - z_0). \tag{11}$$

Observe that the operators $\widetilde{T}(z_0)$ satisfy the same canonical commutation relations as the Heisenberg–Weyl operators $T(z_0)$ namely

$$\widetilde{T}(z_1)\widetilde{T}(z_0) = e^{i\sigma(z_1,z_0)} \widetilde{T}(z_0)\widetilde{T}(z_1) \tag{12}$$

hence they correspond as we will see below to some unitary representation of the Heisenberg (not on $L^2(\mathbb{R}^n)$ but on a closed subspace of $L^2(\mathbb{R}^{2n})$).

For $\phi \in \mathcal{S}(\mathbb{R}^n)$, $||\phi||_{L^2} = 1$, we define an operator $U_\phi : L^2(\mathbb{R}^n) \longrightarrow L^2(\mathbb{R}^{2n})$ by the formula

$$U_\phi\psi(z) = \left(\frac{\pi}{2}\right)^{n/2} W(\psi,\phi)(\tfrac{1}{2}z) \tag{13}$$

where $W(\psi,\phi)$ is the cross-Wigner transform of the pair (ψ,ϕ):

$$W(\psi,\phi)(x,p) = \left(\frac{1}{2\pi}\right)^n \int_{\mathbb{R}^n} e^{-ip\cdot y} \psi(x + \tfrac{1}{2}y) \overline{\phi(x - \tfrac{1}{2}y)} dy.$$

We will call U_ϕ the *wave-packet transform with window* ϕ; is essentially the short-time Fourier transform V_ϕ used in time-frequency analysis and defined by

$$V_\phi(x,\omega) = \int_{\mathbb{R}^n} e^{-2\pi i\omega\cdot t} \psi(t) \overline{\phi(t - x)} dt.$$

When ϕ is a Gaussian both U_ϕ and V_ϕ are closely related to the Bargmann transform [1].

We have (de Gosson [11], Theorem 10.6 p. 312):

Theorem 1. *Let* $\phi \in \mathcal{S}(\mathbb{R}^n)$. *Then:*

(i) U_ϕ *is an isometry of* $L^2(\mathbb{R}^n)$ *on a closed subspace* $\mathcal{H}_\phi$ *of* $L^2(\mathbb{R}^{2n})$;

(ii) *We have* $U_\phi^* U_\phi = I$ *on* $L^2(\mathbb{R}^n)$ *and the operator* $P_\phi = U_\phi U_\phi^*$ *is the orthogonal projection in* $L^2(\mathbb{R}^{2n})$ *onto the space* $\mathcal{H}_\phi$;

(iii) *The intertwining formulae*

$$\widetilde{T}(z_0)U_\phi = U_\phi T(z_0) \quad , \quad \widetilde{A}U_\phi = U_\phi A \tag{14}$$

hold for all $\phi \in \mathcal{S}(\mathbb{R}^n)$.

In particular (iii) implies that we have

$$\left(\tfrac{1}{2}x_j + i\frac{\partial}{\partial p_j}\right) U_\phi\psi = U_\phi(x_j\psi) \tag{15a}$$

$$\left(\tfrac{1}{2}p_j - i\frac{\partial}{\partial x_j}\right) U_\phi\psi = U_\phi(-i\frac{\partial}{\partial x_j}\psi) \tag{15b}$$

for all $\psi \in \mathcal{S}(\mathbb{R}^n)$. This motivates the following notation: if $\widetilde{A}$ is the phase-space operator with symbol a we will write

$$\widetilde{A} = a(\tfrac{1}{2}x + i\tfrac{\partial}{\partial p}+, \tfrac{1}{2}p - i\tfrac{\partial}{\partial x}).$$

1.2. The composition property

We will need the following composition property:

Proposition 2. (i) *Assume that the compose* AB *of the Weyl operators exists and is a Weyl operator. Then* $\widetilde{A}\widetilde{B}$ *exists as well and we have* $\widetilde{A}\widetilde{B} = \widetilde{AB}$.

(ii) *If the Weyl operator* R *has kernel* $K_R \in \mathcal{S}(\mathbb{R}^n \times \mathbb{R}^n)$ *then* $\widetilde{R}$ *has kernel* $K_{\widetilde{R}} \in \mathcal{S}(\mathbb{R}^{2n} \times \mathbb{R}^{2n})$.

Proof. (i) See de Gosson [11], Proposition 10.13, p. 320. (ii) The Weyl symbol r of R is related to the kernel K_R of R by the formula

$$r(x,p) = \int_{\mathbb{R}^n} e^{-ip\cdot y} K(x + \tfrac{1}{2}y, x - \tfrac{1}{2}y)dy$$

hence $r \in \mathcal{S}(\mathbb{R}^{2n})$ if $K_R \in \mathcal{S}(\mathbb{R}^n \times \mathbb{R}^n)$ from which follows that we also have $r_\sigma \in \mathcal{S}(\mathbb{R}^{2n})$ where r_σ is the symplectic Fourier transform of r. We have, using (10),

$$\widetilde{R}\Psi(z) = \left(\frac{1}{2\pi}\right)^n \int_{\mathbb{R}^{2n}} r_\sigma(z_0)e^{-\frac{i}{2}\sigma(z,z_0)}\Psi(z - z_0)dz_0$$

that is, setting $u = z - z_0$:

$$\widetilde{R}\Psi(z) = \left(\frac{1}{2\pi}\right)^n \int_{\mathbb{R}^{2n}} r_\sigma(z - u)e^{-\frac{i}{2}\sigma(z,z-u)}\Psi(u)du.$$

Since $\sigma(z, z-u) = -\sigma(z, u)$ the kernel of $\widetilde{R}$ is thus given by the formula

$$K_{\widetilde{R}}(z,u) = \left(\frac{1}{2\pi}\right)^n e^{\frac{i}{2}\sigma(z,u)} r_\sigma(z-u). \tag{16}$$

The function $(z,u) \longmapsto r_\sigma(z-u)$ being in $\mathcal{S}(\mathbb{R}^{2n} \times \mathbb{R}^{2n})$ so is $K_{\widetilde{R}}$. □

1.3. Symplectic covariance

Since we are in the business of Weyl operators, let us study the symplectic covariance properties of the corresponding phase space operators. Recall that the symplectic group $\mathrm{Sp}(n)$ has a double covering which can be faithfully represented by a group of unitary operators acting on $L^2(\mathbb{R}^{2n})$; that group is called the metaplectic group and we will denote it by $\mathrm{Mp}(n)$. The standard "metaplectic covariance formula" for Weyl calculus reads as follows: for $s \in \mathrm{Sp}(n)$ let S be any of the two operators in $\mathrm{Mp}(n)$ corresponding to s. Then if A has Weyl symbol a the operator SAS^{-1} has Weyl symbol $a \circ s^{-1}$. In de Gosson [8] we proved that if s has no eigenvalue equal to one, then

$$S = \left(\frac{1}{2\pi}\right)^n \frac{i^{\pm\nu}}{\sqrt{|\det(s-I)|}} \int_{\mathbb{R}^{2n}} \exp\left[\frac{i}{2} M_s z^2\right] T(z) dz \tag{17}$$

where $M_S = M_s^T$ is the symplectic Cayley transform defined by

$$M_S = \tfrac{1}{2} J(s+I)(s-I)^{-1}$$

and ν an integer (the "Conley–Zehnder index") that need not preoccupy us here; in addition we showed that every $S \in \mathrm{Mp}(n)$ can be written as the product of exactly two operators of the type above. In view of formula (10) the operator S determines naturally a phase-space operator

$$\widetilde{S} = \left(\frac{1}{2\pi}\right)^n \frac{i^{\pm\nu}}{\sqrt{|\det(s-I)|}} \int_{\mathbb{R}^{2n}} \exp\left[\frac{i}{2} M_s z^2\right] \widetilde{T}(z) dz \tag{18}$$

satisfying the second intertwining relation (14) in Theorem 1.

Proposition 3. *Let $s \in \mathrm{Sp}(n)$ and*

$$\widetilde{A} = a(\tfrac{1}{2}x + i\tfrac{\partial}{\partial p}, \tfrac{1}{2}p - i\tfrac{\partial}{\partial x})$$
$$\widetilde{B} = (a \circ s^{-1})(\tfrac{1}{2}x + i\tfrac{\partial}{\partial p}, \tfrac{1}{2}p - i\tfrac{\partial}{\partial x}).$$

We have $\widetilde{B} = \widetilde{S}\widetilde{A}\widetilde{S}^{-1}$ where $\widetilde{S} \in \mathrm{Mp}(n)$ is any of the two metaplectic operators corresponding to s.

Proof. It is an immediate consequence of the usual symplectic covariance formula

$$(a \circ s^{-1})(x, -i\tfrac{\partial}{\partial x}) = Sa(x, -i\tfrac{\partial}{\partial x})S^{-1}$$

for Weyl operators (see de Gosson [11], Chapter 10, §10.3.3). Alternatively, it follows from the metaplectic covariance relation $\widetilde{S}\widetilde{T}(z)\widetilde{S}^{-1} = \widetilde{T}(sz)$. □

2. Alternative quantizations

2.1. Some physical considerations

For the physical background we refer the reader to Landau and Lifshitz [17] or Messiah [18]. Consider an electron placed in a strong uniform magnetic field $\mathbf{B}$ directed along the z axis: $\mathbf{B} = (0, 0, B_z)$; if $\mathbf{A}$ is a vector potential defined by $\mathbf{B} = \nabla_{\mathbf{r}} \times \mathbf{A}$ ($\mathbf{r} = (x, y, z)$) the Hamiltonian function is

$$\mathcal{H} = \frac{1}{2m}\left(\mathbf{p} - \frac{e}{c}\mathbf{A}\right)^2 \tag{19}$$

with $\mathbf{p} = (p_x, p_y, p_z)$. Choosing the vector potential $\mathbf{A}$ such that $\mathbf{A} = \frac{1}{2}(\mathbf{r} \times \mathbf{B})$ ("symmetric gauge") and disregarding the unessential component p_z, the Hamiltonian H takes the particular form

$$\mathcal{H}_{\mathrm{sym}} = \frac{1}{2m}(p_x^2 + p_y^2) - \omega_L L_z + \frac{m\omega_L^2}{2}(x^2 + y^2)$$

with $\omega_L = eB_z/2mc$ ("Larmor frequency") and $L_z = xp_y - yp_x$ is the angular moment in the z direction. The corresponding quantum operator is given by

$$H_{\mathrm{sym}} = -\frac{\hbar^2}{2m}\Delta + i\hbar\omega_L\left(y\frac{\partial}{\partial x} - x\frac{\partial}{\partial y}\right) + \frac{m\omega_L^2}{2}(x^2 + y^2) \tag{20}$$

where Δ is the Laplacian in the x, y variables. As already observed in the Introduction this operator reduces to Wong's operator (3) if units are chosen appropriately.

Suppose now we choose the vector potential as $\mathbf{A} = (-B_z y, 0, 0)$ (this is called the "Landau gauge" in Physics); then the Hamiltonian function takes the simple form

$$\mathcal{H}_{\mathrm{Lan}} = \frac{1}{2m}\left[\left(p_x + \frac{eB_z}{c}y\right)^2 + p_y^2\right]$$

and the corresponding the quantum operator is

$$H_{\mathrm{Lan}} = -\frac{\hbar^2}{2m}\Delta - i\hbar\omega y\frac{\partial}{\partial x} + \frac{1}{2}m\omega^2 y^2 \tag{21}$$

where $\omega = eB_z/2mc$ is the "cyclotron frequency". It is easy to determine the spectrum of H_{Lan} (which is the same as that of H_{sym} since a change of gauge does not affect the spectrum). Noticing that x does not appear explicitly in the function H_{Lan} the momentum p_x is thus a conserved quantity; setting $\psi(x, y) = e^{\frac{i}{\hbar}p_x x}\phi(y)$ it is easy to check that the eigenvalue problem $H_{\mathrm{Lan}}\psi = E\psi$ reduces to

$$-\frac{\hbar^2}{2m}\frac{d^2\phi}{dy^2} + \frac{1}{2}m\omega^2(y - y_0)^2 = E\psi$$

where $y_0 = -p_x c/eB_z$; but this is just the eigenvalue problem for a translated harmonic oscillator with mass m and frequency ω, whose spectrum consists of the sequence

$$E_N = (N + \tfrac{1}{2})\hbar\omega \quad , \quad N = 0, 1, 2, \ldots.$$

Choosing appropriate units, we recover the spectrum of Wong's operator W as announced in the Introduction.

It turns out that the harmonic oscillator operator ("Hermite operator")

$$H_{\mathrm{Her}} = \frac{\hbar^2}{2m}\frac{\partial^2}{\partial x^2} + \frac{1}{2}m\omega^2 x^2 \tag{22}$$

as well as the operators H_{sym} and H_{Lan} are all obtained from the Hamiltonian function

$$\mathcal{H} = \frac{1}{2m}p^2 + \frac{1}{2}m\omega^2 x^2$$

by applying different quantization rules:

- H_{Her} corresponds to the standard prescription $(x,p) \longrightarrow (x, -i\hbar\frac{\partial}{\partial x})$;
- H_{sym} corresponds to the rule $(x,p) \longrightarrow (i\hbar\frac{\partial}{\partial p} + \frac{1}{2}x, -i\hbar\frac{\partial}{\partial x} + \frac{1}{2}p)$;
- H_{Lan} corresponds to the rule $(x,p) \longrightarrow (i\hbar\frac{\partial}{\partial p} + \frac{1}{2}x, -i\hbar\frac{\partial}{\partial x})$.

Let us generalize this discussion to more general operators.

2.2. Extension; Bopp quantization

As seen above, there is a certain arbitrariness in the definition of our phase-space Weyl operator $\widetilde{A}$ associated with A. If we replace the "quantum translation operator" $\widetilde{T}(z_0)$ defined by (11) by any operator satisfying the canonical commutation relations (12) we will obtain another operator having similar properties as $\widetilde{A}$. Assume for instance that we define (choosing units in which $\hbar = 1$),

$$\widetilde{T}'(z_0)\Psi(z) = e^{i(p_0\cdot x - \frac{1}{2}p_0\cdot x_0)}\Psi(z - z_0)$$

which amounts to extending the Heisenberg–Weyl operator (9) in a trivial way by allowing them to act on phase-space functions. The corresponding phase-space Weyl operator is obtained by replacing definition (10) by the expression

$$\widetilde{A'} = \left(\frac{1}{2\pi}\right)^n \int_{\mathbb{R}^{2n}} a_\sigma(z_0)\widetilde{T}'(z_0)dz_0.$$

This choice corresponds to the quantization rules

$$x_j \longrightarrow X = i\frac{\partial}{\partial p_j} + \frac{1}{2}x_j \ , \ p_j \longrightarrow -i\frac{\partial}{\partial x_j} \tag{23}$$

studied in de Gosson [9]; setting $y = -p$ this quantizes the harmonic oscillator Hamiltonian into the operator

$$W' = -\Delta + x^2 - 2ix\frac{\partial}{\partial y} \tag{24}$$

in place of Wong's operator (3) (see de Gosson [10] for a discussion of other possible quantization compatible with the canonical commutation relations, and their interpretation in terms of classical phases).

It is noteworthy that the quantization rules $(x,p) \longrightarrow (X,P)$ listed above obey the canonical commutation rules $[X,P] = i\hbar$ and thus correspond to different (but of course isomorphic) representations of the Heisenberg group (the rule $(x,p) \longrightarrow (x, -i\hbar\frac{\partial}{\partial x})$ corresponds to the usual Schrödinger representation). There are of course other choices. One easily verifies that for any quadruple of real numbers $(\alpha,\beta,\gamma,\delta)$ such that $\alpha\delta - \beta\gamma = 1$ the operators

$$X = \alpha x + i\beta\hbar\frac{\partial}{\partial p} \quad , P = \gamma p + i\delta\hbar\frac{\partial}{\partial x} \tag{25}$$

satisfy $[X,P] = i\hbar$ and therefore define a *bona fide* quantization rule for which the statements in Theorem 1 remain true for an adequate redefinition $U_\phi^{(\alpha,\beta,\gamma,\delta)}$ of the transform U_ϕ for which the intertwining relations (14) should be replaced by

$$\left(\alpha x + i\beta\hbar\frac{\partial}{\partial p}\right) U_\phi^{(\alpha,\beta,\gamma,\delta)}\psi = U_\phi^{(\alpha,\beta,\gamma,\delta)}(x_j\psi)$$

$$\left(\gamma p + i\delta\hbar\frac{\partial}{\partial x}\right) U_\phi^{(\alpha,\beta,\gamma,\delta)}\psi = U_\phi^{(\alpha,\beta,\gamma,\delta)}(-i\hbar\tfrac{\partial}{\partial x_j}\psi).$$

Applying the rules (25) to the harmonic oscillator Hamiltonian one obtains a whole class of degenerate elliptic operators:

$$\widetilde{A}^{(\alpha,\beta,\gamma,\delta)} = -\frac{1}{2}\left(\delta^2\frac{\partial^2}{\partial x^2} + \beta^2\frac{\partial^2}{\partial p^2}\right) - i\left(\alpha\beta x\frac{\partial}{\partial p} - \delta\gamma p\frac{\partial}{\partial x}\right) + \frac{1}{4}(\alpha^2x^2 + \gamma^2p^2).$$

We notice that the choice $\alpha = \gamma = 1$, $\beta = -\delta = 1/2$ yields the so-called "Bopp quantization" (Bopp [3]) rules

$$X_{\mathrm{Bopp}} = x + i\frac{\hbar}{2}\frac{\partial}{\partial p} \quad , P_{\mathrm{Bopp}} = p - i\frac{\hbar}{2}\frac{\partial}{\partial x}$$

which play an important role in deformation quantization. Still, physically, the phase-space quantization choice

$$\widetilde{X} = \frac{1}{2}x + i\hbar\frac{\partial}{\partial p} \quad , \quad \widetilde{P} = \frac{1}{2}p - i\hbar\frac{\partial}{\partial x} \tag{26}$$

has particular symmetry properties which makes it more attractive; it seems to play a role in the study of quantum gravity (Isidro and de Gosson [15, 16]).

3. Hypoellipticity in the Schwartz space $\mathcal{S}(\mathbb{R}^n)$

Let us now study the question of global hypoellipticity for a class of phase-space operators generalizing those of Wong.

3.1. Global hypoellipticity and Shubin symbols

Let A be a partial differential operator (or more generally, a pseudodifferential operator). One says that A is *hypoelliptic* (in the usual sense) if $A\psi \in C^\infty(\mathbb{R}^n)$ implies that $\psi \in C^\infty(\mathbb{R}^n)$. Recall that the partial differential operator

$$A(x,D) = \sum_{|\alpha|\leq m} a_\alpha(x)D_x^\alpha \quad , \quad a_\alpha \in C^\infty(\mathbb{R}^n)$$

(or, more generally, a classical pseudodifferential operator) is said to be *elliptic* if its principal symbol

$$a_m(x,p) = \sum_{|\alpha|=m} a_\alpha(x)p^\alpha$$

has the property that $a_m(z) = 0$ if and only if $z = 0$. An elliptic operator is hypoelliptic (in the usual sense), as is easily seen by constructing an approximate inverse, or *parametrix* (see for instance Shubin [19] or Trèves [22]). More precisely B (resp. B') is called a left (resp. right) parametrix if

$$BA = I + R \quad (\text{resp. } AB' = I + R')$$

where R and R' are smoothing operators, that is $R, R' : \mathcal{S}'(\mathbb{R}^n) \longrightarrow C^\infty(\mathbb{R}^n)$ (equivalently, R and R' have smooth kernels). The hypoellipticity of an elliptic operator easily follows using the existence of a left parametrix; assume in fact that $A\psi = \phi$ is in $C^\infty(\mathbb{R}^n)$. Then $\psi = B\phi - R\psi$ is also in $C^\infty(\mathbb{R}^n)$: we have $B\phi \in C^\infty(\mathbb{R}^n)$ because $\phi \in C^\infty(\mathbb{R}^n)$ and, on the other hand $R\psi \in C^\infty(\mathbb{R}^n)$ because R is smoothing.

For our purposes the notion of ordinary hypoellipticity as just described is rather useless, because it gives no possibility of controlling the behaviour at infinity. It is preferable to use the notion of global hypoellipticity as introduced by Shubin [19], Corollary 25.1, p. 186 (also Boggiatto *et al.* [2], p. 70). We will say that a linear operator $A : \mathcal{S}'(\mathbb{R}^n) \longrightarrow \mathcal{S}'(\mathbb{R}^n)$ is *globally hypoelliptic* if

$$\psi \in \mathcal{S}'(\mathbb{R}^n) \text{ and } A\psi \in \mathcal{S}(\mathbb{R}^n) \Longrightarrow \psi \in \mathcal{S}(\mathbb{R}^n). \tag{27}$$

Shubin [19] (Chapter IV, §23) has introduced very convenient classes of globally hypoelliptic operators. Let $H\Gamma_\rho^{m_1,m_0}(\mathbb{R}^{2n})$ ($m_0, m_1 \in \mathbb{R}$ and $0 < \rho \leq 1$) be the complex vector space of all functions $a \in C^\infty(\mathbb{R}^n)$ for which there exists $R \geq 0$ such that for $|z| \geq R$ the following estimates hold:

$$C_0|z|^{m_0} \leq |a(z)| \leq C_1|z|^{m_1} \tag{28a}$$

$$|D_z^\alpha a(z)| \leq C_\alpha |a(z)||z|^{-\rho|\alpha|} \tag{28b}$$

with $C_0, C_1, C_\alpha \geq 0$. The main properties we will need are summarized in the following Theorem:

Theorem 4 (Shubin). *Let $a \in H\Gamma_\rho^{m_1,m_0}(\mathbb{R}^{2n})$ and A the Weyl operator with symbol a.*

(i) *There exists a Weyl operator B with symbol $b \in H\Gamma_\rho^{-m_1,-m_0}(\mathbb{R}^{2n})$ such that $BA = I + R_1$ and $AB = I + R_2$ where R_1, R_2 have kernels in $\mathcal{S}(\mathbb{R}^n \times \mathbb{R}^n)$*
(ii) *The operator A is globally hypoelliptic.*

(Note that (ii) immediately follows from (i).) We will call the operator B a *Shubin parametrix* of A.

In the context of Wong's operator W the following example is crucial:

Example 5. *The Hermite operator $-\Delta + |x|^2$ is globally hypoelliptic: it suffices to note that the Weyl symbol of $-\Delta + |x|^2$ is $a(z) = |z|^2$ and thus trivially satisfies the estimates* (28) *with $m_0 = m_1 = 2$, $\rho = 1$.*

In next subsection we generalize this example.

3.2. Main result

We claim that:

Theorem 6. *Assume that the Weyl symbol a of A is in $H\Gamma_{\rho}^{m_1,m_0}(\mathbb{R}^{2n})$ (hence A is a globally hypoelliptic pseudodifferential operator). Then the phase-space operator $\widetilde{A}$ is globally hypoelliptic:*

$$\Psi \in \mathcal{S}'(\mathbb{R}^{2n}) \ , \ \widetilde{A}\Psi \in \mathcal{S}(\mathbb{R}^{2n}) \Longrightarrow \Psi \in \mathcal{S}(\mathbb{R}^{2n}).$$

Proof. Let B be a Shubin parametrix of A: $BA = I + R$ where R is an operator with kernel in $\mathcal{S}(\mathbb{R}^n \times \mathbb{R}^n)$. Writing $\widetilde{A}\Psi = \Psi'$ we have $\Psi = \widetilde{B}\Psi' - \widetilde{R}\Psi$. Clearly $B\Psi' \in \mathcal{S}(\mathbb{R}^{2n})$ hence $\widetilde{B}\Psi' \in \mathcal{S}(\mathbb{R}^{2n})$ so it suffices to show that $\widetilde{R}\Psi \in \mathcal{S}(\mathbb{R}^{2n})$ for all $\Psi \in \mathcal{S}'(\mathbb{R}^{2n})$; for this it suffices to show that the kernel of $\widetilde{R}$ is in $\mathcal{S}(\mathbb{R}^n \times \mathbb{R}^n)$, but this follows from Proposition 2, (ii). □

Quadratic forms on $\mathbb{R}^{2n}$ are very interesting objects: they can be viewed as the Hamiltonian functions generalizing the harmonic oscillator. We recall the following very useful symplectic diagonalization result, which goes back to Williamson [23]: for every real positive-definite symmetric matrix there exists $s \in \operatorname{Sp}(n)$ such that $s^T M s = D$ where D is a diagonal matrix of the type $\begin{pmatrix} \Lambda & 0 \\ 0 & \Lambda \end{pmatrix}$ the diagonal entries of Λ consisting of the moduli $\omega_1, \dots, \omega_n$ of the eigenvalues of JM (these are precisely of the type $\pm i\omega_j$ since JM is equivalent to the antisymmetric matrix $M^{1/2}JM^{1/2}$).

Theorem 6 has the following interesting consequence:

Corollary 7. *Let a be a positive-definite quadratic form on $\mathbb{R}^{2n}$: $a(z) = \frac{1}{2}Mz \cdot z$ with $M = M^T > 0$.*

(i) *The associated phase space operator*

$$\widetilde{A} = a(\tfrac{1}{2}x + i\tfrac{\partial}{\partial p}, \tfrac{1}{2}p - i\tfrac{\partial}{\partial x})$$

obtained by the quantization rule (26) is globally hypoelliptic.

(ii) *There exists $s \in \operatorname{Sp}(n)$ such that the operator $\widetilde{B} = \widetilde{S}\widetilde{A}\widetilde{S}^{-1}$ with symbol $b = a \circ s^{-1}$ is given by the formula*

$$\widetilde{B} = \sum_{j=1}^{n} \frac{\omega_j}{2} \left[\left(\tfrac{1}{2}x + i\tfrac{\partial}{\partial p}\right)^2 + \left(\tfrac{1}{2}p - i\tfrac{\partial}{\partial x}\right)^2 \right].$$

Proof. (i) It suffices to show that $a \in H\Gamma_\rho^{m_1,m_0}(\mathbb{R}^{2n})$ for some m_1, m_0, ρ. Writing $S^T MS = D$ with S and D as above, and ordering the entries of D so that $\omega_1 \leq \cdots \leq \omega_n$ we have

$$\frac{\omega_1}{2}|z|^2 \leq a(z) \leq \frac{\omega_n}{2}|z|^2.$$

We have $|sz| \leq ||s|| \cdot |z|$ and $|s^{-1}z| \leq ||s^{-1}|| \cdot |z|$ hence there exist constants C_0, C_1 such that $C_0|z|^2 \leq a(z) \leq C_1|z|^2$ which is condition (28a) with $m_0 = m_1 = 2$. Let α be a multi-index; if $|\alpha| > 2$ then $|D_z^\alpha a(z)| = 0$. Suppose $|\alpha| \leq 2$; in view of the homogeneity of a there exists $C_\alpha > 0$ such that

$$|D_z^\alpha a(z)| \leq C_\alpha |z|^2 |z|^{-|\alpha|} = C'_\alpha a(z)|z|^{-|\alpha|}$$

so that condition (28b) holds with $\rho = 1$. The statement (ii) is an obvious consequence of the fact that if $a(z) = \frac{1}{2}Mz \cdot z$ then

$$b(z) = a(s^{-1}s) = \sum_{j=1}^{n} \frac{\omega_j}{2}(x_j^2 + p_j^2). \qquad \square$$

(Of course, in the first part of the proof of Corollary 7 we could have used standard diagonalization of the symbol a by orthogonal matrices.)

4. Concluding remarks

A question which poses itself is whether the global hypoellipticity of Section 3 can be generalized to other functional spaces than the Schwartz space $\mathcal{S}(\mathbb{R}^{2n})$. As has been pointed out to me by Franz Luef (NuHAG, Vienna), this might very well be the case if one considers Feichtinger's [6, 7] weighted modulation spaces $M_v^{p,q}$ (also see Gröchenig [13], Chapter 11 for a study of these spaces). The main interest of modulation spaces comes from the fact that they allow a simultaneous control of both local regularity and decay at infinity. Besides their intrinsic interest in Functional Analysis, they play an important role not only in time-frequency analysis (for which they were originally designed), but also in the study of the regularity of the solutions of Schrödinger's equation as we have shown in [12]. We will come back to this important question in a forthcoming paper. In that context let us mention that Dasgupta and Wong [5] have obtained interesting regularity results in Gelfand–Shilov spaces; their work certainly deserves to be extended to the operators we considered in Corollary 7.

Another fact which is certainly worth to be scrutinized is the following. As we pointed out several times in this paper, a change of phase-space quantization seems to correspond to a change of gauge. Is there any "universal rule" behind this property which we only checked for physical operators associated with a magnetic field?

References

[1] V. Bargmann, On a Hilbert Space of Analytic Functions and an Associated Integral Transform. Part II. A Family of Related Function Spaces Application to Distribution Theory, *Communications on Pure and Applied Mathematics* **20**(1) (1967), 1–101.

[2] P. Boggiatto, E. Buzano and L. Rodino, *Global Hypoellipticity and Spectral Theory*, Akademie Verlag, 1996.

[3] F. Bopp, La mécanique quantique est-elle une mécanique statistique particulière? *Ann. Inst. H. Poincaré* **15** (1956), 81–112.

[4] E. Cordero and K. Gröchenig, Symbolic calculus and Fredholm property for localization operators, *J. Fourier Anal. Appl.* **12**(3) (2006), 371–392.

[5] A. Dasgupta and M.W. Wong, Essential self-adjointness and global hypoellipticity of the twisted Laplacian, *Rend. Sem. Mat. Univ. Pol. Torino* **66**(1) (2008), 43–53.

[6] H.G. Feichtinger, On a new Segal algebra, *Monatsh. Math.* **92** (1981), 269–289.

[7] H.G. Feichtinger, Modulation spaces on locally compact abelian groups, *Technical Report, University of Vienna*, 1983.

[8] M. de Gosson, On the Weyl Representation of Metaplectic Operators. *Letters Math. Phys.* **72** (2005), 129–142.

[9] M. de Gosson, Extended Weyl Calculus and Application to the Phase-Space Schrödinger Equation, *J. Phys. A: Math. and General* **38** (2005), L325–L329

[10] M. de Gosson, Symplectically Covariant Schrödinger Equation in Phase Space, *J. Phys. A: Math. Gen.* **38** (2005), 9263–9287

[11] M. de Gosson, *Symplectic Geometry and Quantum Mechanics*, "Operator Theory: Advances and Applications" (Subseries: "Advances in Partial Differential Equations"), Vol. **166** Birkhäuser, Basel, 2006

[12] M. de Gosson, Semi-classical propagation of wavepackets for the phase space Schrödinger equation; interpretation in terms of the Feichtinger algebra, *J. Phys. A: Math. and Theo.* [in press]

[13] K. Gröchenig, *Foundations of Time-Frequency Analysis*, Appl. Numer. Harmon. Anal., Birkhäuser, Boston, MA, 2000.

[14] A. Grossmann, G. Loupias, E.M. Stein, An algebra of pseudo-differential operators and quantum mechanics in phase space, *Ann. Inst. Fourier, Grenoble* **18**(2) (1968), 343–368.

[15] J. Isidro and M. de Gosson, Abelian gerbes as a gauge theory of quantum mechanics on phase space, *J. Phys. A: Math. and Theo.* **40** (2007), 3549–3567.

[16] J. Isidro and M. de Gosson, A gauge theory of quantum mechanics, *Modern Physics Letters A*, **22**(3) (2007), 191–200.

[17] L.D. Landau and E.M. Lifshitz, *Quantum Mechanics: Nonrelativistic Theory*, Pergamon Press, 1997.

[18] A. Messiah, *Quantum Mechanics* (in two volumes), North-Holland Publ. Co., 1991; translated from the French; original title: *Mécanique Quantique*, Dunod, Paris, 1961.

[19] M.A. Shubin, *Pseudodifferential Operators and Spectral Theory*, Springer-Verlag, 1987. [original Russian edition in Nauka, Moskva, 1978.]

[20] G. Torres-Vega and J.H. Frederick, Quantum mechanics in phase space: New approaches to the correspondence principle. *J. Chem. Phys.* **93** (12) (1990), 8862–8874.

[21] G. Torres-Vega and J.H. Frederick, A quantum mechanical representation in phase space, *J. Chem. Phys.* **98**(4) (1993), 3103–3120.

[22] F. Trèves, *Introduction to Pseudo Differential and Fourier Integral Operators*, University Series in Mathematics, Plenum Publ. Co., 1981.

[23] J. Williamson, On the algebraic problem concerning the normal forms of linear dynamical systems, *Amer. J. Math.* **58** (1936), 141–163

[24] M.W. Wong, *Weyl Transforms*. Springer-Verlag, 1998.

[25] M.W. Wong, Weyl transforms and a degenerate elliptic partial differential equation, *Proc. R. Soc. A* **461** (2005), 3863–3870. [doi:10.1098/rspa.2005.1560]

[26] M.W. Wong, Weyl transforms, the heat kernel and Green Function of a degenerate elliptic operator, *Ann. Global Anal. and Geom.* **28** (2005), 271–283.

[27] M.W. Wong, The heat equation for the Hermite operator on the Heisenberg group, *Hokkaido Math. J.* **34** (2005), 393–404.

Maurice de Gosson
Max-Planck-Institut für Mathematik
Pf. 7280
D-53072 Bonn
e-mail: `maurice.degosson@gmail.com`

Current address:
University of Vienna
Faculty of Mathematics, NUHAG
A-1090 Vienna, Austria

Operator Theory:
Advances and Applications, Vol. 189, 15–31

Classes of Degenerate Elliptic Operators in Gelfand–Shilov Spaces

Todor Gramchev, Stevan Pilipović and Luigi Rodino

Abstract. We propose a novel approach for the study of the uniform regularity and the decay at infinity for Shubin type pseudo-differential operators which are globally hypoelliptic but not necessarily globally and even locally elliptic. The basic idea is to use the special role of the Hermite functions for the characterization of inductive and projective Gelfand–Shilov spaces. In this way we transform the problem to infinite dimensional linear systems on S Banach spaces of sequences by using Fourier series expansion with respect to the Hermite functions. As applications of our general results we obtain new theorems for global hypoellipticity for classes of degenerate operators in tensorized generalizations of Shubin spaces and in inductive and projective Gelfand–Shilov spaces.

Mathematics Subject Classification (2000). Primary 47F30; Secondary 46F05, 35B65.

Keywords. Harmonic oscillator type operators, S spaces, global hypoelliptic-ity, Gelfand–Shilov spaces.

1. Introduction

The aim of this paper is to investigate the uniform regularity and the type of exponential decay at infinity for the solutions to linear partial differential equations $Pu = f$ globally defined in $\mathbb{R}^n$, where P is an operator with polynomial coefficients of Shubin type (cf. Shubin [26]) without assuming global ellipticity. We focus our attention on tensor products of harmonic oscillators.

In fact, the globally elliptic Shubin operators generalize the Schrödinger harmonic oscillator operator

$$H = -\Delta + |x|^2, \tag{1.1}$$

T.G and L.R. were partially supported by GNAMPA–INDAM, S.P. was partially supported by the Ministry of Sciences of Serbia, project 144016.

appearing in Quantum Mechanics. The spectrum of H in $L^2(\mathbb{R}^n)$ is discrete with eigenvalues $\lambda = \lambda_k = \sum_{j=1}^n (2k_j + 1)$, $k = (k_1, \dots, k_n) \in \mathbb{Z}_+^n$ while eigenfunctions are the Hermite functions

$$u(x) = H_k(x) = \pi^{-n/4}\left(\prod_{j=1}^n (k_j!)^{-1/2} h_{k_j}(\sqrt{2}x_j)\right) \exp(-|x|^2/2), \qquad (1.2)$$

where $h_r(t)$ stands for the rth Hermite polynomial, e.g., cf. Magnus et al [18]. We refer to the books of Helffer [13] and Boggiatto et al [4] for further generalizations and references concerning spectral theory of globally elliptic operators.

We may distinguish two different aspects. The first is the behaviour at infinity, which in (1.2) is super-exponential of order 2, i.e., for positive constants C, ε:

$$|u(x)| \leq Ce^{-\varepsilon|x|^2}, \qquad x \in \mathbb{R}^n. \qquad (1.3)$$

The main interest here comes historically from Quantum Mechanics, where the exponential decay of eigenfunctions has been intensively studied, see for instance Agmon [2].

As a second aspect, returning to the harmonic oscillator (1.1), we point out the fact that the Hermite functions $H_k(x)$, $k \in \mathbb{Z}_+^n$, (1.2) are not only entire functions, but for every $\gamma \in]0, 1/2[$ there exists $A_\gamma > 0$ such that

$$|\partial_z^\alpha H_k(z)| \leq A_\gamma^{|\alpha|+|k|+1}(\alpha!)^{1/2}e^{-\varepsilon|z|^2}, \quad z \in \mathbb{C}^n, |\mathrm{Im} z| < \gamma|\mathrm{Re} z|, \alpha \in \mathbb{Z}_+^n, \quad (1.4)$$

with $\varepsilon = (1 - 2\gamma)/2$, see [6].

The estimates (1.4) lead in a natural way to the idea that the appropriate functional framework to study (1.3) and (1.4) simultaneously is given by the spaces of Gelfand–Shilov type (cf. the classical book of Gelfand and Shilov [10], see also Mitjagin [19], Avantaggiati [3], Pilipovic [20] and references in these papers). We recall that f belongs to the inductive (respectively, projective) Gelfand–Shilov space $S_\nu^\mu(\mathbb{R}^n)$ (respectively, $\Sigma_\nu^\mu(\mathbb{R}^n)$), $\mu > 0, \nu > 0, \mu+\nu \geq 1$ (respectively, $\mu+\nu > 1$), iff $f \in C^\infty(\mathbb{R}^n)$ and there exists $A > 0$ (respectively, for every $A > 0$) such that we can find $C > 0$ satisfying the following estimates

$$|\partial_x^\alpha f(x)| \leq CA^{-|\alpha|}(\alpha!)^\mu e^{-A|x|^{1/\nu}} \qquad (1.5)$$

for all $x \in \mathbb{R}^n, \alpha \in \mathbb{Z}_+^n$ or, equivalently

$$\sup_{x\in\mathbb{R}^n} |x^\beta \partial_x^\alpha f(x)| \leq CA^{-|\alpha|-|\beta|}(\alpha!)^\mu(\beta!)^\nu, \qquad \alpha, \beta \in \mathbb{Z}_+^n. \qquad (1.6)$$

This definition gives the smallest scale of nontrivial inductive Gelfand–Shilov type space $S_{1-\mu}^\mu(\mathbb{R}^n)$, $\mu \in]0, 1[$ while $\Sigma_\nu^\mu(\mathbb{R}^n)$ contains nonzero functions iff $\mu+\nu > 1$. Especially if $\sigma = \mu = 1/2$, then $\Sigma_{1/2}^{1/2} = \{0\}$. Note that $S_{1/2}^{1/2} \subset \Sigma_\sigma^\sigma \subset S_\sigma^\sigma, \sigma > 1/2$. We call the elements of their respective duals $(S_\nu^\mu(\mathbb{R}^n))'$ and $(\Sigma_\nu^\mu(\mathbb{R}^n))'$ tempered inductive and projective μ-ultradistributions.

The bounds (1.5), (1.6) with $\mu \leq 1$ grant that f extends to $\mathbb{C}^n$ as an entire function, with uniform estimates in neighbourhoods of the real axis, see [10] for precise statements. In particular, for the Hermite functions in (1.2), we can read $u \in S^{1/2}_{1/2}(\mathbb{R}^n)$.

We also mention that projective spaces Σ^μ_μ and their further extensions (cf. [20] for another definition in critical case $\sigma = 1/2$) are used as models in chaos expansions with applications to stochastic differential equations [22].

Recently, Cappiello et al [6], [7] have proved the S^μ_ν–regularity of eigenfunctions to Shubin type partial differential operators in $\mathbb{R}^n$

$$P = \sum_{|\alpha|+|\beta|\leq m} c_{\alpha\beta} x^\beta D^\alpha_x, \tag{1.7}$$

where m is a positive integer, provided P is globally elliptic, namely, there exist $C > 0$ and $R > 0$ such that

$$\left| \sum_{|\alpha|+|\beta|\leq m} c_{\alpha\beta} x^\beta \xi^\alpha \right| \geq C(1 + |x|^2 + |\xi|^2)^{m/2}, \qquad |x| + |\xi| \geq R. \tag{1.8}$$

Global ellipticity in the previous sense implies both local regularity and asymptotic decay of the solutions, namely we have the following basic result (see for example Shubin [26]): $Pu = f \in \mathcal{S}(\mathbb{R}^n)$ for $u \in \mathcal{S}'(\mathbb{R}^n)$ implies actually $u \in \mathcal{S}(\mathbb{R}^n)$. This is what is meant by global hypoellipticity in $S(\mathbb{R}^n)$, and same definition we shall use replacing the Schwartz class $S(\mathbb{R}^n)$ by the Gelfand–Shilov space. The main theorem in [6] states that P is also globally hypoelliptic in the Gelfand–Shilov spaces $S^\mu_\nu(\mathbb{R}^n)$, $\mu, \nu > 0$, $\mu \geq 1/2$, $\nu \geq 1/2$.

In this paper we limit our attention the symmetric Gelfand–Shilov spaces $S^\mu_{\mu'}(\mathbb{R}^n), \mu \geq 1/2$ and $\Sigma^\mu_\mu(\mathbb{R}^n), \mu > 1/2$.

One is led naturally to ask whether we can weaken the global ellipticity condition (1.8) for linear operators with polynomial coefficients. One of the prime motivation for our investigations is the second-order degenerate Shubin type operator in $\mathbb{R}^2$ considered by Wong [27]–[29].

$$\begin{aligned} W &= D^2_{x_1} + D^2_{x_2} + \frac{1}{4}(x_1^2 + x_2^2) - x_2 D_{x_1} + x_1 D_{x_2} \\ &= \left(D_{x_1} - \frac{x_2}{2}\right)^2 + \left(D_{x_2} + \frac{x_1}{2}\right)^2 = -\frac{1}{2}(Z\tilde{Z} + \tilde{Z}Z) \end{aligned} \tag{1.9}$$

where,

$$\begin{aligned} Z &= \frac{\partial}{\partial z} + \frac{1}{2}\bar{z}, & \tilde{Z} &= \frac{\partial}{\partial \bar{z}} - \frac{1}{2}z, & D_{x_k} &= i^{-1}\partial_{x_k}, \\ \frac{\partial}{\partial z} &= \frac{1}{2}(\partial_{x_1} + i\partial_{x_2}), & \frac{\partial}{\partial \bar{z}} &= \frac{1}{2}(\partial_{x_1} - i\partial_{x_2}), & z &= x_1 + ix_2. \end{aligned}$$

We mention also that Popivanov [24] has studied for global hypoellipticity in $S(\mathbb{R}^n)$ another class of second-order degenerate Shubin type operators.

In particular, it is shown in [29] that W is globally hypoelliptic, namely

$$u \in S'(\mathbb{R}^2), \ Wu \in S(\mathbb{R}^2) \Rightarrow u \in S(\mathbb{R}^2). \tag{1.10}$$

Later on, Dasgupta and Wong [9] established global hypoellipticity for W in the critical inductive Gelfand–Shilov class $S^{1/2}_{1/2}(\mathbb{R}^2)$.

We propose a novel approach for dealing with classes of degenerate Shubin type operators and establishing global hypoellipticity both in the Schwartz class and in Gelfand–Shilov classes. As a particular case, we recapture the aforementioned global hypoellipticity results in [29], [9]. The crucial ingredient of our techniques is the use the framework of the so-called S spaces as well as the special role of the Hermite functions for the characterization of spaces of Gelfand–Shilov type ultradifferentiable functions (see Antosik et al [1], Reed and Simon [25], Kashpirovskij [16], Avantaggiati [3], Pilipović [20], [21], Holden, et al. [14], Kamiński et al [15], Langenbruch [17]), Budincevic et al [5].

The paper is organized as follows. In Section 2, we recall Shubin class of spaces ([26]) and (as a novelty) characterize them through Hermite expansions. Moreover, we recall the characterization of Gelfand–Shilov type spaces through Hermite expansions and we derive new characterizations of these spaces through estimates of powers of harmonic oscillators.

In Section 3 we introduce tensor-valued harmonic oscillators, characterize them through Hermite expansions, define tensorized generalization of Sobolev–Shubin spaces and show the main global hypoellipticity results.

Section 4 deals with the degenerate operator W in [29], namely, we reduce W by means of a time frequency analysis conjugation operator to the one dimensional harmonic oscillator (a particular case of tensorized operators studied in the previous section). Hence, as an outcome we give a new proof and extensions to the projective Gelfand–Shilov spaces of the global hypoellipticity results in [29], [9]. Moreover, by intrinsic characterizations of W^{-1}, we show that the loss of derivatives for W is 2 in the scale of Sobolev–Shubin spaces.

2. Hermite expansions for Shubin and Gelfand–Shilov type spaces

Let $s \in \mathbb{R}$. We denote by $Q^s(\mathbb{R}^n)$ the spaces introduced by Shubin [26], namely,

$$Q^s(\mathbb{R}^n) := \{u \in S'(\mathbb{R}^n) : \|u\|_{Q^s} := \|(1+|x|^2-\Delta)^{s/2}u\|_{L^2(\mathbb{R}^n)} < +\infty\}. \quad (2.1)$$

Actually, this space can be defined through another equivalent sequence of norms, see [26], Ch. IV, 25.3, showing that the Hilbert structure induces the dual pairing of Q^s and Q^{-s}. Moreover, every $A \in G^m_\rho$ (cf. [26], Ch. IV) defines a continuous mapping

$$A : Q^s(\mathbb{R}^n) \to Q^{s-m}(\mathbb{R}^n)$$

and

$$A : Q^s(\mathbb{R}^n) \to Q^{s-m-\varepsilon}(\mathbb{R}^n),\ \varepsilon > 0, \text{ is a compact one.}$$

We propose a novel result, a characterization of Q^s be means of the Hermite functions $H_j(x)$, $j \in \mathbb{Z}^n_+$.

We recall that, given a tempered distribution $v \in S'(\mathbb{R}^n)$, or more generally, a tempered μ-ultradistribution $v \in (S^\mu_\mu(\mathbb{R}^n))'$, respectively $(\Sigma^\mu_\mu(\mathbb{R}^n))'$, we can write

$$v_\alpha = (v, H_\alpha), \qquad \alpha \in \mathbb{Z}^n_+, \tag{2.2}$$

where the dual pairing (v, H_α) is understood in the sense of tempered distributions or tempered μ ultradistributions. Here

$$H_\alpha(x) = H_{\alpha_1}(x_1)\dots H_{\alpha_n}(x_n), \qquad \alpha \in \mathbb{Z}^n_+, \tag{2.3}$$

$$H_k(t) = \pi^{-1/4}k!^{-1/2}e^{-t^2/2}h_k(\sqrt{2}t), \qquad k \in \mathbb{Z}_+, \tag{2.4}$$

where

$$h_k(t) = (-1)^k e^{t^2/2}\left(\frac{d}{dt}\right)^k (e^{-t^2/2}) = \sum_{r=0}^{[k/2]}(-1)^r 2^{-r}\frac{k!}{r!(k-2r)!}t^{k-2r} \qquad k \in \mathbb{Z}_+, \tag{2.5}$$

is the kth Hermite polynomial, e.g., see [18]. Clearly,

$$\begin{aligned} H_k(t) &= \pi^{-1/4}k!^{-1/2}\sum_{r=0}^{[k/2]}(-1)^r 2^{k/2-2r}\frac{k!}{r!(k-2r)!}t^{k-2r}e^{-t^2/2} \\ &= \pi^{-1/4}k!^{1/2}\sum_{r=0}^{[k/2]}(-1)^r 2^{k/2-2r}\frac{1}{r!(k-2r)!}t^{k-2r}e^{-t^2/2} \qquad k \in \mathbb{Z}_+. \end{aligned} \tag{2.6}$$

For more details cf. Holden et al [14], see also Pilipovic [20], [15], Budinčević at all [5], Langenbruch [17]. Following [14] (see also Reed and Simon [25]) we introduce an ordering of all n-dimensional multi-indices $\alpha^j = (\alpha^j_1, \dots, \alpha^j_n)$ satisfying

$$0 \le j \le k \Rightarrow |\alpha^j| \le |\alpha^k|. \tag{2.7}$$

Now define $\mathcal{H}_j(x) = H_{\alpha^j}(x)$, $j \in \mathbb{Z}_+$.

Here is the first main result of our paper.

Theorem 2.1. *Let v be a tempered distribution, i.e., $v \in \mathcal{S}'(\mathbb{R}^n)$, so that*

$$v = \sum_{\alpha \in \mathbb{Z}^n_+} v_\alpha H_\alpha(x) \tag{2.8}$$

or

$$v = \sum_{j=0}^{\infty} \tilde{v}_j \mathcal{H}_j(x) \text{ with } \tilde{v}_j = v_{\alpha^j}, \qquad j = 0, 1, \dots, \tag{2.9}$$

Then the following statements are equivalent

i) $$v \in Q^s(\mathbb{R}^n).$$

ii) $$\|\{v_\alpha\}_{\alpha \in \mathbb{Z}^n_+}\|_{l_s} := \left(\sum_{\alpha \in \mathbb{Z}^n_+} |v_\alpha|^2 < \alpha >^s\right)^{1/2} < +\infty. \tag{2.10}$$

iii) $$\|\{\tilde{v}_j\}_{j=0}^\infty\|_{l_{s/n}} := \left(\sum_{j=0}^{\infty} |\tilde{v}_j|^2 < j >^{s/n}\right)^{1/2} < +\infty. \tag{2.11}$$

Let now $\mu \geq 1/2$, respectively $\mu > 1/2$. Then the following properties are equivalent:

a) $v \in S^\mu_\mu(\mathbb{R}^n)$, *respectively* $v \in \Sigma^\mu_\mu(\mathbb{R}^n)$;

b) *there exist* $\varkappa$ *and* $C > 0$, *respectively, for every* $\varkappa > 0$ *there exists* $C > 0$ *such that*

$$|v_\alpha| \leq Ce^{-\varkappa|\alpha|^{1/(2\mu)}}, \qquad \alpha \in \mathbb{Z}^n_+. \tag{2.12}$$

c) *there exist* $\varkappa$ *and* $C > 0$, *respectively, for every* $\varkappa > 0$ *there exists* $C > 0$ *such that*

$$|\tilde{v}_j| \leq Ce^{-\varkappa j^{1/(2n\mu)}}, \qquad j \in \mathbb{Z}_+. \tag{2.13}$$

d) *there exists* $C > 0$ *and* $A > 0$, *respectively, for every* $C > 0$ *there exists* $A > 0$ *such that*

$$\|(|x|^2 - \Delta)^{s/2} u\|_{L^2(\mathbb{R}^n)} \leq AC^s s^{s\mu}, \qquad s \in \mathbb{N}; \tag{2.14}$$

Proof. We will prove the equivalence of i) and ii) since the equivalence of ii) and iii) is clear. Since $v = \sum_{\alpha\in\mathbb{Z}^n_+} v_\alpha H_\alpha$ in $\mathcal{S}'(\mathbb{R}^n)$ and

$$(|x|^2 - \Delta)^{s/2} H_\alpha = (\lambda_\alpha)^{s/2} H_\alpha, \qquad \lambda_\alpha = \sum_{j=1}^n (2\alpha_j + 1),\ \alpha \in \mathbb{Z}^n_+,$$

we have

$$\langle (|x|^2 - \Delta)^{s/2} v, H_\alpha \rangle = \langle v, (|x|^2 - \Delta)^{s/2} H_j \rangle = v_\alpha \lambda_\alpha^{s/2} H_\alpha.$$

By the continuity, this implies

$$(|x|^2 - \Delta)^{s/2} v = \sum_{\alpha\in\mathbb{Z}^n_+} (\lambda_\alpha)^{s/2} v_\alpha H_\alpha$$

in $\mathcal{S}'(\mathbb{R}^n)$ and thus in $L^2(\mathbb{R}^n)$. This gives that $(|x|^2 - \Delta)^{s/2}$ has a left inverse in $L^2(\mathbb{R}^n)$, i.e. $(|x|^2 - \Delta)^{-s/2}(|x|^2 - \Delta)^{s/2} = I$. We have

$$\begin{aligned} \|v\|_{Q^s} &= \|(1 + |x|^2 - \Delta)^{s/2} v\|_{L^2} \\ &= \|(1 + |x|^2 - \Delta)^{s/2}(|x|^2 - \Delta)^{-s/2}(|x|^2 - \Delta)^{s/2} v\|_{L^2} \\ &\leq \|B(s, D_x)(|x|^2 - \Delta)^{s/2} v\|_{L^2}, \end{aligned}$$

where $B(s, D_x)$ is of zero order. Moreover, by the use of the Hermite expansion for v it follows that the operator norm of B satisfies

$$\|B\|_{L^2 \to L^2} \leq \sup_{\alpha\in\mathbb{Z}^n_+} \left(1 + \frac{1}{\sum_{j=1}^n (2\alpha_j + 1)}\right)^{s/2} = \left(1 + \frac{1}{n}\right)^{s/2}$$

which leads

$$\begin{aligned}\|v\|_{Q^s} &\le \left(1+\frac{1}{n}\right)^{s/2} \|(|x|^2-\Delta)^{s/2}v\|_{L^2}\\ &= \left(1+\frac{1}{n}\right)^{s/2} \Big(\sum_{\alpha\in\mathbb{Z}^n_+} \lambda_\alpha^s v_\alpha^2\Big)^{1/2}\\ &\le 2^{s/2}\Big(\sum_{\alpha\in\mathbb{Z}^n_+}^{\infty}\Big(\sum_{i=1}^n(2\alpha_i+1)\Big)^s v_\alpha^2\Big)^{1/2}\\ &\le (4\sqrt{2})^s\sum_{\alpha\in\mathbb{Z}^n_+}\Big(\sum_{i=1}^n(1+\alpha_i^2)\Big)^{s/2} v_\alpha^2)^{1/2}\\ &= (4\sqrt{2})^s\|\{v_\alpha\}_{\alpha\in\mathbb{Z}^n_+}\|_{l_s} \end{aligned} \tag{2.15}$$

We obtain the opposite implication in the same way.

The proof of equivalences of a), b) and c) is well known. (cf. [15], [17], [5]). So we will prove the equivalence of *b*) and *d*) in the inductive case.

Assume b). If $u=\sum_{\alpha\in\mathbb{Z}^n_+} a_\alpha H_\alpha \in S^\mu_\mu(\mathbb{R}^n)$, then by (2.12) we get

$$\begin{aligned}\|(|x|^2-\Delta)^{s/2}u\|_{L^2(\mathbb{R}^n)} &= \Big\|\sum_{\alpha\in\mathbb{Z}^n_+}\Big(\sum_{j=1}^n(2\alpha_j+1)\Big)^{s/2} a_\alpha H_\alpha\Big\|_{L^2(\mathbb{R}^n)}\\ &\le C\sup_{\alpha\in\mathbb{Z}^n_+}\Big(\sum_{j=1}^n(2\alpha_j+1)^{s/2}e^{-(\varkappa-\varepsilon)|\alpha|^{1/(2\mu)}}\Big)\\ &\quad\times\Big\|\sum_{\alpha\in\mathbb{Z}^n_+} e^{-\varepsilon|\alpha|^{1/(2\mu)}}\Big\|_{\ell_2}. \end{aligned} \tag{2.16}$$

Now, calculating sup in (2.16), we obtain (2.14). Suppose now that d) holds. We will prove b). By similar calculations as above we obtain that d) yields

$$\sum_{\alpha\in\mathbb{Z}^n_+}^{\infty}\Big(\sum_{i=1}^n(2\alpha_i+1)\Big)^s\frac{|v_\alpha|^2}{C^{2s}s^{2\mu s}}\le A^2. \tag{2.17}$$

This implies

$$\Big(\sum_{i=1}^n(2\alpha_i+1)\Big)^{s/2}\frac{|v_\alpha|}{C^s s^{\mu s}}\le A \tag{2.18}$$

for all $\alpha\in\mathbb{Z}^n_+$, $s\in\mathbb{N}$. We conclude the proof by observing that there exists $\varkappa>0$ such that

$$\sup_{s\in\mathbb{N}}\frac{(\sum_{i=1}^n(2\alpha_i+1))^{s/2}}{C^{2s}s^{2\mu s}}\le e^{\varkappa|\alpha|^{1/(2\mu)}} \tag{2.19}$$

for all $\alpha\in\mathbb{Z}^n_+$, which implies (2.12), i.e., b) holds. The proof is complete. □

3. Tensor products of multidimensional harmonic oscillators

Consider the operator

$$\mathcal{R} = \prod_1^n \mathcal{R}_i, \text{ where } \mathcal{R}_i = x_i^2 - \frac{\partial^2}{\partial x_i^2}, i = 1, \dots, n \tag{3.1}$$

and $\mathcal{R}^m = \mathcal{R}_1^{m_1} \dots \mathcal{R}_n^{m_n}, m \in \mathbb{Z}_+^n$. This operator is not hypoelliptic in the sense of [26], Definition 25.1.

The next theorem is a simple consequence of Theorem 2.1, d) and the well-known inequality

$$s_1! \dots s_n! \leq (s_1 + \dots + s_n)! \leq C^{s_1+\dots+s_n} s_1! \dots s_n! \text{ for some } C > 0.$$

Theorem 3.1. *Let $u \in \mathcal{S}(\mathbb{R}^n)$. Then $u \in \Sigma_\mu^\mu(\mathbb{R}^n), \mu > 1/2$ (respectively, $u \in S_\mu^\mu(\mathbb{R}^n), \mu \geq 1/2$) iff*

$$\sup_{s \in \mathbb{Z}_+^n} \left(\frac{\|\mathcal{R}^{s/2} u\|_{L^2}}{A^s s!^\sigma} \right) < +\infty$$

for every (respectively, for some) $A > 0$.

Let us introduce somewhat more general version of $\mathcal{R}^m$ which will be instrumental in characterizing the tensorized Shubin spaces.

We take a positive integer $d \leq n$ such that $\mathbb{R}^n$ is decomposed into a direct sum of d linear subspaces

$$\mathbb{R}^n = \mathbb{R}^{n_1} \oplus \dots \oplus \mathbb{R}^{n_d}, \qquad n_1 + \dots + n_d = n. \tag{3.2}$$

Given $\alpha \in \mathbb{Z}_+^n$, by (3.2) we can write

$$\alpha = (\alpha^1, \dots, \alpha^d), \qquad \alpha^k \in \mathbb{Z}_+^{n_k}, \ k = 1, \dots, d, \tag{3.3}$$

and similarly for $x \in \mathbb{R}^n$ we write $x = (x^1, \dots, x^d)$.

We consider the operator

$$\begin{aligned} P_{\vec{m};\lambda}(x, D_x) &= \prod_{k=1}^d (|x^k|^2 - \Delta_{\mathbb{R}^{n_k}})^{m_k} + \lambda \\ &= (|x^1|^2 - \Delta_{\mathbb{R}^{n_1}})^{m_1} \circ \dots \circ (|x^d|^2 - \Delta_{\mathbb{R}^{n_d}})^{m_d} + \lambda, \end{aligned} \tag{3.4}$$

with $\lambda \in \mathbb{C}$ Note that $P_{\vec{m};\lambda}$ is of order $2|\vec{m}| = 2m_1 + \dots + 2m_d$ but it is not even locally elliptic if $d \geq 2$. Indeed, its principal symbol (locally in $\mathbb{R}^n$) is given by

$$P^0_{\vec{m};\lambda}(x, \xi) = \prod_{k=1}^d (|\xi^k|^2)^{m_k}. \tag{3.5}$$

The operator $P_{\vec{m};\lambda}(x, D_x)$ is not globally hypoelliptic in the sense of [26], Definition 25.1. Moreover, the homogeneous operator defined by the principal part (3.5) is not even locally hypoelliptic as constant coefficients operator if $d \geq 2$.

The spectral properties and the action in $L^2(\mathbb{R}^n)$ based spaces of the model operators $P_{\vec{m};\lambda}(x, D_x)$ are analyzed completely in the next theorem, which is the second main result of the paper.

Theorem 3.2. *Let u be a tempered or tempered μ-ultradistribution in $\mathbb{R}^n$. We have*

$$P_{\vec{m};\lambda}(x, D_x)u = \sum_{j\in\mathbb{Z}_+^n} \left(\prod_{k=1}^{d} (2|j^k|+1)^{m_k} + \lambda \right) u_j H_j(x) \tag{3.6}$$

where, recall, $j = (j^1, \ldots, j^d) \in \mathbb{Z}_+^{n_1} \oplus \cdots \oplus \mathbb{Z}_+^{n_d}$.

Suppose now that $Re\lambda \geq 0$. Then

$$P_{\vec{m};\lambda}(x, D_x)u \in L^2(\mathbb{R}^n) \tag{3.7}$$

is equivalent to

$$\sum_{j\in\mathbb{Z}_+^n} \left(\prod_{k=1}^{d} (2|j^k|+1)^{m_k} + \lambda \right)^2 |u_j|^2 < +\infty. \tag{3.8}$$

Moreover, if

$$u \in Q^{2|\vec{m}|}(\mathbb{R}^n) \tag{3.9}$$

then (3.8) holds. Finally, the operator

$$P_{\vec{s}}(x, D_x)u = \prod_{k=1}^{d} (|x^k|^2 - \Delta_{\mathbb{R}^{n_k}})^{s_k/2} u \tag{3.10}$$

is well defined for all $\vec{s} = (s_1, \ldots, s_d) \in \mathbb{R}^d$ by the spectral properties of the powers of the harmonic oscillator, is self-adjoint with the spectrum defined by

$$\operatorname{spec} P_{\vec{s}} = \{\prod_{k=1}^{d} (2|j^k|+1)^{s_k/2} : j = (j^1, \ldots, j^d) \in \mathbb{Z}_+^{n_1} \times \cdots \times \mathbb{Z}_+^{n_d}\}. \tag{3.11}$$

with the Hermite functions forming an orthonormal $L^2(\mathbb{R}^n)$ base, i.e.,

$$P_{\vec{s}}(x, D_x)H_j = \prod_{k=1}^{d} (2|j^k|+1)^{s_k/2} H_j(x). \tag{3.12}$$

Finally, $P_{\vec{s}}(x, D_x)$ acts continuously in the spaces of tempered μ-ultradistributions by the formula

$$P_{\vec{s}}(x, D_x)u = \sum_{j\in\mathbb{Z}_+^n} \left(\prod_{k=1}^{d} (2|j^k|+1)^{s_k/2} \right) u_j H_j(x). \tag{3.13}$$

and is globally $S_\mu^\mu(\mathbb{R}^n)$, (respectively, $\Sigma_\mu^\mu(\mathbb{R}^n)$) hypoelliptic for every $\mu \geq 1/2$ (respectively, $\mu > 1/2$).

Proof. One observes that

$$
\begin{aligned}
&P_{\vec{m};\lambda}(x, D_x)H_j(x) \\
&= \prod_{k=1}^{d}\Big(|x^k|^2 - \Delta_{\mathbb{R}^{n_k}})^{m_k} H_{j^k}(x^k)\Big) + \lambda H_j(x) \qquad (3.14)\\
&= \Big(\prod_{k=1}^{d}(2|j^k|+1)^{m_k} H_{j^k}(x^k)\Big) + \lambda H_j(x) = \Big(\prod_{k=1}^{d}(2|j^k|+1)^{m_k} + \lambda\Big) H_j(x)
\end{aligned}
$$

which leads to (3.6) and the equivalence between (3.7) and (3.8). As it concerns the implication (3.9) $\Longrightarrow$ (3.8) it follows from the fact that one can find $c_1 > 0$ such that

$$\prod_{k=1}^{d}(2t_k+1)^{m_k} \le c_1(t_1 + \cdots + t_d + 1)^{|\vec{m}|} \qquad (3.15)$$

for all $t_k \ge 0$, $k = 1, \ldots, d$. The validity of (3.11) and (3.12) follows from the spectral theory. The global hypoellipticity part follows from the fact that the equation $P_{\vec{s}}(x, D_x)u = f$, in view of the representations,

$$u = \sum_{j\in\mathbb{Z}_+^n} u_j H_j(x), \quad f = \sum_{j\in\mathbb{Z}_+^n} f_j H_j(x)$$

and (3.13), is equivalent to

$$\prod_{k=1}^{d}(2|j^k|+1)^{s_k/2} u_j = f_j, \qquad j \in \mathbb{Z}_+^n, \qquad (3.16)$$

Therefore,

$$u_j = \Big(\prod_{k=1}^{d}(2|j^k|+1)^{s_k/2}\Big)^{-1} f_j, \qquad j \in \mathbb{Z}_+^n, \qquad (3.17)$$

and we may apply the characterization of $S_\mu^\mu(\mathbb{R}^n)$ and $\Sigma_\mu^\mu(\mathbb{R}^n)$ in the first part of this section. The proof is complete. □

Note that (3.9) implies (3.8). This leads naturally to the definition of the following generalization of the Shubin spaces:

$$Q^{\vec{s}}(\mathbb{R}^{n_1} \oplus \cdots \oplus \mathbb{R}^{n_d}) := \{u \in S'(\mathbb{R}^n) : \text{ s.t. } \|P_{\vec{s}}(x, D_x)u\|_{L^2} < +\infty\}, \qquad (3.18)$$

for $\vec{s} = (s_1, \ldots, s_d) \in \mathbb{R}^d$.

Remark 3.3. We note that if $d = 1$ the space $Q^{\vec{s}}((\mathbb{R}^{n_1} \oplus \cdots \oplus \mathbb{R}^{n_d}))$ coincides with the usual Shubin space.

Proposition 3.4. *We have*

$$\bigcap_{\vec{s}\in\mathbb{R}^d} Q^{\vec{s}}(\mathbb{R}^{n_1} \oplus \cdots \oplus \mathbb{R}^{n_d}) = \mathcal{S}(\mathbb{R}^{n_1} \oplus \cdots \oplus \mathbb{R}^{n_d}) = \mathcal{S}(\mathbb{R}^n).$$

The proof is a direct consequence of the definition and Theorem 2.1.

4. Reduction of W to lower-dimensional harmonic oscillators

In this section we transform the operator W in [29] and reduce the problem of the global hypoellipticity in the Schwartz class and the Gelfand–Shilov spaces for W to the one-dimensional harmonic oscillator in $\mathbb{R}^2$, which is a particular example of the tensorized harmonic oscillators in the preceding section.

Define

$$\tilde{W} = KWK, \tag{4.1}$$

where $z \mapsto K(z)$ is the operator of conjugation: $K(z) = \bar{z}$, $z \in \mathbb{C}$. Clearly, $\tilde{W}W = W\tilde{W}$.

By [30], if $j, k \in \mathbb{Z}_+^2$, $x, y \in \mathbb{R}$ and $(H_j)_{j \in \mathbb{Z}_+}$ is the Hermite basis in $L^2(\mathbb{R})$, then

$$e_{j,k}(x,y) = V(H_j(x), H_k(y)) = \mathcal{F}^{-1}(H_j(\cdot + \frac{y}{2})H_k(\cdot - \frac{y}{2}))(x),$$

is an orthonormal basis of $L^2(\mathbb{R}^2)$. Recall,

$$We_{j,k}(x,y) = (2k+1)e_{j,k}(x,y), j,k \in \mathbb{Z}_+, x, y \in \mathbb{R}.$$

It is well known (see Pilipović and Teofanov [23], Gröchenig and Zimmermann [12], Cordero et al [8], Cappiello et al [7], and the references therein) that the Wigner and Fourier transform continuously map $S_\sigma^\sigma(\mathbb{R}^n) \times S_\sigma^\sigma(\mathbb{R}^n)$ into $S_\sigma^\sigma(\mathbb{R}^{2n})$. Moreover, we have:

Proposition 4.1.

$$\tilde{W}e_{j,k} = (2j+1)e_{j,k}, \quad j,k \in \mathbb{Z}_+. \tag{4.2}$$

Proof. It is shown in [29] that

$$Ze_{j,k} = i(2k)^{1/2}e_{j,k-1}, \tilde{Z}e_{j,k} = i(2k+2)^{1/2}e_{j,k+1}. \tag{4.3}$$

Observing that $H_j(-x) = (-1)^j H_j(x), x \in \mathbb{R}, j \in \mathbb{Z}_+$, we have

$$\begin{aligned}
\overline{e_{k,j}(x,y)} = \overline{V(H_k(x), H_j(y))} &= \mathcal{F}\left(H_k\left(\cdot + \frac{y}{2}\right)H_j\left(\cdot - \frac{y}{2}\right)\right)(x) \\
&= \mathcal{F}^{-1}\left(H_k\left(-\cdot + \frac{y}{2}\right)H_j\left(-\cdot - \frac{y}{2}\right)\right)(x) \\
&= (-1)^{j+k}\left\{\mathcal{F}^{-1}\left(H_j\left(\cdot + \frac{y}{2}\right)H_k\left(\cdot - \frac{y}{2}\right)\right)(x)\right\} \\
&= (-1)^{j+k}e_{j,k}(x,y), x, y \in \mathbb{R}.
\end{aligned}$$

Using this we have

$$\tilde{W}e_{j,k} = KW((-1)^{j+k}e_{k,j}) = K((-1)^{j+k}(2j+1)e_{k,j}) = (2j+1)e_{j,k}.$$

The proof is complete. □

Theorem 4.2. *Denote by J a mapping $L^2(\mathbb{R}^2) \to L^2(\mathbb{R}^2)$ defined by*

$$J(H_i(x)H_j(y)) = e_{i,j}(x,y), \; i,j \in \mathbb{N}_0, \; x,y \in \mathbb{R}, \tag{4.4}$$

i.e.,

$$J\Big(\sum_{i,j\in\mathbb{Z}^2_+} a_{i,j}H_i(x)H_j(y)\Big) = \sum_{i,j\in\mathbb{Z}^2_+} a_{i,j}e_{i,j}(x,y), \tag{4.5}$$

for

$$h(x,y) = \sum_{i,j\in\mathbb{Z}^2_+} a_{i,j}H_i(x)H_j(y) \in L^2(\mathbb{R}^2).$$

Then we have:

a) *J is an automorphism of $L^2(\mathbb{R}^2)$ onto itself:*

$$J(f)(x,y) = (2\pi)^{-1}\int_{\mathbb{R}} e^{ix\xi_1} f(\xi_1+\frac{y}{2},\xi_1-\frac{y}{2})d\xi_1, x,y\in\mathbb{R}, \tag{4.6}$$

and

$$J^{-1}(g)(p,q) = \int_{\mathbb{R}} e^{-it\frac{p+q}{2}} g(t,p-q)dt, p,q\in\mathbb{R}. \tag{4.7}$$

b) *With the notation $\mathcal{R}(x) = x^2 + D_x^2$ and I_y the identity mapping, we have*

$$J(I_x\times\mathcal{R}(y))J^{-1} = W,\quad J(\mathcal{R}(x)\times I_y)J^{-1} = \tilde{W}.$$

For any $k=(k_1,k_2)\in\mathbb{Z}^2_+$,

$$J(\mathcal{R}^{k_1}(x)\times\mathcal{R}^{k_2}(y))J^{-1} = \tilde{W}^{k_1}W^{k_2}.$$

Proof. a) It is enough to prove the assertion for

$$f(x,y) = \sum_{(m,n)\in\mathbb{Z}^2_+} a_{m,n}H_m(x)H_n(y) \in L^2(\mathbb{R}^2)\cap\mathcal{S}(\mathbb{R}^2).$$

We have

$$\begin{aligned}
Jf(x,y) &= \sum_{(m,n)\in\mathbb{Z}^2_+} a_{m,n}e_{m,n}(x,y) = \sum_{(r,s)\in\mathbb{Z}^2_+} a_{r,s}\mathcal{F}^{-1}\Big(H_r\Big(\xi_1+\frac{y}{2}\Big)H_s\Big(\xi_1-\frac{y}{2}\Big)\Big)(x)\\
&= \sum_{(m,n)\in\mathbb{Z}^2_+}\Big(\int_{\mathbb{R}}\int_{\mathbb{R}}\sum_{(r,s)\in\mathbb{Z}^2_+} a_{r,s}\mathcal{F}^{-1}\Big(H_r\Big(\xi_1+\frac{\tau}{2}\Big)H_s\Big(\xi_1-\frac{\tau}{2}\Big)\Big)(\eta)H_r(\eta)H_s(\tau)d\eta d\tau\Big)\\
&\qquad\times H_r(x)H_s(y)\\
&= \sum_{(m,n)\in\mathbb{Z}^2_+}\Big(\int_{\mathbb{R}}\int_{\mathbb{R}}\mathcal{F}^{-1}\Big(\sum_{(r,s)\in\mathbb{Z}^2_+} a_{r,s}H_r\Big(\xi_1+\frac{\tau}{2}\Big)H_s\Big(\xi_1-\frac{\tau}{2}\Big)\Big)(\eta)H_r(\eta)H_s(\tau)d\eta d\tau\Big)\\
&\qquad\times H_r(x)H_s(y)\\
&= \sum_{(m,n)\in\mathbb{Z}^2_+}\Big(\int_{\mathbb{R}}\int_{\mathbb{R}}\mathcal{F}^{-1}\Big(f\Big(\xi_1+\frac{\tau}{2},\xi_1-\frac{\tau}{2}\Big)\Big)(\eta)H_r(\eta)H_s(\tau)d\eta d\tau\Big)H_r(x)H_s(y)\\
&= \mathcal{F}^{-1}\Big(f\Big(\xi_1+\frac{y}{2},\xi_1-\frac{y}{2}\Big)\Big)(x).
\end{aligned}$$

This proves the part a).

The assertion b) is an easy consequence by noticing that

$$J(S^{\sigma}_{\sigma}) = S^{\sigma}_{\sigma},\ J(\Sigma^{\sigma}_{\sigma}) = \Sigma^{\sigma}_{\sigma},$$

$\sigma \geq 1/2$, respectively, $\sigma > 1/2$. This completes the proof. □

We may now easily recapture the results of [29] and [9] for global hypoellipticity of W. It is enough to observe that operator J in (4.5) is an automorphism in $\mathcal{S}(\mathbb{R}^2)$, $\mathcal{S}^{\mu}_{\mu}(\mathbb{R}^2)$, $\mu \geq 1/2$, $\Sigma^{\mu}_{\mu}(\mathbb{R}^2)$, $\mu > 1/2$, with extensions to the duals and the inverse J^{-1} defined by (4.7). It follows from Theorem 4.2., b), that W and $\widetilde{W}$ are $\mathcal{S}(\mathbb{R}^2)$, $\mathcal{S}^{\mu}_{\mu}(\mathbb{R}^2)$, $\Sigma^{\mu}_{\mu}(\mathbb{R}^2)$, globally hypoelliptic iff $I_x \times \mathcal{R}(y)$, $\mathcal{R}(x) \times I_y$ are globally hypoelliptic in the same frames. This is granted by the last statement in Proposition 3.4 for $\vec{s} = (0,2)$ and $\vec{s} = (2,0)$ in $\mathbb{R}^2$.

Remark 4.3. We point out that Theorem 4.2 can be extended over $\mathbb{R}^{2n}$ in the two ways. First, by introducing $W_{\text{prod}}(x_1, x_2, D_{x_1}, D_{x_2}, \ldots, x_{2n-1}, x_{2n}, D_{x_{2n-1}}, D_{x_{2n}})$ defined by

$$W(x_1, x_2, D_{x_1}, D_{x_2}) \ldots W(x_{2n-1}, x_{2n}, D_{x_{2n-1}}, D_{x_{2n}}).$$

and $\tilde{W}_{\text{prod}}$ in a similar way, and secondly, introducing

$$W_{\text{sum}}(x, D) = -\Delta + \frac{1}{4}|x|^2 + \sum_{k=1}^{n}(x_{2k-1}D_{x_{2k}} - x_{2k}D_{x_{2k-1}}) \tag{4.8}$$

and $\tilde{W}_{\text{sum}} = KW_{\text{sum}}K$, where K is the operator of conjugation in $\mathbb{C}^n$.

As an alternative approach with respect to the conjugation formula in Theorem 4.2, b) we also can show that the operator W and its polynomial extensions are globally reducible to tensorized harmonic oscillators via a global Fourier integral operator

$$Ev(x) = \int_{\mathbb{R}^2} \exp(i\varphi(x,\eta))\hat{v}(\eta)\, \bar{d}\eta \tag{4.9}$$

where the phase function is given by

$$\varphi(x,\eta) = \eta_1\eta_2 + \frac{x_1x_2}{2} + x_2\eta_1 + x_1\eta_2, \tag{4.10}$$

More details and other generalizations will be presented in a forthcoming paper [11].

We conclude this section by proposing some remarks on the Green function W^{-1} with the kernel (see [29])

$$K_1(\xi,\eta) = (4\pi)^{-1}K_0(\frac{\xi^2+\eta^2}{4}), K_0(t) = \int_0^{\infty} e^{-t\cosh s}ds, t > 0.$$

Proposition 4.4. *The function $K_0(t)$ has the following properties:*

i) *for every $\varepsilon \in]0,1/2[$ there exists $A_\varepsilon > 0$ such that*

$$K_0(t) \leq A_\varepsilon(1+\ln t)e^{-(1/2-\varepsilon)t}, \qquad t > 0 \tag{4.11}$$

ii) *for every* $\varepsilon \in]0,1[$ *there exists* $B_\varepsilon > 0$ *such that*

$$K_0(t) \geq B_\varepsilon (1 + \ln t) e^{-(1/2-\varepsilon)t}, \qquad t > 0. \tag{4.12}$$

iii) *for every* $\varepsilon \in]0,1[$ *there exists* $C_\varepsilon > 0$ *such that*

$$t^{k-1} |K_0^{(k)}(t)| \geq C_\varepsilon^k (k-1)! e^{-(1-\varepsilon)t}, \qquad t > 0. \tag{4.13}$$

iv) *there exists* $C > 0$ *such that*

$$(\tau^2 + \eta^2)^{|k|/2} |K_1^{(k)}(\tau, \eta)| \leq C^{|k|+1} |k|^{|k|}, (\tau, \eta) \in \mathbb{R}^2, k = (k_1, k_2) \in \mathbb{Z}_+^2. \tag{4.14}$$

Proof. We have $\cosh t \leq e^t$ while for every $\varepsilon \in]0,1[$ we can find $\mu_\varepsilon > 0$, $\nu_\varepsilon > 0$ such that

$$\nu_\varepsilon + (1/2 - \varepsilon) e^t \leq \cosh t \leq (1/2 + \varepsilon) e^t + \mu_\varepsilon, \qquad t > 0. \tag{4.15}$$

Therefore, after the change of the variables $re^t = \xi$, we obtain

$$K_0(t) \leq \int_0^{+\infty} e^{-t(1/2-\varepsilon)e^r - t\nu_\varepsilon} dr = e^{-\nu_\varepsilon t} \int_t^{+\infty} \xi^{-1} e^{-(1/2-\varepsilon)\xi} d\xi \tag{4.16}$$

$$K_0(t) \geq \int_0^{+\infty} e^{-t(1/2+\varepsilon)e^r - t\mu_\varepsilon} dr = e^{-\mu_\varepsilon t} \int_t^{+\infty} \xi^{-1} e^{-(1/2+\varepsilon)\xi} d\xi \tag{4.17}$$

The asymptotic analysis of the integral above for $t \searrow 0$ and $t \to +\infty$ yield the desired estimates in i) and ii). Similarly, we obtain that

$$|t^{(s-1)} K_0^{(s)}(t)| = \left| \int_0^{+\infty} e^{-t\cosh r} (\cosh r)^s dr \right|$$
$$= e^{-\nu_\varepsilon t} \int_t^{+\infty} \xi^{s-1} e^{-(1/2-\varepsilon)\xi} d\xi \tag{4.18}$$

and we use estimates for the Euler gamma function. This gives iii) (and smoothing effects of K_1).

iv) Put $g(\tau, \eta) = (4\pi)^{-1} K_0\big(\frac{\tau^2+\eta^2}{4}\big)$. Let $k = (k_1, k_2) \in Z_+^2$. We have

$$(4\pi) g^{(k)}(\tau, \eta) = \int_0^\infty e^{-\frac{\tau^2+\eta^2}{4}} (-1/2)^{k_1+k_2} (\cosh t^{k_1+k_2} (-1/2)^{k_1+k_2} \tau^{k_1} \eta^{k_2}$$
$$+ \cosh t^{k_1+k_2-1} \tau^{k_1} \eta^{k_2-1} + \cosh t^{k_1+k_2-1} \tau^{k_1-1} \eta^{k_2}$$
$$+ \cosh t^{k_1+k_2-2} \tau^{k_1-1} \eta^{k_2-1}) dt.$$

Put $u = e^t$ and write $s = (u^2+1)/u$. We continue

$$(-2)^{|k|} (4\pi) g^{(k)}(\tau, \eta) = \int_1^\infty e^{-s\frac{\tau^2+\eta^2}{4}}$$
$$\big(s^{|k|} \tau^{k_1} \eta^{k_2} + s^{|k|-1} \tau^{k_1} \eta^{k_2-1} + s^{|k|-1} \tau^{k_1-1} \eta^{k_2} + s^{|k|-2} \tau^{k_1-1} \eta^{k_2-1}\big) du/u.$$

We will use $\int_1^\infty (u^2+1)^3/u^4 du < \infty$ and the inequality (for $m > 3$)

$$\sup_s \big\{ s^{m-3} e^{-s(\tau^2+\eta^2)/4} \big\} \leq \frac{(m-3)^{m-3}}{e^{m-3}(\tau^2+\eta^2)^{m-3}}.$$

We continue

$$
\begin{aligned}
2^{|k|}(4\pi)|g^{(k)}(\tau,\eta)| \le\ & C\frac{4^{|k|-3}(|k|-3)^{|k|-3}}{e^{|k|-3}(\tau^2+\eta^2)^{|k|-3}}|\tau|^{k_1}\eta^{k_2} \\
&+\frac{(4^{|k|-4}(|k|-4)^{|k|-4}}{e^{|k|-4}(\tau^2+\eta^2)^{|k|-4}}|\tau|^{k_1}\eta^{k_2-1} \\
&+\frac{4^{|k|-4}(|k|-4)^{|k|-4}}{e^{|k|-4}(\tau^2+\eta^2)^{|k|-4}}|\tau|^{k_1-1}\eta^{k_2} \\
&+\frac{4^{|k|-5}(|k|-5)^{|k|-5}}{e^{|k|-5}(\tau^2+\eta^2)^{|k|-5}}|\tau|^{k_1-1}\eta^{k_2-1}.
\end{aligned}
$$

Now we use

$$
\frac{|\tau|^p|\eta|^q}{(|\tau|^2+|\eta|^2)^{p+q}} \le \frac{1}{(|\tau|^2+|\eta|^2)^{\frac{p+q}{2}}}.
$$

With this and suitable $C>0$ we have (4.14). □

Combining the preceding results, we obtain the main global hypoellipticity results in [29], [9] with some generalizations in the projective Gelfand–Shilov spaces and in the Sobolev–Shubin type spaces in $\mathbb{R}^2$.

Proposition 4.5. *We have*

i) $W^{-1}: Q^s(\mathbb{R}^2) \to Q^s(\mathbb{R}^2)$, $s \in \mathbb{R}$, *is continuous and the loss of derivatives for* W *equals* 2.

ii) *Let* $\sigma \ge 1/2$, *respectively,* $\sigma > 1/2$. *Then*

$$
W : S^\sigma_\sigma(\mathbb{R}^2) \to S^\sigma_\sigma(\mathbb{R}^2), \text{ respectively, } W : \Sigma^\sigma_\sigma(\mathbb{R}^2) \to \Sigma^\sigma_\sigma(\mathbb{R}^2) \tag{4.19}
$$

is an automorphism.

iii) *Let for some* $m \in \mathbb{Z}^n_+$, $W^m u = f \in S^\sigma_\sigma(\mathbb{R}^2)$, $\sigma \ge 1/2$ *(respectively,* $\in \Sigma^\sigma_\sigma(\mathbb{R}^2)$, $\sigma > 1/2$*). Then* $u \in S^\sigma_\sigma(\mathbb{R}^2)$, $\sigma \ge 1/2$ *(respectively,* $\in \Sigma^\sigma_\sigma(R^2)$, $\sigma > 1/2$*).*

Proof. We only explain that the second part of i) follows from Proposition 4.4 iii) with $k=1$ and the definition of K_1 (see also Proposition 4.4 iv)).

Acknowledgment

The research was initiated during the stay of S.P. as a visiting professor in the University of Cagliari in January 2007 and was concluded while T.G. and L.R. visited the University of Novi Sad in December 2007. The authors thank the inviting institutions for the support and the excellent condition for research.

References

[1] P. Antosik, P. Mikusińksi and R. Sikorski, *Theory of Distributions. The Sequential Approach,* Elsevier Scientific Publishing Co., Amsterdam; PWN–Polish Scientific Publishers, Warsaw, 1973.

[2] S. Agmon, *Lectures on Exponential Decay of Second-Order Elliptic Equations: Bounds on Eigenfunctions of N-Body Schrödinger Operators,* Math. Notes **29**, Princeton University Press, Princeton, 1982.

[3] A. Avantaggiati, S-spaces by means of the behaviour of Hermite-Fourier coefficients, *Boll. Un. Mat. Ital.* 4–*A* **6** (1985), 487–495.

[4] P. Boggiatto, E. Buzano and L. Rodino, *Global Hypoellipticity and Spectral Theory.* Math. Res. **92**, Akademie Verlag, Berlin, 1996.

[5] M. Budinčević, D. Perišić and M. Tasković, Gelfand-Shilov spaces, structural and kernel theorems, preprint, arXiv:0706.2268

[6] M. Cappiello, T. Gramchev and L. Rodino, Exponential decay and regularity for SG-elliptic operators with polynomial coefficients, *J. Funct. Anal.* **237** (2006), 634–654.

[7] M. Cappiello, T. Gramchev and L. Rodino, Gelfand-Shilov spaces, pseudo-differential operators and localization operators, in *Modern Trends in Pseudo-Differential Operators,* Editors: J. Toft, M.W. Wong and H. Zhu, Oper. Theory Adv. Appl. **172**, Birkhäuser, Basel, 2007, 297–312.

[8] E. Cordero, S. Pilipović, L. Rodino and N. Teofanov, Localization operators and exponential weights for modulation spaces, *Mediterr. J. Math.* **2** (2005), 381–394.

[9] A. Dasgupta and M.W. Wong, Essential self-adjointness and global hypoellipticity of the twisted Laplacian, *Rend. Sem. Mat. Univ. Pol. Torino* **66** (2008), 43–53.

[10] I.M. Gelfand and G.E. Shilov, *Generalized Functions II,* Academic Press, New York, 1968.

[11] T. Gramchev, S. Pilipović, L. Rodino, in preparation.

[12] K. Gröchenig and G. Zimmermann, Spaces of test functions via the STFT, *J. Funct. Spaces Appl.* **2** (2004), 24–53.

[13] B. Helffer, *Théorie spectrale pour des opérateurs globalement elliptiques.* Astérisque **112**, Société Mathématique de France, Paris, 1984.

[14] H. Holden, B. Oksendal, J. Uboe and T. Zhang, *Stochastic Partial Differential Equations. A Modeling, White Noise Functional Approach,* Probability and its Applications, Birkhäuser Boston, Inc., Boston, MA, 1996.

[15] A. Kamiński, D. Perišić and S. Pilipović, On various integral transformations of tempered ultradistributions, Demonstratio Math. **33** (2000), 641–655.

[16] A.I. Kashpirovskij, Realization of some spaces of S-type as the space of sequences, Vestn. Kiev, Univ. Mat. Meh. **21** (1979), 52–58. (in Russian).

[17] M. Langenbruch, Hermite functions and weighted spaces of generalized functions, Manuscripta Math. **119** (2006), 269–285.

[18] W. Magnus, F. Oberheittinger and F.G. Tricomi, *Higher Transcendental Functions II,* Bateman Project, California Institute of Technology, McGraw–Hill, 1953.

[19] B.S. Mitjagin, Nuclearity and other properties of spaces of type S, *Amer. Math. Soc. Transl. Ser.* 2 **93** (1970), 45–59.

[20] S. Pilipović, Generalization of Zemanian spaces of generalized functions which elements have series expansion, *SIAM J. Math. Anal.* **17** (1986), 477–484.

[21] S. Pilipovic, Tempered ultradistributions, *Boll. Unione Mat. Ital. VII. Ser. B* **2** (1988), 235–251.

[22] S. Pilipović, and D. Seleši, Expansion theorems for generalized random processes, Wick products and applications to stochastic differential equations, *Infin. Dimens. Anal. Quantum Probab. Relat. Top.* **10** (2007), 79–110.

[23] S. Pilipovic and N. Teofanov, *Pseudodifferential operators on ultramodulation spaces, J. Funct. Anal.* **208** (2004), 194–228.

[24] P. Popivanov, A link between between small divisors and smoothness of the solutions of a class of partial differential equations, *Ann. Global Anal. Geom.* **1**, (1983), 77–92.

[25] M. Reed and B. Simon, *Methods of Modern Mathematical Physics I* Academic Press, 1972.

[26] M. Shubin, *Pseudodifferential Operators and Spectral theory.* Springer Series in Soviet Mathematics, Springer-Verlag, Berlin, 1987.

[27] M.W. Wong, Weyl transforms and a degenerate elliptic partial differential equation. *Proc. R. Soc. Lond. Ser. A Math. Phys. Eng. Sci.* **461** (2005), 3863–3870.

[28] M.W. Wong, The heat equation for the Hermite operator on the Heisenberg group, *Hokkaido Math. J.* **34** (2005), 393–404.

[29] M.W. Wong, Weyl transforms, the heat kernel and Green function of a degenerate elliptic operator, *Ann. Global. Anal. Geom.* **28** (2005), 271–283.

[30] M.W. Wong, *Weyl Transforms* Springer-Verlag, Berlin, 1998.

Todor Gramchev
Dipartimento di Matematica e Informatica
Università di Cagliari
Via Ospedale 72
I-09124 Cagliari, Italy
e-mail: `todor@unica.it`

Stevan Pilipović
Department of Mathematics and Informatics
University of Novi Sad
Trg D. Obradovića 4
21000 Novi Sad, Serbia
e-mail: `pilipovic@im.ns.ac.yu`

Luigi Rodino
Dipartimento di Matematica
Università di Torino
Via Carlo Alberto 10
I-10123 Torino, Italy
e-mail: `luigi.rodino@unito.it`

Operator Theory:
Advances and Applications, Vol. 189, 33–42

Weyl Transforms and the Heat Equation for the Sub-Laplacian on the Heisenberg Group

Aparajita Dasgupta and M.W. Wong

Abstract. Using Weyl transforms depending on Planck's constant, we give a new formula for the solution of the initial value problem governed by the sub-Laplacian on the one-dimensional Heisenberg group. Estimates for the solution in terms of the initial value are given.

Mathematics Subject Classification (2000). Primary 47F05, 47G30; Secondary 35H20.

Keywords. Heisenberg group, sub-Laplacian, twisted Laplacian, Fourier–Wigner transforms, Weyl transforms, Hermite functions, heat kernel, semigroup, Sobolev spaces.

1. The Heisenberg group

If we identify $\mathbb{R}^2$ with the complex plane $\mathbb{C}$ via

$$\mathbb{R}^2 \ni (x, y) \leftrightarrow z = x + iy \in \mathbb{C}$$

and let

$$\mathbb{H} = \mathbb{C} \times \mathbb{R},$$

then $\mathbb{H}$ becomes a non-commutative group when equipped with the multiplication $\cdot$ given by

$$(z, t) \cdot (w, s) = \left(z + w, t + s + \frac{1}{4}[z, w]\right), \quad (z, t), (w, s) \in \mathbb{H},$$

where $[z, w]$ is the symplectic form of z and w defined by

$$[z, w] = 2 \operatorname{Im}(z\overline{w}).$$

In fact, $\mathbb{H}$ is a unimodular Lie group on which the Haar measure is just the ordinary Lebesgue measure $dz\, dt$.

This research has been supported by the Natural Sciences and Engineering Research Council of Canada.

Let $\mathfrak{h}$ be the Lie algebra of left-invariant vector fields on $\mathbb{H}$. A basis for $\mathfrak{h}$ is then given by X, Y and T, where

$$X = \frac{\partial}{\partial x} + \frac{1}{2} y \frac{\partial}{\partial t}, \qquad Y = \frac{\partial}{\partial y} - \frac{1}{2} x \frac{\partial}{\partial t},$$

and

$$T = \frac{\partial}{\partial t}.$$

The sub-Laplacian $\mathcal{L}$ on $\mathbb{H}$ is defined by

$$\mathcal{L} = -(X^2 + Y^2).$$

A simple computation gives

$$\mathcal{L} = -\Delta - \frac{1}{4}(x^2 + y^2)\frac{\partial^2}{\partial t^2} + \left(x\frac{\partial}{\partial y} - y\frac{\partial}{\partial x} \right) \frac{\partial}{\partial t},$$

where

$$\Delta = \frac{\partial^2}{\partial x^2} + \frac{\partial^2}{\partial y^2}.$$

Details on the Heisenberg group $\mathbb{H}$ and the sub-Laplacian $\mathcal{L}$ can be found in [9, 10, 11].

The aim of this paper is to give a new formula for the strongly continuous one-parameter semigroup $e^{-u\mathcal{L}}$, $u > 0$, generated by $\mathcal{L}$. Moreover, we use the Sobolev spaces $L^2_s(\mathbb{H})$, $s \in \mathbb{R}$, in [1] to estimate $\|e^{-u\mathcal{L}} f\|_{L^2_s(\mathbb{H})}$, $u > 0$, in terms of $\|f\|_{L^2(\mathbb{H})}$ for all f in $L^2(\mathbb{H})$, and to give an estimate for $\|e^{-u\mathcal{L}} f\|_{L^2(\mathbb{H})}$ in terms of $\|f\|_{L^2_s(\mathbb{H})}$. The function F on $\mathbb{H} \times (0, \infty)$ given by

$$F(z, t, u) = (e^{-u\mathcal{L}} f)(z, t), \quad (z, t) \in \mathbb{H},\ u > 0,$$

is in fact the solution of the initial value problem

$$\begin{cases} \frac{\partial F}{\partial u}(z, t, u) = -(\mathcal{L}F)(z, t, u), & (z, t) \in \mathbb{H},\ u > 0, \\ F(z, t, 0) = f(z, t), & (z, t) \in \mathbb{H}, \end{cases} \tag{1.1}$$

for the sub-Laplacian $\mathcal{L}$.

Using the same techniques in Section 2 of [1], we get for all $f \in L^2(\mathbb{H})$ and $u > 0$,

$$(e^{-u\mathcal{L}} f)(z, t) = (2\pi)^{-1/2} \int_{-\infty}^{\infty} e^{-it\tau} (e^{-uL_\tau} f^\tau)(z)\, d\tau, \quad (z, t) \in \mathbb{H}, \tag{1.2}$$

where L_τ, $\tau > 0$, is the τ-twisted Laplacian given by

$$L_\tau = -\Delta + \frac{1}{4}(x^2 + y^2)\tau^2 - i\left(x\frac{\partial}{\partial y} - y\frac{\partial}{\partial x} \right) \tau$$

and f^τ is the function on $\mathbb{C}$ given by

$$f^\tau(z) = (2\pi)^{-1/2} \int_{-\infty}^{\infty} e^{it\tau} f(z, t)\, dt, \quad z \in \mathbb{C},$$

provided that the integral exists. In fact, $f^\tau(z)$ is the inverse Fourier transform of $f(z,t)$ with respect to t evaluated at τ. In this paper, the nonzero parameter τ can be looked at as Planck's constant.

To put the main results in perspective, let us note that there exists an explicit formula for the heat kernel of $\mathcal{L}$. To recall, for positive time u, the heat kernel of $\mathcal{L}$ is the kernel K_u of the integral operator $e^{-u\mathcal{L}}$ such that

$$e^{-u\mathcal{L}}f = K_u *_{\mathbb{H}} f$$

for all suitable functions f on $\mathbb{H}$, where

$$(K_u *_{\mathbb{H}} f)(z,t) = \int_{\mathbb{H}} K_u((z,t)\cdot(w,s)^{-1})f(w,s)\,dw\,ds, \quad (z,t)\in\mathbb{H},$$

provided that the integral exists. In fact, the formula for K_u, $u>0$, given by

$$K_u(z,t) = \frac{1}{8\pi^2}\int_{-\infty}^{\infty} e^{-it\tau}\frac{\tau}{\sinh(\tau u)}e^{-\frac{1}{4}\tau|z|^2\coth(\tau u)}\,d\tau, \quad (z,t)\in\mathbb{H},$$

dates back to the independent works of Gaveau [4] and Hulanicki [6] in the seventies. Recent works and references on the heat kernel and the heat equation for the sub-Laplacian $\mathcal{L}$ can be found in [8, 9, 10, 11, 12, 13] among others. Although the explicit formula for the heat kernel of $\mathcal{L}$ is interesting in its own right, it is not a tractable tool for obtaining the estimates for the solution of the initial value problem (1.1) that we have in mind. Since the symbol $\sigma(\mathcal{L})$ of $\mathcal{L}$ is given by

$$\sigma(\mathcal{L})(x,y,t;\xi,\eta,\tau) = \left(\xi+\frac{1}{2}y\tau\right)^2+\left(\eta-\frac{1}{2}x\tau\right)^2$$

for all (x,y,t) and (ξ,η,τ) in $\mathbb{R}^3$, it is easy to see that $\mathcal{L}$ is a nowhere elliptic partial differential operator on $\mathbb{R}^3$. So, the standard techniques of strongly continuous one-parameter semigroups generated by strongly elliptic operators in [2, 3, 5, 7, 17] and others are not applicable to the sub-Laplacian $\mathcal{L}$.

For the main results on the estimates in this paper, we use a formula for $e^{-u\mathcal{L}}$ in terms of the τ-Weyl transforms and the τ-Fourier–Wigner transforms of Hermite functions, which we recall in, respectively, Section 2 and Section 3. A result on the L^2-boundedness of τ-Weyl transforms is instrumental in obtaining the estimates.

Basic information on the classical Fourier–Wigner transforms, Wigner transforms and Weyl transforms can be found in [14] among others.

The τ-Weyl transforms and a result on the L^2-boundedness of the τ-Weyl transforms are given in Section 2. The τ-Fourier–Wigner transforms of Hermite functions are recalled in Section 3. A formula for $e^{-uL_\tau}f$, $u>0$, for every function f in $L^2(\mathbb{H})$ is given in Section 4. This formula is an extension of the formula obtained in [16] for $\tau=1$ and gives a formula for $e^{-u\mathcal{L}}$, $u>0$, immediately using the inverse Fourier transform as indicated by (1.2). In Section 5, we use the family $L^2_s(\mathbb{H})$, $s\in\mathbb{R}$, of Sobolev spaces with respect to the center of the Heisenberg group in [1] to obtain Sobolev estimates for $e^{-u\mathcal{L}}f, u>0$, in terms of $\|f\|_{L^2(\mathbb{H})}$, and estimates for $\|e^{-u\mathcal{L}}f\|_{L^2(\mathbb{H})}$, $u>0$, in terms of the Sobolev norms $\|f\|_{L^2_s(\mathbb{H})}$ of f in $L^2_s(\mathbb{H})$.

2. τ-Weyl transforms

Let f and g be functions in $L^2(\mathbb{R})$. Then for τ in $\mathbb{R}\setminus\{0\}$, the τ-Fourier–Wigner transform $V_\tau(f,g)$ is defined by

$$V_\tau(f,g)(q,p) = (2\pi)^{-1/2}|\tau|^{1/2}\int_{-\infty}^{\infty} e^{i\tau qy} f\left(y+\frac{p}{2}\right)\overline{g\left(y-\frac{p}{2}\right)}\,dy \tag{2.1}$$

for all q and p in $\mathbb{R}$. In fact,

$$V_\tau(f,g)(q,p) = |\tau|^{1/2}V(f,g)(\tau q,p), \quad q,p\in\mathbb{R},$$

where $V(f,g)$ is the classical Fourier–Wigner transform of f and g. A proof can be found in [1].

It can be proved that $V_\tau(f,g)$ is a function in $L^2(\mathbb{C})$ and we have the Moyal identity stating that

$$\|V_\tau(f,g)\|_{L^2(\mathbb{C})} = \|f\|_{L^2(\mathbb{R})}\|g\|_{L^2(\mathbb{R})}, \quad \tau\in\mathbb{R}\setminus\{0\}. \tag{2.2}$$

We define the τ-Wigner transform $W_\tau(f,g)$ of f and g by

$$W_\tau(f,g) = V_\tau(f,g)^\wedge. \tag{2.3}$$

Then we have the following connection of the τ-Wigner transform with the usual Wigner transform.

Theorem 2.1. *Let $\tau\in\mathbb{R}\setminus\{0\}$. Then for all functions f and g in $L^2(\mathbb{R})$,*

$$W_\tau(f,g)(x,\xi) = |\tau|^{-1/2}W(f,g)(x/\tau,\xi), \quad x,\xi\in\mathbb{R},$$

where $W(f,g)$ is the classical Wigner transform of f and g.

It is obvious that

$$W_\tau(f,g) = \overline{W_\tau(g,f)}, \quad f,g\in L^2(\mathbb{R}). \tag{2.4}$$

Let $\sigma\in L^p(\mathbb{C})$, $1\le p\le\infty$. Then for all τ in $\mathbb{R}\setminus\{0\}$ and all functions f in the Schwartz space $\mathcal{S}(\mathbb{R})$ on $\mathbb{R}$, we define $W_\sigma^\tau f$ to be the tempered distribution on $\mathbb{R}$ by

$$(W_\sigma^\tau f,g) = (2\pi)^{-1/2}\int_{-\infty}^{\infty}\int_{-\infty}^{\infty}\sigma(x,\xi)W_\tau(f,g)(x,\xi)\,dx\,d\xi \tag{2.5}$$

for all g in $\mathcal{S}(\mathbb{R})$, where (F,G) is defined by

$$(F,G) = \int_{\mathbb{R}^n} F(z)\overline{G(z)}\,dz$$

for all measurable functions F and G on $\mathbb{R}^n$, provided that the integral exists. We call W_σ^τ the τ-Weyl transform associated to the symbol σ. It is easy to see that if σ is a symbol in the Schwartz space $\mathcal{S}(\mathbb{C})$ on $\mathbb{C}$, then $W_\sigma^\tau f$ is a function in $\mathcal{S}(\mathbb{R})$ for all f in $\mathcal{S}(\mathbb{R})$.

We have the following estimate for the norm of the Weyl transform $W_{\hat{\sigma}}^\tau$ in terms of the L^p norm of the symbol σ when $\sigma\in L^p(\mathbb{C})$, $1\le p\le 2$.

Theorem 2.2. *Let $\sigma \in L^p(\mathbb{C})$, $1 \le p \le 2$. Then $W^\tau_{\hat{\sigma}} : L^2(\mathbb{R}) \to L^2(\mathbb{R})$ is a bounded linear operator and*

$$\|W^\tau_{\hat{\sigma}}\|_* \le (2\pi)^{-1/p}|\tau|^{-(1/2)+(1/p)}\|\sigma\|_{L^p(\mathbb{C})},$$

where $\|W^\tau_{\hat{\sigma}}\|_*$ is the operator norm of $W^\tau_{\hat{\sigma}} : L^2(\mathbb{R}) \to L^2(\mathbb{R})$.

Proof. Let f and g be functions in $\mathcal{S}(\mathbb{R})$. Then

$$\begin{aligned}(W^\tau_{\hat{\sigma}} f, g) &= (2\pi)^{-1/2}\int_{\infty}^{\infty}\int_{-\infty}^{\infty} \hat{\sigma}(x,\xi)W_\tau(f,g)(x,\xi)\,dx\,d\xi \\ &= (2\pi)^{-1}|\tau|^{-1/2}\int_{-\infty}^{\infty}\int_{-\infty}^{\infty} \hat{\sigma}(x,\xi)W(f,g)(x/\tau,\xi)\,dx\,d\xi \\ &= (2\pi)^{-1}|\tau|^{1/2}\int_{-\infty}^{\infty}\int_{\infty}^{\infty} \hat{\sigma}(\tau x,\xi)W(f,g)(x,\xi)\,dx\,d\xi.\end{aligned}$$

But

$$\hat{\sigma}(\tau x,\xi) = |\tau|^{-1}\widehat{\sigma_{1/\tau}}(x,\xi), \quad x,\xi \in \mathbb{R},$$

where $\sigma_{1/\tau}$ is the dilation of σ with respect to the first variable by the amount $1/\tau$. More precisely,

$$\sigma_{1/\tau}(q,p) = \sigma(q/\tau, p), \quad q,p \in \mathbb{R}.$$

So,

$$\begin{aligned}(W^\tau_{\hat{\sigma}} f, g) &= (2\pi)^{-1/2}|\tau|^{-1/2}\int_{-\infty}^{\infty}\int_{-\infty}^{\infty} \widehat{\sigma_{1/\tau}}(x,\xi)W(f,g)(x,\xi)\,dx\,d\xi \\ &= |\tau|^{-1/2}(W_{\widehat{\sigma_{1/\tau}}} f, g),\end{aligned}$$

where $W_{\widehat{\sigma_{1/\tau}}}$ is the classical Weyl transform with symbol $\widehat{\sigma_{1/\tau}}$.

Thus, it follows from Theorem 21.1 in [15] that $W^\tau_{\hat{\sigma}} : L^2(\mathbb{R}) \to L^2(\mathbb{R})$ is a bounded linear operator and

$$\|W^\tau_{\hat{\sigma}}\|_* \le |\tau|^{-1/2}(2\pi)^{-1/p}\|\sigma_{1/\tau}\|_{L^p(\mathbb{C})} = (2\pi)^{-1/p}|\tau|^{-(1/2)+(1/p)}\|\sigma\|_{L^p(\mathbb{C})}. \qquad \square$$

3. Fourier–Wigner transforms of Hermite functions

For $\tau \in \mathbb{R}\setminus\{0\}$ and for $k = 0,1,2,\dots$, we define e^τ_k to be the function on $\mathbb{R}$ by

$$e^\tau_k(x) = |\tau|^{1/4}e_k(\sqrt{|\tau|}x), \quad x \in \mathbb{R}.$$

Here, e_k is the Hermite function of order k defined by

$$e_k(x) = \frac{1}{(2^k k!\sqrt{\pi})^{1/2}}e^{-x^2/2}H_k(x), \quad x \in \mathbb{R},$$

where H_k is the Hermite polynomial of degree k given by

$$H_k(x) = (-1)^k e^{x^2/2}\left(\frac{d}{dx}\right)^k (e^{-x^2}), \quad x \in \mathbb{R}.$$

For $j, k = 0, 1, 2, \ldots$, we define $e^\tau_{j,k}$ on $\mathbb{R}^2$ by

$$e^\tau_{j,k} = V_\tau(e^\tau_j, e^\tau_k).$$

The following theorem gives the connection of $\{e^\tau_{j,k} : j, k = 0, 1, 2, \ldots\}$ with $\{e_{j,k} : j, k = 0, 1, 2, \ldots\}$, where

$$e_{j,k} = V(e_j, e_k), \quad j, k = 0, 1, 2, \ldots.$$

A proof can be found in [1].

Theorem 3.1. *For $\tau \in \mathbb{R} \setminus \{0\}$ and for $j, k = 0, 1, 2, \ldots$,*

$$e^\tau_{j,k}(q, p) = |\tau|^{1/2} e_{j,k}\left(\frac{\tau}{\sqrt{|\tau|}} q, \sqrt{|\tau|} p\right), \quad q, p \in \mathbb{R}.$$

Theorem 3.2. $\{e^\tau_{j,k} : j, k = 0, 1, 2, \ldots\}$ *forms an orthonormal basis for $L^2(\mathbb{R}^2)$.*

Theorem 3.2 follows from Theorem 3.1 and Theorem 21.2 in [14] to the effect that $\{e_{j,k} : j, k = 0, 1, 2, \ldots\}$ is an orthonormal basis for $L^2(\mathbb{R}^2)$.

Theorem 3.3. *For $j, k = 0, 1, 2, \ldots$,*

$$L_\tau e^\tau_{j,k} = (2k + 1)|\tau| e^\tau_{j,k}.$$

Theorem 3.3 can be proved using Theorem 3.1 and Theorem 22.2 in [14] telling us that for $j, k = 0, 1, 2, \ldots$, $e_{j,k}$ is an eigenfunction of L_1 corresponding to the eigenvalue $2k + 1$.

4. A formula and an estimate for e^{-uL_τ}, $u > 0$

Let $\tau \in \mathbb{R}\setminus\{0\}$. Then a formula for e^{-uL_τ}, $u > 0$, is given by the following theorem.

Theorem 4.1. *Let $f \in L^2(\mathbb{C})$. Then for $u > 0$,*

$$e^{-uL_\tau} f = (2\pi)^{1/2} \sum_{k=0}^{\infty} e^{-(2k+1)|\tau|u} V_\tau(W^\tau_{\hat{f}} e^\tau_k, e^\tau_k),$$

where the convergence of the series is understood to be in $L^2(\mathbb{C})$.

Proof. Let $f \in L^2(\mathbb{C})$. Then for $u > 0$,

$$e^{-uL_\tau} f = \sum_{k=0}^{\infty} \sum_{j=0}^{\infty} e^{-(2k+1)|\tau|u} (f, e^\tau_{j,k}) e^\tau_{j,k}, \tag{4.1}$$

where the series is convergent in $L^2(\mathbb{C})$. Now, by Plancherel's theorem, (2.1) and (2.3)–(2.5),

$$\begin{aligned}(f, e^\tau_{j,k}) &= \int_{\mathbb{C}} f(z) \overline{V_\tau(e^\tau_j, e^\tau_k)(z)}\, dz = \int_{\mathbb{C}} \hat{f}(\zeta) \overline{V_\tau(e^\tau_j, e^\tau_k)^\wedge(\zeta)}\, d\zeta \\ &= \int_{\mathbb{C}} \hat{f}(\zeta) \overline{W_\tau(e^\tau_j, e^\tau_k)(\zeta)}\, d\zeta = (2\pi)^{1/2} (W_{\hat{f}} e^\tau_k, e^\tau_j)\end{aligned} \tag{4.2}$$

for $j, k = 0, 1, 2, \ldots$.

Similarly, for $j, k = 0, 1, 2, \ldots$, and g in $L^2(\mathbb{C})$, we get

$$(e_{j,k}^\tau, g) = \overline{(g, e_{j,k}^\tau)} = (2\pi)^{1/2}\overline{(W_{\hat{g}}^\tau e_k^\tau, e_j^\tau)} = (2\pi)^{1/2}(e_j^\tau, W_{\hat{g}}^\tau e_k^\tau). \tag{4.3}$$

So, by (4.1)–(4.3), Fubini's theorem and Parseval's identity,

$$\begin{aligned}(e^{-uL_\tau} f, g) &= 2\pi \sum_{k=0}^{\infty} e^{-(2k+1)|\tau|u} \sum_{j=0}^{\infty} (W_{\hat{f}}^\tau e_k^\tau, e_j^\tau)(e_j^\tau, W_{\hat{g}}^\tau e_k^\tau) \\ &= 2\pi \sum_{k=0}^{\infty} e^{-(2k+1)|\tau|u} (W_{\hat{f}}^\tau e_k^\tau, W_{\hat{g}}^\tau e_k^\tau). \end{aligned} \tag{4.4}$$

By Plancherel's theorem and (2.3)–(2.5),

$$\begin{aligned}(W_{\hat{f}}^\tau e_k^\tau, W_{\hat{g}}^\tau e_k^\tau) &= (2\pi)^{-1/2} \overline{\int_{\mathbb{C}} \hat{g}(z) W_\tau(e_k^\tau, W_{\hat{f}}^\tau e_k^\tau)(z)\, dz} \\ &= (2\pi)^{-1/2} \int_{\mathbb{C}} W_\tau(W_{\hat{f}}^\tau e_k^\tau, e_k^\tau)(z) \overline{\hat{g}(z)}\, dz \\ &= (2\pi)^{-1/2} \int_{\mathbb{C}} V_\tau(W_{\hat{f}}^\tau e_k^\tau, e_k^\tau)(z) \overline{g(z)}\, dz \end{aligned} \tag{4.5}$$

for $k = 0, 1, 2, \ldots$. Thus, by (4.4), (4.5) and Fubini's theorem,

$$\begin{aligned}(e^{-uL_\tau} f, g) &= (2\pi)^{1/2} \sum_{k=0}^{\infty} e^{-(2k+1)|\tau|u} (V_\tau(W_{\hat{f}}^\tau e_k^\tau, e_k^\tau), g) \\ &= (2\pi)^{1/2} \left(\sum_{k=0}^{\infty} e^{-(2k+1)|\tau|u} V_\tau(W_{\hat{f}}^\tau e_k^\tau, e_k^\tau), g \right) \end{aligned} \tag{4.6}$$

for all f and g in $L^2(\mathbb{C})$ and $u > 0$. Thus, by (4.6),

$$e^{-uL_\tau} f = (2\pi)^{1/2} \sum_{k=0}^{\infty} e^{-(2k+1)|\tau|u} V_\tau(W_{\hat{f}}^\tau e_k^\tau, e_k^\tau)$$

for all f in $L^2(\mathbb{C})$ and $u > 0$. □

For all τ in $\mathbb{R} \setminus \{0\}$, we have the following estimate for the L^2 norm of $e^{-uL_\tau} f$, $u > 0$, in terms of the L^p norm of f.

Theorem 4.2. *Let $\tau \in \mathbb{R} \setminus \{0\}$. Then for all functions f in $L^p(\mathbb{C})$, $1 \le p \le 2$,*

$$\|e^{-uL_\tau} f\|_{L^2(\mathbb{C})} \le (2\pi)^{-(1/p)+(1/2)} |\tau|^{-(1/2)+(1/p)} \frac{1}{2\sinh(|\tau|u)} \|f\|_{L^p(\mathbb{C})}.$$

Proof. By Theorem 4.1, the Moyal identity (2.2) and the fact that

$$\|e_k^\tau\|_{L^2(\mathbb{R})} = 1, \quad k = 0, 1, 2, \ldots,$$

we get

$$\|e^{-uL_\tau} f\|_{L^2(\mathbb{C})} \le (2\pi)^{1/2} e^{-|\tau|u} \sum_{k=0}^{\infty} e^{-2k|\tau|u} \|W_{\hat{f}}^\tau e_k^\tau\|_{L^2(\mathbb{R})}, \quad u > 0. \tag{4.7}$$

Applying Theorem 2.2 to (4.7), we get

$$\begin{aligned}\|e^{-uL_\tau} f\|_{L^2(\mathbb{C})} &\le (2\pi)^{-(1/p)+(1/2)}|\tau|^{-(1/2)+(1/p)}e^{-|\tau|u}\left(\sum_{k=0}^{\infty} e^{-2k|\tau|u}\right)\|f\|_{L^p(\mathbb{C})}\\ &= (2\pi)^{-(1/p)+(1/2)}|\tau|^{-(1/2)+(1/p)}\frac{1}{2\sinh(|\tau|u)}\|f\|_{L^p(\mathbb{C})},\end{aligned}$$

as asserted. □

5. Sobolev estimates for $e^{-u\mathcal{L}}$, $u > 0$

Let $s \in \mathbb{R}$. Then as in [1], we define $L^2_s(\mathbb{H})$ to be the set of all functions f in $L^2(\mathbb{H})$ such that

$$\int_{\mathbb{C}}\int_{-\infty}^{\infty} |\tau|^{2s}|f^\tau(z)|^2 d\tau\, dz < \infty.$$

For every f in $L^2_s(\mathbb{H})$, we define the norm $\|f\|_{L^2_s(\mathbb{H})}$ by

$$\|f\|^2_{L^2_s(\mathbb{H})} = \int_{\mathbb{C}}\int_{-\infty}^{\infty} |\tau|^{2s}|f^\tau(z)|^2 d\tau\, dz.$$

Then it can be shown easily that $L^2_s(\mathbb{H})$ is an inner product space in which the inner product $(\,,\,)_{L^2_s(\mathbb{H})}$ is given by

$$(f,g)_{L^2_s(\mathbb{H})} = \int_C\int_{-\infty}^{\infty} |\tau|^{2s} f^\tau(z)\overline{g^\tau(z)}\, d\tau\, dz$$

for all f and g in $L^2_s(\mathbb{H})$.

Theorem 5.1. *Let $s \ge 1$. Then for $u > 0$, $e^{-u\mathcal{L}} : L^2(\mathbb{H}) \to L^2_s(\mathbb{H})$ is a bounded linear operator and*

$$\|e^{-u\mathcal{L}} f\|_{L^2_s(\mathbb{H})} \le \frac{c_s}{2u^s}\|f\|_{L^2(\mathbb{H})}, \quad f \in L^2(\mathbb{H}),$$

where

$$c_s = \sup_{\tau\in\mathbb{R}\setminus\{0\}} (|\tau|^s/\sinh|\tau|).$$

Proof. Let $u > 0$. Then, by Fubini's theorem, Plancherel's theorem, (1.2) and Theorem 4.2 with $p = 2$,

$$\begin{aligned}\|e^{-u\mathcal{L}} f\|^2_{L^2_s(\mathbb{H})} &= \int_{\mathbb{C}}\int_{-\infty}^{\infty} |\tau|^{2s}|(e^{-u\mathcal{L}} f)^\tau(z)|^2 d\tau\, dz\\ &= \int_{-\infty}^{\infty} |\tau|^{2s}\left(\int_{\mathbb{C}} |(e^{-u\mathcal{L}} f)^\tau(z)|^2 dz\right) d\tau\\ &= \int_{-\infty}^{\infty} |\tau|^{2s}\left(\int_{\mathbb{C}} |(e^{-uL_\tau} f^\tau)(z)|^2 dz\right) d\tau\\ &= \int_{-\infty}^{\infty} |\tau|^{2s}\|e^{-uL_\tau} f^\tau\|^2_{L^2(\mathbb{C})} d\tau\end{aligned}$$

$$= \frac{1}{4}\left(\int_{-\infty}^{\infty} \frac{|\tau|^{2s}}{\sinh^2(|\tau|u)} \|f^\tau\|^2_{L^2(\mathbb{C})} d\tau\right)$$
$$= \frac{1}{4}\int_{-\infty}^{\infty} \frac{|\tau|^{2s}}{\sinh^2(|\tau|u)} \left(\int_{\mathbb{C}} |f^\tau(z)|^2 dz\right) d\tau$$
$$= \frac{1}{4\,u^{2s+1}}\int_{-\infty}^{\infty} \frac{|\tau|^{2s}}{\sinh^2(|\tau|u)} \left(\int_{\mathbb{C}} |\check{f}(z,\tau/u)|^2 dz\right) d\tau,$$

where $\check{f}$ is the inverse Fourier transform of f with respect to t. So, using a simple change of variable and letting

$$C_s = \sup_{\tau\in\mathbb{R}\setminus\{0\}} (|\tau|^{2s}/\sinh^2|\tau|),$$

we get

$$\|e^{-u\mathcal{L}}f\|^2_{L^2_s(\mathbb{H})} \leq \frac{C_s}{4u^{2s}}\int_{-\infty}^{\infty}\left(\int_{\mathbb{C}} |\check{f}(z,\tau)|^2 dz\right) d\tau = \frac{C_s}{4u^{2s}}\|f\|^2_{L^2(\mathbb{H})}$$

and this completes the proof. □

The following result complements Theorem 5.1.

Theorem 5.2. *Let $s \leq -1$. Then for $u > 0$, $e^{-u\mathcal{L}} : L^2_s(\mathbb{H}) \to L^2(\mathbb{H})$ is a bounded linear operator and*

$$\|e^{-u\mathcal{L}}f\|_{L^2(\mathbb{H})} \leq \frac{c_{-s}}{2u^{-s}}\|f\|_{L^2_s(\mathbb{H})}, \quad f \in L^2_s(\mathbb{H}),$$

where

$$c_{-s} = \sup_{\tau\in\{0\}} (|\tau|^{-s}\sinh|\tau|).$$

The proof of Theorem 5.2 is very similar to that of Theorem 5.1 and is hence omitted.

References

[1] A. Dasgupta and M.W. Wong, Weyl transforms and the inverse of the sub-Laplacian on the Heisenberg group, in *Pseudo-Differential Operators: Partial Differential Equations and Time-Frequency Analysis,* Editors: L. Rodino, B.-W. Schulze and M.W. Wong, Fields Institute Communications, American Mathematical Society, 2007, 27–36.

[2] E.B. Davies, *One-Parameter Semigroups*, Academic Press, 1980.

[3] A. Friedman, *Partial Differential Equations*, Holt, Reinhart and Winston, 1969.

[4] B. Gaveau, Principe de moindre action, propagation de la chaleur et estimées sous elliptiques sur certains groupes nilpotents, *Acta Math.* **139** (1977), 95–153.

[5] E. Hille and R.S. Phillips, *Functional Analysis and Semi-groups*, Third Printing of Revised Version of 1957, American Mathematical Society, 1974.

[6] A. Hulanicki, The distribution of energy in the Brownian motion in the Gaussian field and analytic-hypoellipticity of certain subelliptic operators on the Heisenberg group, *Studia Math.* **56** (1976), 165–173.

[7] G.M. Iancu and M.W. Wong, Global solutions of semilinear heat equations in Hilbert spaces, *Abstr. Appl. Anal.* **1** (1996), 263–276.

[8] J. Kim and M.W. Wong, Positive definite temperature functions on the Heisenberg group, *Appl. Anal.* **85** (2006), 987–1000.

[9] E.M. Stein, *Harmonic Analysis: Real-Variable Methods, Orthogonality, and Oscillatory Integrals,* Princeton University Press, 1993.

[10] S. Thangavelu, *Harmonic Analysis on the Heisenberg Group*, Birkhäuser, 1998.

[11] S. Thangavelu, *An Introduction to the Uncertainty Principle: Hardy's Theorem on Lie Groups,* Birkhäuser, 2004.

[12] S. Thangavelu, A survey of Hardy type theorems, in *Advances in Analysis*, Editors: H.G.W. Begehr, R.P. Gilbert, M.E. Muldoon and M.W. Wong, World Scientific, 2005, 39–70.

[13] J. Tie, The non-isotropic twisted Laplacian on $\mathbb{C}^n$ and the sub-Laplacian on $\mathbf{H}_n$, *Comm. Partial Differential Equations* **31** (2006), 1047–1069.

[14] M.W. Wong, *Weyl Transforms,* Springer-Verlag, 1998.

[15] M.W. Wong, Weyl transforms, the heat kernel and Green function of a degenerate elliptic operator, *Ann. Global Anal. Geom.* **28** (2005), 271–283.

[16] M.W. Wong, The heat equation for the Hermite operator on the Heisenberg group, *Hokkaido Math. J.* **34** (2005), 393–404.

[17] K. Yosida, *Functional Analysis*, Reprint of Sixth Edition, Springer-Verlag, 1995.

Aparajita Dasgupta and M.W. Wong
Department of Mathematics and Statistics
York University
4700 Keele Street
Toronto
Ontario M3J 1P3, Canada
e-mail: adgupta@mathstat.yorku.ca
mwwong@mathstat.yorku.ca

Operator Theory:
Advances and Applications, Vol. 189, 43–65

Construction of the Fundamental Solution and Curvature of Manifolds with Boundary

Chisato Iwasaki

Abstract. The method of construction of the fundamental solution for the heat equations of the initial boundary value problem on manifolds with boundary, which is applicable to calculate traces of operators, is discussed. This shows a relation of the singularity of the fundamental solution and the curvature of manifolds. We may say that this is an extension of a local version of Gauss-Bonnet-Chern theorem for manifolds with boundary.

Mathematics Subject Classification (2000). Primary 58J35; Secondary 58J40, 58J20.

Keywords. Symbolic calculus, fundamental solution, curvature, local index.

1. Introduction

In this paper we give, by means of symbolic calculus of pseudo-differential operators, an extension theorem of a local version of Gauss-Bonnet-Chern theorem on manifolds with boundary given in C. Iwasaki [11].

Let M be a Riemannian manifold and let $\chi(M)$ be the Euler characteristic of M. Let dv and $d\sigma$ be a volume element of M and that of its boundary ∂M respectively. The Gauss-Bonnet-Chern theorem which was proved by S. Chern [2], [3] states, broadly speaking, that topological quantity can be represented by the geometric quantities.

Gauss-Bonnet-Chern Theorem. Let M be a Riemannian manifold of dimension n.

(1) For M without boundary we have

$$\chi(M) = \int_M C_n(x, M)dv.$$

(2) If M has boundary,

$$\chi(M) = \int_M C_n(x, M)dv + \int_{\partial M} D_{n-1}(x)d\sigma.$$

The precise definitions of $C_n(x, M)$ and $D_{n-1}(x)$ are given in Section 3 and Section 5 respectively. We note that if n is odd,

$$D_{n-1}(x) = \frac{1}{2} C_{n-1}(x, \partial M)$$

holds.

Analytical proofs are based on the following formula (cf. V.K. Patodi [15]):

$$\chi(M) = \int_M \sum_{p=0}^{n} (-1)^p \operatorname{tr} e_p(t, x, x) dv, \tag{1.1}$$

where $e_p(t, x, y)$ denotes the kernel of the fundamental solution $E_p(t)$ for the heat equation for Δ_p on differential p-forms $A^p(M) = \Gamma(\wedge^p T^*(M))$ and $\operatorname{tr} e_p(t, x, x)$ means the trace of operator $e_p(t, x, x)$ on $\wedge^p T^*_x(M)$.

If M has no boundary, $E_p(t)$ is the fundamental solution for the Cauchy problem, that is,

$$E_p(t) f(x) = \int_M e_p(t, x, y) \varphi(y) dv_y, \quad \varphi \in A^p(M)$$

satisfies for $0 < T < \infty$

$$\begin{aligned} \left(\frac{d}{dt} + \Delta_p\right) E_p(t) = 0 \quad & \text{in } (0, T) \times M, \\ E_p(0) = I \quad & \text{in } M. \end{aligned}$$

If M has boundary ∂M, $E_p(t)$ satisfies the following equations instead of the above equations

$$\begin{aligned} \left(\frac{d}{dt} + \Delta_p\right) E_p(t) = 0 \quad & \text{in } (0, T) \times M, \\ B_p E_p(t) = 0 \quad & \text{on } (0, T) \times \partial M, \\ E_p(0) = I \quad & \text{in } M, \end{aligned}$$

with some boundary operator B_p (see Section 4).

So, we may say that a local version of Gauss-Bonnet-Chern theorem holds, if we have

$$\sum_{p=0}^{n} (-1)^p \operatorname{tr} e_p(t, x, x) = C_n(x, M) + 0(\sqrt{t}) \tag{1.2}$$

as t tends to 0.

The author has proved (1.2) for $x \in M \backslash \partial M$ and has also shown for $x \in \partial M$

$$\sum_{p=0}^{n} (-1)^p \operatorname{tr} e_p(t, x, x) = 2D_{n-1}(x) \frac{1}{\sqrt{t}} + O(1) \qquad \text{as } t \to 0 \tag{1.3}$$

in [11]. The author has shown a way to constructing the fundamental solution of which the main term is enough to show (1.2) and (1.3). It is constructed by technique of pseudo-differential operators of new weights on symbols.

In this paper, a generalization of a local version of Gauss-Bonnet-Chern theorem on manifolds with boundary is obtained. Before stating our theorems, we introduce some notations.

We denote $\mathcal{I}$ and $\mathcal{I}_0$ the set of index

$$\mathcal{I} = \{I = (i_1, i_2, \dots, i_r) : 0 \le r \le n,\ 1 \le i_1 < \cdots < i_r \le n\},$$
$$\mathcal{I}_0 = \{I = (i_1, i_2, \dots, i_r) : 0 \le r \le n-1,\ 1 \le i_1 < \cdots < i_r \le n-1\}$$

and

$$\binom{a}{b} = 0 \text{ if } a < b, \text{ or } b < 0, \quad \binom{0}{0} = 1.$$

In the rest of this paper fix an integer ℓ such that $0 \le \ell \le n$.

Theorem 1.1 (Main Theorem). *Let M be a Riemannian manifold of dimension n with boundary and let $E_p(t)$ be the fundamental solution on $\Gamma(\wedge^p T^*(M))$. Given arbitrary constants $\{k_j\}_{j=\ell+1,\dots,n}$ define the constants f_p as follows:*

$$f_p = (-1)^p \left\{ \binom{n-p}{n-\ell} + \sum_{j=\max\{p,\ell+1\}}^{n} k_j \binom{n-p}{n-j} \right\} \quad (0 \le p \le n). \tag{1.4}$$

(1) *If $x \in M \backslash \partial M$, we have*

$$\sum_{p=0}^{n} f_p \operatorname{tr} e_p(t,x,x) = C_\ell(x) t^{-\frac{n}{2}+\frac{\ell}{2}} + 0(t^{-\frac{n}{2}+\frac{\ell}{2}+\frac{1}{2}}) \qquad \text{as } t \to 0,$$

where $C_\ell(x)$ is given in Definition 3.4.

(2) *If $x \in \partial M$, we have*

$$\sum_{p=0}^{n} f_p \operatorname{tr} e_p(t,x,x) = 2D_{\ell-1}(x) t^{-\frac{n}{2}+\frac{\ell}{2}-\frac{1}{2}} + 0(t^{-\frac{n}{2}+\frac{\ell}{2}}) \qquad \text{as } t \to 0,$$

where $D_{\ell-1}(x)$ is given in Definition 5.3.

(3) *We have*

$$\int_M \sum_{p=0}^{n} f_p \operatorname{tr} e_p(t,x,x) \psi(x) dv$$
$$= \left(\int_M C_\ell(x,M)\psi(x) dv + \int_{\partial M} D_{\ell-1}(x)\psi(x) d\sigma \right) t^{-\frac{n}{2}+\frac{\ell}{2}} + 0(t^{-\frac{n}{2}+\frac{\ell}{2}+\frac{1}{2}})$$

as $t \to 0$ for any $\psi(x) \in C^\infty(\overline{M})$.

Remark 1.2. Assume $\ell = n$. Then $f_p = (-1)^p$ for all p as in (1.2) and $C_\ell = C_n, D_{\ell-1} = D_{n-1}$. So statements (1) and (2) of Main Theorem are a local version of Gauss-Bonnet-Chern theorem proved in [11]. The statement (3) is the Gauss-Bonnet-Chern theorem.

Remark 1.3. If M has no boundary, $D_{\ell-1}$ vanishes. So (1) of the Main Theorem has been obtained in C. Iwasaki [12].

Remark 1.4. Assume $k_j = 0$ for all j. Then $f_p = (-1)^p\binom{n-p}{n-\ell}$ $(0 \le p \le \ell)$, $f_p = 0(\ell+1 \le p \le n)$. In this case, under the assumption M has no boundary, (1) of the Main Theorem coincides with the result given in [5].

Since B_p is a coercive boundary operator for Δ_p, it is well known that $e_p(t,x,y)$ is regular either $x \neq y$ or $t > 0$. On the diagonal set, $e_p(t,x,y)$ has singularity with respect to t as follows:

$$e_p(t,x,x) \sim c_0(x)t^{-\frac{n}{2}} + c_1(x)t^{-\frac{n}{2}+\frac{1}{2}} + \cdots + \cdots \qquad t \to 0.$$

In [11] it is proved that the singularity of $\sum_{p=0}^{n}(-1)^p \mathrm{tr}\, e_p(t,x,x)$ at any point x in M vanishes by an algebraic theorem on linear spaces stated in H.L. Cycon-R.G. Froese-W. Kirsch-B. Simon [4]. This theorem is given in Section 2 of this paper, in a form suitable for our discussion.

Our point is that one can prove the Main Theorem by only calculating the main term of the symbol of the fundamental solution, introducing a new weight of symbols of pseudo-differential operators.

The plan of this paper is the following. In Section 2 we state some algebraic theorems, which are proved in [12]. In Section 3 the asymptotic expansion of the fundamental solution for the Cauchy problem is discussed. This argument is similar to one of [12]. Our boundary operator B_p will be explained in Section 4. See [11] for the detailed argument. Section 5 is devoted to the proof of (2) and (3) of the Main Theorem, which is an essential part of this paper.

2. Algebraic properties for the calculation of the trace

Let V be a vector space of dimension n with an inner product and let $\wedge^p(V)$ be its anti-symmetric p tensors. Set $\wedge^*(V) = \sum_{p=0}^{n} \wedge^p(V)$. Fix an orthonormal basis $\{v_1, \ldots, v_n\}$ of V. Let a_i^* be a linear transformation on $\wedge^*(V)$ defined by $a_i^* v = v_i \wedge v$ and let a_i be the adjoint operator of a_i^* on $\wedge^*(V)$.

We note that $\{a_i^*, a_j\}_{1\le i,j\le n}$ satisfy the following relations.

$$\begin{aligned} a_i a_j + a_j a_i &= 0, \\ a_i^* a_j^* + a_j^* a_i^* &= 0, \\ a_i a_j^* + a_j^* a_i &= \delta_{ij}. \end{aligned} \tag{2.1}$$

Definition 2.1. Set

$$\beta_\phi = 1, \beta_j = a_j^* a_j - a_j a_j^* \ \text{ for } \ 1 \le j \le n$$

and

$$\beta_I = \beta_{i_1} \cdots \beta_{i_k} \text{for } I = (i_1, \ldots, i_k) \in \mathcal{I}.$$

It is clear that

$$a_k^* a_k = \frac{1}{2}(1+\beta_k), \quad \beta_j\beta_k = \beta_k\beta_j, \quad \beta_j^2 = 1.$$

The following assertion is essentially due to [15].

Proposition 2.2. *We have for any* $I = (i_1, \ldots, i_k) \in \mathcal{I}$ *the following assertions:*

(1) *If* $p < k$
$$\mathrm{tr}[\beta_I a_{j_1} a_{j_2} \cdots a_{j_p} a^*_{h_1} a^*_{h_2} \cdots a^*_{h_p}] = 0.$$

(2) *Suppose* $p = k$ *and* $\{j_1, j_2, \ldots, j_k\} \neq \{i_1, i_2, \ldots, i_k\}$ *or* $\{h_1, h_2, \ldots, h_k\} \neq \{i_1, i_2, \ldots, i_k\}$. *Then*
$$\mathrm{tr}[\beta_I a_{j_1} a_{j_2} \cdots a_{j_p} a^*_{h_1} a^*_{h_2} \cdots a^*_{h_p}] = 0.$$

(3) *Let* π, σ *be elements of the permutation group of degree* k. *Then we have*
$$\mathrm{tr}[\beta_I a^*_{i_{\pi(1)}} a_{i_{\sigma(1)}} a^*_{i_{\pi(2)}} a_{i_{\sigma(2)}} \cdots a^*_{i_{\pi(k)}} a_{i_{\sigma(k)}}] = 2^{n-k}\,\mathrm{sign}(\pi)\,\mathrm{sign}(\sigma).$$

Definition 2.3. Let Ψ_p be the projection of $\wedge^*(V)$ on $\wedge^p(V)$ and let $\Gamma_0 = 1$, $\Gamma_k = \sum_{I \in \mathcal{I}, \sharp(I)=k} \beta_I$.

The following theorem is a key algebraic argument of this paper.

Theorem 2.4. *Let* f_p $(0 \le p \le n)$ *be constants of the form*
$$f_p = (-1)^p \Big\{ \binom{n-p}{n-\ell} + \sum_{j=\max(\ell+1,p)}^{n} k_j \binom{n-p}{n-j} \Big\}$$
with any constants k_j $(\ell + 1 \le j \le n)$. *Then the following equation holds*
$$\sum_{q=0}^{n} f_q \Psi_q = (-1)^\ell 2^{\ell-n} \Gamma_\ell + \sum_{p=\ell+1}^{n} \alpha_p \Gamma_p$$
with α_p $(\ell + 1 \le p \le n)$ *defined by*
$$\alpha_p = (-1)^p \Big\{ 2^{\ell-n} \binom{p}{\ell} + \sum_{j=\ell+1}^{p} \binom{p}{j} k_j \Big\}.$$

Especially

(1) *If* $\ell = n$, *then*
$$\sum_{q=0}^{n} f_q \Psi_q = (-1)^n \Gamma_n$$
holds if and only if
$$f_p = (-1)^p \quad \text{for any } p.$$

(2) *If* $\alpha_p = (-1)^p 2^{\ell-n} \binom{p}{\ell}$ $(\ell + 1 \le p \le n)$,
$$f_p = \begin{cases} (-1)^p \binom{n-p}{n-\ell}, & (0 \le p \le \ell); \\ 0, & (\ell + 1 \le p \le n). \end{cases}$$

(3) *If* $\alpha_p = (-1)^\ell 2^{\ell-n} \binom{p}{\ell}$ $(\ell + 1 \le p \le n)$,
$$f_p = \begin{cases} 0, & (0 \le p \le n - \ell - 1); \\ (-1)^{n-\ell+p} \binom{p}{n-\ell}; & (n - \ell \le p \le n). \end{cases}$$

3. The proof of statement (1) of the Main Theorem

We give a rough review of a proof of the statement of (1) of the Main Theorem. This argument is similar to that of [12].

Let M be a smooth Riemannian manifold of dimension n with a Riemannian metric g. Let $X_1, X_2, \dots, X_n$ be a local orthonormal frame of $T(M)$ in a local path U. And let $\omega^1, \omega^2, \dots, \omega^n$ be its dual. The differential d and its dual ϑ acting on $\Gamma(\wedge^p T^*(M))$ are written as follows,using the Levi-Civita connection ∇ (see Appendix A of [14]):

$$d = \sum_{j=1}^{n} e(\omega^j)\nabla_{X_j}, \qquad \vartheta = -\sum_{j=1}^{n} \imath(X_j)\nabla_{X_j},$$

where we use the following notation.

Notation.

$$e(\omega^j)\omega = \omega^j \wedge \omega, \ \imath(X_j)\omega(Y_1, \dots, Y_{p-1}) = \omega(X_j, Y_1, \dots, Y_{p-1}).$$

Let the function $c_{i,j}^k$ be defined for $1 \leq i, j, k \leq n$ by the identities:

$$\nabla_{X_i} X_j = \sum_{k=1}^{n} c_{i,j}^k X_k$$

and let $R(X, Y)$ be the curvature transformation, that is

$$R(X, Y) = \left[\nabla_X, \nabla_Y\right] - \nabla_{[X,Y]}.$$

Set

$$R(X_i, X_j)X_k = \sum_{m=1}^{n} R_{mkij} X_m \quad 1 \leq i, j, k, \leq n.$$

The Laplacian $\Delta = d\vartheta + \vartheta d$ on $\sum_{p=0}^{n} \Gamma(\wedge^p T^*(M))$ can be expressed by Weitzenböck's formula;

$$\Delta = -\{\sum_{j=1}^{n} \nabla_{X_j}\nabla_{X_j} - \sum_{j=1}^{n} \nabla_{(\nabla_{X_j} X_j)} + \sum_{i,j=1}^{n} e(\omega))\imath(X_j)R(X_i, X_j)\}.$$

We use the following notations in the rest of this paper:

$$\begin{aligned} a_j^* &= e(\omega *), & a_k &= \iota(X_k). \\ a_I &= a_{i_1} a_{i_2} \cdots a_{i_p}, & a_I^* &= a_{i_p}^* \cdots a_{i_1}^*, \\ \omega^I &= \omega^{i_1} \wedge \omega^{i_2} \wedge \cdots \wedge \omega^{i_p} & \text{for } I &= (i_1, i_2, \dots, i_p) \in \mathcal{I}. \end{aligned}$$

From the fact that our connection is the Riemannian connection the coefficients $c_{i,j}^k$ satisfy

$$c_{i,j}^k = -c_{i,k}^j, \ [X_i, X_j] = \sum_{k=1}^{n} (c_{i,j}^k - c_{j,i}^k) X_k. \tag{3.1}$$

By the above notations we have for $\varphi_J\omega^J \in A^*(M)$

$$\nabla_{X_j}(\varphi_J\omega^J) = (X_j\varphi_J)\omega^J - \varphi_J G_J(\omega^J)$$

using

$$\nabla_{X_j}(\omega^i) = -\sum_{k=1}^{n} c_{j,k}^{i}\omega^k, \quad \text{where} \quad G_j = \sum_{k,m=1}^{n} c_{j,k}^{m} a_k^* a_m.$$

We have also

$$R(X_i, X_j)(\varphi_J\omega^J) = -\sum_{k,m=1}^{n} \varphi_J R_{mkij} a_k^* a_m(\omega^J).$$

Then we have the representation in a local chart U

$$\Delta = -\left\{\sum_{j=1}^{n}(X_jI - G_j)^2 - \sum_{i,j=1}^{n} c_{i,i}^{j}(X_jI - G_j) - \sum_{i,j,k,m=1}^{n} R_{mkij} a_i^* a_j a_k^* a_m\right\}$$

on $A^*(M)$. Here I is the identity operator on $\wedge^*(T^*(M))$.

Take a local coordinate $\{x_1, \ldots, x_n\}$ of U. Let $\{\xi_1, \ldots, \xi_n\}$ be its dual. By the above argument we have

Lemma 3.1. *Let us denote the symbol of X_j by α_j, that is,*

$$\alpha_j = \sigma(X_j).$$

Then the symbol of Δ is given by

$$\sigma(\Delta) = -\sum_{j=1}^{n}(\alpha_jI - G_j)^2 + R + r_1,$$

where

$$R = \sum_{i,j,k,m=1}^{n} R_{mkij} a_i^* a_j a_k^* a_m,$$

$$r_1 = \sum_{k,j=1}^{n} i\{\frac{\partial}{\partial\xi_k}\alpha_j\frac{\partial}{\partial x_k}\alpha_j - \frac{\partial}{\partial\xi_k}\alpha_j\frac{\partial}{\partial x_k}G_j\}I + \sum_{j,k=1}^{n} c_{k,k}^{j}(\alpha_jI - G_j).$$

$\mathcal{A} = (\mathcal{A}_{ij})$ denotes a matrix whose elements are $\mathcal{A}_{ij} = a_i^* a_j$ $(1 \leq i,j \leq n)$.

Definition 3.2. A subset K^m of $S_{1,0}^m$ is given by $K^m = \{p(x, \xi : \mathcal{A})$; polynomials with respect to ξ and $\mathcal{A}_{i,j}$, $(i,j = 1,2,\ldots,n)$ of order m with coefficients in $\mathcal{B}(\mathbf{R}^n)\}$. Here $\mathcal{B}(\mathbf{R}^n)$ is the set of smooth functions in $\mathbf{R}^n$ whose all partial derivatives are bounded. We define a pseudo-differential operator $P = p(x, D : \mathcal{A})$ acting on $A^*(M)$ of a symbol $\sigma(P) = p(x, \xi : \mathcal{A}) = \sum_{I,J} p_{I,J}(x,\xi) a_I^* a_J \in K^m$ as follows.

$$p(x, D : \mathcal{A})(\varphi_K\omega^K) = \sum_{I,J} p_{I,J}(x, D)\varphi_K a_I^* a_J(\omega^K).$$

In our case, by Lemma 3.1 the symbol of Δ is of the form $\sigma(\Delta) = r_2 + r_1$, where

$$r_2 = -\sum_{j=1}^{n} (\alpha_j I - G_j)^2 + R.$$

It is clear that r_j belongs to K^j for $j = 1, 2$.

If we review the results in [11], we see the following facts. The above representation of Δ signifies that the fundamental solution$E_p(t)(e_p(t,x,y))$ have a common representation for any p. So we write for short $E(t)(e(t,x,y))$. In our notation, $E_p(t) = \Psi_p E(t)$. It is sufficient to construct an asymptotic solution for the fundamental solution locally because we can reduce the construction to solving an integral equation on the manifold M of Volterra type. Our fundamental solution is of the form

$$E(t) = U(t) + V(t),$$

where $U(t)$ is the fundamental solution for the Cauchy problem and $V(t)$ is smooth with respect t off the boundary. See details in Section 5.

Now consider the following Cauchy problem:

$$\begin{aligned} \left(\frac{d}{dt} + \Delta\right) U(t) &= 0 \qquad \text{in } (0,T) \times \mathbf{R}^n, \\ \lim_{t \to 0} U(t) &= I \qquad \text{in } \mathbf{R}^n. \end{aligned}$$

$U(t)$ constructed in [11] is an integral operator

$$U(t)\varphi(x) = \int_M u(t,x,y)\varphi(y)dv_y, \quad \varphi \in A^*(M)$$

and $U(t)$ has an expansion.

$$U(t) \sim \sum_{j=0} u_j(t,x,D), \tag{3.2}$$

where $u_j(t,x,D)$ are pseudo-differential operators with parameter t.

The following fact is obtained in p. 255 of [11]. The main part of $U(t)$ is obtained as a pseudo-differential operator with symbol $u_0(t,x,\xi) = e^{-r_2 t}$. So we have

$$\begin{aligned} \tilde{u}_0(t,x,x) &= (2\pi)^{-n} \int_{\mathbf{R}^n} u_0(t,x,\xi)d\xi \qquad (3.3) \\ &= \left(\frac{1}{2\sqrt{\pi t}}\right)^n \sqrt{\text{detg}} e^{-tR}\Big(1 + 0(\sqrt{t})\Big). \end{aligned}$$

We shall calculate

$$\text{tr } (\beta_I \tilde{u}_0(t,x,x))dx = \left(\frac{1}{2\sqrt{\pi t}}\right)^n \text{tr } (\beta_I e^{-tR})dv(1 + 0(\sqrt{t})), \tag{3.4}$$

for $I \in \mathcal{I}$, $\sharp(I) = r$.

Using

$$e^{-tR} = \sum_{k=0}^{\infty} \left\{ \frac{(-R)^k}{k!} t^k \right\},$$

and by Proposition 2.2 we have

$$\operatorname{tr}\,(\beta_I e^{-tR}) = \begin{cases} \operatorname{tr}\big(\beta_I(-R)^m\big)\frac{t^m}{m!} + 0(t^{m+1}), & \text{if} \quad r = 2m\ ; \\ 0(t^{\frac{r+1}{2}}), & \text{if} \quad r \text{ is odd.} \end{cases} \tag{3.5}$$

We have the following proposition for the terms of the right-hand side of the above equation.

Proposition 3.3. *For* $I = (i_1, i_2, \ldots, i_r) \in \mathcal{I}$ $(r = 2m)$

$$\operatorname{tr}(\beta_I(-R)^m) = 2^{n-r-m} \sum_{\pi,\sigma \in S_r} \operatorname{sign}(\pi)\operatorname{sign}(\sigma)$$
$$\times R_{i_{\pi(1)} i_{\pi(2)} i_{\sigma(1)} i_{\sigma(2)}} \cdots R_{i_{\pi(r-1)} i_{\pi(r)} i_{\sigma(r-1)} i_{\sigma(r)}}.$$

Definition 3.4.

(1-i) If ℓ is odd, $C_\ell(x, M) = 0$
(1-ii) If ℓ is even $(\ell = 2m)$,

$$C_\ell(x, M) = \sum_{I \in \mathcal{I}, \sharp(I) = \ell} C_I(x, M),$$

where

$$C_I(x, M) = \left(\frac{1}{2\sqrt{\pi}}\right)^n \frac{1}{m!} \left(\frac{1}{2}\right)^m \sum_{\pi,\sigma \in S_\ell} \operatorname{sign}(\pi)\operatorname{sign}(\sigma)$$
$$\times R_{i_{\pi(1)} i_{\pi(2)} i_{\sigma(1)} i_{\sigma(2)}} \cdots\cdots R_{i_{\pi(\ell-1)} i_{\pi(\ell)} i_{\sigma(\ell-1)} i_{\sigma(\ell)}}$$

for $I = (i_1, i_2, \ldots, i_\ell) \in \mathcal{I}$.

By (3.4), (3.5) and Proposition 3.3, we have the following equation for $\sharp(I) = r$

$$\operatorname{tr}\,(\beta_I \tilde{u}_0(t, x, x)) = \begin{cases} 2^{n-r} t^{-\frac{n}{2} + \frac{r}{2}} C_I(x, M) \sqrt{\det g} + 0(t^{-\frac{n}{2} + \frac{r}{2} + 1}), & \text{if} \quad r = 2m\ ; \\ 0(t^{-\frac{n}{2} + \frac{r}{2} + \frac{1}{2}}), & \text{if} \quad r \text{ is odd.} \end{cases} \tag{3.6}$$

Similarly we have

$$\operatorname{tr}\,(\beta_I \tilde{u}_j(t, x, x))dx = 0(t^{-\frac{n}{2} + \frac{r}{2} + \frac{j}{2}}). \tag{3.7}$$

By (3.6) and (3.7) we have

$$\begin{aligned} \operatorname{tr}\,(\beta_I u(t, x, x)) &= \operatorname{tr}\,(\beta_I \tilde{u}_0(t, x, x)) \frac{1}{\sqrt{\det g}} + 0(t^{-\frac{n}{2} + \frac{r}{2} + \frac{1}{2}}) \\ &= \begin{cases} 2^{n-r} t^{-\frac{n}{2} + \frac{r}{2}} C_I(x, M) + 0(t^{-\frac{n}{2} + \frac{r}{2} + \frac{1}{2}}), & \text{if} \quad r = 2m\ ; \\ 0(t^{-\frac{n}{2} + \frac{r}{2} + \frac{1}{2}}), & \text{if} \quad r \text{ is odd.} \end{cases} \end{aligned} \tag{3.8}$$

By Theorem 2.4 we obtain

$$\mathrm{tr}\left(\sum_{p=0}^{n} f_p\Psi_p u(t,x,x)\right) = (-1)^{\ell}2^{\ell-n}\sum_{I\in\mathcal{I},\sharp(I)=\ell}\mathrm{tr}\,(\beta_I u(t,x,x)) + \sum_{p=\ell+1}^{n}\alpha_p \mathrm{tr}\,(\Gamma_p u(t,x,x)).$$

By (3.6) and (3.7) it is easy to see

$$\mathrm{tr}\,(\Gamma_p u(t,x,x)) = 0(t^{-\frac{n}{2}+\frac{\ell}{2}+\frac{1}{2}}) \quad \text{for any } p\geq \ell+1. \tag{3.9}$$

Applying (3.8) and (3.9), we have

$$\mathrm{tr}\left(\sum_{p=0}^{n} f_p\Psi_p u(t,x,x)\right) = C_\ell(x,M)t^{-\frac{n}{2}+\frac{\ell}{2}} + 0(t^{-\frac{n}{2}+\frac{\ell}{2}+\frac{1}{2}}) \quad x\in M.$$

4. The construction of the fundamental solution with boundary

The main part of the construction of the fundamental solution or its asymptotics lies in constructing these in a local chart (cf. C. Iwasaki [10]). So it is enough to show a method of construction of the fundamental solution in $\mathbf{R}^n_+$. The study in [10] is applicable for the construction of the fundamental solution for our initial-boundary value problem. In this case we introduce the symbol class $\mathcal{J}_s$ according to [11], as we used K^m instead of $S^m_{1,0}$ in Section 3.

First of all, we write down the boundary operator B_p for $A^p(M)$ in a local chart. Take a local patch Ω near ∂M such that ∂M is defined by $\{\rho=0\}$ in Ω and $M\cap\Omega\subset\{\rho\geq 0\}$. Assume that $\omega^n = cd\rho$ with some function c on M. Note that X_j $(1\leq j\leq n-1)$ are tangential to ∂M.

The coefficients of the Levi-Civita connection satisfy both (3.1) and the following relations:

$$c^n_{i,j} = c^n_{j,i} \qquad 1\leq i,j\leq n-1.$$

Choose local coordinates $\{x_1,\dots,x_n\}$ in Ω such that $M\cap\Omega = \{(x',x_n);\ x'\in\mathcal{U}, x_n\geq 0\}, \Gamma = \partial M\cap\Omega = \{(x',0); x'\in\mathcal{U}\}$ and

$$X_n|_\Gamma = \frac{d}{dx_n}.$$

The boundary operator B_p is as follows:

$$\varphi\in\mathrm{Dom}(\vartheta), \qquad d\varphi\in\mathrm{Dom}(\vartheta),$$

where

$$\mathrm{Dom}(\vartheta) = \{\varphi = \sum_J \varphi_J\omega^J : \varphi_J|_\Gamma = 0 \text{ if } n\in J\}.$$

Note that $\mathrm{Dom}(\vartheta)$ is the set of all ψ of $A^*(M)$ which satisfy

$$\int_M (d\varphi,\psi)dv = \int_M (\varphi,\vartheta\psi)dv, \text{ for any } \varphi\in A^*(M).$$

We can write our boundary condition in the following form:

$$\varphi_J|_\Gamma = 0 \text{ if } n \in J, \qquad B\Big(\sum_{n\notin J} \varphi_J \omega^J\Big)|_\Gamma = 0,$$

where

$$B = X_n - \sum_{1\le j,k\le n-1} c^j_{n,k} a^*_k a_j - \sum_{1\le j,k\le n-1} c^n_{j,k} a^*_j a_k. \tag{4.1}$$

Definition 4.1.

(1) Set

$$\gamma = \gamma(x' : \mathcal{A}) - G_n|_\Gamma = \sum_{1\le j,k\le n} (c^j_{n,k}|_\Gamma) a^*_k a_j, \tag{4.2}$$

$$b = b(x' : \mathcal{A}) = - \sum_{1\le j,k\le n-1} (c^n_{j,k}|_\Gamma) a^*_j a_k + \sum_{j=n \text{ or } k=n} (c^j_{n,k}|_\Gamma) a^*_k a_j \tag{4.3}$$

(2) Set $\mathcal{P} = a^*_n a_n$, $\mathcal{Q} = a_n a^*_n = I - \mathcal{P}$.

We note that by (4.1)

$$B = \frac{d}{dx_n} - \gamma + b.$$

Let the fundamental solution be of the form $E(t) = U(t) + V(t)$, where $U(t)$ is the fundamental solution for the Cauchy problem. It is sufficient to construct an operator $V(t)$ which satisfies the following problem in $\mathbf{R}^n_+$.

$$\begin{aligned}
\Big(\frac{d}{dt} + \Delta\Big) V(t) &= 0 && \text{in } (0,T)\times \mathbf{R}^n_+,\\
\mathcal{P}V(t) &= -\mathcal{P}U(t) && \text{on } (0,T)\times \mathbf{R}^{n-1}\times\{x_n = 0\},\\
B\mathcal{Q}V(t) &= -B\mathcal{Q}U(t) && \text{on } (0,T)\times \mathbf{R}^{n-1}\times\{x_n = 0\},\\
\lim_{t\to 0} V(t) &= 0 && \text{in } \mathbf{R}^n_+.
\end{aligned}$$

We introduce the following symbol classes $\mathcal{F}_s$ as follows.

Definition 4.2. $\mathcal{F}_s$ is the set of all finite sums of functions of the following form

$$t^d (x_n)^\ell r(x', \xi', \xi_n : A) \ (r \in K^{s+2d+\ell}).$$

We have by Lemma 3.1

$$r_2|_\Gamma = q_2 = (\xi_n + i\gamma)^2 + \beta(x', \xi' : \mathcal{A}) \in \mathcal{F}_2,$$

where

$$\beta = \beta(x', \xi' : \mathcal{A}) = - \sum_{j=1}^{n-1} \Big((\alpha_j|_\Gamma) I - G_j|_\Gamma\Big)^2 + R|_\Gamma.$$

Let $\{\tilde{w}_{j,k}\}$ be symbols defined in Definition 7 of [10] for integers j and non-positive integers k as follows:

$$\tilde{w}_{j,0}(t,\omega) = \frac{1}{\sqrt{\pi}}\left(\frac{1}{2\sqrt{t}}\right)^{j+1} h_j\left(\frac{\omega}{2\sqrt{t}}\right), \text{ if } \quad j \geq 0,$$

$$\tilde{w}_{j,0}(t,\omega) = -\frac{1}{\sqrt{\pi}}\left(\frac{1}{2\sqrt{t}}\right)^{j+1} \int_0^\infty e^{-(\sigma+\frac{\omega}{2\sqrt{t}})^2} \frac{(-\sigma)^{-j-1}}{(-j-1)!} d\sigma, \text{ if } \quad j \leq -1,$$

For $k \leq -1$

$$\begin{aligned}\tilde{w}_{j,k}(t,\omega;b) = &-\frac{1}{\sqrt{\pi}}\left(\frac{1}{2\sqrt{t}}\right)^{j+k+1} \int_0^\infty e^{-(\sigma+\frac{\omega}{2\sqrt{t}})^2+2b\sqrt{t}\sigma} \frac{(-\sigma)^{-k-1}}{(-k-1)!} \\ &\times h_j(\sigma+\frac{\omega}{2\sqrt{t}})d\sigma, \text{ if } j \geq 0;\end{aligned}$$

$$\begin{aligned}\tilde{w}_{j,k}(t,\omega;b) = &\frac{1}{\sqrt{\pi}}\left(\frac{1}{2\sqrt{t}}\right)^{j+k+1} \int_0^\infty \frac{(-\tau)^{-j-1}}{(-j-1)!} d\tau \\ &\times \int_0^\infty e^{-(\sigma+\tau+\frac{\omega}{2\sqrt{t}})^2+2b\sqrt{t}\sigma} \frac{(-\sigma)^{-k-1}}{(-k-1)!} d\sigma, \text{ if } j \leq -1,\end{aligned}$$

where $h_j(\sigma) = \{(\frac{d}{d\sigma})^j e^{-\sigma^2}\}e^{\sigma^2}$.

Definition 4.3. For a pair (j,k) of integer j and nonpositive integer k we define a function

$$\tilde{v}_{j,k}(t,x_n,y_n;b,\gamma) = e^{\gamma(x_n-y_n)}\tilde{w}_{j,k}(t,x_n+y_n;b),$$

with b and γ given in Definition 4.1. An operator $V_{j,k}(t;b,\gamma)$ corresponding to $\tilde{v}_{j,k}$ is defined as follows for a function $\varphi(y_n)$ defined on $\mathbf{R}^1_+$.

$$(V_{j,k}(t;b,\gamma)\varphi)(x_n) = \int_0^\infty \tilde{v}_{j,k}(t,x_n,y_n;b,\gamma)\varphi(y_n)dy_n.$$

The above series of integral operators $V_{j,k}$ has nice properties. See [11]. One of the properties is the following one.

Proposition 4.4. *$V_{j,k}(t;b,\gamma)$ satisfies for $k \leq -1$*

$$\begin{aligned}&\left\{\frac{\partial}{\partial t} - \left(\frac{\partial}{\partial x_n} - \gamma\right)^2\right\} V_{j,k}(t;b,\gamma) = 0 \qquad \text{in } (0,T)\times\bar{\mathbf{R}}^1_+,\\ &\left(\frac{\partial}{\partial x_n} - \gamma\right) V_{j,k}(t;b,\gamma) = V_{j+1,k}(t;b,\gamma) \qquad \text{in } (0,T)\times\bar{\mathbf{R}}^1_+,\\ &BV_{j,k}(t;b,\gamma) = V_{j,k+1}(t;b,\gamma) + Z_{j,k} \qquad \text{in } (0,T)\times\bar{\mathbf{R}}^1_+,\\ &\lim_{t\to+0}(V_{j,k}(t;b,\gamma)\varphi)(x_n) = 0 \qquad \text{in } x_n > 0,\\ &\qquad \text{for } \varphi = \sum_J \varphi_J\omega^J,\ \varphi_J \in C(\bar{\mathbf{R}}^1_+),\end{aligned}$$

where $Z_{j,k}$ *is an operator defined as*

$$Z_{j,k}\varphi = \int_0^\infty [b, \exp\{\gamma(x_n - y_n)\}]\tilde{w}_{j,k}(t, x_n + y_n; b)\varphi(y_n)dy_n.$$

Definition 4.5. (1) $\mathcal{J}_s$ is the set of all finite sums of the functions of the following form

$$g(t, x_n, y_n, x', \xi' : \mathcal{A}) = t^d (x_n)^\ell q(x', \xi' : \mathcal{A})\tilde{v}_{j,k}(t, x_n, y_n; b(x' : \mathcal{A}), \gamma(x' : \mathcal{A})) \times e^{-\beta(x', \xi' : \mathcal{A})t}$$

$(d, \ell, j, k \in \mathbf{Z}, d \geq 0, \ell \geq 0, k < 0, q \in K^m(\mathbf{R}^{n-1})$ with $m = s + 2d + \ell - j - k)$.
For a symbol $g(t, x_n, y_n, x', \xi', \mathcal{A}) \in \mathcal{J}_s$ we define an integral-pseudo-differential operator G as follows:

$$(G\varphi)(t, x', x_n : \mathcal{A}) = \int_0^\infty g(t, x_n, y_n, x', D' : \mathcal{A})\varphi(\cdot, y_n)dy_n.$$

So the kernel $\tilde{g}$ of operator G is given by

$$\tilde{g}(t, x', x_n, y', y_n : \mathcal{A}) = (2\pi)^{-(n-1)} \int_{R^{n-1}} e^{i(x'-y')\bullet\xi'} g(t, x_n, y_n, x', \xi' : \mathcal{A})d\xi'.$$

(2) $\tilde{R}_s$ is the set of all matrices g whose elements belong to $B([0,T] \times [0,\infty) \times [0,\infty); S^m_{1,0}(\mathbf{R}^{n-1}))$ and satisfy the following estimate for any α, β, a, b, k

$$\left\| \left(\frac{\partial}{\partial \xi'}\right)^\alpha \left(\frac{\partial}{\partial x'}\right)^\beta \left(\frac{\partial}{\partial x_n}\right)^a \left(\frac{\partial}{\partial y_n}\right)^b \left(\frac{\partial}{\partial t}\right)^k g \right\|$$
$$\leq C_{\alpha,\beta} \min(|\xi'|^{-|\alpha|}, \sqrt{t}^{|\alpha|}) \left(\frac{1}{\sqrt{t}}\right)^{s+1+2k+a+b} \exp\left(-\delta\frac{(x_n + y_n)^2}{4t} - c_0|\xi'|^2 t\right)$$

for any $\delta < 1$ and some $c_0 > 0$.

By the argument of page 276 of [11] an asymptotic $E_N(t)$ of the fundamental solution for the mixed problem is obtained of the form for any integer N

$$E_N(t) = U_0 + U_1 + \cdots + U_N + V_1 + V_0 + \cdots + V_{-N},$$

where $U_j = u_j(t; x, D)$ are pseudo-differential operators given in (3.2) and V_j are integral operators whose symbols $v_j(t, x_n, y_n, x', \xi' : \mathcal{A}) \in \mathcal{J}_j$.

Theorem 7.2 of [11] asserts that

$$\begin{aligned}
\left(\frac{d}{dt} + \Delta\right) E_N(t) &= 0 \quad \mod \tilde{R}_{-N} \quad && \text{in } (0,T) \times \mathbf{R}^n_+, \\
\mathcal{P}E_N(t) &= 0 \quad \mod \tilde{R}_{-N-2} \quad && \text{on } (0,T) \times \mathbf{R}^{n-1} \times \{x_n = 0\}, \\
B\mathcal{Q}E_N(t) &= 0 \quad \mod \tilde{R}_{-N-1} \quad && \text{on } (0,T) \times \mathbf{R}^{n-1} \times \{x_n = 0\}.
\end{aligned}$$

Especially we have

$$v_1(t, x_n, y_n, x', \xi' : \mathcal{A}) = 2\mathcal{Q}\tilde{v}_{1,-1}(t, x_n, y_n; b, \gamma)e^{-\beta(x', \xi' : \mathcal{A})t}. \tag{4.4}$$

In the rest of this paper $\tilde{v}_j(t,x',x_n,y',y_n : \mathcal{A})$ denotes the kernel of the integral operator V_j, that is,

$$\tilde{v}_j(t,x',x_n,y',y_n : \mathcal{A}) = (2\pi)^{-(n-1)} \int_{R^{n-1}} e^{i(x'-y')\bullet\xi'} v_j(t,x_n,y_n,x',\xi' : \mathcal{A}) d\xi'.$$

5. The proof of Statements (2) and (3) of the Main Theorem

We will calculate $\mathrm{tr}\,(\sum_{p=0}^{n} f_p \Psi_p V_j(t,x,x))(j \le 1)$. We prepare some lemmas and the definition of $D_{\ell-1}(x)$.

Let $\hat{R}(W,Z,X,Y)$ be the Riemannian curvature tensors induced on ∂M. From the equation of Gauss we have

$$R(X_i,X_j,X_k,X_\ell) = \hat{R}(X_i,X_j,X_k,X_\ell) + c^n_{k,j}c^n_{\ell,i} - c^n_{\ell,j}c^n_{k,i}, \\ 1 \le i,j,k,\ell \le n-1 \quad \text{on } \partial M.$$

So it is clear that on ∂M

$$\sum_{1\le i,j,k,\ell\le n-1} R_{ijk\ell}a_i^*a_ja_k^*a_\ell = \sum_{1\le i,j,k,\ell\le n-1} \hat{R}_{ijk\ell}a_i^*a_ja_k^*a_\ell \\ - \sum_{1\le i,j,k,\ell\le n-1} c^n_{i,j}c^n_{k,\ell}a_i^*a_ja_k^*a_\ell + \sum_{1\le j,k,\ell\le n-1} c^n_{j,j}c^n_{k,\ell}a_k^*a_\ell. \tag{5.1}$$

Proposition 5.1. *We have by the Bianchi's identity*

$$\sum_{1\le i,j,k,h\le n-1} \hat{R}_{ijkh}a_i^*a_ja_k^*a_h \\ = \frac{1}{2}\sum_{1\le i,j,k,h\le n-1} \hat{R}_{ikjh}a_i^*a_ja_k^*a_h - \frac{1}{2}\sum_{1\le i,j,k\le n-1} \hat{R}_{kjki}a_j^*a_i. \tag{5.2}$$

Proof. By the properties of the curvature transformation we have

$$\sum_{1\le i,j,k,h\le n-1} \hat{R}_{ijkh}a_i^*a_ja_k^*a_h \\ = \sum_{1\le i,j,k,h\le n-1} \hat{R}_{ihkj}a_i^*a_ha_k^*a_j \\ = \sum_{1\le i,j,k,h\le n-1} \hat{R}_{ihkj}a_i^*a_ja_ha_k^* + \sum_{1\le i,j,h\le n-1} \hat{R}_{ihjj}a_i^*a_h \\ = -\sum_{1\le i,j,k,h\le n-1} \hat{R}_{ihkj}a_i^*a_ja_k^*a_h + \sum_{1\le i,j,h\le n-1} \hat{R}_{ihjj}a_i^*a_h \\ + \sum_{1\le i,j,k\le n-1} \hat{R}_{ikkj}a_i^*a_j.$$

From the above equation it is easy to see

$$\begin{aligned}
&2\sum_{1\le i,j,k,h\le n-1} \hat{R}_{ijkh}a_i^*a_ja_k^*a_h \\
&= \sum_{1\le i,j,k,h\le n-1} (\hat{R}_{ijkh}-\hat{R}_{ihkj})a_i^*a_ja_k^*a_h + \sum_{1\le i,j,k\le n-1}(\hat{R}_{ijkk}+\hat{R}_{ikkj})a_i^*a_j \\
&= \sum_{1\le i,j,k,h\le n-1} \hat{R}_{ikjh}a_i^*a_ja_k^*a_h - \sum_{1\le i,j,k\le n-1}\hat{R}_{ikjk}a_i^*a_j.
\end{aligned}$$

□

Lemma 5.2. *For any $N \in N$ we have $h_N \in K_{-1}$ such that*

$$e^{-\beta t} = (I+h_N)\Pi_{j=1}^{n-1}\exp\{(\alpha_j^\star I - G_j^\star)^2 t\}e^{-R^\star t} \qquad \mathrm{mod}\ \tilde{R}_{-N-1},$$

where $R^\star = R|_{\partial M} = \sum_{1\le i,j,k,\ell\le n} R_{ijk\ell}a_i^*a_ja_k^*a_\ell|_{\partial M}$.

We give the definition of $D_{\ell-1}(x)$ on $x \in \partial M$.

Definition 5.3. For $x \in \partial M$ $D_{\ell-1}(x)$ is defined as follows:

$$D_{\ell-1}(x) = \sum_{I\in\mathcal{I}_0,\sharp(I)=\ell-1} D_I(x),$$

(i) If ℓ is odd ($\ell = 2m+1$),

$$\begin{aligned}
D_I(x) = \frac{1}{2}\left(\frac{1}{2\sqrt{\pi}}\right)^{n-1}\frac{1}{m!}\left(\frac{1}{2}\right)^m \sum_{\pi,\sigma\in S_{\ell-1}} \mathrm{sign}(\pi)\,\mathrm{sign}(\sigma) \\
\times \hat{R}_{i_{\pi(1)}i_{\pi(2)}i_{\sigma(1)}i_{\sigma(2)}}\cdots\cdots\hat{R}_{i_{\pi(\ell-2)}i_{\pi(\ell-1)}i_{\sigma(\ell-2)}i_{\sigma(\ell-1)}}
\end{aligned}$$

for $I = (i_1, i_2, \ldots, i_{\ell-1}) \in \mathcal{I}_0$.

(ii) If ℓ is even ($\ell = 2m$),

$$\begin{aligned}
D_I(x) = \frac{1}{2}\left(\frac{1}{2\sqrt{\pi}}\right)^{n-1}\sum_{k=0}^{m-1}\frac{d_{m-k-1}}{k!}\left(\frac{1}{2}\right)^k \sum_{\pi,\sigma\in S_{\ell-1}} \mathrm{sign}(\pi)\,\mathrm{sign}(\sigma) \\
\times \hat{R}_{i_{\pi(1)}i_{\pi(2)}i_{\sigma(1)}i_{\sigma(2)}}\cdots\quad\cdots\hat{R}_{i_{\pi(2k-1)}i_{\pi(2k)}i_{\sigma(2k-1)}i_{\sigma(2k)}} \\
\times c^n_{\pi(2k+1),\sigma(2k+1)}c^n_{\pi(2k+2),\sigma(2k+2)}\cdots\quad\cdots c^n_{\pi(\ell-1),\sigma(\ell-1)}
\end{aligned}$$

for $I = (i_1, i_2, \ldots, i_{\ell-1}) \in \mathcal{I}_0$ with

$$d_k = \frac{(-1)^k}{k!(k+\frac{1}{2})\sqrt{\pi}}.$$

Now the calculation of $\mathrm{tr}\,\big(\beta_I\tilde{v}_1(t,x',0)\big)$ is obtained as follows.

Lemma 5.4. *Let $\hat{g}$ be the Riemannian metric induced on ∂M and ε be a positive constant.*

(1) *Let $I = (i_1, i_2, \ldots, i_{l-1}, n) \in \mathcal{I}$ and $I_0 = (i_1, i_2, \ldots, i_{l-1}) \in \mathcal{I}_0$.*

(1-i) *We have*

$$\operatorname{tr}\left(\beta_I \tilde{v}_1(t,x',0,x',0)\right) = (-1)^\ell 2^{n-\ell+1} t^{-\frac{n}{2}+\frac{\ell}{2}-\frac{1}{2}} D_{I_0}(x')\sqrt{\det g(x',0)} + 0(t^{-\frac{n}{2}+\frac{\ell}{2}}).$$

(1-ii) *We have for any* $\psi(x',x_n) \in C^\infty(\mathbf{R}^n)$

$$\int_0^\varepsilon \operatorname{tr}\bigl(\beta_I \tilde{v}_1(t,x',x_n,x',x_n)\bigr)\psi(x',x_n)dx_n = (-1)^\ell 2^{n-\ell} t^{-\frac{n}{2}+\frac{\ell}{2}} \psi(x',0) D_{I_0}(x')\sqrt{\det \hat{g}(x')} + 0(t^{-\frac{n}{2}+\frac{\ell}{2}+\frac{1}{2}}).$$

(2) *Let* $I \in \mathcal{I}_0, \sharp(I) = \ell$ *or* $I \in \mathcal{I}, \sharp(I) \geq \ell+1$.

(2-i) *We have*

$$\operatorname{tr}\left(\beta_I \tilde{v}_1(t,x',0,x',0)\right) = 0(t^{-\frac{n}{2}+\frac{\ell}{2}}).$$

(2-ii) *We have for any* $\psi(x',x_n) \in C^\infty(\mathbf{R}^n)$

$$\int_0^\varepsilon \operatorname{tr}\left(\beta_I \tilde{v}_1(t,x',x_n,x',x_n)\right)\psi(x',x_n)dx_n = 0(t^{-\frac{n}{2}+\frac{\ell}{2}+\frac{1}{2}}).$$

Proof. By (4.4) and Lemma 5.2 we have

$$\begin{aligned}\tilde{v}_1(t,x,x) &= 2\left(\frac{1}{2\pi\sqrt{t}}\right)^{n-1}\int_{\mathbf{R}^{n-1}} \mathcal{Q}\tilde{w}_{1,-1}(t,x_n,x_n:b)e^{-\beta(x',\eta':\sqrt{t}\mathcal{A})}d\eta' \\ &= 2\left(\frac{1}{2\sqrt{\pi t}}\right)^{n-1}(1+0(\sqrt{t}))\sqrt{\det g}\,\mathcal{Q}\tilde{w}_{1,-1}(t,x_n,x_n:b)e^{-tR^\star}.\end{aligned}$$

By

$$\tilde{w}_{1,-1}(t,0,0:b) = \frac{1}{2\sqrt{\pi t}} + \frac{b}{\sqrt{t}}\int_0^\infty e^{-\sigma^2+2b\sqrt{t}\sigma}d\sigma = \frac{1}{2\sqrt{t}}\sum_{j=0}^\infty \frac{(b\sqrt{t})^j}{\Gamma(\frac{j}{2}+1)}$$

we have

$$\begin{aligned}\operatorname{tr}\bigl(\beta_I \tilde{v}_1(t,x',0,x',0)\bigr) &= \left(\frac{1}{2\sqrt{\pi t}}\right)^{n-1}(1+0(\sqrt{t}))\sqrt{\det g} \\ &\quad\times \frac{1}{\sqrt{t}}\operatorname{tr}\left(\beta_I \mathcal{Q}\sum_{j=0}^\infty \frac{(b\sqrt{t})^j}{\Gamma(\frac{j}{2}+1)}\sum_{k=0}^\infty \frac{(-tR^\star)^k}{k!}\right).\end{aligned} \tag{5.3}$$

On the other hand, applying Proposition 16′ in [10] we have

$$\begin{aligned}&\int_0^\varepsilon \tilde{v}_{1,-1}(t,x_n,x_n:\gamma,b)\psi(x)dx_n \\ &= \int_0^\varepsilon \tilde{w}_{1,-1}(t,x_n,x_n:b)\psi(x)dx_n \\ &\sim \frac{1}{4}\sum_{j=0}^\infty \frac{(\sqrt{t})^j}{\Gamma(\frac{j}{2}+1)}\sum_{k,i\geq 0,k+i=j} b^k\left(\frac{\partial}{\partial x_n}\right)^i \psi(x',0)2^{-i}.\end{aligned}$$

So the following equality holds:

$$\int_0^{\varepsilon} \operatorname{tr}\left(\beta_I \tilde{v}_1(t,x,x)\right)\psi(x)dx_n = \left(\frac{1}{2\sqrt{\pi t}}\right)^{n-1}(1+0(\sqrt{t}))\sqrt{\det \hat{g}} \tag{5.4}$$
$$\times \frac{1}{2}\operatorname{tr}\left(\beta_I \mathcal{Q}\sum_{j=0}^{\infty}\frac{(\sqrt{t})^j}{\Gamma(\frac{j}{2}+1)}\sum_{k,i\geq 0,k+i=j} b^k\left(\frac{\partial}{\partial x_n}\right)^i \psi(x',0)2^{-i}\sum_{m=0}^{\infty}\frac{(-tR^\star)^m}{m!}\right).$$

Owing to Theorem 3.1 of [11] and its corollary we have

$$\operatorname{tr}\left(\beta_I \mathcal{Q}\sum_{j=0}^{\infty}\frac{(\sqrt{t})^j}{\Gamma(\frac{j}{2}+1)}\sum_{k,i\geq 0,k+i=j} b^k\left(\frac{\partial}{\partial x_n}\right)^i \psi(x',0)2^{-i}\sum_{m=0}^{\infty}\frac{(-tR^\star)^m}{m!}\right)$$
$$= \operatorname{tr}\left(\beta_I \mathcal{Q}\sum_{j=0}^{\infty}\frac{(b\sqrt{t})^j}{\Gamma(\frac{j}{2}+1)}\sum_{k=0}^{\infty}\frac{(-tR^\star)^k}{k!}\right)\psi(x',0)$$
$$+\operatorname{tr}\left(\beta_I \mathcal{Q}\sum_{j=0}^{\infty}\frac{(\sqrt{t})^j}{\Gamma(\frac{j}{2}+1)}\sum_{k\geq 0,i\geq 1,k+i=j} b^k\left(\frac{\partial}{\partial x_n}\right)^i \psi(x',0)2^{-i}\sum_{m=0}^{\infty}\frac{(-tR^\star)^m}{m!}\right).$$

From Proposition 2.2 and the assumption $\sharp(I)\geq \ell$ we have

$$\operatorname{tr}\left(\beta_I \mathcal{Q}\sum_{j=0}^{\infty}\frac{(b\sqrt{t})^j}{\Gamma(\frac{j}{2}+1)}\sum_{k=0}^{\infty}\frac{(-tR^\star)^k}{k!}\right)$$
$$= \operatorname{tr}\left(\beta_I \mathcal{Q}\sum_{j+2k=\ell-1}\frac{b^j}{\Gamma(\frac{j}{2}+1)}\frac{(-R^\star)^k}{k!}\right)t^{\frac{\ell}{2}-\frac{1}{2}}+0(t^{\frac{\ell}{2}})$$

and

$$\operatorname{tr}\left(\beta_I \mathcal{Q}\sum_{j=0}^{\infty}\frac{(\sqrt{t})^j}{\Gamma(\frac{j}{2}+1)}\sum_{k\geq 0,i\geq 1,k+i=j} b^k\left(\frac{\partial}{\partial x_n}\right)^i \psi(x',0)2^{-i}\sum_{m=0}^{\infty}\frac{(-tR^\star)^m}{m!}\right) = 0(t^{\frac{\ell}{2}}).$$

It is sufficient to calculate only

$$X_{\ell,I} = \operatorname{tr}\left(\beta_I \mathcal{Q}\sum_{j+2k=\ell-1}\frac{b^j}{\Gamma(\frac{j}{2}+1)}\frac{(-R^\star)^k}{k!}\right)$$

for the proofs of both (1) and (2) because by (5.3), (5.4) and the above equation we have

$$\operatorname{tr}\left(\beta_I \tilde{v}_1(t,x',0,x',0)\right) = \left(\frac{1}{2\sqrt{\pi}}\right)^{n-1} t^{-\frac{n}{2}+\frac{\ell}{2}-\frac{1}{2}}\sqrt{\det g(x',0)}X_{\ell,I}+0(t^{-\frac{n}{2}+\frac{\ell}{2}})$$

and

$$\int_0^\varepsilon \operatorname{tr}\left(\beta_I \tilde{v}_1(t,x,x)\right)\psi(x)dx_n$$

$$= \frac{1}{2}\left(\frac{1}{2\sqrt{\pi}}\right)^{n-1} t^{-\frac{n}{2}+\frac{\ell}{2}}\sqrt{\det \hat{g}(x')}X_{\ell,I}\psi(x',0) + 0(t^{-\frac{n}{2}+\frac{\ell}{2}+\frac{1}{2}}).$$

We calculate $X_{\ell,I}$ dividing cases into (I) and (II) according to ℓ.

(I) If $\ell = 2m+1$ is odd, then j must be even. Set $j = 2\mu$. We can write

$$X_{2m+1,I} = \sum_{\mu+k=m} \operatorname{tr}\left(\beta_I \mathcal{Q}\frac{b^{2\mu}}{\Gamma(\mu+1)}\frac{(-R^\star)^k}{k!}\right) = \operatorname{tr}\left(\beta_I \mathcal{Q}(b^2 - R^\star)^m\right)\frac{1}{m!}.$$

On the other hand, we have, by (5.1) and the definition of b that is given in (4.3)

$$b^2 - R^\star = \sum_{1\le i,j,k,h\le n-1} \hat{R}_{ijkh}a_i^*a_ja_k^*a_h + \sum_{i=n \text{ or } j=n} A_{ijkh}a_i^*a_ja_k^*a_h \mod K^1$$

with some functions A_{ijkh}. Applying Proposition 5.1, we have

$$b^2 - R^\star = \frac{1}{2}\sum_{1\le i,j,k,h\le n-1} \hat{R}_{ikjh}a_i^*a_ja_k^*a_h + \sum_{i=n \text{ or } j=n} A_{ijkh}a_i^*a_ja_k^*a_h \mod K^1. \tag{5.5}$$

Then it holds with some functions $\{\alpha_{I,J}\}$ and $\{\tilde{\alpha}_{I,J}\}$

$$\begin{aligned}(b^2 - R^\star)^m = &\left(\frac{1}{2}\sum_{1\le i,j,k,h\le n-1} \hat{R}_{ikjh}a_i^*a_ja_k^*a_h\right)^m \\ &+ \sum_{n\in I \text{ or } n\in J, \sharp(I)=2m, \sharp(J)=2m} \alpha_{I,J}a_I^*a_J \\ &+ \sum_{\sharp(I)\le 2m-1, \sharp(J)\le 2m-1} \tilde{\alpha}_{I,J}a_I^*a_J.\end{aligned}$$

Noting $\mathcal{Q} = a_na_n^* = 1 - a_n^*a_n$, we have with some functions $\{\gamma_{I,J}\}$

$$\begin{aligned}\mathcal{Q}(b^2 - R^\star)^m = &-a_n^*a_n\left(\frac{1}{2}\sum_{1\le i,j,k,h\le n-1} \hat{R}_{ikjh}a_i^*a_ja_k^*a_h\right)^m \\ &+ \sum_{\sharp(I)\le 2m, \sharp(J)\le 2m} \gamma_{I,J}a_I^*a_J.\end{aligned}$$

Owing to Proposition 2.2 we have

$$\begin{aligned}m!X_{2m+1,I} &= \operatorname{tr}\left(\beta_I \mathcal{Q}(b^2 - R^\star)^m\right) \\ &= -\operatorname{tr}\left(\beta_I a_n^*a_n\left(\frac{1}{2}\sum_{1\le i,j,k,h\le n-1} \hat{R}_{ikjh}a_i^*a_ja_k^*a_h\right)^m\right).\end{aligned} \tag{5.6}$$

If $I = (i_1, i_2, \ldots, i_{\ell-1}, n)$, then by a method similar to Lemma 6.2 of [11] we have by Proposition 2.5

$$m!X_{2m+1,I} = -2^{n-\ell}\left(\frac{1}{2}\right)^m \sum_{\pi,\sigma\in S_{\ell-1}} \operatorname{sign}(\pi)\operatorname{sign}(\sigma) \times \hat{R}_{i_{\pi(1)}i_{\pi(2)}i_{\sigma(1)}i_{\sigma(2)}} \cdots \hat{R}_{i_{\pi(\ell-2)}i_{\pi(\ell-1)}i_{\sigma(\ell-2)}i_{\sigma(\ell-1)}}.$$

Finally we have

$$X_{2m+1,I} = -2^{n-\ell+1}\Big(\frac{1}{2\sqrt{\pi}}\Big)^{1-n} D_{I_0}. \tag{5.7}$$

If $I \in \mathcal{I}_0$ or $\sharp(I) \geq \ell+1$, then we have by Proposition 2.2 and (5.6) we have $X_{2m+1,I} = 0$.

(II) If $\ell = 2m$ is even, in this case we have

$$X_{2m,I} = \operatorname{tr}\Big(\beta_I \mathcal{Q} \sum_{k=0}^{m-1} \frac{b^{2m-2k-1}}{\Gamma(m-k+\frac{1}{2})} \frac{(-R^\star)^k}{k!}\Big).$$

Set

$$\hat{R} = \sum_{1\leq i,j,k,\ell\leq n-1} \hat{R}_{ikj\ell} a_i^* a_j a_k^* a_\ell.$$

By (5.5) it holds that

$$-R^\star = \frac{1}{2}\hat{R} - b^2 + \sum_{i=n \text{ or } j=n} A_{ijkh} a_i^* a_j a_k^* a_h \bmod K^1.$$

Then we can write with some functions $\{\alpha_{I,J}\}$

$$(-R^\star)^k = \big(\frac{1}{2}\hat{R} - b^2\big)^k + \sum_{n\in I \text{ or } n\in J, \sharp(I)=2k, \sharp(J)=2k} \alpha_{I,J} a_I^* a_J \bmod K^{2k-1}.$$

So we have with some functions $\{\tilde{\alpha}_{I,J}\}$

$$\begin{aligned}
\sum_{k=0}^{m-1} \frac{b^{2m-2k-1}}{\Gamma(m-k+\frac{1}{2})} \frac{(-R^\star)^k}{k!} &= \sum_{k=0}^{m-1} \frac{b^{2m-2k-1}}{\Gamma(m-k+\frac{1}{2})} \sum_{k'+j=k} \frac{(-b^2)^j}{j!} \frac{(\frac{1}{2}\hat{R})^{k'}}{k'!} \\
&\quad + \sum_{\substack{n\in I \text{ or } n\in J,\\ \sharp(I)=2m-1,\\ \sharp(J)=2m-1}} \tilde{\alpha}_{I,J} a_I^* a_J \bmod K^{2m-2} \\
&= \sum_{k=0}^{m-1} d_{m-k-1} b^{2m-2k-1} \frac{(\frac{1}{2}\hat{R})^k}{k!} \\
&\quad + \sum_{\substack{n\in I \text{ or } n\in J,\\ \sharp(I)=2m-1,\\ \sharp(J)=2m-1}} \tilde{\alpha}_{I,J} a_I^* a_J \bmod K^{2m-2}
\end{aligned}$$

with

$$d_k = \sum_{j=0}^{k} \frac{(-1)^j}{j!\Gamma(k-j+\frac{3}{2})} = \frac{(-1)^k}{k!(k+\frac{1}{2})\sqrt{\pi}}.$$

From Proposition 2.2 and the assumption $\sharp(I) \geq \ell$ we have

$$X_{2m.I} = -\mathrm{tr}\left(\beta_I a_n^* a_n \left(\sum_{k=0}^{m-1} d_{m-k-1} b^{2m-2k-1} \frac{(\frac{1}{2}\hat{R})^k}{k!}\right)\right) \tag{5.8}$$

since we obtain

$$\begin{aligned} \mathcal{Q}\Big(\sum_{k=0}^{m-1} \frac{b^{2m-2k-1}}{\Gamma(m-k+\frac{1}{2})} \frac{(-R^\star)^k}{k!}\Big) &= -a_n^* a_n \Big(\sum_{k=0}^{m-1} d_{m-k-1} b^{2m-2k-1} \frac{(\frac{1}{2}\hat{R})^k}{k!}\Big) \\ &\quad + \sum_{\sharp(I)\leq 2m-1, \sharp(J)\leq 2m-1} \tilde{\gamma}_{I,J} a_I^* a_J \end{aligned}$$

with some constants $\{\tilde{\gamma}_{I,J}\}$.

If $I = (i_1, i_2, \ldots, i_{\ell-1}, n)$ we have

$$\begin{aligned} X_{2m.I} = -2^{n-\ell} &\sum_{k=0}^{m-1} (-1)^{2m-2k-1} \left(\frac{1}{2}\right)^k \frac{d_{m-k-1}}{k!} \\ &\times \sum_{\pi,\sigma\in S_{\ell-1}} \mathrm{sign}(\pi)\,\mathrm{sign}(\sigma) \hat{R}_{\pi(1)\pi(2)\sigma(1)\sigma(2)} \cdots \\ &\cdots \hat{R}_{\pi(2k-1)\pi(2k)\sigma(2k-1)\sigma(2k)} c^n_{\pi(2k+1),\sigma(2k+1)} c^n_{\pi(2k+2),\sigma(2k+2)} \cdots \\ &\cdots c^n_{\pi(\ell-1),\sigma(\ell-1)}. \end{aligned}$$

In this case we have

$$X_{2m,I} = 2^{n-\ell+1}\Big(\frac{1}{2\sqrt{\pi}}\Big)^{1-n} D_{I_0}. \tag{5.9}$$

If $I \in \mathcal{I}_0$ or $\sharp(I) \geq \ell+1$, then by (5.8) and Proposition 2.2 we have

$$X_{2m,I} = 0.$$

By (5.7) and (5.9) it holds that

$$X_{\ell,I} = (-1)^\ell 2^{n-\ell+1}\Big(\frac{1}{2\sqrt{\pi}}\Big)^{1-n} D_{I_0}$$

if $\sharp(I) = (i_1, i_2, \ldots, i_{\ell-1}, n)$. So the proof is complete. □

Corollary 5.5. *Let ψ be a smooth function defined in $\mathbf{R}^n$. For any integer N we have*

$$\begin{aligned} &\sum_{p=0}^{n} f_p \mathrm{tr}(\Psi_p \tilde{v}_1)(t,x,x) \\ &= \begin{cases} 0(t^N), & \text{if } x_n \neq 0; \\ 2t^{-\frac{n}{2}+\frac{\ell}{2}-\frac{1}{2}} D_{\ell-1}(x')\sqrt{\det \hat{g}(x',0)} + 0(t^{-\frac{n}{2}+\frac{\ell}{2}}), & \text{if } x_n = 0. \end{cases} \end{aligned}$$

For any positive constant ε *we have*

$$\int_0^{\varepsilon} \sum_{p=0}^{n} f_p \mathrm{tr}(\Psi_p \tilde{v}_1)(t, x', x_n, x', x_n)\psi(x', x_n)dx_n$$
$$= t^{-\frac{n}{2}+\frac{\ell}{2}} D_{\ell-1}(x')\sqrt{\det \hat{g}(x', 0)}\psi(x', 0) + 0(t^{-\frac{n}{2}+\frac{\ell}{2}+\frac{1}{2}}).$$

Proof. By Theorem 2.4 and Lemma 5.4 we have the assertion. □

In order to calculate the trace of $\sum_{p=0}^{n} f_p \Psi_p V_j \quad (j = 0, -1, -2, \ldots, -N)$ we have the following lemma which is obtained by the similar method to Lemma 5.4.

Lemma 5.6. *If* $h \in \mathcal{J}_s$, *then for the kernel* $\tilde{h}$ *of the operator that corresponds to* h *has the following estimates:*
For any integer N *and for any smooth function* ψ *defined in* $\mathbf{R}^n$

$$\sum_{p=0}^{n} f_p \mathrm{tr}(\Psi_p \tilde{h})(t, x, x) = \begin{cases} 0(t^N), & \text{if } x_n \neq 0; \\ 0(t^{-\frac{n}{2}+\frac{\ell}{2}-\frac{s}{2}}), & \text{if } x_n = 0. \end{cases}$$

and for any positive constant ε

$$\int_0^{\varepsilon} \sum_{p=0}^{n} f_p \mathrm{tr}(\Psi_p \tilde{h})(t, x', x_n, x', x_n)\psi(x', x_n)dx_n = 0(t^{-\frac{n}{2}+\frac{\ell}{2}-\frac{s}{2}+\frac{1}{2}}).$$

Corollary 5.7. *Let* ψ *be a smooth function defined in* $\mathbf{R}^n$. *For any integer* N *we have for any* $j \leq 0$

$$\sum_{p=0}^{n} f_p \mathrm{tr}(\Psi_p \tilde{v}_j)(t, x, x) = \begin{cases} 0(t^N), & \text{if } x_n \neq 0; \\ 0(t^{-\frac{n}{2}+\frac{\ell}{2}-\frac{j}{2}}), & \text{if } x_n = 0. \end{cases}$$

and for any positive constant ε

$$\int_0^{\varepsilon} \sum_{p=0}^{n} f_p \mathrm{tr}(\Psi_p \tilde{v}_j)(t, x', x_n, x', x_n)\psi(x', x_n)dx_n = 0(t^{-\frac{n}{2}+\frac{\ell}{2}-\frac{j}{2}+\frac{1}{2}}).$$

Proof of (2) *and* (3) *of the Main Theorem.* If we study the asymptotic behavior of the fundamental solution, it is sufficient to consider the fundamental solution locally by an argument similar to that employed for the case M is closed, which was proved, for example in [11]. In a local patch we have

$$e(t, x, x)dv = \tilde{u}(t, x, x)dx + \tilde{v}(t, x, x)dx,$$
$$\tilde{u}(t, x, y) = \sqrt{\det g(x)}u(t, x, y),$$

where $u(t, x, y)$ is obtained in Section 3.

By Corollary 5.5 and Corollary 5.7 we obtain

Lemma 5.8. *Let* ψ *be a smooth function defined in* $\mathbf{R}^n$. *For any integer* N *we have*

$$\sum_{p=0}^{n} f_p \mathrm{tr}(\Psi_p \tilde{v})(t,x,x) = \begin{cases} 0(t^N), & \text{if } x_n \neq 0; \\ 2t^{-\frac{n}{2}+\frac{\ell}{2}-\frac{1}{2}} D_{\ell-1}(x')\sqrt{\det \hat{g}(x',0)} + 0(t^{-\frac{n}{2}+\frac{\ell}{2}}), & \text{if } x_n = 0. \end{cases}$$

and for any positive constant ε

$$\int_0^{\varepsilon} \sum_{p=0}^{n} f_p \mathrm{tr}(\Psi_p \tilde{v})(t, x', x_n, x', x_n)\psi(x', x_n) dx_n$$
$$= t^{-\frac{n}{2}+\frac{\ell}{2}} D_{\ell-1}(x')\sqrt{\det \hat{g}(x', 0)}\psi(x', 0) + 0(t^{-\frac{n}{2}+\frac{\ell}{2}+\frac{1}{2}}).$$

From Lemma 5.8 and

$$e_p(t, x, x) = \Psi_p(\tilde{u}(t, x, x) + \tilde{v}(t, x, x))\frac{1}{\sqrt{\det g}}$$

it is easy to show for any N

$$\sum_{p=0}^{n} f_p \mathrm{tr}\, e_p(t, x, x) = \frac{1}{\sqrt{\det g}} \sum_{p=0}^{n} f_p \mathrm{tr}(\Psi_p \tilde{u})(t, x, x) + 0(t^N) \quad x \in M \backslash \partial M,$$
$$= \sum_{p=0}^{n} f_p \mathrm{tr}(\Psi_p u)(t, x, x) + 0(t^N) \quad x \in M \backslash \partial M,$$
$$= C_\ell(x) t^{-\frac{n}{2}+\frac{\ell}{2}} + 0(t^{-\frac{n}{2}+\frac{\ell}{2}+\frac{1}{2}}) \quad x \in M \backslash \partial M,$$
$$\sum_{p=0}^{n} f_p \mathrm{tr}\, e_p(t, x, x) = 2D_{\ell-1}(x) t^{-\frac{n}{2}+\frac{\ell}{2}-\frac{1}{2}} + 0(t^{-\frac{n}{2}+\frac{\ell}{2}}) \quad x \in \partial M.$$

and

$$\int_M \sum_{p=0}^{n} f_p \mathrm{tr}\, e_p(t, x, x)\psi(x) dv$$
$$= \Big(\int_M C_\ell(x)\psi(x) dv + \int_{\partial M} D_{\ell-1}(x')\psi(x') d\sigma\Big) t^{-\frac{n}{2}+\frac{\ell}{2}} + 0(t^{-\frac{n}{2}+\frac{\ell}{2}+\frac{1}{2}}).$$

So the proof is complete. □

References

[1] N. Berline, E. Getzler and M. Vergne, *Heat Kernels and Dirac Operators,* Springer-Verlag, 1992.

[2] S. Chern, A simple intrinsic proof of the Gauss Bonnet formula for closed Riemannian manifolds, *Ann. Math.* **45** (1944), 747–752.

[3] S. Chern, On the curvature integral in a Riemannian manifold, *Ann. Math.* **46** (1945), 674–684.

[4] H.L. Cycon, R.G. Froese, W. Kirsch and B. Simon, *Schrödinger Operators,* Texts and Monographs in Physics, Springer, 1987.

[5] P. Günther and R. Schimming, Curvature and spectrum of compact Riemannian manifolds, *J. Diff. Geom.* **12** (1977), 599–618.

[6] E. Getzler, The local Atiyah-Singer index theorem, in *Critical Phenomena, Random Systems, Gauge Theories*, Editors: K. Sterwalder and R. Stora Les Houches, North-Holland, 1984, 967–974.

[7] E. Getzler, A short proof of the local Atiyah-Singer index theorem, *Topology* **25** (1986), 111–117.

[8] P.B. Gilkey, *Invariance Theory, the Heat Equation, and the Atiyah-Singer Index Theorem,* Publish or Perish, Inc., 1984.

[9] P.B. Gilkey, *The boundary integrand in the formula for the signature and Euler characteristic of a Riemannian manifold with boundary, Adv. Math.* **15** (1975), 334–360.

[10] C. Iwasaki, The asymptotic expansion of the fundamental solution for initial-boundary value problems and its application, *Osaka J. Math.* **31** (1994), 663–728.

[11] C. Iwasaki, A proof of the Gauss-Bonnet-Chern theorem by the symbol calculus of pseudo-differential operators, *Japanese J. Math* **21** (1995), 235–285.

[12] C. Iwasaki, Symbolic calculus of pseudo-differential operators and curvature of manifolds, in *Modern Trends in Pseudo-Differential Operators*, Editors: J. Toft, M.W. Wong and H. Zhu, Birkhäuser, 2007, 51–66.

[13] S. Kobayashi and K. Nomizu, *Foundations of Differential Geometry I, II*, John Wiley & Sons, 1963.

[14] S. Murakami, *Manifolds,* Kyoritsusshuppan, 1969 (in Japanese).

[15] V.K. Patodi, Curvature and the eigenforms of the Laplace operator, *J. Diff. Geom.* **5** (1971), 233–249.

Chisato Iwasaki
Department of Mathematics
University of Hyogo
2167 Shosha
Himeji Hyogo 671-2201, Japan
e-mail: iwasaki@sci.u-hyogo.ac.jp

Operator Theory:
Advances and Applications, Vol. 189, 67–106

Operators with Corner-Degenerate Symbols

Jamil Abed and Bert-Wolfgang Schulze

Abstract. We establish elements of a new approach to ellipticity and parametrices within operator algebras on a manifold with higher singularities, only based on some general axiomatic requirements on parameter-dependent operators in suitable scales of spaces. The idea is to model an iterative process with new generations of parameter-dependent operator theories, together with new scales of spaces that satisfy analogous requirements as the original ones, now on a corresponding higher level.

The "full" calculus is voluminous; so we content ourselves here with some typical aspects such as symbols in terms of order reducing families, classes of relevant examples, and operators near a corner point.

Mathematics Subject Classification (2000). Primary 35S35; Secondary 35J70.

Keywords. Pseudo-differential operators, weighted spaces, edge- and corner-degenerate symbols, ellipticity near corner singularities.

Introduction

This paper is aimed at studying operators with certain degenerate operator-valued amplitude functions, motivated by the iterative calculus of pseudo-differential operators on manifolds with higher singularities. Here, in contrast to [38], [39], we develop the aspect of symbols, based on "abstract" reductions of orders which makes the approach transparent from a new point of view. To illustrate the idea, let us first consider, for example, the Laplacian on a manifold with conical singularities (say, without boundary). In this case the ellipticity does not only refer to the "standard" principal homogeneous symbol but also to the so-called conormal symbol. The latter one, contributed by the conical point, is operator-valued and singles out the weights in Sobolev spaces, where the operator has the Fredholm property. Another example of ellipticity with different principal symbolic components is the case of boundary value problems. The boundary (say, smooth), interpreted as an edge, contributes the operator-valued boundary (or edge) symbol which is responsible for the nature of boundary conditions (for instance, of Dirichlet or Neumann type in the case of the Laplacian). In general, if the configuration has polyhedral singularities of order k, we have to expect a principal symbolic hierarchy of length $k+1$, with components contributed by the various strata. In order to characterise the solvability of elliptic equations, especially, the regularity of solutions in suitable scales of spaces, it is adequate to embed the problem in a pseudo-differential calculus, and to construct a parametrix. For higher singularities this is a program of tremendous complexity. It is therefore advisable to organise general elements of the calculus by means of an axiomatic framework which contains the typical features, such as the cone- or edge-degenerate behaviour of symbols but ignores the (in general) huge tail of $k-1$ iterative steps to reach the singularity level k.

The "concrete" (pseudo-differential) calculus of operators on manifolds with conical or edge singularities may be found in several papers and monographs, see, for instance, [29], [33], [32], [5]. Operators on manifolds of singularity order 2 are studied in [34], [38], [17], [7]. Theories of that kind are also possible for boundary value problems with the transmission property at the (smooth part of the) boundary, see, for instance, [28], [13], [9]. This is useful in numerous applications, for instance, to models of elasticity or crack theory, see [13], [8]. Elements of operator structures on manifolds with higher singularities are developed, for instance, in [37], [1]. The nature of such theories depends very much on specific assumptions on the degeneracy of the involved symbols. There are worldwide different schools studying operators on singular manifolds, partly motivated by problems of geometry, index theory, and topology, see, for instance, Melrose [18], Melrose and Piazza [19], Nistor [24], Nazaikinskij, Savin, Sternin [20], [21], [22], and many others. We do not study here operators of "multi-Fuchs" type, often associated with "corner manifolds". Our operators are of a rather different behaviour with respect to the degeneracy of symbols. Nevertheless the various theories have intersections and common sources, see the paper of Kondratyev [14] or papers and monographs of other representatives of a corresponding Russian school, see, for instance, [26], [27].

Let us briefly recall a few basic facts on operators on manifolds with conical singularities or edges.

Let M be a manifold with conical singularity $v \in M$, i.e., $M \setminus \{v\}$ is smooth, and M is close to v modelled on a cone $X^{\Delta} := (\overline{\mathbb{R}}_+ \times X)/(\{0\} \times X)$ with base X, where X is a closed compact C^∞ manifold. We then have differential operators of order $\mu \in \mathbb{N}$ on $M \setminus \{v\}$, locally near v in the splitting of variables $(r, x) \in \mathbb{R}_+ \times X$ of the form

$$A := r^{-\mu} \sum_{j=0}^{\mu} a_j(r) \left(-r\frac{\partial}{\partial r}\right)^j \tag{0.1}$$

with coefficients $a_j \in C^\infty(\overline{\mathbb{R}}_+, \mathrm{Diff}^{\mu-j}(X))$ (here $\mathrm{Diff}^\nu(\cdot)$ denotes the space of all differential operators of order ν on the manifold in parentheses, with smooth coefficients). Observe that when we consider a Riemannian metric on $\mathbb{R}_+ \times X := X^\wedge$ of the form $dr^2 + r^2 g_X$, where g_X is a Riemannian metric on X, then the associated Laplace-Beltrami operator is just of the form (0.1) for $\mu = 2$. For such operators we have the homogeneous principal symbol $\sigma_\psi(A) \in C^\infty(T^*(M \setminus \{v\}) \setminus 0)$, and locally near v in the variables (r,x) with covariables (ρ,ξ) the function

$$\tilde{\sigma}_\psi(A)(r,x,\rho,\xi) := r^\mu \sigma_\psi(A)(r,x,r^{-1}\rho,\xi)$$

which is smooth up to $r=0$. If a symbol (or an operator function) contains r and ρ in the combination $r\rho$ we speak of degeneracy of Fuchs type.

It is interesting to ask the nature of an operator algebra that contains Fuchs type differential operators of the from (0.1) on X^Δ, together with the parametrices of elliptic elements. An analogous problem is meaningful on M. Answers may be found in [33], including the tools of the resulting so-called cone algebra. As noted above the ellipticity close to the tip $r=0$ is connected with a second symbolic structure, namely, the conormal symbol

$$\sigma_{\mathrm{c}}(A)(w) := \sum_{j=0}^{\mu} a_j(0)w^j : H^s(X) \to H^{s-\mu}(X) \tag{0.2}$$

which is a family of operators, depending on $w \in \Gamma_{\frac{n+1}{2}-\gamma}$, $\Gamma_\beta := \{w \in \mathbb{C} : \mathrm{Re}w = \beta\}$, $n = \dim X$. Here $H^s(X)$ are the standard Sobolev spaces of smoothness $s \in \mathbb{R}$ on X. Ellipticity of A with respect to a weight $\gamma \in \mathbb{R}$ means that (0.2) is a family of isomorphisms for all $w \in \Gamma_{\frac{n+1}{2}-\gamma}$.

The ellipticity on the infinite cone X^Δ refers to a further principal symbolic structure, to be observed when $r \to \infty$. The behaviour in that respect is not symmetric under the substitution $r \to r^{-1}$. The present axiomatic approach will refer to "abstract" corners represented by $r \to 0$. The considerations are based on specific insight on families of reductions of orders in given scales of spaces (in the simplest case $H^s(X), s \in \mathbb{R}$, when the corner is a conical singularity). In order to motivate our general constructions we briefly recall the form of corner operators of second generation.

First, a differential operator on an open stretched wedge $\mathbb{R}_+ \times X \times \Omega \ni (r,x,y)$, $\Omega \subseteq \mathbb{R}^q$ open, is called edge-degenerate, if it has the form

$$A = r^{-\mu} \sum_{j+|\alpha|\le\mu} a_{j\alpha}(r,y)\left(-r\frac{\partial}{\partial r}\right)^j (rD_y)^\alpha, \tag{0.3}$$

$a_{j\alpha} \in C^\infty(\overline{\mathbb{R}}_+ \times \Omega, \mathrm{Diff}^{\mu-(j+|\alpha|)}(X))$. Observe that (0.3) can be written in the form $A = r^{-\mu}\mathrm{Op}_{r,y}(p)$ for an operator-valued symbol p of the form $p(r,y,\rho,\eta) = \tilde{p}(r,y,r\rho,r\eta)$ and $\tilde{p}(r,y,\tilde{\rho},\tilde{\eta}) \in C^\infty(\overline{\mathbb{R}}_+ \times \Omega, L^\mu_{\mathrm{cl}}(X;\mathbb{R}^{1+q}_{\tilde{\rho},\tilde{\eta}}))$,

$$\mathrm{Op}_{r,y}(p)u(r,y) = \iint e^{i(r-r')\rho+i(y-y')\eta}p(r,y,\rho,\eta)u(r',y')dr'dy'đ\rho đ\eta.$$

Here $L^\mu_{\mathrm{cl}}(X;\mathbb{R}^l_\lambda)$ means the space of classical parameter-dependent pseudo-differential operators on X of order μ, with parameter $\lambda \in \mathbb{R}^l$, that is, locally on X the operators are given in terms of amplitude functions $a(x,\xi,\lambda)$, where (ξ,λ) is treated as an $(n+l)$-dimensional covariable, and we have $L^{-\infty}(X;\mathbb{R}^l) := \mathcal{S}(\mathbb{R}^l, L^{-\infty}(X))$ with $L^{-\infty}(X)$ being the (Fréchet) space of smoothing operators on X.

Let $\mathrm{Diff}^{\mu}_{\mathrm{deg}}(M)$ for a manifold M with edge Y denote the space of all differential operators on $M \setminus Y$ of order μ that are locally near Y in the splitting of variables $(r, x, y) \in \mathbb{R}_+ \times X \times \Omega$ of the form (0.3). If we replace in the definition the edge covariable η by $(\eta, \lambda) \in \mathbb{R}^{q+l}$ $(q = \dim Y)$ we obtain parameter-dependent families of operators in $\mathrm{Diff}^{\mu}_{\mathrm{deg}}(M)$. Similarly as (0.1) an operator of the form

$$A := t^{-\mu} \sum_{j=0}^{\mu} a_j(t) \left(-t\frac{\partial}{\partial t}\right)^j$$

is called corner degenerate if $a_j \in C^\infty(\overline{\mathbb{R}}_+, \mathrm{Diff}^{\mu-j}_{\mathrm{deg}}(M))$, $j = 0, 1, \dots, \mu$. The corner conormal symbol $\sigma_{\mathrm{c}}(A)(z) = \sum_{j=0}^{\mu} a_j(0) z^j$, $z \in \Gamma_{\frac{\dim M+1}{2}-\delta}$ for a corner weight $\delta \in \mathbb{R}$, is just a parameter-dependent family in $\mathrm{Diff}^{\mu}_{\mathrm{deg}}(M)$ with parameter $\mathrm{Im} z$ on the indicated weight line. The program to study such operators close to the tip $t \to 0$ (see [1], [7]) is just a concrete realisation of the present theory.

This paper is organised as follows. In Chapter 1 we introduce spaces of symbols based on families of reductions of orders in given scales of (analogues of Sobolev) spaces.

Chapter 2 is devoted to the specific effects of an axiomatic calculus near the tip of the corner. The corner axis is represented by a real axis $\mathbb{R} \ni r$, and the operators take values in vector-valued analogues of Sobolev spaces in r.

As indicated above, our results are designed as a step of a larger concept of abstract edge and corner theories, organised in an iterative manner. The full calculus employs the one for $r \to \infty$ as a counterpart of our Mellin operators on $\mathbb{R}_+$ near $r = 0$. However, the continuation of the calculus in that sense needs more space than available in the present note.

1. Symbols associated with order reductions

1.1. Scales and order reducing families

Let $\mathfrak{E}$ denote the set of all families $\mathcal{E} = (E^s)_{s\in\mathbb{R}}$ of Hilbert spaces with continuous embeddings $E^{s'} \hookrightarrow E^s$, $s' \geq s$, so that $E^\infty := \bigcap_{s\in\mathbb{R}} E^s$ is dense in every E^s, $s \in \mathbb{R}$ and that there is a dual scale $\mathcal{E}^* = (E^{*s})_{s\in\mathbb{R}}$ with a non-degenerate sesquilinear pairing $(.,.)_0 : E^0 \times E^{*0} \to \mathbb{C}$, such that $(.,.)_0 : E^\infty \times E^{*\infty} \to \mathbb{C}$, extends to a non-degenerate sesquilinear pairing

$$E^s \times E^{*-s} \to \mathbb{C}$$

for every $s \in \mathbb{R}$, where $\sup_{f\in E^{*-s}\setminus\{0\}} \frac{|(u,f)_0|}{\|f\|_{E^{*-s}}}$ and $\sup_{g\in E^s\setminus\{0\}} \frac{|(g,v)_0|}{\|g\|_{E^s}}$ are equivalent norms in the spaces E^s and E^{*-s}, respectively; moreover, if $\mathcal{E} = (E^s)_{s\in\mathbb{R}}$, $\widetilde{\mathcal{E}} = (\widetilde{E}^s)_{s\in\mathbb{R}}$ are two scales in consideration and $a \in \mathcal{L}^\mu(\mathcal{E}, \widetilde{\mathcal{E}}) := \bigcap_{s\in\mathbb{R}} \mathcal{L}(E^s, \widetilde{E}^{s-\mu})$, for some $\mu \in \mathbb{R}$, then

$$\sup_{s\in[s',s'']} \|a\|_{s,s-\mu} < \infty$$

for every $s' \leq s''$; here $\|.\|_{s,\tilde{s}} := \|.\|_{\mathcal{L}(E^s,\widetilde{E}^{\tilde{s}})}$. Later on, in the case $s = \tilde{s} = 0$ we often write $\|.\| := \|.\|_{0,0}$.

Let us say that a scale $\mathcal{E} \in \mathfrak{E}$ is said to have the compact embedding property, if the embeddings $E^{s'} \hookrightarrow E^s$ are compact when $s' > s$.

Remark 1.1. Every $a \in \mathcal{L}^\mu(\mathcal{E}, \widetilde{\mathcal{E}})$ has a formal adjoint $a^* \in \mathcal{L}^\mu(\widetilde{\mathcal{E}}^*, \mathcal{E}^*)$, obtained by $(au, v)_0 = (u, a^*v)_0$ for all $u \in E^\infty, v \in \widetilde{E}^{*\infty}$.

Remark 1.2. The space $\mathcal{L}^\mu(\mathcal{E}, \widetilde{\mathcal{E}})$ is Fréchet in a natural way for every $\mu \in \mathbb{R}$.

Definition 1.3. A system $(b^\mu(\eta))_{\mu\in\mathbb{R}}$ of operator functions $b^\mu(\eta) \in C^\infty(\mathbb{R}^q, \mathcal{L}^\mu(\mathcal{E},\mathcal{E}))$ is called an order reducing family of the scale $\mathcal{E}$, if $b^\mu(\eta) : E^s \to E^{s-\mu}$ is an isomorphism for every $s, \mu \in \mathbb{R}$, $\eta \in \mathbb{R}^q$, $b^0(\eta) = \mathrm{id}$ for every $\eta \in \mathbb{R}^q$, and

(i) $D_\eta^\beta b^\mu(\eta) \in C^\infty(\mathbb{R}^q, \mathcal{L}^{\mu-|\beta|}(\mathcal{E}, \mathcal{E}))$ for every $\beta \in \mathbb{N}^q$;

(ii) for every $s \in \mathbb{R}, \beta \in \mathbb{N}^q$ we have

$$\max_{|\beta|\leq k} \sup_{\substack{\eta\in\mathbb{R}^q \\ s\in[s',s'']}} \|b^{s-\mu+|\beta|}(\eta)\{D_\eta^\beta b^\mu(\eta)\}b^{-s}(\eta)\|_{0,0} < \infty$$

for all $k \in \mathbb{N}$, and for all real $s' \leq s''$.

(iii) for every $\mu, \nu \in \mathbb{R}$, $\nu \geq \mu$, we have

$$\sup_{s\in[s',s'']} \|b^\mu(\eta)\|_{s,s-\nu} \leq c\langle\eta\rangle^B$$

for all $\eta \in \mathbb{R}^q$ and $s' \leq s''$ with constants $c(\mu, \nu, s), B(\mu, \nu, s) > 0$, uniformly bounded in compact s-intervals and compact μ, ν-intervals for $\nu \geq \mu$; moreover, for every $\mu \leq 0$ we have

$$\|b^\mu(\eta)\|_{0,0} \leq c\langle\eta\rangle^\mu$$

for all $\eta \in \mathbb{R}^q$ with constants $c > 0$, uniformly bounded in compact μ-intervals, $\mu \leq 0$.

Clearly the operators b^μ in (iii) for $\nu \geq \mu$ or $\mu \leq 0$, are composed with a corresponding embedding operator.

In addition we require that the operator families $(b^\mu(\eta))^{-1}$ are equivalent to $b^{-\mu}(\eta)$, according to the following notation. Another order reducing family $(b_1^\mu(\eta))_{\mu\in\mathbb{R}}, \eta \in \mathbb{R}^q$, in the scale $\mathcal{E}$ is said to be equivalent to $(b^\mu(\eta))_{\mu\in\mathbb{R}}$, if for every $s \in \mathbb{R}, \beta \in \mathbb{N}^q$, there are constants $c = c(\beta, s)$ such that

$$\|b_1^{s-\mu+|\beta|}(\eta)\{D_\eta^\beta b^\mu(\eta)\}b_1^{-s}(\eta)\|_{0,0} \leq c,$$

$$\|b^{s-\mu+|\beta|}(\eta)\{D_\eta^\beta b_1^\mu(\eta)\}b^{-s}(\eta)\|_{0,0} \leq c,$$

for all $\eta \in \mathbb{R}^q$, uniformly in $s \in [s', s'']$ for every $s' \leq s''$.

Remark 1.4. Parameter-dependent theories of operators are common in many concrete contexts. For instance, if Ω is an (open) C^∞ manifold, there is the space $L_{\mathrm{cl}}^\mu(\Omega, \mathbb{R}^q)$ of parameter-dependent pseudo-differential operators on Ω of order $\mu \in \mathbb{R}$, with parameter $\eta \in \mathbb{R}^q$, where the local amplitude functions $a(x, \xi, \eta)$

are classical symbols in $(\xi, \eta) \in \mathbb{R}^{n+q}$, treated as covariables, $n = \dim \Omega$, while $L^{-\infty}(\Omega, \mathbb{R}^q)$ is the space of Schwartz functions in $\eta \in \mathbb{R}^q$ with values in $L^{-\infty}(\Omega)$, the space of smoothing operators on Ω. Later on we will also consider specific examples with more control on the dependence on η, namely, when $\Omega = M \setminus \{v\}$ for a manifold M with conical singularity v.

Example. Let X be a closed compact C^∞ manifold, $E^s := H^s(X), s \in \mathbb{R}$, the scale of classical Sobolev spaces on X and $b^\mu(\eta) \in L^\mu_{\mathrm{cl}}(X; \mathbb{R}^q_\eta)$ a parameter-dependent elliptic family that induces isomorphisms $b^\mu(\eta) : H^s(X) \to H^{s-\mu}(X)$ for all $s \in \mathbb{R}$. Then for $\nu \geq \mu$ we have

$$\|b^\mu(\eta)\|_{\mathcal{L}(H^s(X), H^{s-\nu}(X))} \leq c\langle\eta\rangle^{\pi(\mu,\nu)}$$

for all $\eta \in \mathbb{R}^q$, uniformly in $s \in [s', s'']$ for arbitrary s', s'', as well as in compact μ- and ν-intervals for $\nu \geq \mu$, where

$$\pi(\mu, \nu) := \max(\mu, \mu - \nu) \tag{1.1}$$

with a constant $c = c(\mu, \nu, s', s'') > 0$. Observe that $\sup_{\xi \in \mathbb{R}^p} \frac{\langle\xi,\eta\rangle^\mu}{\langle\xi\rangle^\nu} \leq \langle\eta\rangle^{\pi(\mu,\nu)}$ for all $\eta \in \mathbb{R}^q$.

Remark 1.5. Let $b^s(\tilde{\tau}, \tilde{\eta}) \in L^\mu_{\mathrm{cl}}(X; \mathbb{R}^{1+q}_{\tilde{\tau},\tilde{\eta}})$ be an order reducing family as in the above example, now with the parameter $(\tilde{\tau}, \tilde{\eta}) \in \mathbb{R}^{1+q}$ rather than η, and of order $s \in \mathbb{R}$. Then, setting $b^s(t, \tau, \eta) := b^s(t\tau, t\eta)$ the expression

$$\Big\{ \int \|[t]^{-s}\mathrm{Op}_t(b^s)(\eta^1)u\|^2_{L^2(X)} dt \Big\}^{\frac{1}{2}}$$

for $\eta^1 \in \mathbb{R}^q \setminus \{0\}$, $|\eta^1|$ sufficiently large, is a norm on the space $\mathcal{S}(\mathbb{R}, C^\infty(X))$. Let $H^s_{\mathrm{cone}}(\mathbb{R} \times X)$ denote the completion of $\mathcal{S}(\mathbb{R}, C^\infty(X))$ in this norm. Observe that this space is independent of the choice of η^1, $|\eta^1|$ sufficiently large. For reference below we also form weighted variants $H^{s;g}_{\mathrm{cone}}(\mathbb{R} \times X) := \langle t\rangle^{-g} H^s_{\mathrm{cone}}(\mathbb{R} \times X)$, $g \in \mathbb{R}$, and set

$$H^{s;g}_{\mathrm{cone}}(\mathbb{R}_+ \times X) := H^{s;g}_{\mathrm{cone}}(\mathbb{R} \times X)|_{\mathbb{R}_+ \times X}. \tag{1.2}$$

As is known, cf. [13], the spaces $H^{s;g}_{\mathrm{cone}}(\mathbb{R} \times X)$ are weighted Sobolev spaces in the calculus of pseudo-differential operators on $\mathbb{R}_+ \times X$ with $|t| \to \infty$ being interpreted as a conical exit to infinity.

Another feature of order reducing families, known, for instance, in the case of the above example, is that when $U \subseteq \mathbb{R}^p$ is an open set and $m(y) \in C^\infty(U)$ a strictly positive function, $m(y) \geq c$ for $c > 0$ and for all $y \in U$, the family $b^s_1(y, \eta) := b^s(m(y)\eta)$, $s \in \mathbb{R}$, is order reducing in the sense of Definition 1.3 and equivalent to $b(\eta)$ for every $y \in U$, uniformly in $y \in K$ for any compact subset $K \subset U$. A natural requirement is that when $m > 0$ is a parameter, there is a constant $M = M(s', s'') > 0$ such that

$$\|b^s(\eta)b^{-s}(m\eta)\|_{0,0} \leq c \max(m, m^{-1})^M \tag{1.3}$$

for every $s \in [s', s'']$, $m \in \mathbb{R}_+$, and $\eta \in \mathbb{R}^q$.

We now turn to another example of an order reducing family, motivated by the calculus of pseudo-differential operators on a manifold with edge (here in "abstract" form), where all the above requirements are satisfied, including the latter one.

Definition 1.6.

(i) If H is a Hilbert space and $\kappa := \{\kappa_\lambda\}_{\lambda\in\mathbb{R}_+}$ a group of isomorphisms $\kappa_\lambda : H \to H$, such that $\lambda \to \kappa_\lambda h$ defines a continuous function $\mathbb{R}_+ \to H$ for every $h \in H$, and $\kappa_\lambda \kappa_\rho = \kappa_{\lambda\rho}$ for $\lambda, \rho \in \mathbb{R}$, we call κ a group action on H.

(ii) Let $\mathcal{H} = (H^s)_{s\in\mathbb{R}} \in \mathfrak{C}$ and assume that H^0 is endowed with a group action $\kappa = \{\kappa_\lambda\}_{\lambda\in\mathbb{R}_+}$ that restricts (for $s > 0$) or extends (for $s < 0$) to a group action on H^s for every $s \in \mathbb{R}$. In addition, we assume that κ is a unitary group action on H^0. We then say that $\mathcal{H}$ is endowed with a group action.

If H and κ are as in Definition 1.6 (i), it is known that there are constants $c, M > 0$, such that

$$\|\kappa_\lambda\|_{\mathcal{L}(H)} \le c \max(\lambda, \lambda^{-1})^M \tag{1.4}$$

for all $\lambda \in \mathbb{R}_+$.

Let $\mathcal{W}^s(\mathbb{R}^q, H)$ denote the completion of $\mathcal{S}(\mathbb{R}^q, H)$ with respect to the norm

$$\|u\|_{\mathcal{W}^s(\mathbb{R}^q,H)} := \left\{ \int \langle\eta\rangle^{2s} \|\kappa^{-1}_{\langle\eta\rangle}\hat{u}(\eta)\|^2_H d\eta \right\}^{\frac{1}{2}};$$

$\hat{u}(\eta) = \mathcal{F}_{y\to\eta} u(\eta)$ is the Fourier transform in $\mathbb{R}^q$. The space $\mathcal{W}^s(\mathbb{R}^q, H)$ will be referred to as edge space on $\mathbb{R}^q$ of smoothness $s \in \mathbb{R}$ (modelled on H). Given a scale $\mathcal{H} = (H^s)_{s\in\mathbb{R}} \in \mathfrak{C}$ with group action we have the edge spaces

$$W^s := \mathcal{W}^s(\mathbb{R}^q, H^s), \ s \in \mathbb{R}.$$

If necessary we also write $\mathcal{W}^s(\mathbb{R}^q, H^s)_\kappa$. The spaces form again a scale $\mathcal{W} := (W^s)_{s\in\mathbb{R}} \in \mathfrak{C}$.

For purposes below we now formulate a class of operator-valued symbols

$$S^\mu(U \times \mathbb{R}^q; H, \widetilde{H})_{\kappa,\tilde{\kappa}} \tag{1.5}$$

for open $U \subseteq \mathbb{R}^p$ and Hilbert spaces H and $\widetilde{H}$, endowed with group actions $\kappa = \{\kappa_\lambda\}_{\lambda\in\mathbb{R}_+}, \tilde{\kappa} = \{\tilde{\kappa}_\lambda\}_{\lambda\in\mathbb{R}_+}$, respectively, as follows. The space (1.5) is defined to be the set of all $a(y, \eta) \in C^\infty(U \times \mathbb{R}^q, \mathcal{L}(H, \widetilde{H}))$ such that

$$\sup_{(y,\eta)\in K\times\mathbb{R}^q} \langle\eta\rangle^{-\mu+|\beta|} \|\tilde{\kappa}^{-1}_{\langle\eta\rangle}\{D^\alpha_y D^\beta_\eta a(y,\eta)\}\kappa_{\langle\eta\rangle}\|_{\mathcal{L}(H,\tilde{H})} < \infty \tag{1.6}$$

for every compact $K \subset U, \alpha \in \mathbb{N}^p, \beta \in \mathbb{N}^q$.

Remark 1.7. Analogous symbols can also be defined in the case when $\widetilde{H}$ is a Fréchet space with group action, i.e., $\widetilde{H}$ is written as a projective limit of Hilbert spaces $\widetilde{H}_j, j \in \mathbb{N}$, with continuous embeddings $\widetilde{H}_j \hookrightarrow \widetilde{H}_0$, where the group action on $\widetilde{H}_0$ restricts to group actions on $\widetilde{H}_j$ for every j. Then $S^\mu(U \times \mathbb{R}^q; H, \widetilde{H}) := \varprojlim_{j\in\mathbb{N}} S^\mu(U \times \mathbb{R}^q; H, \widetilde{H}_j)$.

Consider an operator function $p(\xi,\eta) \in C^\infty(\mathbb{R}^{p+q}_{\xi,\eta}, \mathcal{L}^\mu(\mathcal{H},\mathcal{H}))$ that represents a symbol

$$p(\xi,\eta) \in S^\mu(\mathbb{R}^{p+q}_{\xi,\eta}; H^s, H^{s-\mu})_{\kappa,\kappa}$$

for every $s \in \mathbb{R}$, such that $p(\xi,\eta) : H^s \to H^{s-\mu}$ is a family of isomorphisms for all $s \in \mathbb{R}$, and the inverses $p^{-1}(\xi,\eta)$ represent a symbol

$$p^{-1}(\xi,\eta) \in S^{-\mu}(\mathbb{R}^{p+q}_{\xi,\eta}; H^s, H^{s+\mu})_{\kappa,\kappa}$$

for every $s \in \mathbb{R}$. Then $b^\mu(\eta) := \mathrm{Op}_x(p)(\eta)$ is a family of isomorphisms

$$b^\mu(\eta) : W^s \to W^{s-\mu}, \ \eta \in \mathbb{R}^q,$$

with the inverses $b^{-\mu}(\eta) := \mathrm{Op}_x(p^{-1})(\eta)$.

Proposition 1.8.

(i) *We have*

$$\|b^\mu(\eta)\|_{\mathcal{L}(W^0,W^0)} \leq c\langle\eta\rangle^\mu \tag{1.7}$$

for every $\mu \leq 0$, *with a constant* $c(\mu) > 0$.

(ii) *For every* $s, \mu, \nu \in \mathbb{R}$, $\nu \geq \mu$, *we have*

$$\|b^\mu(\eta)\|_{\mathcal{L}(W^s,W^{s-\nu})} \leq c\langle\eta\rangle^{\pi(\mu,\nu)+M(s)+M(s-\mu)} \tag{1.8}$$

for all $\eta \in \mathbb{R}^q$, *with a constant* $c(\mu, s) > 0$, *and* $M(s) \geq 0$ *defined by*

$$\|\kappa_\lambda\|_{\mathcal{L}(H^s,H^s)} \leq c\lambda^{M(s)} \text{ for all } \lambda \geq 1.$$

Proof. (i) Let us check the estimate (1.7). For the computations we denote by $j : H^{-\mu} \hookrightarrow H^0$ the embedding operator. We have for $u \in W^0$

$$\begin{aligned}
\|b^\mu(\eta)u\|^2_{W^0} &= \int \|jp(\xi,\eta)(\mathcal{F}u)(\xi)\|^2_{H^0} d\xi \\
&= \int \|\kappa^{-1}_{\langle\xi,\eta\rangle} j \kappa_{\langle\xi,\eta\rangle} \kappa^{-1}_{\langle\xi,\eta\rangle} p(\xi,\eta) \kappa_{\langle\xi,\eta\rangle} \kappa^{-1}_{\langle\xi,\eta\rangle} (\mathcal{F}u)(\xi)\|^2_{H^0} d\xi \\
&\leq \int \|\kappa^{-1}_{\langle\xi,\eta\rangle} j \kappa_{\langle\xi,\eta\rangle}\|^2_{\mathcal{L}(H^{-\mu},H^0)} \|\kappa^{-1}_{\langle\xi,\eta\rangle} p(\xi,\eta) \kappa_{\langle\xi,\eta\rangle} \kappa^{-1}_{\langle\xi,\eta\rangle} (\mathcal{F}u)(\xi)\|^2_{H^{-\mu}} d\xi \\
&\leq c \int \|\kappa^{-1}_{\langle\xi,\eta\rangle} p(\xi,\eta) \kappa_{\langle\xi,\eta\rangle}\|^2_{\mathcal{L}(H^0,H^{-\mu})} \|\kappa^{-1}_{\langle\xi,\eta\rangle} (\mathcal{F}u)(\xi)\|^2_{H^0} d\xi \\
&\leq c \sup_{\xi\in\mathbb{R}^p} \langle\xi,\eta\rangle^{2\mu} \|u\|^2_{W^0}.
\end{aligned}$$

Thus $\|b^\mu(\eta)\|_{\mathcal{L}(W^0,W^0)} \leq c \sup_{\xi\in\mathbb{R}^p} \langle\xi,\eta\rangle^\mu \leq c\langle\eta\rangle^\mu$, since $\mu \leq 0$.

(ii) Let $j : H^{s-\mu} \hookrightarrow H^{s-\nu}$ denote the canonical embedding. For every fixed $s \subset \mathbb{R}$ we have

$$\begin{aligned}
\|b^{\mu}(\eta)u\|^2_{W^{s-\nu}} &= \int \langle\xi\rangle^{2(s-\nu)} \|\kappa^{-1}_{\langle\xi\rangle} jp(\xi,\eta)(\mathcal{F}_{x\to\xi}u)(\xi)\|^2_{H^{s-\nu}} d\xi \\
&= \int \langle\xi\rangle^{2(s-\nu)} \|\kappa^{-1}_{\langle\xi\rangle} jp(\xi,\eta)\kappa_{\langle\xi\rangle}\langle\xi\rangle^{-s}\langle\xi\rangle^{s}\kappa^{-1}_{\langle\xi\rangle}(\mathcal{F}_{x\to\xi}u)(\xi)\|^2_{H^{s-\nu}} d\xi \\
&= \sup_{\xi\in\mathbb{R}^p} \langle\xi\rangle^{-2\nu} \|\kappa^{-1}_{\langle\xi\rangle} jp(\xi,\eta)\kappa_{\langle\xi\rangle}\|^2_{\mathcal{L}(H^s,H^{s-\nu})} \int \langle\xi\rangle^{2s} \|\kappa^{-1}_{\langle\xi\rangle}\mathcal{F}_{x\to\xi}u(\xi)\|^2_{H^s} d\xi.
\end{aligned}$$

For the first factor on the right we obtain

$$\begin{aligned}
&\|\kappa^{-1}_{\langle\xi\rangle}\big(jp(\xi,\eta)\big)\kappa_{\langle\xi\rangle}\|_{\mathcal{L}(H^s,H^{s-\nu})} \\
&\qquad \le \|\kappa^{-1}_{\langle\xi\rangle} j\kappa_{\langle\xi\rangle}\|_{\mathcal{L}(H^{s-\mu},H^{s-\nu})} \|\kappa^{-1}_{\langle\xi\rangle} p(\xi,\eta)\kappa_{\langle\xi\rangle}\|_{\mathcal{L}(H^s,H^{s-\mu})} \\
&\qquad\qquad \le c\|\kappa^{-1}_{\langle\xi\rangle} p(\xi,\eta)\kappa_{\langle\xi\rangle}\|_{\mathcal{L}(H^s,H^{s-\mu})}
\end{aligned}$$

with a constant $c > 0$. We employed here that $\|\kappa^{-1}_{\langle\xi\rangle} j\kappa_{\langle\xi\rangle}\|_{\mathcal{L}(H^{s-\mu},H^{s-\nu})} \le c$ for all $\xi \in \mathbb{R}^p$. Moreover,

$$\begin{aligned}
&\|\kappa^{-1}_{\langle\xi\rangle} p(\xi,\eta)\kappa_{\langle\xi\rangle}\|_{\mathcal{L}(H^s,H^{s-\mu})} \\
&\le \|\kappa^{-1}_{\langle\xi\rangle}\kappa_{\langle\xi,\eta\rangle}\|_{\mathcal{L}(H^{s-\mu},H^{s-\mu})} \|\kappa^{-1}_{\langle\xi,\eta\rangle} p(\xi,\eta)\kappa_{\langle\xi,\eta\rangle}\|_{\mathcal{L}(H^s,H^{s-\mu})} \|\kappa^{-1}_{\langle\xi,\eta\rangle}\kappa_{\langle\xi\rangle}\|_{\mathcal{L}(H^s,H^s)} \\
&\le c\langle\xi,\eta\rangle^{\mu} \|\kappa_{\langle\xi,\eta\rangle\langle\xi\rangle^{-1}}\|_{\mathcal{L}(H^{s-\mu},H^{s-\mu})} \|\kappa_{\langle\xi,\eta\rangle^{-1}\langle\xi\rangle}\|_{\mathcal{L}(H^s,H^s)} \\
&\le c\langle\xi,\eta\rangle^{\mu} \left(\frac{\langle\xi,\eta\rangle}{\langle\xi\rangle}\right)^{M(s-\mu)+M(s)}.
\end{aligned}$$

As usual, $c > 0$ denotes different constants (they may also depend on s); the numbers $M(s)$, $s \in \mathbb{R}$, are determined by the estimates

$$\|\kappa_\lambda\|_{\mathcal{L}(H^s,H^s)} \le c\lambda^{M(s)} \text{ for all } \lambda \ge 1.$$

We obtain altogether that

$$\begin{aligned}
\|b^{\mu}(\eta)\|_{\mathcal{L}(W^s,W^{s-\nu})} &\le c \sup_{\xi\in\mathbb{R}^n} \frac{\langle\xi,\eta\rangle^{\mu}}{\langle\xi\rangle^{\nu}} \Big(\frac{\langle\xi,\eta\rangle}{\langle\xi\rangle}\Big)^{M(s-\mu)+M(s)} \\
&\le c\langle\eta\rangle^{\pi(\mu,\nu)+M(s-\mu)+M(s)}.
\end{aligned}$$

□

It can be proved that the operators in Proposition 1.8 also have the uniformity properties with respect to s, μ, ν in compact sets, imposed in Definition 1.3.

1.2. Symbols based on order reductions

We now turn to operator-valued symbols, referring to scales

$$\mathcal{E} = (E^s)_{s\in\mathbb{R}}, \widetilde{\mathcal{E}} = (\widetilde{E}^s)_{s\in\mathbb{R}} \in \mathfrak{C}.$$

For purposes below we slightly generalise the concept of order reducing families by replacing the parameter space $\mathbb{R}^q \ni \eta$ by $\mathbb{H} \ni \eta$, where

$$\mathbb{H} := \{\eta = (\eta', \eta'') \in \mathbb{R}^{q'+q''} : q = q' + q'', \eta'' \neq 0\}. \tag{1.9}$$

In other words for every $\mu \in \mathbb{R}$ we fix order-reducing families $b^{\mu}(\eta)$ and $\tilde{b}^{\mu}(\eta)$ in the scales $\mathcal{E}$ and $\widetilde{\mathcal{E}}$, respectively, where η varies over $\mathbb{H}$, and the properties of Definition 1.3 are required for all $\eta \in \mathbb{H}$. In many cases we may admit the case $\mathbb{H} = \mathbb{R}^q$ as well.

Definition 1.9. By $S^{\mu}(U \times \mathbb{H}; \mathcal{E}, \widetilde{\mathcal{E}})$ for open $U \subseteq \mathbb{R}^p, \mu \in \mathbb{R}$, we denote the set of all $a(y,\eta) \in C^{\infty}(U \times \mathbb{H}, \mathcal{L}^{\mu}(\mathcal{E}, \widetilde{\mathcal{E}}))$ such that

$$D_y^{\alpha} D_{\eta}^{\beta} a(y,\eta) \in C^{\infty}(U \times \mathbb{H}, \mathcal{L}^{\mu-|\beta|}(\mathcal{E}, \widetilde{\mathcal{E}})), \tag{1.10}$$

and for every $s \in \mathbb{R}$ we have

$$\max_{|\alpha|+|\beta| \le k} \sup_{\substack{y \in K, \eta \in \mathbb{H}, |\eta| \ge h \\ s \in [s', s'']}} \|\tilde{b}^{s-\mu+|\beta|}(\eta)\{D_y^{\alpha} D_{\eta}^{\beta} a(y,\eta)\} b^{-s}(\eta)\|_{0,0} \tag{1.11}$$

is finite for all compact $K \subset U$, $k \in \mathbb{N}, h > 0$.

Let $S^{\mu}(\mathbb{H}; \mathcal{E}, \widetilde{\mathcal{E}})$ denote the subspace of all elements of $S^{\mu}(U \times \mathbb{H}; \mathcal{E}, \widetilde{\mathcal{E}})$ that are independent of y.

Observe that when $(b^{\mu}(\eta))_{\mu \in \mathbb{R}}$ is an order reducing family parametrised by $\eta \in \mathbb{H}$ then we have

$$b^{\mu}(\eta) \in S^{\mu}(\mathbb{H}; \mathcal{E}, \mathcal{E}) \tag{1.12}$$

for every $\mu \in \mathbb{R}$.

Remark 1.10. The space $S^{\mu}(U \times \mathbb{H}; \mathcal{E}, \widetilde{\mathcal{E}})$ is Fréchet with the semi-norms

$$a \to \max_{|\alpha|+|\beta| \le k} \sup_{\substack{(y,\eta) \in K \times \mathbb{H}, |\eta| \ge h \\ s \in [s', s'']}} \|\tilde{b}^{s-\mu+|\beta|}(\eta)\{D_y^{\alpha} D_{\eta}^{\beta} a(y,\eta)\} b^{-s}(\eta)\|_{0,0} \tag{1.13}$$

parametrised by compact $K \subset U$, $s \in \mathbb{Z}$, $\alpha \in \mathbb{N}^p$, $\beta \in \mathbb{N}^q$, $h > 0$, which are the best constants in the estimates (1.11). We then have

$$S^{\mu}(U \times \mathbb{H}; \mathcal{E}, \widetilde{\mathcal{E}}) = C^{\infty}(U, S^{\mu}(\mathbb{H}; \mathcal{E}, \widetilde{\mathcal{E}})) = C^{\infty}(U) \hat{\otimes}_{\pi} S^{\mu}(\mathbb{H}; \mathcal{E}, \widetilde{\mathcal{E}}).$$

We will also employ other variants of such symbols, for instance, when $\Omega \subseteq \mathbb{R}^m$ is an open set,

$$S^{\mu}(\overline{\mathbb{R}}_+ \times \Omega \times \mathbb{H}; \mathcal{E}, \widetilde{\mathcal{E}}) := C^{\infty}(\overline{\mathbb{R}}_+ \times \Omega, S^{\mu}(\mathbb{H}; \mathcal{E}, \widetilde{\mathcal{E}})).$$

In order to emphasize the similarity of our considerations for $\mathbb{H}$ with the case $\mathbb{H} = \mathbb{R}^q$ we often write again $\mathbb{R}^q$ and later on tacitly use the corresponding results for $\mathbb{H}$ in general.

Remark 1.11. Let $a(y,\eta) \in S^{\mu}(U \times \mathbb{R}^q)$ be a polynomial in η of order μ and $\mathcal{E} = (E^s)_{s \in \mathbb{R}}$ a scale and identify $D_y^{\alpha} D_{\eta}^{\beta} a(y,\eta)$ with $\big(D_y^{\alpha} D_{\eta}^{\beta} a(y,\eta)\big)\iota$ with the embedding $\iota : E^s \to E^{s-\mu+|\beta|}$. Then we have

$$\begin{aligned} &\|b^{s-\mu+|\beta|}(\eta)\big(D_y^{\alpha} D_{\eta}^{\beta} a(y,\eta)\big) b^{-s}(\eta)\|_{0,0} \\ &\qquad \le |D_y^{\alpha} D_{\eta}^{\beta} a(y,\eta)| \|b^{-\mu+|\beta|}(\eta)\|_{0,0} \le c\langle\eta\rangle^{\mu-|\beta|}\langle\eta\rangle^{-\mu+|\beta|} = c \end{aligned}$$

for all $\beta \in \mathbb{N}^q$, $|\beta| \le \mu$, $y \in K \subset U$, K compact (see Definition 1.3 (iii)). Thus $a(y,\eta)$ is canonically identified with an element of $S^{\mu}(U \times \mathbb{R}^q; \mathcal{E}, \mathcal{E})$.

Proposition 1.12. *We have*

$$S^{-\infty}(U \times \mathbb{R}^q; \mathcal{E}, \widetilde{\mathcal{E}}) := \bigcap_{\mu \in \mathbb{R}} S^{\mu}(U \times \mathbb{R}^q; \mathcal{E}, \widetilde{\mathcal{E}}) = C^{\infty}(U, \mathcal{S}(\mathbb{R}^q, \mathcal{L}^{-\infty}(\mathcal{E}, \widetilde{\mathcal{E}}))).$$

Proof. Let us show the assertion for y-independent symbols; the y-dependent case is then straightforward. For notational convenience we set $\widetilde{\mathcal{E}} = \mathcal{E}$; the general case is analogous. First let $a(\eta) \in S^{-\infty}(\mathbb{R}^q; \mathcal{E}, \mathcal{E})$, which means that $a(\eta) \in C^{\infty}(\mathbb{R}^q, \mathcal{L}^{-\infty}(\mathcal{E}, \mathcal{E}))$ and

$$\|b^{s+N}(\eta)\{D_\eta^\beta a(\eta)\} b^{-s}(\eta)\|_{0,0} < c \tag{1.14}$$

for all $s \in \mathbb{R}$, $N \in \mathbb{N}$, $\beta \in \mathbb{N}^q$ and show that

$$\sup_{\eta \in \mathbb{R}^q} \|\langle\eta\rangle^M D_\eta^\beta a(\eta)\|_{s,t} < \infty \tag{1.15}$$

for every $s, t \in \mathbb{R}$, $M \in \mathbb{N}$, $\beta \in \mathbb{N}^q$. To estimate (1.15) it is enough to assume $t > 0$. We have

$$\|\langle\eta\rangle^M D_\eta^\beta a(\eta)\|_{s,t} = \|b^{-kt}(\eta) b^{kt}(\eta)\langle\eta\rangle^M D_\eta^\beta a(\eta) b^{-s}(\eta) b^{s}(\eta)\|_{s,t} \tag{1.16}$$

for every $k \in \mathbb{N}$, $k \geq 1$, it is sufficient to show that the right-hand side is uniformly bounded in $\eta \in \mathbb{R}^q$ for sufficiently large choice of k. The right-hand side of (1.16) can be estimated by

$$\|b^{-t}(\eta)\|_{0,t} \|b^{(1-k)t}(\eta)\|_{0,0} \|b^{kt}(\eta) D_\eta^\beta a(\eta) b^{-s}(\eta)\|_{0,0} \|b^{s}(\eta)\|_{s,0}.$$

Using $\|b^{kt}(\eta) D_\eta^\beta a(\eta) b^{-s}(\eta)\|_{0,0} \leq c$, which is true by assumption and the estimates

$$\|b^{s}(\eta)\|_{s,0} \leq c\langle\eta\rangle^B, \quad \|b^{-t}(\eta)\|_{0,t} \leq c\langle\eta\rangle^{B'},$$

with different $B, B' \in \mathbb{R}$ and $\|b^{(1-k)t}(\eta)\|_{0,0} \leq c\langle\eta\rangle^{(1-k)t}$ (see Definition 1.3 (iii)) we obtain altogether

$$\|\langle\eta\rangle^M D_\eta^\beta a(\eta)\|_{s,t} \leq c\langle\eta\rangle^{M+B+B'+(1-k)t}$$

for some $c > 0$. Choosing k large enough it follows that the exponent on the right-hand side is < 0, i.e., we obtain uniform boundedness in $\eta \in \mathbb{R}^q$.

To show the reverse direction suppose that $a(\eta)$ satisfies (1.15), and let $\beta \in \mathbb{N}^q$, $M, s, t \in \mathbb{R}$ be arbitrary. We have

$$\begin{aligned} &\|b^{t}(\eta) D_\eta^\beta a(\eta) b^{-s}(\eta)\|_{0,0} \leq \\ &\qquad \|b^{t}(\eta)\langle\eta\rangle^{-M}\|_{t,0} \|\langle\eta\rangle^{2M} D_\eta^\beta a(\eta)\|_{s,t} \|\langle\eta\rangle^{-M} b^{-s}(\eta)\|_{0,s}. \end{aligned} \tag{1.17}$$

Now using (1.15) and the estimates

$$\|b^{t}(\eta)\langle\eta\rangle^{-M}\|_{t,0} \leq c\langle\eta\rangle^{A-M}, \quad \|\langle\eta\rangle^{-M} b^{-s}(\eta)\|_{0,s} \leq c\langle\eta\rangle^{A'-M},$$

with constants $A, A' \in \mathbb{R}$, we obtain

$$\|b^{t}(\eta) D_\eta^\beta a(\eta) b^{-s}(\eta)\|_{0,0} \leq c\langle\eta\rangle^{A+A'-2M}.$$

Choosing M large enough we get uniform boundedness of (1.17) in $\eta \in \mathbb{R}^q$ which completes the proof. □

Proposition 1.13. *Let $a(y,\eta)\in S^{\mu}(U\times\mathbb{R}^q;\mathcal{E},\widetilde{\mathcal{E}})$ and $\mu\le 0$. Then we have*

$$\|a(y,\eta)\|_{0,0}\le c\langle\eta\rangle^{\mu}$$

for all $y\in K\subset U$, K compact, $\eta\in\mathbb{R}^q$, with a constant $c=c(s,K)>0$.

Proof. For simplicity we consider the y-independent case. It is enough to show that $\|a(\eta)u\|_{\widetilde{E}^0}\le c\langle\eta\rangle^{\mu}\|u\|_{E^0}$ for all $u\in E^{\infty}$. Let $j:E^{-\mu}\to E^0$ denote the embedding operator. We then have

$$\begin{aligned}\|a(\eta)u\|_{\widetilde{E}^0}&=\|a(\eta)b^{-\mu}(\eta)jb^{\mu}(\eta)u\|_{\widetilde{E}^0}\\&\le\|a(\eta)b^{-\mu}(\eta)\|_{\mathcal{L}(E^0,\widetilde{E}^0)}\|jb^{\mu}(\eta)u\|_{E^0}\le c\langle\eta\rangle^{\mu}\|u\|_{E^0}.\end{aligned}$$

□

Proposition 1.14. *A symbol $a(y,\eta)\in S^{\mu}(U\times\mathbb{R}^q;\mathcal{E},\widetilde{\mathcal{E}})$, $\mu\in\mathbb{R}$, satisfies the estimates*

$$\|a(y,\eta)\|_{s,s-\nu}\le c\langle\eta\rangle^{A} \tag{1.18}$$

for every $\nu\ge\mu$, for every $y\in K\subset U$, K compact, $\eta\in\mathbb{R}^q, s\in\mathbb{R}$, with constants $c=c(s,\mu,\nu)>0, A=A(s,\mu,\nu,K)>0$ that are uniformly bounded when s,μ,ν vary over compact sets, $\nu\ge\mu$.

Proof. For simplicity we consider again the y-independent case. Let $j:\widetilde{E}^{s-\mu}\hookrightarrow\widetilde{E}^{s-\nu}$ be the embedding operator. Then we have

$$\begin{aligned}\|a(\eta)\|_{s,s-\nu}&=\|j\tilde{b}^{-s+\mu}(\eta)\tilde{b}^{s-\mu}(\eta)a(\eta)b^{-s}(\eta)b^{s}(\eta)\|_{s,s-\nu}\\&\le\|j\tilde{b}^{-s+\mu}(\eta)\|_{0,s-\nu}\|\tilde{b}^{s-\mu}(\eta)a(\eta)b^{-s}(\eta)\|_{0,0}\|b^{s}(\eta)\|_{s,0}.\end{aligned}$$

Applying (1.11) and Definition 1.3 (iii) we obtain (1.18) with $A=B(-s+\mu,-s+\nu,0)+B(s,s,0)$, together with the uniform boundedness of the involved constants.

□

Also here it can be proved that the involved constants in Propositions 1.13, 1.14 are uniform in compact sets with respect to s,μ,ν.

Proposition 1.15. *The symbol spaces have the following properties:*

(i) *$S^{\mu}(U\times\mathbb{R}^q;\mathcal{E},\widetilde{\mathcal{E}})\subseteq S^{\mu'}(U\times\mathbb{R}^q;\mathcal{E},\widetilde{\mathcal{E}})$ for every $\mu'\ge\mu$;*

(ii) *$D_y^{\alpha}D_{\eta}^{\beta}S^{\mu}(U\times\mathbb{R}^q;\mathcal{E},\widetilde{\mathcal{E}})\subseteq S^{\mu-|\beta|}(U\times\mathbb{R}^q;\mathcal{E},\widetilde{\mathcal{E}})$ for every $\alpha\in\mathbb{N}^p,\beta\in\mathbb{N}^q$;*

(iii) *$S^{\mu}(U\times\mathbb{R}^q;\mathcal{E}_0,\widetilde{\mathcal{E}})S^{\nu}(U\times\mathbb{R}^q;\mathcal{E},\mathcal{E}_0)\subseteq S^{\mu+\nu}(U\times\mathbb{R}^q;\mathcal{E},\widetilde{\mathcal{E}})$ for every $\mu,\nu\in\mathbb{R}$ (the notation on the left-hand side of the latter relation means the space of all (y,η)-wise compositions of elements in the respective factors).*

Proof. For simplicity we consider symbols with constant coefficients. Let us write $\|\cdot\|:=\|\cdot\|_{0,0}$, etc.

(i) $a(\eta)\in S^{\mu}(\mathbb{R}^q;\mathcal{E},\widetilde{\mathcal{E}})$ means (1.10) and (1.11); this implies

$$\begin{aligned}&\|\tilde{b}^{s-\mu'+|\beta|}(\eta)\{D_{\eta}^{\beta}a(\eta)\}b^{-s}(\eta)\|=\|\tilde{b}^{\mu-\mu'}(\eta)\tilde{b}^{s-\mu+|\beta|}(\eta)\{D_{\eta}^{\beta}a(\eta)\}b^{-s}(\eta)\|\\&\le c\langle\eta\rangle^{\mu-\mu'}\|\tilde{b}^{s-\mu+|\beta|}(\eta)\{D_{\eta}^{\beta}a(\eta)\}b^{-s}(\eta)\|\le c\|\tilde{b}^{s-\mu+|\beta|}(\eta)\{D_{\eta}^{\beta}a(\eta)\}b^{-s}(\eta)\|.\end{aligned}$$

We employed $\mu-\mu'\le 0$ and the property (iii) in Definition 1.3.

(ii) The estimates (1.10) can be written as

$$\|\tilde{b}^{s-(\mu-|\beta|)}(\eta)\{D_\eta^\beta a(\eta)\}b^{-s}(\eta)\| \leq c$$

which just means that $D_\eta^\beta a(\eta) \in S^{\mu-|\beta|}(\mathbb{R}^q; \mathcal{E}, \widetilde{\mathcal{E}})$.

(iii) Given $a(\eta) \in S^\mu(\mathbb{R}^q; \mathcal{E}_0, \widetilde{\mathcal{E}}), \tilde{a}(\eta) \in S^\nu(\mathbb{R}^q; \mathcal{E}, \mathcal{E}_0)$ we have (with obvious meaning of notation)

$$\|\tilde{b}_0^{s-\nu+|\gamma|}(\eta)\{D_\eta^\gamma \tilde{a}(\eta)\}b^{-s}(\eta)\| \leq c, \ \|\tilde{b}^{s-\mu+|\delta|}(\eta)\{D_\eta^\delta a(\eta)\}b_0^{-s}(\eta)\| \leq c$$

for all $\gamma, \delta \in \mathbb{N}^q$. If $\alpha \in \mathbb{N}^q$ is any multi-index, $D_\eta^\alpha(a\tilde{a})(\eta)$ is a linear combination of compositions $D_\eta^\delta a(\eta) D_\eta^\gamma \tilde{a}(\eta)$ with $|\gamma| + |\delta| = |\alpha|$. It follows that

$$\begin{aligned}
&\|\tilde{b}^{s-(\mu+\nu)+|\alpha|}(\eta) D_\eta^\delta a(\eta)\{D_\eta^\gamma \tilde{a}(\eta)\}b^{-s}(\eta)\| \\
&\quad = \|\tilde{b}^{s-(\mu+\nu)+|\alpha|}(\eta) D_\eta^\delta a(\eta) b_0^{-s+\nu-|\gamma|}(\eta) b_0^{s-\nu+|\gamma|}(\eta) D_\eta^\gamma \tilde{a}(\eta) b^{-s}(\eta)\| \\
&\quad \leq \|\tilde{b}^{t-\mu+|\alpha|-|\gamma|}(\eta) D_\eta^\delta a(\eta) b_0^{-t}(\eta)\| \ \|b_0^{s-\nu+|\gamma|}(\eta) D_\eta^\gamma \tilde{a}(\eta) b^{-s}(\eta)\| \quad (1.19)
\end{aligned}$$

for $t = s - \nu + |\gamma|$; the right-hand side is bounded in η, since $|\alpha| - |\gamma| = |\delta|$. □

Remark 1.16. Observe from (1.19) that the semi-norms of compositions of symbols can be estimated by products of semi-norms of the factors.

1.3. An example from the parameter-dependent cone calculus

We now construct a specific family of reductions of orders between weighted spaces on a compact manifold M with conical singularity v, locally near v modelled on a cone

$$X^\Delta := (\overline{\mathbb{R}}_+ \times X)/(\{0\} \times X)$$

with a smooth compact manifold X as base. The parameter η will play the role of covariables of the calculus of operators on a manifold with edge; that is why we talk about an example from the edge calculus. The associated "abstract" cone calculus according to what we did so far in the Sections 1.1 and 1.2 and then below in Chapter 2 will be a contribution to the calculus of corner operators of second generation. It will be convenient to pass to the stretched manifold $\mathbb{M}$ associated with M which is a compact C^∞ manifold with boundary $\partial\mathbb{M} \cong X$ such that when we squeeze down $\partial\mathbb{M}$ to a single point v we just recover M. Close to $\partial\mathbb{M}$ the manifold $\mathbb{M}$ is equal to a cylinder $[0,1) \times X \ni (t,x)$, a collar neighbourhood of $\partial\mathbb{M}$ in M. A part of the considerations will be performed on the open stretched cone $X^\wedge := \mathbb{R}_+ \times X \ni (t,x)$ where we identify $(0,1) \times X$ with the interior of the collar neighbourhood (for convenience, without indicating any pull backs of functions or operators with respect to that identification). Let $\widetilde{M} := 2\mathbb{M}$ be the double of $\mathbb{M}$ (obtained by gluing together two copies $\mathbb{M}_\pm$ of $\mathbb{M}$ along the common boundary $\partial\mathbb{M}$, where we identify $\mathbb{M}$ with $\mathbb{M}_+$); then $\widetilde{M}$ is a closed compact C^∞ manifold. On the space M we have a family of weighted Sobolev spaces $H^{s,\gamma}(M), s, \gamma \in \mathbb{R}$, that may be defined as

$$H^{s,\gamma}(M) := \{\sigma u + (1-\sigma)v : u \in \mathcal{H}^{s,\gamma}(X^\wedge), v \in H^s_{\mathrm{loc}}(M \setminus \{v\})\},$$

where $\sigma(t)$ is a cut-off function (i.e., $\sigma \in C_0^\infty(\overline{\mathbb{R}}_+), \sigma \equiv 1$ near $t = 0$), $\sigma(t) = 0$ for $t > 2/3$. Here $\mathcal{H}^{s,\gamma}(X^\wedge)$ is defined to be the completion of $C_0^\infty(X^\wedge)$ with respect to the norm

$$\left\{ \frac{1}{2\pi i} \int_{\Gamma_{\frac{n+1}{2}-\gamma}} \|b^\mu_{\mathrm{base}}(\mathrm{Im}\, w)(\mathcal{M}u)(w)\|^2_{L^2(X)} dw \right\}^{\frac{1}{2}}, \tag{1.20}$$

$n = \dim X$, where $b^\mu_{\mathrm{base}}(\tau) \in L^\mu_{\mathrm{cl}}(X; \mathbb{R}_\tau)$ is a family of reductions of order on X, similarly as in the example in Section 1.1 (in particular, $b^s_{\mathrm{base}}(\tau) : H^s(X) \to H^0(X) = L^2(X)$ is a family of isomorphisms). Moreover, $\mathcal{M}$ is the Mellin transform, $(\mathcal{M}u)(w) = \int_0^\infty t^{w-1}u(t)dt$, $w \in \mathbb{C}$ the complex Mellin covariable, and

$$\Gamma_\beta := \{w \in \mathbb{C} : \mathrm{Re}\, w = \beta\}$$

for any real β. From $t^\delta \mathcal{H}^{s,\gamma}(X^\wedge) = \mathcal{H}^{s,\gamma+\delta}(X^\wedge)$ for all $s, \gamma, \delta \in \mathbb{R}$ it follows the existence of a strictly positive function $\mathrm{h}^\delta \in C^\infty(M \setminus \{v\})$, such that the operator of multiplication by h^δ induces an isomorphism

$$\mathrm{h}^\delta : H^{s,\gamma}(M) \to H^{s,\gamma+\delta}(M) \tag{1.21}$$

for every $s, \gamma, \delta \in \mathbb{R}$.

Moreover, again according to the example in Section 1.1, now for any smooth compact manifold $\widetilde{M}$ we have an order reducing family $\tilde{b}(\eta)$ in the scale of Sobolev spaces $H^s(\widetilde{M}), s \in \mathbb{R}$. More generally, we employ parameter-dependent families $\tilde{a}(\eta) \in L^\mu_{\mathrm{cl}}(\widetilde{M}; \mathbb{R}^q)$. The symbols $a(\eta)$ that we want to establish in the scale $H^{s,\gamma}(M)$ on our compact manifold M with conical singularity v will be essentially (i.e., modulo Schwartz functions in η with values in globally smoothing operators on M) constructed in the form

$$a(\eta) := \sigma a_{\mathrm{edge}}(\eta)\tilde{\sigma} + (1 - \sigma)a_{\mathrm{int}}(\eta)(1 - \tilde{\tilde{\sigma}}), \tag{1.22}$$

$a_{\mathrm{int}}(\eta) := \tilde{a}(\eta)|_{\mathrm{int}\mathbb{M}}$, with cut-off functions $\sigma(t), \tilde{\sigma}(t), \tilde{\tilde{\sigma}}(t)$ on the half-axis, supported in $[0, 2/3)$, with the property

$$\tilde{\tilde{\sigma}} \prec \sigma \prec \tilde{\sigma}$$

(here $\sigma \prec \tilde{\sigma}$ means the $\tilde{\sigma}$ is equal to 1 in a neighbourhood of supp σ).

The "edge" part of (1.22) will be defined in the variables $(t, x) \in X^\wedge$. Let us choose a parameter-dependent elliptic family of operators of order μ on X

$$\tilde{p}(t, \tilde{\tau}, \tilde{\eta}) \in C^\infty(\overline{\mathbb{R}}_+, L^\mu_{\mathrm{cl}}(X; \mathbb{R}^{1+q}_{\tilde{\tau},\tilde{\eta}})).$$

Setting

$$p(t, \tau, \eta) := \tilde{p}(t, t\tau, t\eta) \tag{1.23}$$

we have what is known as an edge-degenerate family of operators on X. We now employ the following Mellin quantisation theorem.

Definition 1.17. Let $M^\mu_{\mathcal{O}}(X; \mathbb{R}^q)$ defined as the set of all $h(z, \eta) \in \mathcal{A}(\mathbb{C}, L^\mu_{\mathrm{cl}}(X; \mathbb{R}^q))$ such that $h(\beta + i\tau, \eta) \in L^\mu_{\mathrm{cl}}(X; \mathbb{R}^{1+q}_{\tau,\eta})$ for every $\beta \in \mathbb{R}$, uniformly in compact β-intervals (here $\mathcal{A}(\mathbb{C}, E)$ with any Fréchet space E denotes the space of all E-valued holomorphic functions in $\mathbb{C}$, in the Fréchet topology of uniform convergence on compact sets).

Observe that also $M^{\mu}_{\mathcal{O}}(X;\mathbb{R}^q)$ is a Fréchet space in a natural way. Given an $f(t,t',z,\eta) \in C^{\infty}(\mathbb{R}_+ \times \mathbb{R}_+, L^{\mu}_{\mathrm{cl}}(X;\Gamma_{\frac{1}{2}-\gamma} \times \mathbb{R}^q))$ we set

$$\mathrm{op}^{\gamma}_{M}(f)(\eta)u(r) := \int_{\mathbb{R}}\int_0^{\infty}\left(\frac{t}{t'}\right)^{-(\frac{1}{2}-\gamma+i\tau)} f\left(t,t',\frac{1}{2}-\gamma+i\tau,\eta\right)u(t')\frac{dt'}{t'}\,đ\tau,$$

$đ\tau := (2\pi)^{-1}d\tau$, which is regarded as a (parameter-dependent) weighted pseudo-differential operator with symbol f, referring to the weight $\gamma \in \mathbb{R}$. There exists an element

$$\tilde{h}(t,z,\tilde{\eta}) \in C^{\infty}(\overline{\mathbb{R}}_+, M^{\mu}_{\mathcal{O}}(X;\mathbb{R}^q_{\tilde{\eta}})) \tag{1.24}$$

such that, when we set

$$h(t,z,\eta) := \tilde{h}(t,z,t\eta) \tag{1.25}$$

we have

$$\mathrm{op}^{\gamma}_{M}(h)(\eta) = \mathrm{Op}_t(p)(\eta) \tag{1.26}$$

mod $L^{-\infty}(X^{\wedge};\mathbb{R}^q_{\eta})$, for every weight $\gamma \in \mathbb{R}$. Observe that when we set

$$p_0(t,\tau,\eta) := \tilde{p}(0,t\tau,t\eta),\ h_0(t,z,\eta) := \tilde{h}(0,z,t\eta)$$

we also have $\mathrm{op}^{\gamma}_{M}(h_0)(\eta) = \mathrm{Op}_t(p_0)(\eta) \bmod L^{-\infty}(X^{\wedge};\overline{\mathbb{R}}_+)$, for all $\gamma \in \mathbb{R}$.

Let us now choose cut-off functions $\omega(t), \tilde{\omega}(t), \tilde{\tilde{\omega}}(t)$ such that $\tilde{\tilde{\omega}} \prec \omega \prec \tilde{\omega}$.

Fix the notation $\omega_{\eta}(t) := \omega(t[\eta])$, and form the operator function

$$\begin{aligned} a_{\mathrm{edge}}(\eta) := {} & \omega_{\eta}(t)t^{-\mu}\mathrm{op}^{\gamma-\frac{n}{2}}_{M}(h)(\eta)\tilde{\omega}_{\eta}(t) \\ & + t^{-\mu}(1-\omega_{\eta}(t))\mathrm{Op}_t(p)(\eta)(1-\tilde{\tilde{\omega}}_{\eta}(t)) + m(\eta) + g(\eta). \end{aligned} \tag{1.27}$$

Here $m(\eta)$ and $g(\eta)$ are smoothing Mellin and Green symbols of the edge calculus. The definition of $m(\eta)$ is based on smoothing Mellin symbols $f(z) \in M^{-\infty}(X;\Gamma_{\beta})$. Here $M^{-\infty}(X;\Gamma_{\beta})$ is the subspace of all $f(z) \in L^{-\infty}(X;\Gamma_{\beta})$ such that for some $\varepsilon > 0$ (depending on f) the function f extends to an

$$l(z) \in \mathcal{A}(U_{\beta,\varepsilon}, L^{-\infty}(X))$$

where $U_{\beta,\varepsilon} := \{z \in \mathbb{C} : |\mathrm{Re} z - \beta| < \varepsilon\}$ and

$$l(\delta + i\tau) \in L^{-\infty}(X;\mathbb{R}_{\tau})$$

for every $\delta \in (\beta-\varepsilon, \beta+\varepsilon)$, uniformly in compact subintervals. By definition we then have $f(\beta+i\tau) = l(\beta+i\tau)$; for brevity we often denote the holomorphic extension l of f again by f. For $f \in M^{-\infty}(X;\Gamma_{\frac{n+1}{2}-\gamma})$ we set

$$m(\eta) := t^{-\mu}\omega_{\eta}\mathrm{op}^{\gamma-\frac{n}{2}}_{M}(f)\tilde{\omega}_{\eta}$$

for any cut-off functions $\omega, \tilde{\omega}$.

In order to explain the structure of $g(\eta)$ in (1.27) we first introduce weighted spaces on the infinite stretched cone $X^\wedge = \mathbb{R}_+ \times X$, namely,

$$\mathcal{K}^{s,\gamma;g}(X^\wedge) := \omega \mathcal{H}^{s,\gamma}(X^\wedge) + (1-\omega) H^{s;g}_{\text{cone}}(X^\wedge) \tag{1.28}$$

for any $s, \gamma, g \in \mathbb{R}$, and a cut-off function ω, see (1.20) which defines $\mathcal{H}^{s,\gamma}(X^\wedge)$ and the formula (1.2). Moreover, we set $\mathcal{K}^{s,\gamma}(X^\wedge) := \mathcal{K}^{s,\gamma;0}(X^\wedge)$. The operator families $g(\eta)$ are so-called Green symbols in the covariable $\eta \in \mathbb{R}^q$, defined by

$$g(\eta) \in S^{\mu}_{\text{cl}}(\mathbb{R}^q_\eta; \mathcal{K}^{s,\gamma;g}(X^\wedge), \mathcal{S}^{\gamma-\mu+\varepsilon}(X^\wedge)), \tag{1.29}$$

$$g^*(\eta) \in S^{\mu}_{\text{cl}}(\mathbb{R}^q_\eta; \mathcal{K}^{s,-\gamma+\mu;g}(X^\wedge), \mathcal{S}^{-\gamma+\varepsilon}(X^\wedge)), \tag{1.30}$$

for all $s, \gamma, g \in \mathbb{R}$, where g^* denotes the η-wise formal adjoint with respect to the scalar product of $\mathcal{K}^{0,0;0}(X^\wedge) = r^{-\frac{n}{2}} L^2(\mathbb{R}_+ \times X)$ and $\varepsilon = \varepsilon(g) > 0$. Here

$$\mathcal{S}^\beta(X^\wedge) := \omega \mathcal{K}^{\infty,\beta}(X^\wedge) + (1-\omega)\mathcal{S}(\overline{\mathbb{R}}_+, C^\infty(X))$$

for any cut-off function ω. The notion of operator-valued symbols in (1.29), (1.30) refers to (1.5) in its generalisation to Fréchet spaces $\widetilde{H}$ (rather than Hilbert spaces) with group actions (see Remark 1.7) that is in the present case given by

$$\kappa_\lambda : u(t,x) \to \lambda^{\frac{n+1}{2}+g} u(\lambda t, x), \ \lambda \in \mathbb{R}_+ \tag{1.31}$$

$n = \dim X$, both in the spaces $\mathcal{K}^{s;\gamma,g}(X^\wedge)$ and $\mathcal{S}^{\gamma-\mu+\varepsilon}(X^\wedge)$.

The following theorem is crucial for proving that our new order reduction family is well defined. Therefore we will sketch the main steps of the proof, which is based on the edge calculus. Various aspects of the proof can be found in the literature, for example in Kapanadze and Schulze [12, Proposition 3.3.79], Schrohe and Schulze [31], Harutyunyan and Schulze [9]. Among the tools we have the pseudo-differential operators on $X^\wedge$ interpreted as a manifold with conical exit to infinity $r \to \infty$; the general background may be found in Schulze [36]. The calculus of such exit operators goes back to Parenti [25], Cordes [3], Shubin [42], and others.

Theorem 1.18. *We have*

$$\sigma a_{\text{edge}}(\eta)\tilde{\sigma} \in S^\mu(\mathbb{R}^q; \mathcal{K}^{s,\gamma;g}(X^\wedge), \mathcal{K}^{s-\mu,\gamma-\mu;g}(X^\wedge)) \tag{1.32}$$

for every $s, g \in \mathbb{R}$, more precisely,

$$D^\beta_\eta\{\sigma a_{\text{edge}}(\eta)\tilde{\sigma}\} \in S^{\mu-|\beta|}(\mathbb{R}^q; \mathcal{K}^{s,\gamma;g}(X^\wedge), \mathcal{K}^{s-\mu+|\beta|,\gamma-\mu;g}(X^\wedge)) \tag{1.33}$$

for all $s, g \in \mathbb{R}$ and all $\beta \in \mathbb{N}^q$. (The spaces of symbols in (1.32), (1.33) *refer to the group action* (1.31)).

Proof. To prove the assertions it is enough to consider the case without $m(\eta)+g(\eta)$, since the latter sum maps to $\mathcal{K}^{\infty,\gamma;g}(X^\wedge)$ anyway. The first part of the Theorem is known, see, for instance, [9] or [4]. Concerning the relation (1.33) we write

$$\sigma a_{\text{edge}}(\eta)\tilde{\sigma} = \sigma\{a_{\text{c}}(\eta) + a_\psi(\eta)\}\tilde{\sigma} \tag{1.34}$$

with

$$\begin{aligned} a_{\mathrm{c}}(\eta) &:= t^{-\mu}\omega_\eta \mathrm{op}_M^{\gamma-\frac{n}{2}}(h)(\eta)\tilde{\omega}_\eta, \\ a_\psi(\eta) &:= t^{-\mu}(1-\omega_\eta)\mathrm{Op}_t(p)(\eta)(1-\tilde{\tilde{\omega}}_\eta) \end{aligned}$$

and it suffices to take the summands separately. In order to show (1.33) we consider, for instance, the derivative $\partial/\partial\eta_j =: \partial_j$ for some $1 \le j \le q$. By iterating the process we then obtain the assertion. We have

$$\partial_j\sigma\{a_{\mathrm{c}}(\eta)+a_\psi(\eta)\}\tilde{\sigma} = \sigma\{\partial_j a_{\mathrm{c}}(\eta)+\partial_j a_\psi(\eta)\}\tilde{\sigma} = b_1(\eta)+b_2(\eta)+b_3(\eta)$$

with

$$\begin{aligned} b_1(\eta) &:= \sigma t^{-\mu}\Big\{\omega_\eta \mathrm{op}_M^{\gamma-\frac{n}{2}}(h)(\eta)\partial_j\tilde{\omega}_\eta + (1-\omega_\eta)\mathrm{Op}_t(p)(\eta)\partial_j(1-\tilde{\tilde{\omega}}_\eta)\Big\}\tilde{\sigma}, \\ b_2(\eta) &:= \sigma t^{-\mu}\Big\{\omega_\eta \mathrm{op}_M^{\gamma-\frac{n}{2}}(\partial_j h)(\eta)\tilde{\omega}_\eta + (1-\omega_\eta)\mathrm{Op}_t(\partial_j p)(\eta)(1-\tilde{\tilde{\omega}}_\eta)\Big\}\tilde{\tilde{\sigma}}, \\ b_3(\eta) &:= \sigma t^{-\mu}\Big\{(\partial_j\omega_\eta)\mathrm{op}_M^{\gamma-\frac{n}{2}}(h)(\eta)\tilde{\omega}_\eta + (\partial_j(1-\omega_\eta))\mathrm{Op}_t(p)(\eta)(1-\tilde{\tilde{\omega}}_\eta)\Big\}\tilde{\sigma}. \end{aligned}$$

In $b_1(\eta)$ we can apply a pseudo-locality argument which is possible since $\partial_j\tilde{\omega}_\eta \equiv 0$ on $\operatorname{supp}\omega_\eta$ and $\partial_j(1-\tilde{\tilde{\omega}}_\eta) \equiv 0$ on $\operatorname{supp}(1-\omega_\eta)$; this yields (together with similar considerations as for the proof of (1.32))

$$b_1(\eta) \in S^{\mu-1}(\mathbb{R}^q; \mathcal{K}^{s,\gamma;g}(X^\wedge), \mathcal{K}^{\infty,\gamma-\mu;g}(X^\wedge)).$$

Moreover we obtain

$$b_2(\eta) \in S^{\mu-1}(\mathbb{R}^q; \mathcal{K}^{s,\gamma;g}(X^\wedge), \mathcal{K}^{s-\mu+1,\gamma-\mu;g}(X^\wedge))$$

since $\partial_j h$ and $\partial_j p$ are of order $\mu-1$ (again combined with arguments for (1.32)). Concerning $b_3(\eta)$ we use the fact that there is a $\psi \in C_0^\infty(\mathbb{R}_+)$ such that $\psi \equiv 1$ on $\operatorname{supp}\partial_j\omega$, $\tilde{\omega}-\psi \equiv 0$ on $\operatorname{supp}\partial_j\omega$ and $(1-\tilde{\tilde{\omega}})-\psi \equiv 0$ on $\operatorname{supp}\partial_j\omega$. Thus, when we set $\psi_\eta(t) := \psi(t[\eta])$, we obtain $b_3(\eta) := c_3(\eta)+c_4(\eta)$ with

$$\begin{aligned} c_3(\eta) &:= \sigma t^{-\mu}\Big\{(\partial_j\omega_\eta)\mathrm{op}_M^{\gamma-\frac{n}{2}}(h)(\eta)\psi_\eta - (\partial_j\omega_\eta)\mathrm{Op}_t(p)(\eta)\psi_\eta\Big\}\tilde{\sigma}, \\ c_4(\eta) &:= \sigma t^{-\mu}\Big\{(\partial_j\omega_\eta)\mathrm{op}_M^{\gamma-\frac{n}{2}}(h)(\eta)[\tilde{\omega}_\eta-\psi_\eta] - (\partial_j\omega_\eta)\mathrm{Op}_t(p)(\eta)[(1-\tilde{\tilde{\omega}}_\eta)-\psi_\eta]\Big\}\tilde{\sigma}. \end{aligned}$$

Here, using $\partial_j\omega_\eta = (\omega')_\eta\partial_j(t[\eta])$ which yields an extra power of t on the left of the operator, together with pseudo-locality, we obtain

$$c_4(\eta) \in S^{\mu-1}(\mathbb{R}^q; \mathcal{K}^{s,\gamma;g}(X^\wedge), \mathcal{K}^{\infty,\gamma-\mu;g}(X^\wedge)).$$

To treat $c_3(\eta)$ we employ that both $\partial_j\omega_\eta$ and ψ_η are compactly supported on $\mathbb{R}_+$. Using the property (1.26), we have

$$\begin{aligned} c_3(\eta) &= \sigma t^{-\mu}(\partial_j\omega_\eta)\{\mathrm{op}_M^{\gamma-\frac{n}{2}}(h)(\eta) - \mathrm{Op}_t(p)(\eta)\}\psi_\eta\tilde{\sigma} \\ &\qquad \in S^{\mu-1}(\mathbb{R}^q; \mathcal{K}^{s,\gamma;g}(X^\wedge), \mathcal{K}^{\infty,\gamma-\mu;g}(X^\wedge)). \qquad \square \end{aligned}$$

Definition 1.19. A family of operators $c(\eta) \in \mathcal{S}(\mathbb{R}^q, \bigcap_{s\in\mathbb{R}} \mathcal{L}(H^{s,\gamma}(M), H^{\infty,\delta}(M)))$ is called a smoothing element in the parameter-dependent cone calculus on M associated with the weight data $(\gamma, \delta) \in \mathbb{R}^2$, written $c \in C_G(M, (\gamma, \delta); \mathbb{R}^q)$, if there is an $\varepsilon = \varepsilon(c) > 0$ such that

$$c(\eta) \in \mathcal{S}(\mathbb{R}^q, \mathcal{L}(H^{s,\gamma}(M), H^{\infty,\delta+\varepsilon}(M))),$$
$$c^*(\eta) \in \mathcal{S}(\mathbb{R}^q, \mathcal{L}(H^{s,-\delta}(M), H^{\infty,-\gamma+\varepsilon}(M)));$$

for all $s \in \mathbb{R}$; here c^* is the η-wise formal adjoint of c with respect to the $H^{0,0}(M)$-scalar product.

The η-wise kernels of the operators $c(\eta)$ are in $C^\infty((M \setminus \{v\}) \times (M \setminus \{v\}))$. However, they are of flatness ε in the respective distance variables to v, relative to the weights δ and γ, respectively. Let us look at a simple example to illustrate the structure. We choose elements $k \in \mathcal{S}(\mathbb{R}^q, H^{\infty,\delta+\varepsilon}(M))$, $k' \in \mathcal{S}(\mathbb{R}^q, H^{\infty,-\gamma+\varepsilon}(M))$ and assume for convenience that k and k' vanish outside a neighbourhood of v, for all $\eta \in \mathbb{R}^q$. Then with respect to a local splitting of variables (t, x) near v we can write $k = k(\eta, t, x)$ and $k' = k'(\eta, t', x')$, respectively. Set

$$c(\eta)u(t,x) := \iint k(\eta, t, x) k'(\eta, t', x') u(t', x') t'^n dt' dx'$$

with the formal adjoint

$$c^*(\eta)v(t', x') := \iint \overline{k'(\eta, t', x') k(\eta, t, x)} v(t, x) t^n dt dx.$$

Then $c(\eta)$ is a smoothing element in the parameter-dependent cone calculus.

By $C^\mu(M, (\gamma, \gamma - \mu); \mathbb{R}^q)$ we denote the set of all operator families

$$a(\eta) = \sigma a_{\mathrm{edge}}(\eta)\tilde{\sigma} + (1 - \sigma)a_{\mathrm{int}}(\eta)(1 - \tilde{\tilde{\sigma}}) + c(\eta) \tag{1.35}$$

where a_{edge} is of the form (1.27), $a_{\mathrm{int}} \in L^\mu_{\mathrm{cl}}(M \setminus \{v\}; \mathbb{R}^q)$, while $c(\eta)$ is a parameter-dependent smoothing operator on M, associated with the weight data $(\gamma, \gamma - \mu)$.

Theorem 1.20. *Let M be a compact manifold with a conical singularity. Then the η-dependent families* (1.22) *which define continuous operators*

$$a(\eta) : H^{s,\gamma}(M) \to H^{s-\nu,\gamma-\nu}(M) \tag{1.36}$$

for all $s \in \mathbb{R}$, $\nu \geq \mu$, have the properties:

$$\|a(\eta)\|_{\mathcal{L}(H^{s,\gamma}(M), H^{s-\nu,\gamma-\nu}(M))} \leq c\langle\eta\rangle^B \tag{1.37}$$

for all $\eta \in \mathbb{R}^q$, and $s \in \mathbb{R}$, with constants $c = c(\mu, \nu, s) > 0$, $B = B(\mu, \nu, s)$, and, when $\mu \leq 0$

$$\|a(\eta)\|_{\mathcal{L}(H^{0,0}(M), H^{0,0}(M))} \leq c\langle\eta\rangle^\mu \tag{1.38}$$

for all $\eta \in \mathbb{R}$, $s \in \mathbb{R}$, with constants $c = c(\mu, s) > 0$.

Proof. The result is known for the summand $(1 - \sigma)a_{\mathrm{int}}(\eta)(1 - \tilde{\tilde{\sigma}})$ as we see from the example in Section 1.1. Therefore, we may concentrate on

$$p(\eta) := \sigma a_{\mathrm{edge}}(\eta)\tilde{\sigma} : H^{s,\gamma}(M) \to H^{s-\nu,\gamma-\nu}(M).$$

To show (1.37) we pass to

$$\sigma a_{\mathrm{edge}}(\eta)\tilde{\sigma} : \mathcal{K}^{s,\gamma}(X^\wedge) \to \mathcal{K}^{s-\nu,\gamma-\nu}(X^\wedge).$$

Then Theorem 1.18 shows that we have symbolic estimates, especially

$$\|\kappa_{\langle\eta\rangle}^{-1} p(\eta)\kappa_{\langle\eta\rangle}\|_{\mathcal{L}(\mathcal{K}^{s,\gamma}(X^\wedge),\mathcal{K}^{s-\mu,\gamma-\mu}(X^\wedge))} \le c\langle\eta\rangle^{\mu}.$$

We have

$$\|p(\eta)\|_{\mathcal{L}(\mathcal{K}^{s,\gamma}(X^\wedge),\mathcal{K}^{s-\nu,\gamma-\nu}(X^\wedge))} \le \|p(\eta)\|_{\mathcal{L}(\mathcal{K}^{s,\gamma}(X^\wedge),\mathcal{K}^{s-\mu,\gamma-\mu}(X^\wedge))},$$

and

$$\begin{aligned}\|p(\eta)\|_{\mathcal{L}(\mathcal{K}^{s,\gamma}(X^\wedge),\mathcal{K}^{s-\mu,\gamma-\mu}(X^\wedge))} &= \|\kappa_{\langle\eta\rangle}\kappa_{\langle\eta\rangle}^{-1} p(\eta)\kappa_{\langle\eta\rangle}\kappa_{\langle\eta\rangle}^{-1}\|_{\mathcal{L}(\mathcal{K}^{s,\gamma}(X^\wedge),\mathcal{K}^{s-\mu,\gamma-\mu}(X^\wedge))} \\ \le \|\kappa_{\langle\eta\rangle}\|_{\mathcal{L}(\mathcal{K}^{s-\mu,\gamma-\mu}(X^\wedge),\mathcal{K}^{s-\mu,\gamma-\mu}(X^\wedge))} &\|\kappa_{\langle\eta\rangle}^{-1} p(\eta)\kappa_{\langle\eta\rangle}\|_{\mathcal{L}(\mathcal{K}^{s-\mu,\gamma-\mu}(X^\wedge),\mathcal{K}^{s,\gamma}(X^\wedge))} \\ &\|\kappa_{\langle\eta\rangle}^{-1}\|_{\mathcal{L}(\mathcal{K}^{s,\gamma}(X^\wedge),\mathcal{K}^{s,\gamma}(X^\wedge))} \le c\langle\eta\rangle^{\mu+\widetilde{M}+M}.\end{aligned}$$

Here we used that $\kappa_{\langle\eta\rangle}, \kappa_{\langle\eta\rangle}^{-1}$ satisfy estimates like (1.4).

For (1.38) we employ that κ_λ is operating as a unitary group on $\mathcal{K}^{0,0}(X^\wedge)$. This gives us

$$\begin{aligned}\|p(\eta)\|_{\mathcal{L}(\mathcal{K}^{0,0}(X^\wedge),\mathcal{K}^{0,0}(X^\wedge))} &= \|\kappa_{\langle\eta\rangle}^{-1} p(\eta)\kappa_{\langle\eta\rangle}\|_{\mathcal{L}(\mathcal{K}^{0,0}(X^\wedge),\mathcal{K}^{0,0}(X^\wedge))} \\ &\le \|\kappa_{\langle\eta\rangle}^{-1} p(\eta)\kappa_{\langle\eta\rangle}\|_{\mathcal{L}(\mathcal{K}^{0,0}(X^\wedge),\mathcal{K}^{-\mu,-\mu}(X^\wedge))} \le c\langle\eta\rangle^{\mu}. \quad \square\end{aligned}$$

Theorem 1.21. *For every $k \in \mathbb{Z}$ there exists an $f_k(z) \in M^{-\infty}(X;\Gamma_{\frac{n+1}{2}-\gamma})$ such that for every cut-off functions $\omega, \tilde{\omega}$ the operator*

$$A := 1 + \omega \mathrm{op}_M^{\gamma-\frac{n}{2}}(f_k)\tilde{\omega} : H^{s,\gamma}(M) \to H^{s,\gamma}(M) \tag{1.39}$$

is Fredholm and of index k, for all $s \in \mathbb{R}$.

Proof. We employ the result (cf. [35]) that for every $k \in \mathbb{Z}$ there exists an $f_k(z)$ such that

$$\widetilde{A} := 1 + \omega \mathrm{op}_M^{\gamma-\frac{n}{2}}(f_k)\tilde{\omega} : \mathcal{K}^{s,\gamma}(X^\wedge) \to \mathcal{K}^{s,\gamma}(X^\wedge) \tag{1.40}$$

is Fredholm of index k. Recall that the proof of the latter result follows from a corresponding theorem in the case $\dim X = 0$. The Mellin symbol f_k is constructed in such a way that $1+f_k(z) \neq 0$ for all $z \in \Gamma_{\frac{1}{2}-\gamma}$ and the argument of $1+f_k(z)|_{\Gamma_{\frac{1}{2}-\gamma}}$ varies from 1 to $2\pi k$ when $z \in \Gamma_{\frac{1}{2}-\gamma}$ goes from $\mathrm{Im} z = -\infty$ to $\mathrm{Im} z = +\infty$. The choice of $\omega, \tilde{\omega}$ is unessential; so we assume that $\omega, \tilde{\omega} \equiv 0$ for $r \ge 1-\varepsilon$ with some $\varepsilon > 0$. Let us represent the cone $\widetilde{M} := X^\Delta$ as a union of $\big([0, 1+\frac{\varepsilon}{2}) \times X\big)/(\{0\} \times X) =: \widetilde{M}_-$ and $(1-\frac{\varepsilon}{2}, \infty) \times X =: \widetilde{M}_+$. Then

$$\widetilde{A}|_{\widetilde{M}_-} = 1 + \omega \mathrm{op}_M^{\gamma-\frac{n}{2}}(f_k)\tilde{\omega}, \; \widetilde{A}|_{\widetilde{M}_+} = 1. \tag{1.41}$$

Moreover, without loss of generality, we represent M as a union $\big([0, 1+\frac{\varepsilon}{2}) \times X\big)/(\{0\} \times X) \cup M_+$ where M_+ is an open C^∞ manifold which intersects $\big([0, 1+$

$\frac{\varepsilon}{2}) \times X)/(\{0\} \times X) =: M_-$ in a cylinder of the form $(1 - \frac{\varepsilon}{2}, 1 + \frac{\varepsilon}{2}) \times X$. Let B denote the operator on M, defined by

$$B_- := A|_{M_-} = 1 + \omega \mathrm{op}_M^{\gamma - \frac{n}{2}}(f_k)\tilde{\omega}, \ B_+ := A|_{M_+} = 1 \tag{1.42}$$

We are then in a special situation of cutting and pasting of Fredholm operators. We can pass to manifolds with conical singularities N and $\widetilde{N}$ by setting

$$N = \widetilde{M_-} \cup M_+, \ \widetilde{N} = M_- \cup \widetilde{M_+}$$

and transferring the former operators in (1.41), (1.42) to N and $\widetilde{N}$, respectively, by gluing together the $\pm$ pieces of $\widetilde{A}$ and A to belong to $\widetilde{M_\pm}$ and $M_\pm$ to corresponding operators $\widetilde{B}$ on $\widetilde{N}$ and B on N. We then have the relative index formula

$$\mathrm{ind} A - \mathrm{ind} B = \mathrm{ind} \widetilde{A} - \mathrm{ind} \widetilde{B} \tag{1.43}$$

(see [23]). In the present case $\widetilde{A}$ and $\widetilde{M}$ are the same as B and N where $\widetilde{B}$ and $\widetilde{N}$ are the same as A and M. It follows that

$$\mathrm{ind} \widetilde{A} - \mathrm{ind} \widetilde{B} = \mathrm{ind} B - \mathrm{ind} A. \tag{1.44}$$

From (1.43), (1.44) it follows that $\mathrm{ind} A = \mathrm{ind} B = \mathrm{ind} \widetilde{A}$. □

Theorem 1.22. *There is a choice of m and g such that the operators (1.22) form a family of isomorphisms*

$$a(\eta) : H^{s,\gamma}(M) \to H^{s-\mu,\gamma-\mu}(M) \tag{1.45}$$

for all $s \in \mathbb{R}$ and all $\eta \in \mathbb{R}^q$.

Proof. We choose a function

$$p(t, \tau, \eta, \zeta) := \tilde{p}(t\tau, t\eta, \zeta)$$

similarly as (1.23) where $\tilde{p}(\tilde{\tau}, \tilde{\eta}, \zeta) \in L^\mu_{\mathrm{cl}}(X; \mathbb{R}^{1+q+l}_{\tilde{\tau},\tilde{\eta},\zeta}), l \geq 1$, is parameter-dependent elliptic with parameters $\tilde{\tau}, \tilde{\eta}, \zeta$. For purposes below we specify $\tilde{p}(\tilde{\tau}, \tilde{\eta}, \zeta)$ in such a way that the parameter-dependent homogeneous principal symbol in $(x, \tilde{\tau}, \xi, \tilde{\eta}, \zeta)$ for $(\tilde{\tau}, \xi, \tilde{\eta}, \zeta) \neq 0$ is equal to

$$(|\tilde{\tau}|^2 + |\xi|^2 + |\tilde{\eta}|^2 + |\zeta|^2)^{\frac{\mu}{2}}.$$

We now form an element

$$\tilde{h}(z, \tilde{\eta}, \zeta) \in M^\mu_{\mathcal{O}}(X; \mathbb{R}^{q+l}_{\tilde{\eta},\zeta})$$

analogously as (1.24) such that

$$h(t, z, \eta, \zeta) := \tilde{h}(z, t\eta, \zeta)$$

satisfies

$$\mathrm{op}^\gamma_M(h)(\eta, \zeta) = \mathrm{Op}_t(p)(\eta, \zeta)$$

$\mathrm{mod}\, L^{-\infty}(X^\wedge; \mathbb{R}^{q+l}_{\eta,\zeta})$. For every fixed $\zeta \in \mathbb{R}^l$ this is exactly as before, but in this way we obtain corresponding ζ-dependent families of such objects. It follows

$$\sigma b_{\mathrm{edge}}(\eta, \zeta)\tilde{\sigma} := t^{-\mu}\sigma \left\{ \omega_\eta \mathrm{op}_M^{\gamma - \frac{n}{2}}(h)(\eta, \zeta)\tilde{\omega}_\eta + \chi_\eta \mathrm{Op}_t(p)(\eta, \zeta)\tilde{\chi}_\eta \right\} \tilde{\sigma}$$

with

$$\chi_\eta(t) := 1 - \omega_\eta(t), \ \tilde{\chi}_\eta(t) := 1 - \tilde{\tilde{\omega}}_\eta(t).$$

Let us form the principal edge symbol

$$\sigma_\wedge(\sigma b_{\mathrm{edge}}\tilde{\sigma})(\eta, \zeta) = t^{-\mu}\left\{\omega_{|\eta|}\mathrm{op}_M^{\gamma-\frac{n}{2}}(h)(\eta, \zeta)\tilde{\omega}_{|\eta|} + \chi_{|\eta|}\mathrm{Op}_t(p)(\eta, \zeta)\tilde{\chi}_{|\eta|}\right\}$$

for $|\eta| \neq 0$. The latter is interpreted as a family of continuous operators

$$\sigma_\wedge(\sigma b_{\mathrm{edge}}\tilde{\sigma})(\eta, \zeta) : \mathcal{K}^{s,\gamma;g}(X^\wedge) \to \mathcal{K}^{s-\mu,\gamma-\mu;g}(X^\wedge) \tag{1.46}$$

which is elliptic as a family of classical pseudo-differential operators on $X^\wedge$. In addition it is exit elliptic on $X^\wedge$ with respect to the conical exit of $X^\wedge$ to infinity. In order that (1.46) is Fredholm for the given weight $\gamma \in \mathbb{R}$ and all $s, g \in \mathbb{R}$ it is necessary and sufficient that the subordinate conormal symbol

$$\sigma_c\sigma_\wedge(\sigma b_{\mathrm{edge}}\tilde{\sigma})(z, \zeta) : H^s(X) \to H^{s-\mu}(X)$$

is a family of isomorphisms for all $z \in \Gamma_{\frac{n+1}{2}-\gamma}$. This is standard information from the calculus on the stretched cone $X^\wedge$. By definition the conormal symbol is just

$$\tilde{h}(0, z, 0, \zeta) : H^s(X) \to H^{s-\mu}(X). \tag{1.47}$$

Since by construction $\tilde{h}(\beta + i\tau, 0, \zeta)$ is parameter-dependent elliptic on X with parameters $(\tau, \zeta) \in \mathbb{R}^{1+l}$, for every $\beta \in \mathbb{R}$ (uniformly in finite β-intervals) there is a $C > 0$ such that (1.47) becomes bijective whenever $|\tau, \zeta| > C$. In particular, choosing ζ large enough it follows the bijectivity for all $\tau \in \mathbb{R}$, i.e., for all $z \in \Gamma_{\frac{n+1}{2}-\gamma}$. Let us fix ζ^1 in that way and write again

$$p(t, \tau, \eta) := \tilde{p}(t\tau, t\eta, \zeta^1), \ h(t, z, \eta) := \tilde{h}(z, t\eta, \zeta^1).$$

We are now in the same situation we started with, but we know in addition that (1.46) is a family of Fredholm operators of a certain index, say, $-k$ for some $k \in \mathbb{Z}$. With the smoothing Mellin symbol $f_k(z)$ as in (1.40) we now form the composition

$$\sigma b_{\mathrm{edge}}(\eta)\tilde{\sigma}(1 + \omega_\eta \mathrm{op}_M^{\gamma-\frac{n}{2}}(f_k)\tilde{\omega}_\eta) \tag{1.48}$$

which is of the form

$$\sigma b_{\mathrm{edge}}(\eta)\tilde{\sigma} + \omega_\eta \mathrm{op}_M^{\gamma-\frac{n}{2}}(f)\tilde{\omega}_\eta + g(\eta) \tag{1.49}$$

for another smoothing Mellin symbol $f(z)$ and a certain Green symbol $g(\eta)$. Here, by a suitable choice of $\omega, \tilde{\omega}$, without loss of generality we assume that $\sigma \equiv 1$ and $\tilde{\sigma} \equiv 1$ on $\operatorname{supp}\omega_\eta \cup \operatorname{supp}\tilde{\omega}_\eta$, for all $\eta \in \mathbb{R}^q$. Since (1.48) is a composition of parameter-dependent cone operators the associated edge symbol is equal to

$$F(\eta) := \sigma_\wedge(\sigma b_{\mathrm{edge}}\tilde{\sigma})(\eta)(1 + \omega_{|\eta|}\mathrm{op}_M^{\gamma-\frac{n}{2}}(f_k)\tilde{\omega}_{|\eta|}) : \mathcal{K}^{s,\gamma}(X^\wedge) \to \mathcal{K}^{s-\mu,\gamma-\mu}(X^\wedge) \tag{1.50}$$

which is a family of Fredholm operators of index 0. By construction (1.50) depends only on $|\eta|$. For $\eta \in S^{q-1}$ we now add a Green operator g_0 on $X^\wedge$ such that

$$F(\eta) + g_0(\eta) : \mathcal{K}^{s,\gamma}(X^\wedge) \to \mathcal{K}^{s-\mu,\gamma-\mu}(X^\wedge)$$

is an isomorphism; it is known that such g_0 (of finite rank) exists (for $N = \dim\ker F(\eta)$ it can be written in the form $g_0 u := \sum_{j=1}^{N}(u, v_j)w_j$, where $(\cdot,\cdot)$ is the $\mathcal{K}^{0,0}(X^\wedge)$-scalar product and $(v_j)_{j=1,\dots,N}$ and $(w_j)_{j=1,\dots,N}$ are orthonormal systems of functions in $C_0^\infty(X^\wedge)$). Setting

$$g(\eta) := \sigma\vartheta(\eta)|\eta|^\mu \kappa_{|\eta|} g_0 \kappa_{|\eta|}^{-1}\tilde{\sigma}$$

with an excision function $\vartheta(\eta)$ in $\mathbb{R}^q$ we obtain a Green symbol with $\sigma_\wedge(g)(\eta) = |\eta|^\mu \kappa_{|\eta|} g_0 \kappa_{|\eta|}^{-1}$ and hence

$$\sigma_\wedge(F(\eta) + g(\eta)) : \mathcal{K}^{s,\gamma}(X^\wedge) \to \mathcal{K}^{s-\mu,\gamma-\mu}(X^\wedge)$$

is a family of isomorphisms for all $\eta \in \mathbb{R}^q \setminus \{0\}$. Setting

$$a_{\mathrm{edge}}(\eta) := t^{-\mu}\left[\omega_\eta \mathrm{op}_M^{\gamma-\frac{n}{2}}(h)(\eta)\tilde{\omega}_\eta + \chi_\eta \mathrm{Op}_t(p)(\eta)\tilde{\chi}_\eta\right]\left(1 + \omega_\eta \mathrm{op}_M^{\gamma-\frac{n}{2}}(f_k)\tilde{\omega}_\eta\right) + |\eta|^\mu \vartheta(\eta)\kappa_{|\eta|} g_0 \kappa_{|\eta|}^{-1} \quad (1.51)$$

we obtain an operator family

$$\sigma a_{\mathrm{edge}}(\eta)\tilde{\sigma} = F(\eta) + g(\eta)$$

as announced before. Next we choose a parameter-dependent elliptic $a_{\mathrm{int}}(\eta) \in L_{\mathrm{cl}}^\mu(M\setminus\{v\}; \mathbb{R}_\eta^q)$ such that its parameter-dependent homogeneous principal symbol close to $t = 0$ (in the splitting of variables (t, x)) is equal to

$$(|\tau|^2 + |\xi|^2 + |\eta|^2)^{\frac{\mu}{2}}.$$

Then we form

$$a(\eta) := \sigma a_{\mathrm{edge}}(\eta)\tilde{\sigma} + (1-\sigma)a_{\mathrm{int}}(\eta)(1-\tilde{\tilde{\sigma}})$$

with $\sigma, \tilde{\sigma}, \tilde{\tilde{\sigma}}$ as in (1.22). This is now a parameter-dependent elliptic element of the cone calculus on M with parameter $\eta \in \mathbb{R}^q$. It is known, see the explanations after this proof, that there is a constant $C > 0$ such that the operators (1.45) are isomorphisms for all $|\eta| \geq C$. Now, in order to construct $a(\eta)$ such that (1.45) are isomorphisms for all $\eta \in \mathbb{R}^q$ we simply perform the construction with $(\eta, \lambda) \in \mathbb{R}^{q+r}, r \geq 1$ in place of η, then obtain a family $a(\eta, \lambda)$ and define $a(\eta) := a(\eta, \lambda^1)$ with a $\lambda^1 \in \mathbb{R}^r, |\lambda^1| \geq C$. □

Let us now give more information on the above-mentioned space

$$C^\mu(M, \boldsymbol{g}; \mathbb{R}^q), \ \boldsymbol{g} = (\gamma, \gamma - \mu),$$

of parameter-dependent cone operators on M of order $\mu \in \mathbb{R}$, with the weight data $\boldsymbol{g}$. The elements $a(\eta) \in C^\mu(M, \boldsymbol{g}; \mathbb{R}^q)$ have a principal symbolic hierarchy

$$\sigma(a) := (\sigma_\psi(a), \sigma_\wedge(a)) \quad (1.52)$$

where $\sigma_\psi(a)$ is the parameter-dependent homogeneous principal symbol of order μ, defined through $a(\eta) \in L_{\mathrm{cl}}^\mu(M \setminus \{v\}; \mathbb{R}^q)$. This determines the reduced symbol

$$\tilde{\sigma}_\psi(a)(t, x, \tau, \xi, \eta) := t^\mu \sigma_\psi(a)(t, x, t^{-1}\tau, \xi, t^{-1}\eta)$$

given close to v in the splitting of variables (t, x) with covariables (τ, ξ). By construction $\tilde{\sigma}_\psi(a)$ is smooth up to $t = 0$. The second component $\sigma_\wedge(a)(\eta)$ is defined as

$$\sigma_\wedge(a)(\eta) := t^{-\mu}\omega_{|\eta|}\mathrm{op}_M^{\gamma-\frac{n}{2}}(h_0)(\eta)\tilde{\omega}_{|\eta|} \\ + t^{-\mu}(1-\omega_{|\eta|})\mathrm{Op}_t(p_0)(\eta)(1-\tilde{\tilde{\omega}}_{|\eta|}) + \sigma_\wedge(m+g)(\eta)$$

where $\sigma_\wedge(m+g)(\eta)$ is just the (twisted) homogeneous principal symbol of $m+g$ as a classical operator-valued symbol.

The element $a(\eta)$ of $C_G(M, \boldsymbol{g}; \mathbb{R}^q)$ represents families of continuous operators

$$a(\eta) : H^{s,\gamma}(M) \to H^{s,\gamma-\mu}(M) \tag{1.53}$$

for all $s \in \mathbb{R}$.

Definition 1.23. An element $a(\eta) \in C^\mu(M, \boldsymbol{g}; \mathbb{R}^q)$ is called elliptic, if

(i) $\sigma_\psi(a)$ never vanishes as a function on $T^*((M \setminus \{v\}) \times \mathbb{R}^q) \setminus 0$ and if $\tilde{\sigma}_\psi(a)$ does not vanish for all (t, x, τ, ξ, η), $(\tau, \xi, \eta) \neq 0$, up to $t = 0$;

(ii) $\sigma_\wedge(a)(\eta) : \mathcal{K}^{s,\gamma}(X^\wedge) \to \mathcal{K}^{s-\mu,\gamma-\mu}(X^\wedge)$ is a family of isomorphisms for all $\eta \neq 0$, and any $s \in \mathbb{R}$.

Theorem 1.24. *If $a(\eta) \in C^\mu(M, \boldsymbol{g}; \mathbb{R}^q)$, $\boldsymbol{g} = (\gamma, \gamma - \mu)$ is elliptic, there exists an element $a^{(-1)}(\eta) \in C^{-\mu}(M, \boldsymbol{g}^{-1}; \mathbb{R}^q)$ $\boldsymbol{g}^{-1} := (\gamma - \mu, \gamma)$, such that*

$$1 - a^{(-1)}(\eta)a(\eta) \in C_G(M, \boldsymbol{g}_l; \mathbb{R}^q), \quad 1 - a(\eta)a^{(-1)}(\eta) \in C_G(M, \boldsymbol{g}_r; \mathbb{R}^q),$$

where $\boldsymbol{g}_l := (\gamma, \gamma)$, $\boldsymbol{g}_r := (\gamma - \mu, \gamma - \mu)$.

The proof employs known elements of the edge symbolic calculus (cf. [36]); so we do not recall the details here. Let us only note that the inverses of $\sigma_\psi(a), \tilde{\sigma}_\psi(a)$ and $\sigma_\wedge(a)$ can be employed to construct an operator family $b(\eta) \in C^{-\mu}(M, \boldsymbol{g}^{-1}; \mathbb{R}^q)$ such that

$$\sigma_\psi(a^{(-1)}) = \sigma_\psi(b), \quad \tilde{\sigma}_\psi(a^{(-1)}) = \tilde{\sigma}_\psi(b), \quad \sigma_\wedge(a^{(-1)}) = \sigma_\wedge(b).$$

This gives us $1 - b(\eta)a(\eta) =: c_0(\eta) \in C^{-1}(M, \boldsymbol{g}_l; \mathbb{R}^q)$, and a formal Neumann series argument allows us to improve $b(\eta)$ to a left parametrix $a^{(-1)}(\eta)$ by setting $a^{(-1)}(\eta) := \left(\sum_{j=0}^\infty c_0^j(\eta)\right) b(\eta)$ (using the existence of the asymptotic sum in $C^0(M, \boldsymbol{g}; \mathbb{R}^q)$). In a similar manner we can construct a right parametrix, i.e., $a^{(-1)}(\eta)$ is as desired.

Corollary 1.25. *If $a(\eta)$ is as in Theorem 1.24, then (1.53) is a family of Fredholm operators of index 0, and there is a constant $C > 0$ such that the operators (1.53) are isomorphisms for all $|\eta| \geq C$, $s \in \mathbb{R}$.*

Corollary 1.26. *If we perform the construction of Theorem 1.24 with the parameter $(\eta, \lambda) \in \mathbb{R}^{q+l}$, $l \geq 1$, rather than η, Corollary 1.25 yields that $a(\eta, \lambda)$ is invertible for all $\eta \in \mathbb{R}^q$, $|\lambda| \geq C$. Then, setting $a(\eta) := a(\eta, \lambda^1)$, $|\lambda^1| \geq C$ fixed, we obtain $a^{-1}(\eta) \in C^{-\mu}(M, \boldsymbol{g}^{-1}; \mathbb{R}^q)$.*

Observe that the operator functions of Theorem 1.20 refer to scales of spaces with two parameters, namely, $s \in \mathbb{R}$, the smoothness, and $\gamma \in \mathbb{R}$, the weight. Compared with Definition 1.9 we have here an additional weight. There are two ways to make the different view points compatible. One is to apply weight reducing isomorphisms

$$\mathrm{h}^{-\mu} : H^{s,\gamma}(M) \to H^{s,\gamma-\mu}(M) \tag{1.54}$$

in (1.21). Then, passing from

$$a(\eta) : H^{s,\gamma}(M) \to H^{s-\mu,\gamma-\mu}(M) \tag{1.55}$$

to

$$b^{\mu}(\eta) := \mathrm{h}^{-\gamma+\mu} a(\eta) \mathrm{h}^{\gamma} : H^{s,0}(M) \to H^{s-\mu,0}(M) \tag{1.56}$$

we obtain operator functions between spaces only referring to s but with properties as required in Definition 1.9 (which remains to be verified).

Remark 1.27. The spaces $E^s := H^{s,0}(M)$, $s \in \mathbb{R}$, form a scale with the properties at the beginning of Section 1.1.

Another way is to modify the abstract framework by admitting scales $E^{s,\gamma}$ rather than E^s, where in general γ may be in $\mathbb{R}^k$ (which is motivated by the higher corner calculus). We do not study the second possibility here but we only note that the variant with $E^{s,\gamma}$-spaces is very similar to the one without γ.

Let us now look at operator functions of the form (1.56).

Theorem 1.28. *The operators* (1.56) *constitute an order reducing family in the spaces* $E^s := H^{s,0}(M)$, *where the properties* (i)–(iii) *of Definition* 1.3 *are satisfied.*

Proof. In this proof we concentrate on the properties of our operators for every fixed s, μ, ν with $\nu \geq \mu$. The uniformity of the involved constants can easily be deduced; however, the simple (but lengthy) considerations will be left out.

(i) We have to show that

$$D_\eta^\beta b^\mu(\eta) = D_\eta^\beta \{\mathrm{h}^{-\gamma+\mu} a(\eta) \mathrm{h}^\gamma\} \in C^\infty(\mathbb{R}^q, \mathcal{L}(E^s, E^{s-\mu+|\beta|}))$$

for all $s \in \mathbb{R}$, $\beta \in \mathbb{N}^q$. According to (1.22) the operator function is a sum of two contributions. The second summand

$$(1-\sigma)\mathrm{h}^{-\gamma+\mu} a_{\mathrm{int}}(\eta) \mathrm{h}^\gamma (1-\tilde{\tilde{\sigma}})$$

is a parameter-dependent family in $L^\mu_{\mathrm{cl}}(2\mathrm{M}; \mathbb{R}^q)$ and obviously has the desired property. The first summand is of the form

$$\sigma \mathrm{h}^{-\gamma+\mu} \{a_{\mathrm{edge}}(\eta) + m(\eta) + g(\eta)\} \mathrm{h}^\gamma \tilde{\sigma}.$$

From the proof of Theorem 1.20 we have

$$D_\eta^\beta \sigma a_{\mathrm{edge}}(\eta) \tilde{\sigma} \in S^{\mu-|\beta|}(\mathbb{R}^q; \mathcal{K}^{s,\gamma;g}(X^\wedge), \mathcal{K}^{s-\mu+|\beta|,\gamma-\mu;g}(X^\wedge))$$

for every $\beta \in \mathbb{N}^q$. In particular, these operator functions are smooth in η and the derivatives improve the smoothness in the image by $|\beta|$. This gives us the desired property of $\sigma \mathrm{h}^{-\gamma+\mu} a_{\mathrm{edge}}(\eta) \mathrm{h}^\gamma \tilde{\sigma}$. The C^∞ dependence of $m(\eta) + g(\eta)$ in η is clear (those are operator-valued symbols), and they map to $\mathcal{K}^{\infty,\gamma-\mu;g}(X^\wedge)$

anyway. Therefore, the desired property of $\sigma \mathrm{h}^{-\gamma+\mu}\{m(\eta)+g(\eta)\}\mathrm{h}^{\gamma}\tilde{\sigma}$ is satisfied as well.

(ii) This property essentially corresponds to the fact that the product in consideration close to the conical point is a symbol in η of order zero and that the group action in $\mathcal{K}^{0,0}(X^{\wedge})$-spaces is unitary. Outside the conical point the boundedness is as in the example in Section 1.1.

(iii) The proof of this property close to the conical point is of a similar structure as Proposition 1.8, since our operators are based on operator-valued symbols referring to spaces with group action. The contribution outside the conical point is as in the same example in Section 1.1. □

Remark 1.29. For $E^s := \mathcal{H}^{s,0}(M)$, $s \in \mathbb{R}$, $\mathcal{E} = (E^s)_{s\in\mathbb{R}}$, the operator functions $b^{\mu}(\eta)$ of the form (1.56) belong to $S^{\mu}(\mathbb{R}^q;\mathcal{E},\mathcal{E})$ (see the notation after Definition 1.9).

2. Operators referring to a corner point

2.1. Weighted spaces

Let $\mathcal{E} = (E^s)_{s\in\mathbb{R}} \in \mathfrak{E}$ be a scale and $(b^{\mu}(\varrho))_{\mu\in\mathbb{R}}$, $\varrho \in \mathbb{R}$, be an order reducing family (see Definition 1.3 with $q = 1$). We define a new scale of spaces adapted to the Mellin transform and the approach of the cone calculus. In the following definition the Mellin transform refers to the variable $r \in \mathbb{R}_+$, i.e., $M = M_{r\to w}$.

Definition 2.1. For every $s, \gamma \in \mathbb{R}$ we define the space $\mathcal{H}^{s,\gamma}(\mathbb{R}_+,\mathcal{E})$ to be the completion of $C_0^{\infty}(\mathbb{R}_+, E^{\infty})$ with respect to be norm

$$\|u\|_{\mathcal{H}^{s,\gamma}(\mathbb{R}_+,\mathcal{E})} = \left\{\frac{1}{2\pi i}\int_{\Gamma_{\frac{d+1}{2}-\gamma}} \|b^s(\mathrm{Im}w)(Mu)(w)\|^2_{E^0}dw\right\}^{\frac{1}{2}} \tag{2.1}$$

for a $d = d_{\mathcal{E}} \in \mathbb{N}$. The Mellin transform M in (2.1) is interpreted as the weighted Mellin transform $M_{\gamma-\frac{d}{2}}$.

The role of $d_{\mathcal{E}}$ is an extra information, given together with the scale $\mathcal{E}$. In the example $\mathcal{E} = (H^s(X))_{s\in\mathbb{R}}$ for a closed compact C^{∞} manifold X we set $d_{\mathcal{E}} := \dim X$.

Observe that when we replace the order reducing family in (2.1) by an equivalent one the resulting norm is equivalent to (2.1).

By virtue of the identity

$$r^{\beta}\mathcal{H}^{s,\gamma}(\mathbb{R}_+,\mathcal{E}) = \mathcal{H}^{s,\gamma+\beta}(\mathbb{R}_+,\mathcal{E})$$

for every $s, \gamma, \beta \in \mathbb{R}$, it is often enough to refer the considerations to one particular weight, or to set

$$d_{\mathcal{E}} = 0. \tag{2.2}$$

For simplicity we now assume (2.2).

Let us consider Definition 1.9 for the case $U = \mathbb{R}$, $q = 1$, and denote the covariable now by $\varrho \in \mathbb{R}$. Set

$$S^\mu(\overline{\mathbb{R}}_+ \times \overline{\mathbb{R}}_+ \times \mathbb{R}; \mathcal{E}, \widetilde{\mathcal{E}}) := S^\mu(\mathbb{R} \times \mathbb{R} \times \mathbb{R}; \mathcal{E}, \widetilde{\mathcal{E}})|_{\overline{\mathbb{R}}_+ \times \overline{\mathbb{R}}_+ \times \mathbb{R}}$$

and

$$S^\mu(\overline{\mathbb{R}}_+ \times \overline{\mathbb{R}}_+ \times \Gamma_\delta; \mathcal{E}, \widetilde{\mathcal{E}}) := \{a(r, r', w) \in C^\infty(\overline{\mathbb{R}}_+ \times \overline{\mathbb{R}}_+ \times \Gamma_\delta, \mathcal{L}^\mu(\mathcal{E}, \widetilde{\mathcal{E}})) \\ : a(r, r', \delta + i\varrho) \in S^\mu(\overline{\mathbb{R}}_+ \times \overline{\mathbb{R}}_+ \times \mathbb{R}_\rho; \mathcal{E}, \widetilde{\mathcal{E}})\}$$

for any $\delta \in \mathbb{R}$. The subspaces of r'-independent ((r, r')-independent) symbols are denoted by $S^\mu(\overline{\mathbb{R}}_+ \times \mathbb{R}; \mathcal{E}, \widetilde{\mathcal{E}})$ $(S^\mu(\mathbb{R}; \mathcal{E}, \widetilde{\mathcal{E}}'))$ and $S^\mu(\overline{\mathbb{R}}_+ \times \Gamma_\delta; \mathcal{E}, \widetilde{\mathcal{E}})$ $(S^\mu(\Gamma_\delta; \mathcal{E}, \widetilde{\mathcal{E}}))$, respectively.

Given an element $f(r, r', w) \in S^\mu(\overline{\mathbb{R}}_+ \times \overline{\mathbb{R}}_+ \times \Gamma_{\frac{1}{2}-\gamma}; \mathcal{E}, \widetilde{\mathcal{E}})$ we set

$$\mathrm{op}_M^\gamma(f)u(r) = \frac{1}{2\pi} \int \int_0^\infty \left(\frac{r}{r'}\right)^{-(\frac{1}{2}-\gamma+i\varrho)} f\left(r, r', \frac{1}{2} - \gamma + i\varrho\right) u(r') \frac{dr'}{r'} d\varrho. \quad (2.3)$$

Let, for instance, f be independent of r'. Then (2.3) induces a continuous operator

$$\mathrm{op}_M^\gamma(f) : C_0^\infty(\mathbb{R}_+, E^\infty) \to C^\infty(\mathbb{R}_+, \widetilde{E}^\infty). \quad (2.4)$$

In fact, we have $\mathrm{op}_M^\gamma(f) = M_{\gamma, w\to r}^{-1} f(r, w) M_{\gamma, r'\to w}$. The weighted Mellin transform M_γ induces a continuous operator

$$M_\gamma : C_0^\infty(\mathbb{R}_+, E^s) \to \mathcal{S}(\Gamma_{\frac{1}{2}-\gamma}, E^s)$$

for every $s \in \mathbb{R}$. The subsequent multiplication of $M_\gamma u(w)$ by $f(r, w)$ gives rise to an element in $C^\infty(\mathbb{R}_+, \mathcal{S}(\Gamma_{\frac{1}{2}-\gamma}, \widetilde{E}^{s-\mu}))$, and then it follows easily that $\mathrm{op}_M^\gamma(f)u \in C^\infty(\mathbb{R}_+, \widetilde{E}^{s-\mu})$. We now formulate a continuity result, first for the case of symbols with constant coefficients.

Theorem 2.2. *For every $f(w) \in S^\mu(\Gamma_{\frac{1}{2}-\gamma}; \mathcal{E}, \widetilde{\mathcal{E}})$ the operator (2.4) extends to a continuous operator*

$$\mathrm{op}_M^\gamma(f) : \mathcal{H}^{s,\gamma}(\mathbb{R}_+, \mathcal{E}) \to \mathcal{H}^{s-\mu,\gamma}(\mathbb{R}_+, \widetilde{\mathcal{E}}) \quad (2.5)$$

for every $s \in \mathbb{R}$. Moreover, $f \to \mathrm{op}_M^\gamma(f)$ induces a continuous operator

$$S^\mu(\Gamma_{\frac{1}{2}-\gamma}; \mathcal{E}, \widetilde{\mathcal{E}}) \to \mathcal{L}(\mathcal{H}^{s,\gamma}(\mathbb{R}_+, \mathcal{E}), \mathcal{H}^{s-\mu,\gamma}(\mathbb{R}_+, \widetilde{\mathcal{E}})) \quad (2.6)$$

for every $s \in \mathbb{R}$.

Proof. We have

$$\begin{aligned} &\|\mathrm{op}_M^\gamma(f)u\|^2_{\mathcal{H}^{s-\mu,\gamma}(\mathbb{R}_+, \widetilde{\mathcal{E}})} \\ &= \int_{\mathbb{R}} \|\tilde{b}^{s-\mu}(\varrho) M_\gamma \left(M_\gamma^{-1} f\left(\frac{1}{2} - \gamma + i\varrho\right)\right) (M_\gamma u) \left(\frac{1}{2} - \gamma + i\varrho\right) \|^2_{\widetilde{E}^0} d\varrho \\ &= \int_{\mathbb{R}} \|\tilde{b}^{s-\mu}(\varrho) f\left(\frac{1}{2} - \gamma + i\varrho\right) b^{-s}(\varrho) b^s(\varrho)(M_\gamma u) \left(\frac{1}{2} - \gamma + i\varrho\right) \|^2_{\widetilde{E}^0} d\varrho \\ &\le c^2 \|u\|^2_{\mathcal{H}^{s,\gamma}(\mathbb{R}_+, \mathcal{E})} \end{aligned}$$

with

$$c = \sup_{\varrho \in \mathbb{R}} \|\tilde{b}^{s-\mu}(\varrho) f\left(\frac{1}{2} - \gamma + i\varrho\right) b^{-s}(\varrho)\|_{\mathcal{L}(E^0, \widetilde{E}^0)}$$

which is finite for every $s \in \mathbb{R}$ (cf. the estimates (1.10)). Thus we have proved the continuity both of (2.5) and (2.6). □

In order to generalise Theorem 2.2 to symbols with variable coefficients we impose conditions of reasonable generality that allow us to reduce the arguments to a vector-valued analogue of Kumano-go's technique.

Given a Fréchet space V with a countable system of semi-norms $(\pi_\iota)_{\iota \in \mathbb{N}}$ that defines its topology, we denote by

$$C_B^\infty(\mathbb{R}_+ \times \mathbb{R}_+, V)$$

the set of all $u(r, r') \in C^\infty(\mathbb{R}_+ \times \mathbb{R}_+, V)$ such that

$$\sup_{r, r' \in \mathbb{R}_+} \pi_\iota\Big((r\partial_r)^k (r'\partial_{r'})^{k'} u(r, r')\Big) < \infty$$

for all $k, k' \in \mathbb{N}$. In a similar manner by $C_B^\infty(\mathbb{R}_+, V)$ we denote the set of such functions that are independent of r'.

Moreover, we set

$$S_B^\mu(\mathbb{R}_+ \times \mathbb{R}_+ \times \Gamma_{\frac{1}{2}-\gamma}; \mathcal{E}, \widetilde{\mathcal{E}}) := C_B^\infty(\mathbb{R}_+ \times \mathbb{R}_+, S^\mu(\Gamma_{\frac{1}{2}-\gamma}; \mathcal{E}, \widetilde{\mathcal{E}}))$$

and, similarly, $S_B^\mu(\mathbb{R}_+ \times \Gamma_{\frac{1}{2}-\gamma}; \mathcal{E}, \widetilde{\mathcal{E}}) := C_B^\infty(\mathbb{R}_+, S^\mu(\Gamma_{\frac{1}{2}-\gamma}; \mathcal{E}, \widetilde{\mathcal{E}}))$.

Theorem 2.3. *For every $f(r, w) \in S_B^\mu(\mathbb{R}_+ \times \Gamma_{\frac{1}{2}-\gamma}; \mathcal{E}, \widetilde{\mathcal{E}})$ the operator $\mathrm{op}_M^\gamma(f)$ induces a continuous mapping*

$$\mathrm{op}_M^\gamma(f) : \mathcal{H}^{s,\gamma}(\mathbb{R}_+, \mathcal{E}) \to \mathcal{H}^{s-\mu,\gamma}(\mathbb{R}_+, \widetilde{\mathcal{E}}),$$

and $f \to \mathrm{op}_M^\gamma(f)$ a continuous operator

$$S_B^\mu(\mathbb{R}_+ \times \Gamma_{\frac{1}{2}-\gamma}; \mathcal{E}, \widetilde{\mathcal{E}}) \to \mathcal{L}(\mathcal{H}^{s,\gamma}(\mathbb{R}_+, \mathcal{E}), \mathcal{H}^{s-\mu,\gamma}(\mathbb{R}_+, \widetilde{\mathcal{E}}))$$

for every $s \in \mathbb{R}$.

Parallel to the spaces of Definition 2.1 it also makes sense to consider their "cylindrical" analogue, defined as follows.

Definition 2.4. Let $(b^s(\eta))_{s \in \mathbb{R}}$, be an order reducing family as in Definition 1.3. For every $s \in \mathbb{R}$ we define the space $H^s(\mathbb{R}^q, \mathcal{E})$ to be the completion of $C_0^\infty(\mathbb{R}^q, E^\infty)$ with respect to the norm

$$\|u\|_{H^s(\mathbb{R}^q, \mathcal{E})} := \Big\{\int_{\mathbb{R}^q} \|b^s(\eta)(\mathcal{F}u)(\eta)\|_{E^0}^2 d\eta\Big\}^{\frac{1}{2}}.$$

Clearly, similarly as above, with a symbol $a(y,y',\eta) \in S^\mu(\mathbb{R}^q \times \mathbb{R}^q \times \mathbb{R}^q; \mathcal{E}, \widetilde{\mathcal{E}})$ (when we impose a suitable control with respect to the dependence on y' for large $|y'|$) we can associate a pseudo-differential operator

$$\mathrm{Op}_y(a)u(y) = \iint e^{i(y-y')\eta} a(y,y',\eta)u(y')dy'đ\eta.$$

In particular, if $a = a(\eta)$ has constant coefficients, then we obtain a continuous operator

$$\mathrm{Op}_y(a) : H^s(\mathbb{R}^q, \mathcal{E}) \to H^{s-\mu}(\mathbb{R}^q, \widetilde{\mathcal{E}})$$

for every $s \in \mathbb{R}$. In the case of variable coefficients we need some precautions on the nature of symbols. This will be postponed for the moment.

We are mainly interested in the case $q = 1$. Consider the transformation

$$(S_\gamma u)(y) := e^{-(\frac{1}{2}-\gamma)y} u(e^{-y})$$

from functions in $r \in \mathbb{R}_+$ to functions in $y \in \mathbb{R}$. We then have the identity

$$(M_\gamma u)\left(\frac{1}{2} - \gamma + i\varrho\right) = (\mathcal{F} S_\gamma u)(\varrho)$$

with $\mathcal{F}$ being the one-dimensional Fourier transform. This gives us

$$\left\{\frac{1}{2\pi}\int_{\mathbb{R}} \|b^s(\eta)(\mathcal{F}S_\gamma u)(\eta)\|^2_{E^0} d\eta\right\}^{\frac{1}{2}} = \|S_\gamma u\|_{H^s(\mathbb{R},\mathcal{E})} = \|u\|_{\mathcal{H}^{s,\gamma}(\mathbb{R}_+,\mathcal{E})},$$

i.e., S_γ induces an isomorphism

$$S_\gamma : \mathcal{H}^{s,\gamma}(\mathbb{R}_+, \mathcal{E}) \to H^s(\mathbb{R}, \mathcal{E}).$$

Remark 2.5. By reformulating the expression (2.3) we obtain

$$\mathrm{op}_M^\gamma(f)u(r) = \frac{1}{2\pi}\iint e^{(\frac{1}{2}-\gamma+i\varrho)(\log r' - \log r)} f\left(r, r', \frac{1}{2} - \gamma + i\varrho\right) u(r')\frac{dr'}{r'}d\varrho.$$

Substituting $r = e^{-y}$, $r' = e^{-y'}$ gives us

$$\begin{aligned}\mathrm{op}_M^\gamma(f)u(r) =& \frac{1}{2\pi}\iint e^{i(y-y')\varrho} e^{(\frac{1}{2}-\gamma)(y-y')} f(e^{-y},\\ & e^{-y'}, \frac{1}{2} - \gamma + i\varrho)u(e^{-y'})dy'd\varrho = \mathrm{Op}_y(g_\gamma)v(y)\end{aligned}$$

with $v(y) := u(e^{-y})$ and $g_\gamma(y,y',\varrho) := e^{(\frac{1}{2}-\gamma)(y-y')} f(e^{-y}, e^{-y'}, \frac{1}{2} - \gamma + i\varrho)$.

In other words, if $\chi : \mathbb{R}_+ \to \mathbb{R}$ is defined by $\chi(r) = -\log r =: y$, we have $(\chi^* v)(r) = v(-\log r)$ or $((\chi^{-1})^* u)(g) = u(e^{-y})$ and

$$\mathrm{op}_M^\gamma(f) = \chi^* \mathrm{Op}_y(g_\gamma)(\chi^{-1})^*.$$

Thus $\mathrm{Op}_y(g_\gamma)$ is the operator push forward of $\mathrm{op}_M^\gamma(f)$ under χ.

2.2. Mellin quantisation and kernel cut-off

The axiomatic cone calculus that we develop here is a substructure of the general calculus of operators with symbols in $a(r,\rho) \in S^\mu(\mathbb{R}_+ \times \mathbb{R}; \mathcal{E}, \widetilde{\mathcal{E}})$ of the form $a(r,\rho) = \tilde{a}(r, r\rho), \tilde{a}(r,\tilde{\rho}) \in S^\mu(\overline{\mathbb{R}}_+ \times \mathbb{R}_{\tilde{\rho}}; \mathcal{E}, \widetilde{\mathcal{E}})$ (up to a weight factor and modulo smoothing operators) with a special control near $r = 0$ via Mellin quantisation. By $L^{-\infty}(\mathbb{R}_+; \mathcal{E}, \widetilde{\mathcal{E}}; \mathbb{R}^q)$ we denote the space of all Schwartz functions in $\eta \in \mathbb{R}^q$ with values in operators

$$C_0^\infty(\mathbb{R}_+, E^{-\infty}) \to C^\infty(\mathbb{R}_+, \widetilde{E}^\infty).$$

We then define

$$\begin{aligned} L^\mu(\mathbb{R}_+; \mathcal{E}, \widetilde{\mathcal{E}}; \mathbb{R}^q) = \{\mathrm{Op}_r(a)(\eta) + C(\eta) :& \\ a(r,\rho,\eta) \in S^\mu(\mathbb{R}_+ \times \mathbb{R}^{1+q}_{\rho,\eta}; \mathcal{E}, \widetilde{\mathcal{E}}), C(\eta) \in L^{-\infty}(\mathbb{R}_+; \mathcal{E}, \widetilde{\mathcal{E}}; \mathbb{R}^q)\}.& \end{aligned}$$

Our next objective is to formulate a Mellin quantisation result of symbols

$$a(r,\rho,\eta) = \tilde{a}(r, r\rho, r\eta), \quad \tilde{a}(r,\tilde{\rho},\tilde{\eta}) \in S^\mu(\overline{\mathbb{R}}_+ \times \mathbb{R}^{1+q}_{\tilde{\rho},\tilde{\eta}}; \mathcal{E}, \widetilde{\mathcal{E}}) \tag{2.7}$$

(see Remark 1.10).

Definition 2.6. By $M^\mu_{\mathcal{O}}(\mathcal{E},\widetilde{\mathcal{E}};\mathbb{R}^q_{\tilde{\eta}})$ we denote the set of all $h(z,\tilde{\eta}) \in \mathcal{A}(\mathbb{C}, S^\mu(\mathbb{R}^q_{\tilde{\eta}};\mathcal{E},\widetilde{\mathcal{E}}))$ such that

$$h(\beta + i\rho, \tilde{\eta}) \in S^\mu(\mathbb{R}_\rho \times \mathbb{R}^q_{\tilde{\eta}}; \mathcal{E}, \widetilde{\mathcal{E}})$$

for every $\beta \in \mathbb{R}$, uniformly in compact β-intervals. For $q = 0$ we simply write $M^\mu_{\mathcal{O}}(\mathcal{E}, \widetilde{\mathcal{E}})$.

Theorem 2.7. *For every symbol $a(r,\rho,\eta)$ of the form* (2.7) *there exists an $\tilde{h}(r,z,\tilde{\eta}) \in C^\infty(\overline{\mathbb{R}}_+, M^\mu_{\mathcal{O}}(\mathcal{E},\widetilde{\mathcal{E}};\mathbb{R}^q))$ such that for $h(r,z,\eta) := \tilde{h}(r,z,r\eta)$ and every $\delta \in \mathbb{R}$ we have*

$$\mathrm{op}^\delta_M(h)(\eta) = \mathrm{Op}_r(a)(\eta)$$

modulo operators in $L^{-\infty}(\mathbb{R}_+; \mathcal{E}, \widetilde{\mathcal{E}}; \mathbb{R}^q)$.

This result in the context of operator-valued symbols based on order reductions is mentioned here for completeness. It is contained in a joint work (in preparation) of the second author with C.-I. Martin (Potsdam) and N. Rablou (Göttingen). It extends a corresponding result of the edge symbolic calculus, see [6, Theorem 3.2]. More information in that case is given in [15, Chapter 4]. Here we adapt some part of this approach to realise the kernel cut-off principle that allows us to recognize how many parameter-dependent meromorphic Mellin symbols exist.

Definition 2.8. $S^\mu(\boldsymbol{C} \times \mathbb{R}^q;\mathcal{E},\widetilde{\mathcal{E}})$ denotes the space of all $h(\zeta,\eta) \in \mathcal{A}(\mathbb{C}, S^\mu(\mathbb{R}^q_\eta;\mathcal{E},\widetilde{\mathcal{E}}))$ such that

$$h(\rho + i\delta, \eta) \in S^\mu(\mathbb{R}^{1+q}_{\rho,\eta}; \mathcal{E}, \widetilde{\mathcal{E}})$$

for every $\delta \in \mathbb{R}$, uniformly in compact δ-intervals.

Clearly the space $S^\mu(\boldsymbol{C}\times\mathbb{R}^q;\mathcal{E},\widetilde{\mathcal{E}})$ is a generalisation of $M^\mu_{\mathcal{O}}(\mathcal{E},\widetilde{\mathcal{E}})$, however, with an interchanged role of real and imaginary part of the complex covariable. To produce elements of $S^\mu(\boldsymbol{C}\times\mathbb{R}^q;\mathcal{E},\widetilde{\mathcal{E}})$ we consider a so-called kernel cut-off operator

$$V: C_0^\infty(\mathbb{R})\times S^\mu(\mathbb{R}^{1+q};\mathcal{E},\widetilde{\mathcal{E}})\to S^\mu(\boldsymbol{C}\times\mathbb{R}^q;\mathcal{E},\widetilde{\mathcal{E}})$$

transforming an arbitrary element $a(\rho,\eta)\in S^\mu(\mathbb{R}^{1+q};\mathcal{E},\widetilde{\mathcal{E}})$ into $\big(V(\varphi)a\big)(\zeta,\eta)\in S^\mu(\boldsymbol{C}\times\mathbb{R}^q;\mathcal{E},\widetilde{\mathcal{E}})$ for any $\varphi\in C_0^\infty(\mathbb{R})$. It will be useful to admit φ to belong to the space $C_{\mathrm{b}}^\infty(\mathbb{R}):=\{\varphi\in C^\infty(\mathbb{R}_\theta):\sup_{\theta\in\mathbb{R}}|D_\theta^k\varphi(\theta)|<\infty \text{ for every } k\in\mathbb{N}\}$. We set

$$\big(V(\varphi)a\big)(\rho,\eta):=\iint e^{-i\theta\tilde{\rho}}\varphi(\theta)a(\rho-\tilde{\rho},\eta)d\theta đ\tilde{\rho}, \tag{2.8}$$

interpreted as an oscillatory integral (see also [16]). We now prove the following result:

Theorem 2.9. *The kernel cut-off operator* $V:(\varphi,a)\to V(\varphi)a$ *defines a bilinear and continuous mapping*

$$V: C_{\mathrm{b}}^\infty(\mathbb{R})\times S^\mu(\mathbb{R}^{1+q};\mathcal{E},\widetilde{\mathcal{E}})\to S^\mu(\mathbb{R}^{1+q};\mathcal{E},\widetilde{\mathcal{E}}), \tag{2.9}$$

and $\big(V(\varphi)a\big)(\rho,\eta)$ *admits an asymptotic expansion*

$$\big(V(\varphi)a\big)(\rho,\eta)\sim\sum_{k=0}^\infty\frac{(-1)^k}{k!}D_\theta^k\varphi(0)\partial_\rho^k a(\rho,\eta). \tag{2.10}$$

Proof. First note that the mapping

$$(\varphi,a)\to\varphi(\theta)a(\rho-\tilde{\rho},\eta),$$
$$C_{\mathrm{b}}^\infty(\mathbb{R})\times S^\mu(\mathbb{R}^{1+q}_{\rho,\eta};\mathcal{E},\widetilde{\mathcal{E}})\to C^\infty(\mathbb{R}^q_{\rho,\eta},S_{\mathrm{b}}^\mu(\mathbb{R}_\theta\times\mathbb{R}_{\tilde{\rho}};\mathcal{E},\widetilde{\mathcal{E}}))$$

for $S_{\mathrm{b}}^\mu(\mathbb{R}_\theta\times\mathbb{R}_{\tilde{\rho}};\mathcal{E},\widetilde{\mathcal{E}}):=C_{\mathrm{b}}^\infty(\mathbb{R}_\theta,S^\mu(\mathbb{R}_{\tilde{\rho}};\mathcal{E},\widetilde{\mathcal{E}}))$ is bilinear and continuous. For the proof of the continuity of (2.9) it suffices to verify that $\big(V(\varphi)a\big)(\rho,\eta)\in S^\mu(\mathbb{R}^{1+q};\mathcal{E},\widetilde{\mathcal{E}})$ and then to apply the closed graph theorem. By virtue of

$$D^\beta_{\rho,\eta}\big(V(\varphi)a\big)(\rho,\eta)=\big(V(\varphi)(D^\beta_{\rho,\eta}a)\big)(\rho,\eta)$$

for every $\beta\in\mathbb{N}^{1+q}$ we only have to check that for every $s\in\mathbb{R}$

$$\|\tilde{b}^{s-\mu}(\rho,\eta)\big(V(\varphi)a\big)(\rho,\eta)b^{-s}(\rho,\eta)\|_{\mathcal{L}(E^0,\widetilde{E}^0)}\le c \tag{2.11}$$

for all $(\rho,\eta)\in\mathbb{R}^{1+q}$, with a constant $c=c(s)>0$. We regularise the oscillatory integral (2.8)

$$\big(V(\varphi)a\big)(\rho,\eta)=\iint e^{-i\theta\tilde{\rho}}\langle\theta\rangle^{-2}\{(1-\partial_\theta^2)^N\varphi(\theta)\}a_N(\rho,\tilde{\rho},\eta)d\theta đ\tilde{\rho}$$

for

$$a_N(\rho,\tilde{\rho},\eta):=(1-\partial_{\tilde{\rho}}^2)\{\langle\tilde{\rho}\rangle^{-2N}a(\rho-\tilde{\rho},\eta)\}. \tag{2.12}$$

The function (2.12) is a linear combination of terms

$$(\partial_{\tilde{\rho}}^j\langle\tilde{\rho}\rangle^{-2N})(\partial_\rho^k a)(\rho-\tilde{\rho},\eta) \text{ for } 0\le j,k\le 2.$$

We have

$$\|\tilde{b}^{s-\mu}(\rho,\eta)\Big\{\iint e^{-i\theta\tilde{\rho}}\langle\theta\rangle^{-2}(1-\partial_\theta^2)^N\varphi(\theta)(\partial_{\tilde{\rho}}^j\langle\tilde{\rho}\rangle^{-2N})$$
$$(\partial_\rho^k a)(\rho-\tilde{\rho},\eta)d\theta đ\tilde{\rho}\Big\}b^{-s}(\rho,\eta)\|_{\mathcal{L}(E^0,\tilde{E}^0)} \tag{2.13}$$
$$= \|\iint \tilde{b}^{s-\mu}(\rho,\eta)\tilde{b}^{-s+\mu}(\rho-\tilde{\rho},\eta)\tilde{b}^{s-\mu}(\rho-\tilde{\rho},\eta)\{e^{-i\theta\tilde{\rho}}\langle\theta\rangle^{-2}(1-\partial_\theta^2)^N\varphi(\theta)$$
$$(\partial_{\tilde{\rho}}^j\langle\tilde{\rho}\rangle^{-2N})(\partial_\rho^k a)(\rho-\tilde{\rho},\eta)\}b^{-s}(\rho-\tilde{\rho},\eta)b^{s}(\rho-\tilde{\rho},\eta)b^{-s}(\rho,\eta)d\theta đ\tilde{\rho}\|_{\mathcal{L}(E^0,E^0)}$$
$$\le c\iint \|\tilde{b}^{s-\mu}(\rho,\eta)\tilde{b}^{-s+\mu}(\rho-\tilde{\rho},\eta)\|_{\mathcal{L}(\tilde{E}^0,\tilde{E}^0)}\|\tilde{b}^{s-\mu}(\rho-\tilde{\rho},\eta)(\partial_{\tilde{\rho}}^j\langle\tilde{\rho}\rangle^{-2N})$$
$$(\partial_\rho^k a)(\rho-\tilde{\rho},\eta)b^{-s}(\rho-\tilde{\rho},\eta)\|_{\mathcal{L}(E^0,\tilde{E}^0)}\ \|b^{s}(\rho-\tilde{\rho},\eta)b^{-s}(\rho,\eta)\|_{\mathcal{L}(E^0,E^0)}đ\tilde{\rho}.$$

For the norms under the integral we apply the Taylor expansion

$$b^s(\rho-\tilde{\rho},\eta)=\sum_{m=0}^{M}\frac{1}{m!}(\partial_\rho^m b^s)(\rho,\eta)(-\tilde{\rho})^m$$
$$+\frac{\tilde{\rho}}{M!}^{M+1}\int_0^1(1-t)^M(\partial_\rho^{M+1}b^s)(\rho-t\tilde{\rho},\eta)dt.$$

This yields

$$\|b^s(\rho-\tilde{\rho},\eta)b^{-s}(\rho,\eta)\|_{\mathcal{L}(E^0,E^0)}$$
$$\le\sum_{m=0}^{M}\frac{1}{m!}\langle\tilde{\rho}\rangle^m\|(\partial_\rho^m b^s)(\rho,\eta)b^{-s}(\rho,\eta)\|_{\mathcal{L}(E^0,E^0)}$$
$$+\frac{\langle\tilde{\rho}\rangle^{M+1}}{M!}\int_0^1(1-t)^M\|(\partial_\rho^{M+1}b^s)(\rho-t\tilde{\rho},\eta)b^{-s}(\rho,\eta)\|_{\mathcal{L}(E^0,E^0)}dt.$$

By virtue of (1.12), Proposition 1.15 and Proposition 1.13 we obtain

$$\|(\partial_\rho^m b^s)(\rho,\eta)b^{-s}(\rho,\eta)\|_{\mathcal{L}(E^0,E^0)}\le c\langle\rho,\eta\rangle^{-m}.$$

Moreover, using Definition 1.3 (iii), it follows that

$$\|(\partial_\rho^{M+1}b^s)(\rho-t\tilde{\rho},\eta)b^{-s}(\rho-t\tilde{\rho},\eta)b^{s}(\rho-t\tilde{\rho},\eta)b^{-s}(\rho,\eta)\|_{\mathcal{L}(E^0,E^0)}$$
$$\le c\|(\partial_\rho^{M+1}b^s)(\rho-t\tilde{\rho},\eta)b^{-s}(\rho-t\tilde{\rho},\eta)\|_{\mathcal{L}(E^0,E^0)}$$
$$\|b^s(\rho-t\tilde{\rho},\eta)\|_{\mathcal{L}(E^s,E^0)}\|b^{-s}(\rho,\eta)\|_{\mathcal{L}(E^0,E^s)}$$
$$\le\langle\rho-t\tilde{\rho},\eta\rangle^{B_1(s)}\langle\rho,\eta\rangle^{B_2(s)}$$

with certain $B_i(s)$, $i=1,2$. We thus obtain

$$\|b^s(\rho-\tilde{\rho},\eta)b^{-s}(\rho,\eta)\|_{\mathcal{L}(E^0,E^0)}\le c\langle\tilde{\rho}\rangle^{M+1}(\sup_{|t|\le1}\langle\rho-t\tilde{\rho},\eta\rangle^{-(M+1)+B_1(s)})\langle\rho,\eta\rangle^{B_2(s)}.$$

By Peetre's inequality for $L\ge0$ we have $\sup_{|t|\le1}\langle\rho-t\tilde{\rho},\eta\rangle^{-L}\le c\langle\tilde{\rho}\rangle^L\langle\rho,\eta\rangle^{-L}$. Thus choosing M so large that

$$-(M+1)+B_1(s)\le0,\ -(M+1)+B_1(s)+B_2(s)\le0,$$

it follows that

$$\begin{aligned} &\|b^s(\rho-\tilde{\rho},\eta)b^{-s}(\rho,\eta)\|_{\mathcal{L}(E^0,E^0)} \\ &\leq c\langle\tilde{\rho}\rangle^{M+1}\langle\tilde{\rho}\rangle^{M+1-B_1(s)}\langle\rho,\eta\rangle^{-(M+1)+B_1(s)+B_2(s)} \leq c\langle\tilde{\rho}\rangle^{A(s)} \end{aligned} \tag{2.14}$$

for $A(s) := 2(M+1) - B_2(s)$.

In a similar manner we can show that

$$\|\tilde{b}^{s-\mu}(\rho,\eta)\tilde{b}^{-s+\mu}(\rho-\tilde{\rho},\eta)\|_{\mathcal{L}(\tilde{E}^0,\tilde{E}^0)} \leq c\langle\tilde{\rho}\rangle^{\widetilde{A}(s)} \tag{2.15}$$

for some $\widetilde{A}(s) \in \mathbb{R}$. Applying (2.14) and (2.15) in the estimate (2.13) it follows that

$$\|\tilde{b}^{s-\mu}(\rho,\eta)\big(V(\varphi)a\big)(\rho,\eta)b^{-s}(\rho,\eta)\|_{\mathcal{L}(E^0,\tilde{E}^0)} \leq c\sum_{0\leq j\leq 2}\int|\partial^j_{\tilde{\rho}}\langle\tilde{\rho}\rangle^{-2N}|\langle\tilde{\rho}\rangle^{A(s)+\widetilde{A}(s)}đ\tilde{\rho}. \tag{2.16}$$

Since $N \in \mathbb{N}$ can be chosen as large as we want, it follows that the right-hand side of (2.16) is finite for an appropriate N. This completes the proof of (2.11). The relation (2.10) immediately follows by applying the Taylor expansion of φ at 0. □

Theorem 2.10. *The kernel cut-off operator $V : (\varphi, a) \to V(\varphi)a$ defines a bilinear and continuous mapping*

$$V : C_0^\infty(\mathbb{R}) \times S^\mu(\mathbb{R}^{1+q};\mathcal{E},\widetilde{\mathcal{E}}) \to S^\mu(\boldsymbol{C}\times\mathbb{R}^q;\mathcal{E},\widetilde{\mathcal{E}}). \tag{2.17}$$

Proof. Writing

$$\big(V(\varphi)a\big)(\rho,\eta) = \int e^{-i\theta\rho}\varphi(\theta)\{\int e^{i\theta\rho'}a(\rho',\eta)đ\rho'\}d\theta$$

we see that $\big(V(\varphi)a\big)(\rho,\eta)$ is the Fourier transform of a distribution

$$\varphi(\theta)\int e^{i\theta\rho'}a(\rho',\eta)đ\rho' \in \mathcal{S}'(\mathbb{R}_\theta,\mathcal{L}^\mu(\mathcal{E},\widetilde{\mathcal{E}}))$$

with compact support. This extends to a holomorphic $\mathcal{L}^\mu(\mathcal{E},\widetilde{\mathcal{E}})$-valued function in $\zeta = \rho + i\delta$, given by

$$\big(V(\varphi)a\big)(\rho+i\delta,\eta) = \big(V(\varphi_\delta)a\big)(\rho,\eta)$$

for $\varphi_\delta(\theta) := e^{\theta\delta}\varphi(\theta)$. Theorem 2.9 gives us $\big(V(\varphi)a\big)(\rho+i\delta,\eta) \in S^\mu(\mathbb{R}^{1+q};\mathcal{E},\widetilde{\mathcal{E}})$ for every $\delta \in \mathbb{R}$. By virtue of the continuity of $\delta \to \varphi_\delta$, $\mathbb{R} \to C_0^\infty(\mathbb{R})$ and of the continuity of (2.9) it follows that (2.17) induces a continuous mapping

$$V : C_0^\infty(\mathbb{R}) \times S^\mu(\mathbb{R}^{1+q};\mathcal{E},\widetilde{\mathcal{E}}) \to S^\mu(I_\delta\times\mathbb{R}^q;\mathcal{E},\widetilde{\mathcal{E}}),$$

$I_\delta := \{\zeta \in \mathbb{C} : \mathrm{Im}\zeta = \delta\}$ which is uniform in compact δ-intervals. The closed graph theorem gives us also the continuity of (2.17) with respect to the Fréchet topology of $S^\mu(\boldsymbol{C}\times\mathbb{R}^q;\mathcal{E},\widetilde{\mathcal{E}})$. □

2.3. Meromorphic Mellin symbols and operators with asymptotics

As an ingredient of our cone algebra we now study meromorphic Mellin symbols, starting from $M^{\mu}_{\mathcal{O}}(\mathcal{E},\widetilde{\mathcal{E}})$ (see Definition 2.6 for $q=0$).

Theorem 2.11. *$h\in M^{\mu}_{\mathcal{O}}(\mathcal{E},\widetilde{\mathcal{E}})$ and $h|_{\Gamma_\beta}\in S^{\mu-\varepsilon}(\Gamma_\beta;\mathcal{E},\widetilde{\mathcal{E}})$ for some $\varepsilon>0$ entails $h\in M^{\mu-\varepsilon}_{\mathcal{O}}(\mathcal{E},\widetilde{\mathcal{E}})$.*

Proof. The ideas of the proof are similar to the case of the cone calculus with smooth base X and the scales $\big(H^s(X)\big)_{s\in\mathbb{R}}$ (see, e.g., the thesis of Seiler [40]). □

Proposition 2.12. *Let $h(w)\in M^{\mu}_{\mathcal{O}}(\mathcal{E}_0,\widetilde{\mathcal{E}})$, $f(w)\in M^{\nu}_{\mathcal{O}}(\mathcal{E},\mathcal{E}_0)$; then for pointwise composition we have $h(w)f(w)\in M^{\mu+\nu}_{\mathcal{O}}(\mathcal{E},\widetilde{\mathcal{E}})$.*

Proof. The proof is obvious. □

Definition 2.13. An element $h(w)\in M^{\mu}_{\mathcal{O}}(\mathcal{E},\widetilde{\mathcal{E}})$ is called elliptic, if for some $\beta\in\mathbb{R}$ the operators $h(\beta+i\rho):E^s\to\widetilde{E}^{s-\mu}$ are invertible for all $s\in\mathbb{R}$ and $h^{-1}(\beta+i\rho)\in S^{-\mu}(\mathbb{R}_\rho;\widetilde{\mathcal{E}},\mathcal{E})$.

Theorem 2.14. *Let $h\in M^{\mu}_{\mathcal{O}}(\mathcal{E},\widetilde{\mathcal{E}})$ be elliptic. Then,*

$$h(w):E^s\to\widetilde{E}^{s-\mu}\tag{2.18}$$

is a holomorphic family of Fredholm operators of index zero for $s\in\mathbb{R}$. There is a set $D\subset\mathbb{C}$, with $D\cap\{c\le\mathrm{Re}\,w\le c'\}$ finite for every $c\le c'$, such that the operators (2.18) are invertible for all $w\in\mathbb{C}\setminus D$.

Proof. By assumption we have $g:=(h|_{\Gamma_\beta})^{-1}\in S^{-\mu}(\Gamma_\beta;\widetilde{\mathcal{E}},\mathcal{E})$. Applying a version of the kernel cut-off construction, now referring to parallels of the imaginary axis rather than the real axis, with a function $\psi\in C_0^\infty(\mathbb{R}_+)$, $\psi\equiv 1$ near 1, we obtain a continuous operator

$$V(\psi):S^{-\mu}(\Gamma_\beta;\widetilde{\mathcal{E}},\mathcal{E})\to M^{-\mu}_{\mathcal{O}}(\widetilde{\mathcal{E}},\mathcal{E})$$

where $V(\psi)g|_{\Gamma_\beta}=g \mod S^{-\infty}(\Gamma_\beta;\widetilde{\mathcal{E}},\mathcal{E})$. Setting $h^{(-1)}(w):=V(\psi)g$ we obtain $h^{(-1)}(w)\in M^{-\mu}_{\mathcal{O}}(\widetilde{\mathcal{E}},\mathcal{E})$, and from Proposition 2.12 it follows that

$$h(w)h^{(-1)}(w)\in M^{0}_{\mathcal{O}}(\widetilde{\mathcal{E}},\widetilde{\mathcal{E}}),\ h^{(-1)}(w)h(w)\in M^{0}_{\mathcal{O}}(\mathcal{E},\mathcal{E})$$

and

$$h(w)h^{(-1)}(w)|_{\Gamma_\beta}-1\in S^{-\infty}(\Gamma_\beta;\widetilde{\mathcal{E}},\widetilde{\mathcal{E}}),h^{(-1)}(w)h(w)|_{\Gamma_\beta}-1\in S^{-\infty}(\Gamma_\beta;\mathcal{E},\mathcal{E}),\tag{2.19}$$

for every $\beta\in\mathbb{R}$, and hence

$$h(w)h^{(-1)}(w)=1+m(w),\ h^{(-1)}(w)h(w)=1+l(w)\tag{2.20}$$

for certain $m(w)\in M^{-\infty}_{\mathcal{O}}(\widetilde{\mathcal{E}},\widetilde{\mathcal{E}})$, $l(w)\in M^{-\infty}_{\mathcal{O}}(\mathcal{E},\mathcal{E})$. For every $s\in\mathbb{R}$ and every fixed $w\in\mathbb{C}$ the operators

$$m(w):\widetilde{E}^s\to\widetilde{E}^\infty,\ l(w):E^s\to E^\infty$$

are continuous. Therefore, since the scales have the compact embedding property, from (2.20) we obtain that $h^{(-1)}(w)$ is a two-sided parametrix of $h(w)$ for every w, i.e., the operators (2.18) are Fredholm. Since $h(w) \in \mathcal{A}(\mathbb{C}, \mathcal{L}^{\mu}(E^s, E^{s-\mu}))$ is continuous in $w \in \mathbb{C}$ we have ind $h(w_1) =$ ind $h(w_2)$ for every $w_1, w_2 \in \mathbb{C}$. However, since $h(w) = 0$ consists of invertible operators on the line Γ_β it follows that ind $h(w) = 0$ for all $w \in \mathbb{C}$. Finally, from the relations (2.18) we see that for every $c \leq c'$ there is an $L(c, c') > 0$ such that the operators (2.18) are invertible for all $w \in \mathbb{C}$ with $|\text{Im } w| \geq L(c, c')$, $c \leq \text{Re } w \leq c'$. Then a general result on holomorphic Fredholm families tells us that the strip $c \leq \text{Re } w \leq c'$ contains at most finitely many points where (2.18) is not invertible. Those points just constitute the set D, it is also independent of s, since ker $h(w)$ is independent of s as we easily see from (2.20) and the smoothing remainders; then vanishing of the index shows that the invertibility holds exactly when ker $h(w) = 0$. □

Theorem 2.15. *The ellipticity of h with respect to Γ_β as in Definition* 2.13 *entails the ellipticity with respect to Γ_δ for all $\delta \in \mathbb{R}$ such that $\Gamma_\delta \cap D = \emptyset$. In that sense Definition* 2.13 *is independent of the choice of β.*

Proof. Let us apply the kernel cut-off operator $V(\psi_\varepsilon)$, where $\psi_\varepsilon \in C_0^\infty(\mathbb{R}_+)$ is of the form $\psi_\varepsilon(t) = \psi(\varepsilon t)$, $\varepsilon > 0$, for some cut-off function ψ. Then, setting

$$V(\psi_\varepsilon)(h^{-1}(\beta + i\rho)) =: f_\varepsilon \in M_{\mathcal{O}}^{-\mu}(\widetilde{\mathcal{E}}, \mathcal{E})$$

we have $f_\varepsilon|_{\Gamma_\beta} \in S^{-\mu}(\Gamma_\beta; \widetilde{\mathcal{E}}, \mathcal{E})$ and $f_\varepsilon|_{\Gamma_\beta} \to h^{-1}(\beta + i\rho)$ as $\varepsilon \to 0$ in the topology of $S^{-\mu}(\Gamma_\beta; \widetilde{\mathcal{E}}, \mathcal{E})$. This shows us that $f_{\varepsilon_1}|_{\Gamma_\beta}$ is pointwise invertible when $\varepsilon_1 > 0$ is sufficiently small. Let us set $h^{(-1)}(w) = f_{\varepsilon_1}(w)$. According to Proposition 2.12 we have $g(w) := h^{(-1)}(w)h(w) \in M_{\mathcal{O}}^0(\mathcal{E}, \mathcal{E})$ and by construction

$$g|_{\Gamma_\beta} = 1 + l \text{ for some } l \in S^{-\infty}(\Gamma_\beta; \mathcal{E}, \mathcal{E}).$$

Then Theorem 2.11 yields $g = 1 \bmod M_{\mathcal{O}}^{-\infty}(\mathcal{E}, \mathcal{E})$. It follows that

$$h^{-1}|_{\Gamma_\delta} h|_{\Gamma_\delta} = 1 + l_\delta \text{ for some } l_\delta \in S^{-\infty}(\Gamma_\delta; \mathcal{E}, \mathcal{E})$$

and hence, since $h|_{\Gamma_\delta}$ is pointwise invertible,

$$h^{(-1)}|_{\Gamma_\delta} = (1 + l_\delta)\left(h|_{\Gamma_\delta}\right)^{-1}.$$

which yields

$$\left(h|_{\Gamma_\delta}\right)^{-1} = (1 + l_\delta)^{-1} h^{(-1)}|_{\Gamma_\delta}. \tag{2.21}$$

From Proposition 1.12 we know that $l_\delta \in \mathcal{S}(\Gamma_\delta, \mathcal{L}^{-\infty}(\mathcal{E}, \mathcal{E}))$ and it is also clear that $(1 + l_\delta)^{-1} = 1 + m_\delta$ for some $m_\delta \in \mathcal{S}(\Gamma_\delta, \mathcal{L}^{-\infty}(\mathcal{E}, \mathcal{E}))$. Then Proposition 1.15 (iii) shows that $\left(h|_{\Gamma_\delta}\right)^{-1} \in S^{-\mu}(\Gamma_\delta; \widetilde{\mathcal{E}}, \mathcal{E})$. □

A sequence

$$R = \{(p_j, m_j, L_j)\}_{j \in \mathbb{Z}}$$

is called a discrete asymptotic type of Mellin symbols, if $p_j \in \mathbb{C}$, $m_j \in \mathbb{N}$, and $L_j \subset \mathcal{L}^{-\infty}(\mathcal{E}, \widetilde{\mathcal{E}})$ is a finite-dimensional subspace of finite rank operators; moreover, $\pi_{\mathbb{C}} R := (p_j)_{j \in \mathbb{Z}}$ is assumed to intersect the strips $\{w \in \mathbb{C} : c_1 \leq \text{Re} w \leq c_2\}$ in a finite set, for every $c_1 \leq c_2$. Let $M_R^{-\infty}(\mathcal{E},\widetilde{\mathcal{E}})$ denote the space of all functions $m \in \mathcal{A}(\mathbb{C} \setminus \pi_{\mathbb{C}} R, \mathcal{L}^{-\infty}(\mathcal{E},\widetilde{\mathcal{E}}))$ which are meromorphic with poles at the points p_j of multiplicity $m_j + 1$ and Laurent coefficients at $(w - p_j)^{-(k+1)}$ belonging to L_j for $0 \leq k \leq m_j$, and $\chi(w)m(w)|_{\Gamma_\delta} \in \mathcal{S}(\Gamma_\delta; \mathcal{E}, \widetilde{\mathcal{E}})$ for every $\delta \in \mathbb{R}$, uniformly in compact δ-intervals, where χ is any $\pi_{\mathbb{C}} R$-excision function. Moreover, we set

$$M_R^\mu(\mathcal{E},\widetilde{\mathcal{E}}) := M_{\mathcal{O}}^\mu(\mathcal{E},\widetilde{\mathcal{E}}) + M_R^{-\infty}(\mathcal{E},\widetilde{\mathcal{E}}). \tag{2.22}$$

Theorem 2.16. *Let $h \in M_R^\mu(\mathcal{E}_0,\widetilde{\mathcal{E}})$, $f \in M_S^\nu(\mathcal{E},\mathcal{E}_0)$ with asymptotic types R, S and orders $\mu, \nu \in \mathbb{R}$, then we have $hf \in M_P^{\mu+\nu}(\mathcal{E},\widetilde{\mathcal{E}})$ with some resulting asymptotic type P.*

Proof. The proof of this result is analogous to the one in the "concrete" cone calculus, see [33]. □

Proposition 2.17. *For every $m \in M_R^{-\infty}(\mathcal{E},\mathcal{E})$ there exists an $m^{(-1)} \in M_S^{-\infty}(\mathcal{E},\mathcal{E})$ with another asymptotic type S such that*

$$\big(1 + m(w)\big)\big(1 + m^{(-1)}(w)\big) = 1.$$

For the proof we employ the following Lemma.

Lemma 2.18. *Let E be a Banach space, $U \subseteq \mathbb{C}$ open, $0 \in U$, and let $h \in \mathcal{A}(U, \mathcal{L}(E))$ be an element such that $h(w) = 0$ on a closed space of finite codimension $F \subseteq E$. Moreover, let $a_1, \ldots, a_N \in \mathcal{L}(E)$ be operators of finite rank, for some $N \in \mathbb{N} \setminus \{0\}$. Then there is a $\delta > 0$ such that the meromorphic $\mathcal{L}(E)$-valued function*

$$f(w) = 1 + h(w) + \sum_{j=1}^{N} a_j w^{-j}$$

is invertible for all $0 < |w| < \delta$.

Proof of Proposition 2.17. First observe that if $m \in \mathcal{L}^{-\infty}(\mathcal{E},\mathcal{E})$ is an operator such that

$$1 + m : E^s \to E^s$$

is invertible for all $s \in \mathbb{R}$, we can define an operator $g \in \mathcal{L}^0(\mathcal{E},\mathcal{E})$ such that $(1 + m)(1 + g) = 1$. This gives us $1 + m + g + mg = 1$, and $m, mg \in \mathcal{L}^{-\infty}(\mathcal{E},\mathcal{E})$ implies $g = -m(1 + g) \in \mathcal{L}^{-\infty}(\mathcal{E},\mathcal{E})$.

Moreover, our operator function $1 + m$ is holomorphic in $\mathbb{C} \setminus \pi_{\mathbb{C}} R$. Then $g = (1+m)^{-1} - 1$ is holomorphic in $\mathbb{C} \setminus D$ with values in $L^{-\infty}(\mathcal{E},\mathcal{E})$, where $D \subseteq \mathbb{C}$ is a countable set such that $\{w \in \mathbb{C} : c_1 \leq \mathrm{Re} w \leq c_2\} \cap \{w \in \mathbb{C} : \mathrm{dist}(w, \pi_{\mathbb{C}} R) > \varepsilon\} \cap D$ is finite for every $c_1 \leq c_2$ and $\varepsilon > 0$. If $\chi(w)$ is any D-excision function, then we have a representation $(1 + m)^{-1} = \sum_{j=0}^{\infty} (-1)^j m^j$ as a convergent series of functions with values in $\mathcal{L}(E^s, E^s)$ uniformly in $w \in \mathbb{C}$ for $c \leq \mathrm{Re} w \leq c'$, $|\mathrm{Im} w| \geq C$ for every $c \leq c'$ and $C > 0$ sufficiently large. In a similar manner we obtain convergence of all w-derivatives of $\sum_{j=0}^{\infty} (-1)^j m^j$ in a set of such a structure. Thus, from $g = -m(1 + g) = -m(1 + m)^{-1}$ and the Schwartz property of m for large $|\mathrm{Im} w|$, uniformly in finite strips $c \leq \mathrm{Re} w \leq c'$, we obtain the same of g itself. It remains to show that g is meromorphic with poles at the points of D, that D has no accumulation points at $\pi_{\mathbb{C}}R$, and that the Laurent coefficients are of the desired kind, namely, to belong to $L^{-\infty}(\mathcal{E},\mathcal{E})$ and to be of finite rank. Let us verify that there are no accumulation points of the singularities of $(1+m(w))^{-1}$. Let w_0 be a pole of m, i.e., $w_0 \in \pi_{\mathbb{C}}R$. Then we can write

$$1+m(w) = 1+m_0(w)+\sum_{k=1}^{K} b_k(w-w_0)^{-k}$$

with suitable $K \in \mathbb{N}$, m_0 holomorphic in a neighbourhood of w_0 and $\mathcal{L}^{-\infty}(\mathcal{E},\mathcal{E})$-valued, with finite rank operators b_k. Note that $m_0 \not\equiv -1$. Setting $n(w) := \sum_{k=1}^{K} b_k(w-w_0)^{-k}$ we have

$$1+m(w) = (1+m_0(w))(1+(1+m_0(w))^{-1}n(w)).$$

Since m_0 is holomorphic near w_0 and $1+m_0(w)$ a Fredholm family, the singularities of $(1+m_0(w))^{-1}$ form a countable discrete set; therefore there is a $\delta > 0$ such that $(1+m_0(w))^{-1}$ exists for all w such that $0 < |w-w_0| < \delta$. Moreover, $(1+m_0(w))^{-1}n(w)$ can be written in the form $h(w)+\sum_{j=1}^{N} a_j(w-w_0)^{-j}$ with a suitable h which is holomorphic near w_0 and finite rank operators a_j, $1 \le j \le N$. The operator $(1+m_0(w))^{-1}n(w)$ vanishes on the space $F := \bigcap_{k=1}^{K} \ker b_k$ which is of finite codimension. Setting $M := \bigcap_{j=1}^{N} \ker a_j$ it follows that $h(w)u = 0$ for all $u \in M \cap F$; the latter space is also of finite codimension. Lemma 2.18 then shows that $1+(1+m_0(w))^{-1}n(w)$ is invertible in $0 < |w-w_0| < \delta$ for a suitable $\delta > 0$. □

Theorem 2.19. *Let $h \in M_{\mathcal{O}}^{\mu}(\mathcal{E},\widetilde{\mathcal{E}})$ be elliptic, then there is an $f \in M_S^{-\mu}(\widetilde{\mathcal{E}},\mathcal{E})$ with asymptotic type S such that $hf = 1$.*

Proof. Let $h^{(-1)}(w) \in M_{\mathcal{O}}^{\mu}(\mathcal{E},\widetilde{\mathcal{E}})$ be as in the proof of Theorem 2.14. Then we have the relation (2.20). By virtue of Proposition 2.17 there exists a $g \in M_P^{-\infty}(\mathcal{E},\mathcal{E})$ for some asymptotic type P such that $(1+m(w))(1+g(w)) = 1$. This yields $h(w)f(w) = 1$ for $f := h^{(-1)}(1+g)$ which belongs to $M_S^{-\mu}(\widetilde{\mathcal{E}},\mathcal{E})$, according to Theorem 2.16. In a similar manner we find an $\tilde{f} \in M_{\widetilde{S}}^{-\mu}(\widetilde{\mathcal{E}},\mathcal{E})$ such that $\tilde{f}(w)h(w) = 1$. This implies $f = \tilde{f}$. □

Definition 2.20. A $g \in M_R^{\mu}(\mathcal{E},\widetilde{\mathcal{E}})$ is said to be elliptic, if there is a $\beta \in \mathbb{R}$ such that $(g|_{\Gamma_\beta})^{-1} \in S^{-\mu}(\mathbb{R}_\rho;\widetilde{\mathcal{E}},\mathcal{E})$.

Theorem 2.21. *If $g \in M_R^{\mu}(\mathcal{E},\widetilde{\mathcal{E}})$ is elliptic, there is an $f \in M_S^{-\mu}(\widetilde{\mathcal{E}},\mathcal{E})$ such that $gf = 1$.*

Proof. Applying a kernel cut-off argument to $(g|_{\Gamma_\beta})^{-1}$ we can find an $h^{(-1)} \in M_{\mathcal{O}}^{-\mu}(\widetilde{\mathcal{E}},\mathcal{E})$ such that $h^{(-1)}|_{\Gamma_\beta} - (g|_{\Gamma_\beta})^{-1} \in S^{-\infty}(\Gamma_\beta;\widetilde{\mathcal{E}},\mathcal{E})$. By definition we have $g = g_0 + g_1$ for certain $g_0 \in M_{\mathcal{O}}^{\mu}(\mathcal{E},\widetilde{\mathcal{E}})$, $g_1 \in M_R^{-\infty}(\mathcal{E},\widetilde{\mathcal{E}})$. Then $h^{(-1)}g_0|_{\Gamma_\beta} = 1 \bmod S^{-\infty}(\Gamma_\beta;\mathcal{E},\mathcal{E})$ implies $h^{(-1)}g_0 = 1 \bmod M_{\mathcal{O}}^{-\infty}(\mathcal{E},\mathcal{E})$ (see Theorem 2.11). If follows that $h^{(-1)}g = 1 + m$ for some $m \in M_R^{-\infty}(\mathcal{E}, \mathcal{E})$ with an asymptotic type R (see Theorem 2.16). Thus Proposition 2.17 gives us $g^{-1} = (1+m)^{-1}h^{(-1)} \in M_S^{-\mu}(\widetilde{\mathcal{E}}, \mathcal{E})$ with some asymptotic type S. □

Parallel to the spaces of Mellin symbols (2.22) we now introduce subspaces of $\mathcal{H}^{s,\gamma}(\mathbb{R}_+, \mathcal{E})$ with discrete asymptotics. We consider a sequence

$$P := \{(p_j, m_j)\}_{0 \le j \le N} \tag{2.23}$$

with $N \in \mathbb{N} \cup \{+\infty\}$, $m_j \in \mathbb{N}$, $0 \le j \le N$.

A sequence (2.23) is said to be a discrete asymptotic type, associated with weight data (γ, Θ) (with a weight $\gamma \in \mathbb{R}$ and a weight interval $\Theta = (\vartheta, 0]$, $-\infty \le \vartheta \le 0$), if

$$\pi_{\mathbb{C}} P := \{p_j\}_{0 \le j \le N} \subset \{w \in \mathbb{C} : \frac{d+1}{2} - \gamma + \vartheta < \operatorname{Re} w < \frac{d+1}{2} - \gamma\},$$

and $\pi_{\mathbb{C}} P$ is finite when ϑ is finite, and $\operatorname{Re} p_j \to -\infty$ as $j \to \infty$ when $\vartheta = -\infty$ and $N = +\infty$. We will say that P satisfies the shadow condition, if $(p, m) \in P$ implies $(p - j, m) \in P$ for all $j \in \mathbb{N}$ with $\frac{d+1}{2} - \gamma + \vartheta < \operatorname{Re}(p - j) < \frac{d+1}{2} - \gamma$.

If Θ is finite we define the (finite-dimensional) space

$$\mathcal{S}_P(\mathbb{R}_+, \mathcal{E}) := \Big\{ \sum_{j=0}^{N} \sum_{k=0}^{m_j} \omega(r) c_{jk} r^{-p_j} \log^k r : \ c_{jk} \in E^\infty, \ 0 \le k \le m_j, \ 0 \le j \le N \Big\}$$

with some fixed cut-off function ω on the half-axis. We then have

$$\mathcal{S}_P(\mathbb{R}_+, \mathcal{E}) \subset \mathcal{H}^{\infty,\gamma}(\mathbb{R}_+, \mathcal{E}).$$

Moreover, we set

$$\mathcal{H}_\Theta^{s,\gamma}(\mathbb{R}_+, \mathcal{E}) := \varprojlim_{m \in \mathbb{N}} \{\omega \mathcal{H}^{s,\gamma - \vartheta - \frac{1}{m+1}}(\mathbb{R}_+, \mathcal{E}) + (1 - \omega)\mathcal{H}^{s,\gamma}(\mathbb{R}_+, \mathcal{E})\}$$

endowed with the Fréchet topology of the projective limit, and

$$\mathcal{H}_P^{s,\gamma}(\mathbb{R}_+, \mathcal{E}) := \mathcal{H}_\Theta^{s,\gamma}(\mathbb{R}_+, \mathcal{E}) + \mathcal{S}_P(\mathbb{R}_+, \mathcal{E})$$

as a direct sum of Fréchet spaces.

In order to formulate spaces with discrete asymptotics of type P in the case $\Theta = (-\infty, 0]$ we form $P_k := \{(p, m) \in P : \operatorname{Re} p > \frac{d+1}{2} - \gamma - (k+1)\}$ for any $k \in \mathbb{N}$. From the above construction we have the spaces $\mathcal{H}_{P_k}^{s,\gamma}(\mathbb{R}_+, \mathcal{E})$ together with continuous embeddings

$$\mathcal{H}_{P_{k+1}}^{s,\gamma}(\mathbb{R}_+, \mathcal{E}) \hookrightarrow \mathcal{H}_{P_k}^{s,\gamma}(\mathbb{R}_+, \mathcal{E}), \ k \in \mathbb{N}.$$

We then define

$$\mathcal{H}_P^{s,\gamma}(\mathbb{R}_+, \mathcal{E}) := \varprojlim_{k \in \mathbb{N}} \mathcal{H}_{P_k}^{s,\gamma}(\mathbb{R}_+, \mathcal{E}) \tag{2.24}$$

in the corresponding Fréchet topology of the projective limit.

Remark 2.22. The relation $u \in \mathcal{H}_P^{s,\gamma}(\mathbb{R}_+, \mathcal{E})$ with P being associated with (γ, Θ), $\Theta = (-\infty, 0]$, is equivalent with the existence of (unique) coefficients $c_{jk} \in E^\infty$, $0 \leq k \leq m_j$, such that for every $l \in \mathbb{R}_+$ there is an $N = N(l) \in \mathbb{N}$ with

$$\omega(r)\Big(u(r,x) - \sum_{j=0}^{N} \sum_{k=0}^{m_j} c_{jk} r^{-p_j} \log^k r\Big) \in \mathcal{H}^{s,\gamma+l}(\mathbb{R}_+, \mathcal{E}).$$

Similarly as in the "concrete" cone calculus (see [33]) we have the following continuity result:

Theorem 2.23. *For every $f \in M_R^\mu(\mathcal{E}, \widetilde{\mathcal{E}})$ the operator* (2.5) *restricts to a continuous operator*

$$\mathrm{op}_M^\gamma(f) : \mathcal{H}_P^{s,\gamma}(\mathbb{R}_+, \mathcal{E}) \to \mathcal{H}_Q^{s-\mu,\gamma}(\mathbb{R}_+, \widetilde{\mathcal{E}})$$

for every $s \in \mathbb{R}$ and every asymptotic type P with some resulting Q.

The case of Mellin symbols with variable coefficients is also of interest in the corner calculus. It is then adequate to assume $f(r,w) \in C^\infty(\overline{\mathbb{R}}_+, M_R^\mu(\mathcal{E}, \widetilde{\mathcal{E}}))$ and to consider operators $\omega \mathrm{op}_M^\gamma(f)\tilde{\omega}$ in combination with cut-off functions $\omega(r), \tilde{\omega}(r)$. Those induce continuous operators $\mathcal{H}_P^{s,\gamma}(\mathbb{R}_+, \mathcal{E}) \to \mathcal{H}_Q^{s-\mu,\gamma}(\mathbb{R}_+, \widetilde{\mathcal{E}})$ as well.

References

[1] D. Calvo, C.-I. Martin and B.-W. Schulze, Symbolic structures on corner manifolds, in "Microlocal Analysis and Asymptotic Analysis", RIMS Conf. dedicated to L. Boutet de Monvel, Kyoto, August 2004, Keio University, Tokyo, 2005, 22–35.

[2] D. Calvo and B.-W. Schulze, Operators on corner manifolds with exits to infinity, *J. Partial Differential Equations* **19**(2) (2006), 147–192.

[3] H.O. Cordes, *A global parametrix for pseudo-differential operators over* $\mathbb{R}^n$ *with applications*, preprint, SFB 72, *Universität Bonn*, 1976.

[4] N. Dines, Ellipticity of a class of corner operators, in *Pseudo-Differential Operators: Partial Differential Equations and Time-frequency Analysis*, Editors: L. Rodino, B.-W. Schulze and M.W. Wong, Fields Institute Communications Series **52**, American Mathematical Society, 2007, 131–169.

[5] Ju.V. Egorov and B.-W. Schulze, *Pseudo-Differential Operators, Singularities, Applications*, Operator Theory, Advances and Applications **93**, Birkhäuser Verlag, Basel, 1997.

[6] J.B. Gil, B.-W. Schulze and J. Seiler, Cone pseudodifferential operators in the edge symbolic calculus, *Osaka J. Math.* **37** (2000), 219–258.

[7] G. Harutyunyan and B.-W. Schulze, The relative index for corner singularities, Integral Equations Operators Theory **54**(3) (2006), 385–426.

[8] G. Harutyunyan and B.-W. Schulze, The Zaremba problem with singular interfaces as a corner boundary value problem, *Potential Analysis* **25**(4) (2006), 327–369.

[9] G. Harutyunyan and B.-W. Schulze, *Elliptic Mixed, Transmission and Singular Crack Problems*, European Mathematical Society, Zürich, 2008.

[10] G. Harutyunyan, B.-W. Schulze and I. Witt, Boundary value problems in the edge pseudo-differential calculus, preprint 2000/10, Institut für Mathematik, *Universität Potsdam*, 2000.

[11] I.L. Hwang, The L^2-boundedness of pseudodifferential operators, *Trans. Amer. Math. Soc.* **302** (1987), 55–76.

[12] D. Kapanadze and B.-W. Schulze, Pseudo-differential crack theory, *Mem. Diff. Equ. Math. Phys.* **22** (2001), 3–76.

[13] D. Kapanadze and B.-W. Schulze, *Crack Theory and Edge Singularities*, Kluwer Academic Publishers, Dordrecht, 2003.

[14] V.A. Kondratyev, Boundary value problems for elliptic equations in domains with conical points, *Trudy Mosk. Mat. Obshch.* **16** (1967), 209–292.

[15] T. Krainer The calculus of Volterra Mellin Pseudodifferential Operators with Operator-valued Symbols, in *Advances in Partial Differential Equations* (*Parabolicity, Volterra Calculus and Conical Singularities*) Editors: S. Albeverio, M. Demuth, E. Schrohe and B.-W. Schulze, Operator Theory: Advances and Applications **138** Birkhäuser Verlag, Basel, 2002, 47–91.

[16] H. Kumano-go, *Pseudo-Differential Operators*, The MIT Press, Cambridge, Massachusetts and London, England, 1981.

[17] L. Maniccia and B.-W. Schulze, An algebra of meromorphic corner symbols, *Bull. des Sciences Math.* **127**(1) (2003), 55–99.

[18] R.B. Melrose, *The Atiyah-Patodi-Singer Index Theorem*, Research Notes in Mathematics, A.K. Peters, Wellesley, 1993.

[19] R.B. Melrose and P. Piazza, Analytic K-theory on manifolds with corners, *Adv. Math.* **92**(1) (1992), 1–26.

[20] V. Nazaikinskij, A. Savin and B. Sternin, Elliptic theory on manifolds with corners: I. dual manifolds and pseudodifferential operators, preprint, arXiv: Math. OA/0608353v1.

[21] V. Nazaikinskij, A. Savin and B. Sternin, Elliptic theory on manifolds with corners: II. homotopy classification and K-homology, preprint, arXiv: Math. KT/0608354v1.

[22] V. Nazaikinskij, A. Savin and B. Sternin, On the homotopy classification of elliptic operators on stratified manifolds, preprint, arXiv: Math. KT/0608332v1.

[23] V. Nazaikinskij and B.Ju. Sternin, The index locality principle in elliptic theory, *Func. Anal. Appl.* **35** (2001), 37–52.

[24] V. Nistor, Pseudodifferential operators on non-compact manifolds and analysis on polyhedral domains, in *Spectral Geometry of Manifolds with Boundary and Decomposition of Manifolds*, Contemp. Math. **366**, Amer. Math. Soc., Providence, RI, 2005, 307–328.

[25] C. Parenti, Operatori pseudo-differenziali in $\mathbb{R}^n$ e applicazioni, *Annali Mat. Pura Appl.* **93**(4) (1972), 359–389.

[26] B.A. Plamenevskij, On the boundedness of singular integrals in spaces with weight, *Mat. Sb.* **76**(4) (1968), 573–592.

[27] B.A. Plamenevskij, *Algebras of Pseudo-Differential Operators*, Nauka, Moscow, 1986.

[28] S. Rempel and B.-W. Schulze, Mellin symbolic calculus and asymptotics for boundary value problems, *Seminar Analysis 1984/1985* (1985), 23–72.

[29] S. Rempel and B.-W. Schulze, Complete Mellin and Green symbolic calculus in spaces with conormal asymptotics, *Ann. Global Anal. Geom.* **4**(2) (1986), 137–224.

[30] S. Rempel and B.-W. Schulze, *Asymptotics for Elliptic Mixed Boundary Problems* (*Pseudo-Differential and Mellin Operators in Spaces with Conormal Singularity*), Math. Res. **50**, Akademie-Verlag, Berlin, 1989.

[31] E. Schrohe and B.-W. Schulze, Edge-degenerate boundary value problems on cones, in *Evolution Equations and Their Applications in Physical and Life Sciences,* Proc. Bad Herrenalb (Karlsruhe), 2000.

[32] B.-W. Schulze, Pseudo-differential operators on manifolds with edges, in *Symp. Partial Differential Equations, Holzhau* 1988, Teubner-Texte zur Mathematik **112**, Teubner, Leibzig, 1989, 259–287.

[33] B.-W. Schulze, *Pseudo-Differential Operators on Manifolds with singularities*, North-Holland, Amsterdam, 1991.

[34] B.-W. Schulze, The Mellin pseudo-differential calculus on manifolds with corners, in *Symp. Analysis in Domains and on Manifolds with Singularities, Breitenbrunn* 1990, Teubner-Texte zur Mathematik, **131**, Teubner, Leibzig, 1992, 208–289.

[35] B.-W. Schulze, *Pseudo-Differential Boundary Value Problems, Conical Singularities, and Asymptotics* Akademie-Verlag, Berlin, 1994.

[36] B.-W. Schulze, *Boundary Value Problems and Singular Pseudo-Differential Operators*, J. Wiley, Chichester, 1998.

[37] B.-W. Schulze, Operator algebras with symbol hierarchies on manifolds with singularities, in *Advances in Partial Differential Equations* (*Approaches to Singular Analysis*), Editors: J. Gil, D. Grieser and M. Lesch, Operator Theory: Advances and Applications, Birkhäuser Verlag, Basel, 2001, 167–207.

[38] B.-W. Schulze, Operators with symbol hierarchies and iterated asymptotics, *Publ. RIMS, Kyoto University* **38**(4) (2002), 735–802.

[39] B.-W. Schulze, The structure of operators on manifolds with polyhedral singularities, preprint 2006/05, Institut für Mathematik, *Universität Potsdam*, 2006, arXiv: Math. AP/0610618.

[40] J. Seiler, *Pseudodifferential Calculus on Manifolds with Non-Compact Edges*, Ph.D. Thesis, University of Potsdam, 1998.

[41] J. Seiler, Continuity of edge and corner pseudo-differential operators, *Math. Nachr.* **205** (1999), 163–182.

[42] M.A. Shubin, Pseudodifferential operators in $\mathbb{R}^n$, *Dokl. Akad. Nauk SSSR* **196** (1971), 316–319.

Jamil Abed and Bert-Wolfgang Schulze
Universität Potsdam
Institut für Mathematik
am Neuen Palais 10
D-14469 Potsdam, Germany
e-mail: abed@math.uni-potsdam.de
schulze@math.uni-potsdam.de

Operator Theory:
Advances and Applications, Vol. 189, 107–116

Ellipticity of Fredholm Pseudo-Differential Operators on $L^p(\mathbb{R}^n)$

Aparajita Dasgupta

Abstract. Based on the works of Grieme and Schulze [8], it is shown in this paper that if a pseudo-differential operator with symbol in S^m, $-\infty < m < \infty$, is Fredholm on $L^p(\mathbb{R}^n), 1 < p < \infty$, then the pseudo-differential operator is elliptic. The basic idea is to construct an isometric operator R_λ, $\lambda \in \mathbb{R} \setminus \{0\}$, on $L^p(\mathbb{R}^n)$ in order to prove the ellipticity of the Fredholm pseudo-differential operator with symbol in S^0. This result is then generalized for arbitrary symbol classes.

Mathematics Subject Classification (2000). Primary 35S05, 47G30; Secondary 47A53.

Keywords. Pseudo-differential operators, SG pseudo-differential operators, Sobolev spaces, Fredholm operators and elliptic pseudo-differential operators.

1. Introduction

For $m \in (-\infty, \infty)$, let S^m be the set of all functions in $C^\infty(\mathbb{R}^{2n})$ such that for all multi-indices α and β, there exists a positive constant $C_{\alpha,\beta}$ for which

$$|(D_x^\alpha D_\xi^\beta \sigma)(x, \xi)| \leq C_{\alpha,\beta}(1 + |\xi|)^{m-|\beta|}, \quad x, \xi \in \mathbb{R}^n.$$

Let $\sigma \in S^m$. Then we define the corresponding pseudo-differential operator T_σ on the Schwartz space $\mathcal{S}$ by

$$(T_\sigma \phi)(x) = (2\pi)^{-n/2} \int_{\mathbb{R}^n} e^{ix\cdot\xi} \sigma(x, \xi) \hat{\phi}(\xi)\, d\xi, \quad x \in \mathbb{R}^n.$$

It can be proved easily that $T_\sigma : \mathcal{S} \to \mathcal{S}$ is a continuous linear mapping. In fact, as is shown in Theorem 11.9 in the book [20] by Wong, T_σ can be extended to a bounded linear operator from $H^{s,p}$ into $H^{s-m,p}$ for $-\infty < s < \infty$ and $1 < p < \infty$.

This research has been supported by the Natural Sciences and Engineering Research Council of Canada.

The aim of this paper is to prove that if $T_\sigma : H^{s,p} \to H^{s-m,p}$ is Fredholm for some s and p with $-\infty < s < \infty$ and $1 < p < \infty$, then σ must be elliptic in the sense that there exist positive constants C and R such that

$$|\sigma(x,\xi)| \geq C(1+|\xi|)^m, \quad x,\xi \in \mathbb{R}^n.$$

That Fredholmness does not follow from ellipticity can be seen from the Laplacian Δ on $\mathbb{R}^n$. It is well known that Δ is elliptic, but Δ is not a Fredhlom operator from $H^{2,p}$ into $L^p(\mathbb{R}^n)$ for $1 < p < \infty$.

In this section we give an introduction to the SG pseudo-differential operators, which are also known as pseudo-differential operators of global type. They are also known as pseudo-differential operators with exit behavior, as in [7, 16]. A discussion of SG pseudo-differential operators and related topics can also be found in [3, 4, 10, 11, 12, 13, 16] and the references therein.

The main techniques that we use in this paper come from the study of SG pseudo-differential operators in the Ph.D. thesis [8] of Grieme, which we now recall.

Let $m_1, m_2 \in (-\infty, \infty)$. Then S^{m_1,m_2} is the set of all functions in $C^\infty(\mathbb{R}^{2n})$ such that for all multi-indices α and β, there exist positive constants $C_{\alpha,\beta}$ for which

$$|(D_x^\alpha D_\xi^\beta \sigma)(x,\xi)| \leq C_{\alpha,\beta} \langle x\rangle^{m_2-|\alpha|} \langle \xi\rangle^{m_1-|\beta|}, \; x,\xi \in \mathbb{R}^n,$$

where $\langle . \rangle$ denotes the function on $\mathbb{R}^N$ given by

$$\langle z\rangle = (1+|z|^2)^{1/2}, \; z \in \mathbb{R}^N,$$

for every positive integer N. For all $\sigma \in S^{m_1,m_2}$ we denote

$$p_{\alpha,\beta}^{m_1,m_2}(\sigma) = \sup_{(x,\xi)\in\mathbb{R}^{2n}} \left\{ |D_x^\alpha D_\xi^\beta \sigma(x,\xi)| \langle x\rangle^{-m_2+|\alpha|} \langle \xi\rangle^{-m_1+|\beta|} \right\},$$

for all multi-indices α and β. A function in S^{m_1,m_2} is said to be a SG symbol of orders m_1, m_2. It is clear that if $\sigma \in S^{m_1,m_2}$, then $\sigma \in S^{m_1}$, where S^{m_1} is the class of symbols of classical pseudo-differential operators studied exclusively in the book [18] by Wong. For $\sigma \in S^{m_1,m_2}$, we then define the pseudo-differential operator T_σ with symbol σ by

$$(T_\sigma \phi)(x) = (2\pi)^{-n/2} \int_{\mathbb{R}^n} e^{ix\cdot\xi} \sigma(x,\xi) \hat{\phi}(\xi) d\xi, \; x \in \mathbb{R}^n,$$

for all functions ϕ in the Schwartz space $\mathcal{S}$, where

$$\hat{\phi}(\xi) = (2\pi)^{-n/2} \int_{\mathbb{R}^n} e^{-ix\cdot\xi} \phi(x) dx, \; \xi \in \mathbb{R}^n.$$

It can be easily shown that $T_\sigma : \mathcal{S} \to \mathcal{S}$ is a continuous linear mapping. Further information and formal properties of SG pseudo-differential operators can be found in [6]. The aim of this paper is to prove ellipticity of Fredholm pseudo-differential operators on $L^p(\mathbb{R}^n)$. In Section 2, the L^p-Sobolev spaces for SG pseudo-differential operators is discussed. In Section 3, the elliptic and Fredholm SG pseudo-differential operators are recalled and the Fredholmness of elliptic SG pseudo-differential operators is given. Detail works on this can be found in [6]. In Section 4, an isometric operator R_λ on $L^p(\mathbb{R}^n)$ is introduced in order to prove the ellipticity of the Fredholm pseudo-differential operator T_σ, with symbol $\sigma \in S^0$. Later this result is generalized for symbols in arbitrary symbol classes S^m, $m \in (-\infty, \infty)$, which again in turn used to prove the ellipticity of the Fredholm SG pseudo-differential operators on $L^p(\mathbb{R}^n)$. Similar results for SG pseudo-differential operators on $L^2(\mathbb{R}^n)$ can be found in [8].

2. Sobolev spaces

For $s_1, s_2 \in (-\infty, \infty)$ let J_{s_1,s_2} be the Bessel potential of orders s_1, s_2 defined by

$$J_{s_1,s_2} = T_{\sigma_{s_1,s_2}},$$

where

$$\sigma_{s_1,s_2}(x,\xi) = \langle x\rangle^{-s_2}\langle\xi\rangle^{-s_1}, \ (x,\xi) \in \mathbb{R}^n.$$

Obviously then $\sigma_{s_1,s_2} \in S^{-s_1,-s_2}$. It can be shown easily that for $-\infty < s_1, s_2 < \infty$, the mapping $J_{s_1,s_2} : \mathcal{S}' \to \mathcal{S}'$ is a bijection and

$$J_{s_1,s_2}^{-1} = J_{-s_1,0}J_{0,-s_2} \tag{2.1}$$

and J_{s_1,s_2}^{-1} is a pseudo-differential operator of orders $-s_1, -s_2$.

For $1 < p < \infty$ and $-\infty < s_1, s_2 < \infty$, we define the L^p-Sobolev space $H^{s_1,s_2,p}$ of orders s_1, s_2 by

$$H^{s_1,s_2,p} = \left\{u \in \mathcal{S}' : J_{-s_1,-s_2}u \in L^p(\mathbb{R}^n)\right\}.$$

Then $H^{s_1,s_2,p}$ is a Banach space in which the norm $\|.\|_{s_1,s_2,p}$ is given by

$$\|u\|_{s_1,s_2,p} = \|J_{s_1,s_2}u\|_{L^p(\mathbb{R}^n)}, \ u \in H^{-s_1,-s_2,p},$$

where $\|.\|_{L^p(\mathbb{R}^n)}$ is the norm in $L^p(\mathbb{R}^n)$. Obviously, $H^{0,0,p} = L^p(\mathbb{R}^n)$. Then we have,

Theorem 2.1. *Let $\sigma \in S^{m_1,m_2}$, $-\infty < m_1, m_2 < \infty$. Then for $1 < p < \infty$ and $-\infty < s_1, s_2 < \infty$, $T_\sigma : H^{s_1,s_2,p} \to H^{s_1-m_1,s_2-m_2,p}$ is a bounded linear operator.*

3. Elliptic and Fredholm SG pseudo-differential operators

Let $\sigma \in S^{m_1,m_2}$, $-\infty < m_1, m_2 < \infty$. Then σ is said to be elliptic if there exists positive constants C and R such that

$$|\sigma(x,\xi)| \geq C\langle x\rangle^{m_2}\langle\xi\rangle^{m_1}, \ |x|^2 + |\xi|^2 \geq R.$$

Then the pseudo-differential operator T_σ with the elliptic symbol σ is called the elliptic operator.

Theorem 3.1. *Let $\sigma \in S^{m_1,m_2}$, $-\infty < m_1, m_2 < \infty$, be elliptic. Then there exists a symbol τ in $S^{-m_1,-m_2}$ such that*

$$T_\sigma T_\tau = I + S \quad \text{and} \quad T_\tau T_\sigma = I + R,$$

where R and S are infinitely smoothing in the sense that they are SG pseudo-differential operators with symbols in $\bigcap_{k_1,k_2\in\mathbb{R}} S^{k_1,k_2}$.

The SG pseudo-differential operator T_τ in Theorem 3.1 is known as a parametrix of T_σ.

Now let us first recall from the book [14] by Schechter, a closed linear operator A from a complex Banach space X into a complex Banach space Y with dense domain $D(A)$ is said to be Fredholm if the range $R(A)$ of A is a closed subspace of Y, the null space $N(A)$ of A and the null space $N(A^t)$ of the true adjoint A^t of A are finite dimensional. For a Fredholm operator A, the index $i(A)$ of A is defined by

$$i(A) = \dim N(A) - \dim N(A^t).$$

The following theorem which is known as the Atkinson theorem also gives another equivalent definition of the Fredholm operators.

Theorem 3.2. *Let A be a closed linear operator from a complex Banach space X into a complex Banach space Y with a dense domain $D(A)$. Then A is Fredholm if and only if we can find a bounded linear operator $B : Y \to X$, a compact operator $K_1 : X \to X$ and a compact operator $K_2 : Y \to Y$ such that $BA = I + K_1$ on $D(A)$ and $AB = I + K_2$ on Y.*

All the results on Fredholm operators hitherto can be found in the book [14] by Schechter. The first result in this section is the following theorem which can be found in [6].

Theorem 3.3. *Let $\sigma \in S^{m_1,m_2}$, $m_1, m_2 > 0$, be elliptic. Then for $1 < p < \infty$, $T_{\sigma,0}$ is a Fredholm operator on $L^p(\mathbb{R}^n)$ with domain $H^{m_1,m_2,p}$. Furthermore, if $\sigma \in S^{0,0}$ is elliptic, then the bounded linear operator $T_\sigma : L^p(\mathbb{R}^n) \to L^p(\mathbb{R}^n)$ is Fredholm.*

4. Ellipticity of Fredholm pseudo-differential operators

Here we want to prove that the Fredholm pseudo-differential operators is elliptic on $L^p(\mathbb{R}^n)$ and for that we need some technical preparations.

Definition 4.1. *Let $\lambda > 0$, $x_0, \xi_0 \in \mathbb{R}^n \cup \{0\}$ and $\tau \in \mathbb{R}^+ \cup \{0\}$. For every $1 < p < \infty$ we define an operator, $R_\lambda : L^p(\mathbb{R}^n) \to L^p(\mathbb{R}^n)$ by,*

$$(R_\lambda(x_0, \xi_0)u)(x) = \lambda^{\tau n/p} e^{i\lambda x\xi_0} u(\lambda^\tau (x - x_0)).$$

Then it is easy to prove,

Theorem 4.2. *The operator $R_\lambda(x_0, \xi_0) : L^p(\mathbb{R}^n) \to L^p(\mathbb{R}^n)$ is an isometrical invertible operator and the inverse is given by*

$$(R_\lambda^{-1}(x_0, \xi_0)u)(x) = \lambda^{-\tau n/p} e^{-i\lambda(x_0+\lambda^{-\tau}x)\xi_0} u(x_0 + \lambda^{-\tau} x).$$

Proof. Fix $x_0, \xi_0 \in \mathbb{R}^n$ and $R_\lambda := R_\lambda(x_0, \xi_0)$. Let $u \in L^p(\mathbb{R}^n)$. Then

$$\|R_\lambda u\|_p^p = \int_{\mathbb{R}^n} |R_\lambda u(x)|^p dx = \int_{\mathbb{R}^n} \lambda^{\tau n} |u(\lambda^\tau(x - x_0))|^p dx. \tag{4.2}$$

Substituting $y = \lambda^\tau(x - x_0)$ in (4.2), we have

$$\|R_\lambda u\|_p = \|u\|_p.$$

So R_λ is an isometric operator on $L^p(\mathbb{R}^n)$. Now let

$$R_\lambda u(x) = v(x).$$

Then from the definition (4.1),

$$v(x) = \lambda^{\tau n/p} e^{i\lambda x \xi_0} u(\lambda^\tau(x - x_0)). \tag{4.3}$$

Again substituting $\lambda^\tau(x - x_0) = y$ in (4.3), we have

$$u(y) = \lambda^{-\tau n/p} e^{-i\lambda(x_0 + \lambda^{-\tau} y)\xi_0} v(x_0 + \lambda^{-\tau} y).$$

Then

$$R_\lambda^{-1} v(y) = \lambda^{-\tau n/p} e^{-i\lambda(x_0 + \lambda^{-\tau} y)\xi_0} v(x_0 + \lambda^{-\tau} y).$$ □

Theorem 4.3. *For all $u \in L^p(\mathbb{R}^n)$ and $v \in L^{p'}(\mathbb{R}^n)$, where $1/p + 1/p' = 1$, $1 < p < \infty$ and $\tau > 0$, $(R_\lambda(x_0, \xi_0)u, v)_2 \to 0$ as $\lambda \to \infty$.*

Proof. Fix $x_0, \xi_0 \in \mathbb{R}^n$ and let $R_\lambda := R_\lambda(x_0, \xi_0)$. Furthermore, let $u, v \in C_0^\infty(\mathbb{R}^n)$. Then we have

$$\begin{aligned}
|(R_\lambda u, v)| &\le \lambda^{\tau n/p} \int_{\mathbb{R}^n} |u(\lambda^\tau(x - x_0))| |\overline{v(x)}| dx \\
&= \lambda^{\tau n/p} \int_{\mathbb{R}^n} \lambda^{-\tau n} |u(y)| |\overline{v(x_0 + \lambda^{-\tau} y)}| dy \\
&= \lambda^{-\tau n/p'} \int_{\mathbb{R}^n} |\overline{v(x_0 + \lambda^{-\tau} y)}| |u(y)| dy \\
&\le \lambda^{-\tau n/p'} \sup_{x \in \mathbb{R}^n} |v(y)| \int_{\mathbb{R}^n} |u(y)| dy.
\end{aligned} \tag{4.4}$$

So then $(R_\lambda u, v) \to 0$ as $\lambda \to \infty$ for all $u, v \in C_0^\infty(\mathbb{R}^n)$. Now $C_0^\infty(\mathbb{R}^n)$ is dense in $L^p(\mathbb{R}^n)$, so for $\epsilon > 0$ there exists $J \in \mathbb{N}$ such that,

$$|(R_\lambda u, v) - (R_\lambda u_j, v_j)| \le \epsilon/2,$$

for all $j \ge J$, where $u_j, v_j \in C_0^\infty(\mathbb{R}^n)$ and $u \in L^p(\mathbb{R}^n), v \in L^{p'}(\mathbb{R}^n)$. Then again using the fact that $(R_\lambda u_j, v_j) \to 0$ as $\lambda \to \infty$ for $u_j, v_j \in C_0^\infty(\mathbb{R}^n)$ the result follows immediately. □

Lemma 4.4. *For every T_σ, $\sigma \in S^{m_1, m_2}$,*

$$R_\lambda^{-1}(x_0, \xi_0) T_\sigma R_\lambda(x_0, \xi_0) = T_{\sigma_\lambda} \tag{4.5}$$

holds with

$$\sigma_\lambda(x, \eta) = \sigma(x_0 + \lambda^{-\tau} x, \lambda \xi_0 + \lambda^\tau \eta). \tag{4.6}$$

Moreover if $\sigma \in S^{0,0}$, $\lambda \geq 1$ *and* $0 \leq \tau \leq 1/2$, *then the estimate*

$$|\partial_x^\alpha \partial_\xi^\beta \sigma_\lambda(x,\eta)| \leq C p_{\alpha,\beta}(\sigma) \frac{\langle \eta \rangle^{|\beta|}}{|\xi_0|^{|\beta|}} \lambda^{-\tau|\alpha|} \lambda^{-(1-2\tau)|\beta|} \tag{4.7}$$

holds for all $x, \eta \in \mathbb{R}^n$ *and a suitable constant* $c > 0$. *Here* $p_{\alpha,\beta}$ *denotes the corresponding semi-norm on* $S^{0,0}$.

Proof. Fix $x_0, \xi_0 \in \mathbb{R}^n$ and let $R_\lambda := R_\lambda(x_0, \xi_0)$. We first write $R_\lambda^{-1} T_\sigma R_\lambda u(x)$ in the form

$$e^{-i\lambda(x_0+\lambda^{-\tau}x)\xi_0}(2\pi)^{-n} \iint e^{i\lambda(x_0+\lambda^{-\tau}x-y)\xi} \sigma(x_0+\lambda^{-\tau}x, \xi) e^{i\lambda y \xi_0} u(\lambda^\tau(y-x_0)) dy d\xi.$$

Then substituting $z := \lambda^\tau(y - x_0)$, the above expression takes the form

$$\lambda^{-\tau n}(2\pi)^{-n} \iint e^{i\lambda^{-\tau}(\xi-\lambda\xi_0)(x-z)} \sigma(x_0 + \lambda^{-\tau}x, \xi) u(z) dz d\xi. \tag{4.8}$$

Now if we set $\eta := \lambda^{-\tau}(\xi - \lambda\xi_0)$ in (4.8), the expression above turns to

$$(2\pi)^{-n} \iint e^{i(x-z)\eta} \sigma(x_0 + \lambda^{-\tau}x, \lambda\xi_0 + \lambda^\tau \eta) u(z) dz d\xi. \tag{4.9}$$

Then by (4.6) the expression in (4.9) equals to $T_{\sigma_\lambda} u(x)$. Hence (4.5) holds. Now then by using (4.4), chain rule and Peetre's inequality we have,

$$\begin{aligned} |\partial_x^\alpha \partial_\xi^\beta \sigma_\lambda(x,\eta)| &= |\partial_x^\alpha \partial_\xi^\beta \sigma(x_0 + \lambda^{-\tau}x, \lambda\xi_0 + \lambda^\tau\eta)| \\ &= |(\partial_x^\alpha \partial_\xi^\beta \sigma)(x_0 + \lambda^{-\tau}x, \lambda\xi_0 + \lambda^\tau\eta) \lambda^{-\tau|\alpha|} \lambda^{\tau|\beta|}| \\ &\leq c p_{\alpha,\beta}(\sigma) |\xi_0|^{|\beta|} \langle \lambda\xi_0 + \lambda^\tau \eta \rangle^{-|\beta|} \lambda^{-\tau|\alpha|} \lambda^{\tau|\beta|} \\ &\leq c p_{\alpha,\beta}(\sigma) |\xi_0|^{-|\beta|} \langle \eta \rangle^{|\beta|} \lambda^{\frac{(4\tau-2)|\beta|}{2}} \lambda^{-\tau|\alpha|}. \end{aligned} \tag{4.10}$$

Hence (4.7) is proved. □

The next lemma is an immediate consequence of Lemma 4.4 and Taylor's formula.

Lemma 4.5. *Let* $\sigma \in S^{0,0}$, $\lambda \geq 1$, *and* $0 < \tau < 1/2$. *Then*

- *For* $|\alpha| + |\beta| > 0$,

$$\partial_x^\alpha \partial_\xi^\beta \sigma_\lambda(x,\eta) \to 0$$

  as $\lambda \to \infty$, *uniformly on every compact* $K \subset \mathbb{R}^{2n}$.
- *If* $\lambda \to \infty$, *then*

$$\sigma_\lambda(x,\eta) - \sigma_\lambda(0,0) \to 0$$

  uniformly on every compact $K \subset \mathbb{R}^{2n}$.

Lemma 4.4 and Lemma 4.5 are also valid for $\sigma \in S^m$, $m \in (-\infty, \infty)$, moreover for $\sigma \in S^0$ (4.7) is also true. And in that case $p_{\alpha,\beta}$ is the corresponding semi-norm on the space S^0.

Remark 4.6. *Let $\sigma \in S^0$ and $0 < \tau < 1/2$. Then*

$$A_\lambda := \{\sigma_\lambda : \lambda \geq 1\} \subseteq C^\infty(\mathbb{R}^{2n})$$

is a bounded set in $C^\infty(\mathbb{R}^{2n})$. Now since in $C^\infty(\mathbb{R}^{2n})$ every bounded sequence has a convergent subsequence so without any loss of generality let $\{\sigma_{k_n}\}$ be a convergent subsequence of $\{\sigma_{\lambda_n}\}$ and

$$\sigma_{\lambda_{k_n}} \to \sigma_\infty$$

as $n \to \infty$ in $C^\infty(\mathbb{R}^{2n})$ under the topology of corresponding semi-norms given by

$$p_N(\sigma_\lambda) = \sup_{0 \leq |\alpha| \leq N} \sup_{x \in K_N} \{|D^\alpha \sigma_\lambda(x)| : \lambda \geq 1\},$$

where $K_i \subseteq K_{i+1}$ are compact sets in $\mathbb{R}^{2n}$ and $\mathbb{R}^{2n} = \cup_{i=1}^\infty K_i$. Then from Lemma 4.5 we have

$$\partial_x^\alpha \partial_\eta^\beta \sigma_\infty = 0,$$

in every compact set $K \subseteq \mathbb{R}^{2n}$. Hence σ_∞ is a constant in $\mathbb{R}^{2n}$.

Lemma 4.7. *Let $\sigma \in S^0$ and $0 < \tau < 1/2$. Moreover, let $\sigma_\infty \in \mathbb{C}$ and $\lambda_k \to \infty$ such that*

$$\sigma_{\lambda_k} \to \sigma_\infty$$

in $C^\infty(\mathbb{R}^{2n})$. Then

$$R_{\lambda_k}^{-1} T_\sigma R_{\lambda_k} u \to \sigma_\infty u \tag{4.11}$$

in $L^p(\mathbb{R}^n)$ for all u in $L^p(\mathbb{R}^n)$.

Proof. Fix $x_0, \xi_0 \in \mathbb{R}^n$ and let $R_{\lambda_k} := R_{\lambda_k}(x_0, \xi_0)$. Now since

$$\|R_{\lambda_k}^{-1} T_\sigma R_{\lambda_k}\| \leq \|R_{\lambda_k}^{-1}\| \|T_\sigma\| \|R_{\lambda_k}\| = \|T_\sigma\|,$$

so it is sufficient to show (4.11) on a dense subset of $L^p(\mathbb{R}^n)$. Let $u \in \mathcal{S}$. Then

$$|e^{ix\eta} \sigma_\lambda(x, \eta) \hat{u}(\eta)| \leq c p_{0,0}(\sigma) |\hat{u}(\eta)|. \tag{4.12}$$

By the dominated convergence theorem and (4.11) we have

$$(R_{\lambda_k}^{-1} T_\sigma R_{\lambda_k}) u(x) \to \sigma_\infty u(x), \tag{4.13}$$

pointwise for all $x \in \mathbb{R}^n$ as $\lambda_k \to \infty$. Moreover for all $l \in \mathbb{N}$, using (4.5), (4.6) and (4.7) and by integration by parts,

$$|\langle x \rangle^{2l} R_{\lambda_k}^{-1} T_\sigma R_{\lambda_k} u(x)| \leq C \tag{4.14}$$

for all $\lambda \geq 1$. Hence

$$|(R_{\lambda_k}^{-1} T_\sigma R_{\lambda_k})|^p \leq C' \langle x \rangle^{-2lp}. \tag{4.15}$$

Now if, $2lp > n$ then $\langle x \rangle^{-2lp} \in L^1(\mathbb{R}^n)$. So

$$|(R_{\lambda_k}^{-1} T_\sigma R_{\lambda_k} - \sigma_\infty) u(x)| \leq C_1 \langle x \rangle^{-2lp}. \tag{4.16}$$

Now then (4.11) follows from (4.14), (4.17) and the dominated convergence theorem. □

The following theorem is one of the main theorems of this paper.

Theorem 4.8. *Let $\sigma \in S^0$ be such that*

$$T_\sigma : L^p(\mathbb{R}^n) \to L^p(\mathbb{R}^n),$$

$1 < p < \infty$, is a Fredholm operator. Then there are constants $C > 0$ and $R > 0$ such that, $|\sigma(x,\xi)| \geq C$ for all $x \in \mathbb{R}^n$ and $|\xi| \geq R$.

Proof. Since T_σ is a Fredholm operator so there exist operators $S \in B(L^p(\mathbb{R}^n))$ and $K \in \mathcal{K}(L^p(\mathbb{R}^n))$ such that $ST_\sigma = I + K$, K is compact. Consider the set,

$$M = \left\{\xi \in \mathbb{R}^n : \left(\exists x \in \mathbb{R}^n : |\sigma(x,\xi)| \leq \frac{1}{2\|S\|}\right)\right\}.$$

Now if M is bounded then the proof is obvious. Suppose M is not bounded. Then there exists a sequence (x_k, ξ_k) in $\mathbb{R}^{2n}$ such that $|\xi_k| \to \infty$ and $|\sigma(x_k,\xi_k)| \leq 1/2\|S\|$. Without loss of generality we get an $\sigma_\infty \in \mathbb{C}$ such that,

$$\sigma(x_k,\xi_k) \to \sigma_\infty, \ |\sigma_\infty| \leq \frac{1}{2\|S\|} \tag{4.17}$$

as $|\xi_k| \to \infty$. Let

$$\lambda_k := |\xi_k|, R_k := R_{\lambda_k}\left(x_k, \frac{\xi_k}{|\xi_k|}\right). \tag{4.18}$$

Then from Lemma 4.4

$$R_k^{-1} T_\sigma R_k = T_{\sigma_k} \tag{4.19}$$

$$\sigma_k(x,\eta) = \sigma(x_k + \lambda_k^\tau x, \xi_k + \lambda_k^\tau \eta). \tag{4.20}$$

Now since (4.7) holds uniformly for all $(x_0,\xi_0) \in \mathbb{R}^n \times \{\xi : |\xi| = 1\}$, we have

$$|\partial_x^\alpha \partial_\eta^\beta \sigma_k(x,\eta)| \leq c p_{\alpha,\beta}(\sigma) \langle\eta\rangle^{|\beta|} \lambda_k^{-\tau|\alpha|} \lambda_k^{-(1-2\tau)|\beta|} \tag{4.21}$$

We denote

$$\sigma_k(0,0) = \sigma(x_k,\xi_k) = \sigma_k^\infty. \tag{4.22}$$

Then using Taylor's formula it can be shown,

$$\sigma_k(x,\eta) - \sigma_k^\infty \to 0 \tag{4.23}$$

uniformly for (x,η) in a compact set $K \subset \mathbb{R}^{2n}$ as $k \to \infty$. Now let $u \in \mathcal{S}$. Using the dominated convergence theorem and proceeding similarly as Lemma 4.7 we have,

$$|(R_k^{-1} T_\sigma R_k u(x) - \sigma_k^\infty u(x))|^p \leq C \langle x\rangle^{-2lp}, \tag{4.24}$$

where $l \in \mathbb{N}$. So $\langle x\rangle^{-2lp} \in L^p(\mathbb{R}^n)$ when $2lp > n$. Again by the dominated convergence theorem,

$$R_k^{-1} T_\sigma R_k u(x) \to \sigma_k^\infty u(x) \tag{4.25}$$

in $L^p(\mathbb{R}^n)$ as $k \to \infty$. By (4.22) and (4.26), we have

$$R_k^{-1} T_\sigma R_k u(x) \to \sigma_\infty u(x) \tag{4.26}$$

in $L^p(\mathbb{R}^n)$ as $k \to \infty$. For $u \in L^p(\mathbb{R}^n)$, $u \neq 0$,

$$\begin{aligned} 0 < \|u\|_p = \|(ST_\sigma + K)R_k u\|_p &\leq \|ST_\sigma R_k u\|_p + \|KR_k u\|_p \\ &\leq \|R_k^{-1} T_\sigma R_k u\|_p \|S\| + \|KR_k u\|_p. \end{aligned} \tag{4.27}$$

Now by Lemma 4.7 and using the fact that K is a compact operator and Lemma 4.3 it follows that $\|KR_k u\|_p \to 0$ as $k \to \infty$. Then from (4.28)

$$\|u\|_p \le \|S\| |\sigma_\infty| \|u\|_p. \tag{4.28}$$

Then, (4.18) and (4.29) yield the contradiction

$$\frac{1}{\|S\|} \le |\sigma_\infty| \le \frac{1}{2\|S\|},$$

which completes the proof. □

The above theorem (4.8) can be generalized for all values of $m \in (-\infty, \infty)$.

Theorem 4.9. *Let $\sigma \in S^m$, $-\infty < m < \infty$ and let $T_\sigma : H^{s,p} \to H^{s-m,p}$ a Fredholm operator. Then T_σ is an elliptic operator.*

Proof. $T_\sigma : H^{s,p} \to H^{s-m,p}$, $J_{-s} : H^{s,p} \to L^p(\mathbb{R}^n)$ and $J_{m-s} : H^{s-m,p} \to L^p(\mathbb{R}^n)$ are bounded linear operators. Here J_s, the Bessel potential, is a pseudo-differential operator with symbol $\sigma_s \in S^{-s}$ with $\sigma_s(\xi) = \langle\xi\rangle^{-s}$. Let

$$J_{m-s} T_\sigma J_s = T_\tau. \tag{4.29}$$

Then

$$T_\tau : L^p(\mathbb{R}^n) \to L^p(\mathbb{R}^n),$$

where $\tau \in S^0$. Now since J_s is one-one and onto so J_s is Fredholm and elliptic for all $s \in (-\infty, \infty)$. So by Theorem 3.3 T_τ is elliptic. By (4.30) and the fact that J_s, $s \in \mathbb{R}^n$, is bijective it follows immediately that T_σ is elliptic. □

The following theorem which is a simple consequence of Theorem 3.3 and the main theorems of this paper, Theorem 4.9 and Theorem 4.8, proves the ellipticity of the Fredholm SG pseudo-differential operators on $L^p(\mathbb{R}^n)$.

Theorem 4.10. *Let $\sigma \in S^{m_1,m_2}$, $m_1, m_2 \in (-\infty, \infty)$ and let*

$$T_\sigma : H^{s_1,s_2,p} \to H^{s_1-m_1,s_2-m_2,p}$$

is a Fredholm operator. Then T_σ is elliptic. Moreover if, $\sigma \in S^{0,0}$ then

$$T_\sigma : L^p(\mathbb{R}^n) \to L^p(\mathbb{R}^n)$$

is Fredholm if and only if it is elliptic.

References

[1] F.V. Atkinson, The normal solubility of linear equations in normed spaces, *Mat. Sbornik N.S* **28**(70) (1951), 3–14. (in Russian)

[2] P. Boggiatto, E. Buzano and L. Rodino, *Global Hypoellipticity and Spectral Theory*, Akademie-Verlag, 1996.

[3] M. Cappiello and L. Rodino, SG-pseudo-differential operators and Gelfand–Shilov spaces, *Rocky Mountain J. Math.* **36** (2006), 1117–1148.

[4] H.O. Cordes, *The Technique of Pseudodifferential Operators*, Cambridge University Press, 1995.

[5] S. Coriasco and L. Rodino, Cauchy problem for SG-hyperbolic equations with constant multiplicities, *Ricerche Mat.* Suppl. Vol. **XLVIII** (1999), 25–43.

[6] A. Dasgupta and M.W. Wong, Spectral theory of SG pseudo-differential operators on $L^p(\mathbb{R}^n)$, *Studia Math.* **187** (2008), 185–197.

[7] Y.V. Egorov and B.-W. Schulze, *Pseudo-Differential Operators, Singularities, Applications,* Birkhäuser, 1997.

[8] U. Grieme, *Pseudo-Differential Operators with Operator-Valued Symbols on Non-Compact Manifolds,* Ph.D. Thesis, Universität Potsdam, 1998.

[9] V.V. Grushin, Pseudo-differential operators on $\mathbb{R}^n$ with bounded symbols, *Funct. Anal. Appl.* **4** (1970), 202–212.

[10] G. Harutyunyan and B.-W. Schulze, *Elliptic Mixed, Transmission and Singular Crack Problems*, EMS Tracts in Mathematics **4**, European Mathematical Society, 2008.

[11] F. Nicola, K-theory of SG-pseudo-differential algebras, *Proc. Amer. Math. Soc.* **131** (2003), 2841–2848.

[12] F. Nicola and L. Rodino, SG pseudo-differential operators and weak hyperbolicity, *Pliska Stud. Math. Bulgar.* **15** (2002), 5–19.

[13] C. Parenti, Operatori pseudo-differentiali in $\mathbb{R}^n$ e applicazioni, *Ann. Mat. Pura Appl.* **93** (1972), 359–389.

[14] M. Schechter, *Spectra of Partial Differential Operators,* Second Edition, North-Holland, 1986.

[15] M. Schechter, *Principles of Functional Analysis,* Second Edition, American Mathematical Society, 2002.

[16] B.-W. Schulze, *Boundary Value Problems and Singular Pseudo-Differential Operators,* Wiley, 1998.

[17] M. Shubin, *Pseudodifferential Operators and Spectral Theory,* Springer-Verlag, 1987.

[18] M.W. Wong, Fredholm pseudo-differential operators on weighted Sobolev spaces, *Ark. Mat.* **21** (1983), 271–282.

[19] M.W. Wong, Spectral theory of pseudo-differential operators, *Adv. in Appl. Math.* **15** (1994), 437–451.

[20] M.W. Wong, *An Introduction to Pseudo-Differential Operators,* Second Edition, World Scientific, 1999.

[21] M.W. Wong, *M-elliptic pseudo-differential operators on* $L^p(\mathbb{R}^n)$, *Math. Nachr.* **279** (2006), 319–326.

Aparajita Dasgupta
Department of Mathematics and Statistics
York University
4700 Keele Street
Toronto, Ontario M3J 1P3
Canada
e-mail: adgupta@mathstat.yorku.ca

Operator Theory:
Advances and Applications, Vol. 189, 117–136

Hyperbolic Systems with Discontinuous Coefficients: Generalized Wavefront Sets

Michael Oberguggenberger

Abstract. We study linear hyperbolic systems of pseudo-differential equations with nonsmooth, possibly discontinuous symbols and distributional data. This note initiates an approach whereby solutions are constructed in the dual of the Colombeau algebra of generalized functions. We compute the generalized wave front set of the solution to a transport equation with discontinuous propagation speed and delta functions as initial data. The generalized wave front set turns out to have a more refined and informative structure than the wavefront set of the corresponding distributional limit.

Mathematics Subject Classification (2000). Primary 35D05, 35S10, 46F30; Secondary 35D10, 35L45.

Keywords. Pseudo-differential equations, algebras of generalized functions, duality methods, propagation of singularities.

1. Introduction

In this note, we study symmetric hyperbolic $(n \times n)$-systems of differential or pseudo-differential equations

$$\begin{aligned} \partial_t u(t,x) &= A(t,x,\mathrm{D}_x)\, u(t,x) + f(t,x), \quad x \in \mathbb{R}^d, t > 0, \\ u(0,x) &= g(x), \quad x \in \mathbb{R}^d. \end{aligned} \tag{1.1}$$

We are interested in a framework that allows distributional data and non-smooth, e.g., discontinuous coefficients simultaneously. Simple examples show that in such a situation, solutions in the sense of distributions may fail to exist. For this reason, we take recourse to the theory of generalized functions of Colombeau [1, 2]. The matrix of operators in (1.1) is either given by

$$A(t,x,\mathrm{D}_x) = \sum_{j=1}^{d} A_j(t,x)\mathrm{D}_{x_j} + A_0(t,x), \tag{1.2}$$

where the entries of $A_0, \dots, A_d$ are elements of the Colombeau algebra $\mathcal{G}([0,\infty) \times \mathbb{R}^d)$, or by a matrix of first-order pseudo-differential operators with Colombeau symbols.

It has been known since about two decades that hyperbolic systems of the type (1.1), (1.2) have unique solutions in the Colombeau algebra $\mathcal{G}([0,\infty)\times\mathbb{R}^d)$, see [15, 17, 18]; the case of a scalar first-order hyperbolic pseudo-differential equation was settled by [12]. By now, microlocal ellipticity is well established for operators with Colombeau coefficients or symbols [7, 8, 14]. However, the propagation of singularities along bicharacteristics for solutions in $\mathcal{G}([0,\infty)\times\mathbb{R}^d)$ has been a notoriously difficult problem. Positive results have been obtained for solutions in $\mathcal{G}([0,\infty)\times\mathbb{R}^d)$ to systems with classical, smooth coefficients [8] and in the special case of a transport equation in conservative form [9, 13]. The general question still remains open.

In [4], Garetto introduced the continuous dual of a Colombeau algebra and proved microlocal elliptic regularity results in [5]. The dual framework suggests itself as ideally suited for addressing the problem of propagation of singularities for (1.1) in the Colombeau setting. Indeed, one may use the Colombeau algebra for accommodating regularizations of the coefficients (or symbols) of the operators, while keeping the initial data and the forcing terms as distributions (now viewed as elements of the dual of the Colombeau algebra). It is the purpose of this note to show that this program works (and does so with much more ease than remaining on the level of the algebras).

Thus, we will first show that system (1.1) is solvable in the dual space, the space of $\widetilde{\mathbb{C}}$-linear, continuous functionals $\mathcal{L}(\mathcal{G}_c([0,\infty)\times\mathbb{R}^d),\widetilde{\mathbb{C}})$ and second, indicate in an explicit example of a transport equation that the Colombeau generalized wave front set can be calculated and exhibits a finer structure than the classical wave front set of the distributional limit of the solution.

The plan of the paper is as follows. In Section 2, we begin by recalling basic results from Colombeau algebras and their duals. In Section 3, we discuss the Fourier transform of $\widetilde{\mathbb{C}}$-linear functionals, the definition of their wave front set, and the notion of generalized pseudo-differential operators acting on them. It is worthwhile to point out that there are two types of regularity for elements of the dual: to belong to the Colombeau algebra $\mathcal{G}$ or to belong to the subalgebra $\mathcal{G}^\infty$ of regular elements. In Section 4 we show that system (1.1) is solvable in the dual of the Colombeau algebra under certain hypotheses on the symbols and data. We discuss the question of uniqueness and show that $\mathcal{G}$-regular solutions are unique. Section 5 is devoted to an explicit calculation of the wave front set of a (generic) scalar transport equation with discontinuous coefficient and initial data given by a delta function. We exemplify the solution concept and show that the fiber of the wave front set above the kink in the characteristic curve carrying the singularities is smaller than the wave front set of the distributional limit and reflects the shape of the kink. The general question of propagation of singularities along bicharacteristics remains open; we hope that this note will spark interest in the duality methods and lead to progress on the problem.

2. Colombeau theory: Basic notions

In this section we collect various basic notions from the theory of generalized functions of Colombeau [1, 2] which we need in this paper. For more details we refer, e.g., to [10].

Let Ω be an open subset of $\mathbb{R}^d$. The starting point of the theory as we use it here are families $(u_\varepsilon)_{\varepsilon\in(0,1]}$ of smooth functions $u_\varepsilon \in \mathcal{C}^\infty(\Omega)$ for $0 < \varepsilon \leq 1$. The following subalgebras are singled out:
Moderate families, denoted by $\mathcal{E}_{\mathrm{M}}(\Omega)$, are defined by the property:

$$\forall K \Subset \Omega\, \forall \alpha \in \mathbb{N}_0^d\, \exists p \geq 0 : \sup_{x\in K} |\partial^\alpha u_\varepsilon(x)| = \mathcal{O}(\varepsilon^{-p}) \text{ as } \varepsilon \to 0. \tag{2.1}$$

Null or *negligible families*, denoted by $\mathcal{N}(\Omega)$, are defined by the property:

$$\forall K \Subset \Omega\, \forall \alpha \in \mathbb{N}_0^d\, \forall q \geq 0 : \sup_{x\in K} |\partial^\alpha u_\varepsilon(x)| = \mathcal{O}(\varepsilon^{q}) \text{ as } \varepsilon \to 0. \tag{2.2}$$

The null families form a differential ideal in the collection of moderate families. The *Colombeau algebra* is the factor algebra

$$\mathcal{G}(\Omega) = \mathcal{E}_{\mathrm{M}}(\Omega)/\mathcal{N}(\Omega).$$

Restrictions of the elements of $\mathcal{G}(\Omega)$ to open subsets of Ω are defined on representatives. Thus the notion of *support* of an element is meaningful; actually, $\Omega \to \mathcal{G}(\Omega)$ is a sheaf of differential algebras on $\mathbb{R}^d$. We denote the compactly supported Colombeau generalized functions by $\mathcal{G}_{\mathrm{c}}(\Omega)$. The space of compactly supported distributions $\mathcal{E}'(\Omega)$ is imbedded in $\mathcal{G}(\Omega)$ by convolution:

$$\iota : \mathcal{E}'(\Omega) \to \mathcal{G}(\Omega),\ \iota(w) = \text{ class of } ((w * \varphi_\varepsilon)|_\Omega)_{\varepsilon\in(0,1]}, \tag{2.3}$$

where

$$\varphi_\varepsilon(x) = \varepsilon^{-d}\varphi(x/\varepsilon) \tag{2.4}$$

is obtained by scaling a fixed test function $\varphi \in \mathcal{S}(\mathbb{R}^d)$ of integral one with all moments vanishing. By the sheaf property, this can be extended in a unique way to an imbedding of the space of distributions $\mathcal{D}'(\Omega)$ in $\mathcal{G}(\Omega)$. This imbedding renders $\mathcal{C}^\infty(\Omega)$ a faithful subalgebra. In fact, given $f \in \mathcal{C}^\infty(\Omega)$, one can define a corresponding element of $\mathcal{G}(\Omega)$ by the constant imbedding $\sigma(f) =$ class of $(f_\varepsilon)_{\varepsilon\in(0,1]}$ with $f_\varepsilon \equiv f$ for all ε. Then $\iota(f) = \sigma(f)$ in $\mathcal{G}(\Omega)$.

Families $(r_\varepsilon)_{\varepsilon\in(0,1]}$ of complex numbers such that $|r_\varepsilon| = \mathcal{O}(\varepsilon^{-p})$ as $\varepsilon \to 0$ for some $p \geq 0$ are called *moderate*, those for which $|r_\varepsilon| = \mathcal{O}(\varepsilon^q)$ for every $q \geq 0$ are termed *negligible*. The ring $\widetilde{\mathbb{C}}$ of Colombeau generalized numbers is obtained by factoring moderate families of complex numbers with respect to negligible families. When Ω is connected, $\widetilde{\mathbb{C}}$ coincides with the *ring of constants* in the differential algebra $\mathcal{G}(\Omega)$.

Regularity theory for linear equations has been based on the subalgebra $\mathcal{G}^\infty(\Omega)$ of *regular generalized functions* in $\mathcal{G}(\Omega)$. It is defined by those elements which have a representative satisfying

$$\forall K \Subset \Omega\, \exists p \geq 0\, \forall \alpha \in \mathbb{N}_0^d : \sup_{x\in K} |\partial^\alpha u_\varepsilon(x)| = \mathcal{O}(\varepsilon^{-p}) \text{ as } \varepsilon \to 0. \tag{2.5}$$

Observe the change of quantifiers with respect to formula (2.1); locally, all derivatives of a regular generalized function have the same order of growth in $\varepsilon > 0$. It holds that that (see [18])

$$\mathcal{G}^\infty(\Omega) \cap \mathcal{D}'(\Omega) = \mathcal{C}^\infty(\Omega).$$

For $K \Subset \Omega$, $\alpha \in \mathbb{N}_0^d$, $p \geq 0$, denote by $V_{K,\alpha,p} \subset \mathcal{G}(\Omega)$ the set of elements u of $\mathcal{G}(\Omega)$ such that

$$\sup_{x \in K} |\partial^\alpha u_\varepsilon(x)| = \mathcal{O}(\varepsilon^p) \text{ as } \varepsilon \to 0$$

for some – and hence all – representatives $(u_\varepsilon)_{\varepsilon \in (0,1]}$ of u. The so-called *sharp topology* on $\mathcal{G}(\Omega)$ is obtained by taking the collection of all $V_{K,\alpha,p}$ as a base of neighborhoods of zero. Similarly, the ring $\widetilde{\mathbb{C}}$ of Colombeau generalized numbers may be topologized by means of the neighborhoods V_p, $p \geq 0$, defined by those moderate elements that satisfy $|r_\varepsilon| = \mathcal{O}(\varepsilon^p)$. With this topology, $\widetilde{\mathbb{C}}$ turns into an ultrametric, complete topological ring and $\mathcal{G}(\Omega)$ becomes an ultrametric, complete topological differential algebra and a topological module over $\widetilde{\mathbb{C}}$. Further, $\mathcal{G}_c(\Omega)$ can be endowed with the structure of an inductive limit of topological $\widetilde{\mathbb{C}}$-modules. The theory of topological $\widetilde{\mathbb{C}}$-modules has been developed in detail by [3].

Duality theory. Following [3, 4], we introduce the dual $\mathcal{L}(\mathcal{G}_c(\Omega), \widetilde{\mathbb{C}})$ of continuous $\widetilde{\mathbb{C}}$-linear forms on $\mathcal{G}_c(\Omega)$ (with values in $\widetilde{\mathbb{C}}$). As usual, we denote the action of an element u of the dual on an element $\psi \in \mathcal{G}_c(\Omega)$ by $\langle u, \psi \rangle \in \widetilde{\mathbb{C}}$. Equipped with the corresponding weak topology, $\mathcal{L}(\mathcal{G}_c(\Omega), \widetilde{\mathbb{C}})$, $\Omega \subset \mathbb{R}^d$, forms a sheaf of topological $\widetilde{\mathbb{C}}$-modules on $\mathbb{R}^d$.

The Colombeau algebra $\mathcal{G}(\Omega)$ is imbedded in $\mathcal{L}(\mathcal{G}_c(\Omega), \widetilde{\mathbb{C}})$ by means of the assignment

$$\langle u, \psi \rangle = \text{class of } \Big(\int u_\varepsilon(x) \psi_\varepsilon(x)\,\mathrm{d}x \Big)_{\varepsilon \in (0,1]}$$

in $\widetilde{\mathbb{C}}$ where $(u_\varepsilon)_{\varepsilon \in (0,1]}$ and $(\psi_\varepsilon)_{\varepsilon \in (0,1]}$ are representatives of u and ψ. What is more important, $\mathcal{D}'(\Omega)$ can be directly imbedded in $\mathcal{L}(\mathcal{G}_c(\Omega), \widetilde{\mathbb{C}})$ by

$$\langle w, \psi \rangle = \text{class of } \big(\langle w, \psi_\varepsilon \rangle \big)_{\varepsilon \in (0,1]}$$

for $w \in \mathcal{D}'(\Omega)$, $\psi \in \mathcal{G}_c(\Omega)$. More generally, moderate families of distributions $(w_\varepsilon)_{\varepsilon \in (0,1]}$ define elements of $\mathcal{L}(\mathcal{G}_c(\Omega), \widetilde{\mathbb{C}})$. A family of distributions is called *moderate*, if for all compact subsets K of Ω:

$$\exists j \in \mathbb{N}_0 \, \exists p \geq 0 : \; |\langle w_\varepsilon, \varphi \rangle| \leq \varepsilon^{-p} \sup_{x \in K, |\alpha| \leq j} |\partial^\alpha \varphi(x)| \tag{2.6}$$

for all $\varphi \in \mathcal{D}(\Omega)$ with support in K (and sufficiently small ε). Elements of $\mathcal{L}(\mathcal{G}_c(\Omega), \widetilde{\mathbb{C}})$ defined in this way (by a representing family of distributions) are called *basic functionals*. Observe that, in particular, functionals defined by elements of $\mathcal{G}(\Omega)$ are basic; the action on a test function does not depend on the choice of representative.

The imbeddings $\mathcal{G}^\infty(\Omega) \to \mathcal{G}(\Omega) \to \mathcal{L}(\mathcal{G}_c(\Omega), \widetilde{\mathbb{C}})$ are sheaf imbeddings. Therefore, given $u \in \mathcal{L}(\mathcal{G}_c(\Omega), \widetilde{\mathbb{C}})$, it makes sense to consider the largest open subset ω of Ω on which u is defined by an element of $\mathcal{G}(\omega)$. The complement of this set is called the $\mathcal{G}$*-singular support* of u. In the same way, the $\mathcal{G}^\infty$*-singular support* can be introduced.

Finally, the theory can be carried over to generalized functions on the closed half-space $[0,\infty)\times\mathbb{R}^d$ as well. The representatives of the elements of the Colombeau algebra $\mathcal{G}([0,\infty)\times\mathbb{R}^d)$ satisfy estimates (2.1), respectively (2.2) on compact subsets of $[0,\infty)\times\mathbb{R}^d$, i.e., up to the boundary $\{t=0\}$. Similarly, $\mathcal{G}_c([0,\infty)\times\mathbb{R}^d)$ denotes the elements of $\mathcal{G}([0,\infty)\times\mathbb{R}^d)$ whose support is a compact subset of the closed half-space $[0,\infty)\times\mathbb{R}^d$. The restriction of an element $u \in \mathcal{G}_c([0,\infty)\times\mathbb{R}^d)$ to the boundary $\{t=0\}$ is defined on representatives by

$$u|_{\{t=0\}} = \text{ class of } (u_\varepsilon(0,\cdot))_{\varepsilon\in(0,1]}.$$

The space of $\widetilde{\mathbb{C}}$-linear functionals on $[0,\infty)\times\mathbb{R}^d$ is defined as the dual space $\mathcal{L}(\mathcal{G}_c([0,\infty)\times\mathbb{R}^d), \widetilde{\mathbb{C}})$.

3. Fourier transform and pseudo-differential operators

The space of compactly supported functionals from $\mathcal{L}(\mathcal{G}_c(\Omega), \widetilde{\mathbb{C}})$ can be identified with $\mathcal{L}(\mathcal{G}(\Omega), \widetilde{\mathbb{C}})$. It is tempting to define the Fourier transform of an element $u \in \mathcal{L}(\mathcal{G}(\mathbb{R}^d), \widetilde{\mathbb{C}})$ as a map from $\mathbb{R}^d$ into $\widetilde{\mathbb{C}}$ by

$$\xi \to \langle u, \mathrm{e}_\xi\rangle, \qquad \mathrm{e}_\xi(x) = \mathrm{e}^{-\mathrm{i}x\xi}.$$

However, we need to be a bit more precise about the range of the Fourier transform. Thus we introduce the subalgebra $\mathcal{E}_\tau(\mathbb{R}^d)$ of $\mathcal{E}_{\mathrm{M}}(\mathbb{R}^d)$ as those families $(u_\varepsilon)_{\varepsilon\in(0,1]}$ of smooth functions which satisfy

$$\forall\alpha \in \mathbb{N}_0^d\, \exists p \geq 0 : \sup_{\xi\in\mathbb{R}^d} \langle\xi\rangle^{-p}|\partial^\alpha u_\varepsilon(\xi)| = \mathcal{O}(\varepsilon^{-p}) \text{ as } \varepsilon \to 0; \tag{3.1}$$

as usual, we let $\langle\xi\rangle = (1+|\xi|^2)^{1/2}$. The ideal $\mathcal{N}_\tau(\mathbb{R}^d)$ in $\mathcal{E}_\tau(\mathbb{R}^d)$ is constituted by the families with the property

$$\forall\alpha \in \mathbb{N}_0^d\, \exists p \geq 0\, \forall q \geq 0 : \sup_{\xi\in\mathbb{R}^d} \langle\xi\rangle^{-p}|\partial^\alpha u_\varepsilon(\xi)| = \mathcal{O}(\varepsilon^{q}) \text{ as } \varepsilon \to 0.$$

The algebra $\mathcal{G}_\tau(\mathbb{R}^d) = \mathcal{E}_\tau(\mathbb{R}^d)/\mathcal{N}_\tau(\mathbb{R}^d)$ is called the algebra of *tempered* Colombeau generalized functions.

Now let w be a basic functional in $\mathcal{L}(\mathcal{G}(\mathbb{R}^d), \widetilde{\mathbb{C}})$. There exists a representing family of compactly supported distributions $(w_\varepsilon)_{\varepsilon\in(0,1]}$ and a compact set $K \subset \mathbb{R}^d$ such that an estimate of type (2.6) holds for all $\varphi \in \mathcal{C}^\infty(\mathbb{R}^d)$. It follows that the family of smooth functions $\xi \to \langle w_\varepsilon, \mathrm{e}_\xi\rangle$, $\varepsilon \in (0,1]$, satisfies property (3.1). Actually, one may take the same exponent p from (2.6) for all α. This will serve as the definition of the Fourier transform.

Definition 3.1. The Fourier transform of a basic functional $w \in \mathcal{L}(\mathcal{G}(\mathbb{R}^d), \widetilde{\mathbb{C}})$ is the element $\mathcal{F}w \in \mathcal{G}_\tau(\mathbb{R}^d)$ defined by the representing family $(\xi \to \langle w_\varepsilon, \mathrm{e}_\xi\rangle)_{\varepsilon\in(0,1]}$.

We are ready to define the generalized wave front set of a basic functional. First, let $v \in \mathcal{G}_\tau(\mathbb{R}^d)$ and $\xi_0 \in \mathbb{R}^d \setminus 0$. Then ξ_0 is called a *$\mathcal{G}$-regular direction* for v, if there is a representative $(v_\varepsilon)_{\varepsilon\in(0,1]}$ and an open cone $\Gamma \subset \mathbb{R}^d \setminus 0$ about ξ_0 such that

$$\forall l \geq 0\, \exists p \geq 0 : \sup_{\xi\in\Gamma} \langle\xi\rangle^l |v_\varepsilon(\xi)| = \mathcal{O}(\varepsilon^{-p}) \text{ as } \varepsilon \to 0. \tag{3.2}$$

If instead we have that

$$\exists p \geq 0\, \forall l \geq 0 : \sup_{\xi\in\Gamma} \langle\xi\rangle^l |v_\varepsilon(\xi)| = \mathcal{O}(\varepsilon^{-p}) \text{ as } \varepsilon \to 0, \tag{3.3}$$

then ξ_0 is called a *$\mathcal{G}^\infty$-regular direction* for v.

Definition 3.2. Let w be a basic functional in $\mathcal{L}(\mathcal{G}_\mathrm{c}(\Omega), \widetilde{\mathbb{C}})$, $(x_0, \xi_0) \in \Omega \times \mathbb{R}^d \setminus 0$. Then w is called *microlocally $\mathcal{G}$-regular*, respectively *microlocally $\mathcal{G}^\infty$-regular* at (x_0, ξ_0), if there is a cut-off function $\phi \in \mathcal{D}(\Omega)$, $\phi(x_0) \neq 0$, such that ξ_0 is a $\mathcal{G}$-regular direction, respectively $\mathcal{G}^\infty$-regular direction for $\mathcal{F}(\phi w)$. The collection of all (x_0, ξ_0) at which w is not microlocally regular forms the *$\mathcal{G}$-wave front set* $\mathrm{WF}_\mathcal{G}(w)$, respectively *$\mathcal{G}^\infty$-wave front set* $\mathrm{WF}_{\mathcal{G}^\infty}(w)$.

It is clear that in case w is defined by an element of $\mathcal{G}(\Omega)$, then $\mathrm{WF}_\mathcal{G}(w)$ is empty and $\mathrm{WF}_{\mathcal{G}^\infty}(w)$ coincides with the previously defined generalized wave front set $\mathrm{WF}_g(w)$ (see [14, 16]). Also, if w is given by a single distribution, both $\mathrm{WF}_\mathcal{G}(w)$ and $\mathrm{WF}_{\mathcal{G}^\infty}(w)$ coincide with the distributional wave front set $\mathrm{WF}(w)$, because $\mathcal{F}(\phi w)$ does not depend on ε. The following result is due to [5] who proved it by means of a pseudo-differential characterization of the generalized wave front set. For the purpose of illustration, we will provide a direct proof.

Proposition 3.3. *Let w be a basic functional in $\mathcal{L}(\mathcal{G}_\mathrm{c}(\Omega), \widetilde{\mathbb{C}})$. Then the projection of $\mathrm{WF}_\mathcal{G}(w)$ to Ω coincides with the $\mathcal{G}$-singular support of w, and the projection of $\mathrm{WF}_{\mathcal{G}^\infty}(w)$ yields the $\mathcal{G}^\infty$-singular support of w.*

Proof. We begin with the global argument. Suppose that $v_\varepsilon = \mathcal{F}(\phi w_\varepsilon)$ satisfies the estimate (3.2) on $\Gamma = \mathbb{R}^d \setminus 0$. We have to produce a representative $(u_\varepsilon)_{\varepsilon\in(0,1]}$ of an element $u \in \mathcal{G}(\Omega)$ such that

$$\int u(x)\psi(x)\,\mathrm{d}x = \langle \phi w, \psi\rangle$$

in $\widetilde{\mathbb{C}}$ for all $\psi \in \mathcal{G}_\mathrm{c}(\Omega)$. To this end, we take $\rho \in \mathcal{D}(\mathbb{R}^d)$, $\int \rho(x)\,\mathrm{d}x = 1$, set $\rho^\varepsilon(x) = \mathrm{e}^{1/\varepsilon}\rho(\mathrm{e}^{1/\varepsilon}x)$ and define $u_\varepsilon = (\phi w_\varepsilon) * \rho^\varepsilon$. Clearly, u_ε is smooth. Due to the sequence of estimates (3.2) on the Fourier transform of ϕw_ε, we have that $\partial^\alpha u_\varepsilon$ is bounded for all $\alpha \in \mathbb{N}_0^d$ and sufficiently small ε (depending on α), and

$$\|\partial^\alpha u_\varepsilon\|_{\mathrm{L}^\infty(\mathbb{R}^d)} \leq \|\partial^\alpha(\phi w_\varepsilon)\|_{\mathrm{L}^\infty(\mathbb{R}^d)} \|\rho\|_{\mathrm{L}^1(\mathbb{R}^d)} = \mathcal{O}(\varepsilon^{-p}) \text{ as } \varepsilon \to 0$$

for some $p \geq 0$. Thus $(u_\varepsilon)_{\varepsilon\in(0,1]}$ is moderate and qualifies as a representative of an element $u \in \mathcal{G}(\Omega)$. It follows from the mean value theorem that the family $\psi_\varepsilon * \check{\rho}^\varepsilon$ is a representative of ψ as well, where $\check{}$ denotes inflection. Thus

$$\int u_\varepsilon(x)\psi_\varepsilon(x)\,\mathrm{d}x - \langle \phi w_\varepsilon, \psi_\varepsilon\rangle = \langle \phi w_\varepsilon, \psi_\varepsilon * \check{\rho}^\varepsilon - \psi_\varepsilon\rangle$$

is a negligible family of numbers, as desired. In other words, if the $\mathcal{G}$-wave front set of a basic functional is empty, then it is given by an element of $\mathcal{G}(\Omega)$. The microlocal version is obtained by the usual argument, piecing together finitely many cones and convolving the corresponding finitely many cut-off functions. The assertion about the $\mathcal{G}^\infty$-singular support is proved in the same way. □

Generalized pseudo-differential operators. Recall that the Hörmander class $S^m(\Omega\times\mathbb{R}^d)$ of symbols of order m is defined as the set of functions $a \in \mathcal{C}^\infty(\Omega\times\mathbb{R}^d)$ such that

$$|a|^{(m)}_{K,\alpha,\beta} = \sup_{x\in K,\xi\in\mathbb{R}^d} \langle\xi\rangle^{-m+|\alpha|}|\partial_\xi^\alpha\partial_x^\beta a(x,\xi)| < \infty \tag{3.4}$$

for all compact subsets K of Ω and all $\alpha,\beta \in \mathbb{N}_0^d$. Equipped with the seminorms (3.4), $S^m(\Omega\times\mathbb{R}^d)$ is a Fréchet space. We will later make use of time-dependent symbols $t \to a(t)$ which are given by $\mathcal{C}^\infty$-maps of $t \in [0,T]$ with values in $S^m(\Omega\times\mathbb{R}^d)$, equipped with the seminorms

$$|a|^{(m)}_{T,k,K,\alpha,\beta} = \sup_{t\in[0,T]} |\partial_t^k a(t)|^{(m)}_{K,\alpha,\beta}. \tag{3.5}$$

A family of symbols $(a_\varepsilon)_{\varepsilon\in(0,1]}$ from $S^m(\Omega\times\mathbb{R}^d)$ will be termed *moderate*, if

$$\forall K \Subset \Omega\,\forall\alpha,\beta\in\mathbb{N}_0^d\,\exists p\geq 0 :\ |a_\varepsilon|^{(m)}_{K,\alpha,\beta} = \mathcal{O}(\varepsilon^{-p}) \text{ as } \varepsilon\to 0.$$

Similarly, the family is called *negligible*, if

$$\forall K \Subset \Omega\,\forall\alpha,\beta\in\mathbb{N}_0^d\,\forall q\geq 0 :\ |a_\varepsilon|^{(m)}_{K,\alpha,\beta} = \mathcal{O}(\varepsilon^{q}) \text{ as } \varepsilon\to 0.$$

The space $\widetilde{S}^m(\Omega\times\mathbb{R}^d)$ of *generalized symbols* is the factor space of moderate modulo negligible symbols. In a similar way, *generalized amplitudes* are defined as generalized symbols on $\Omega\times\Omega\times\mathbb{R}^d$ and represented by families of the form $a_\varepsilon(x,y,\xi)$, $\varepsilon\in(0,1]$.

Given a generalized amplitude a and an element u of $\mathcal{G}_c(\Omega)$, the integral

$$v_\varepsilon(x) = \int_{\mathbb{R}^d}\int_\Omega \mathrm{e}^{\mathrm{i}(x-y)\xi} a_\varepsilon(x,y,\xi)u_\varepsilon(y)\,\mathrm{d}y\,đ\xi$$

can be interpreted as an oscillatory integral (inserting the operator $\langle\xi\rangle^{-N}\langle \mathrm{D}_y\rangle^N$ enforces convergence) and is easily seen to produce a family $(v_\varepsilon)_{\varepsilon\in(0,1]}$ of smooth functions that belong to $\mathcal{E}_\mathrm{M}(\Omega)$. Changing representatives of a or u results in a family that differs from $(v_\varepsilon)_{\varepsilon\in(0,1]}$ by an element of $\mathcal{N}(\Omega)$. Therefore, the assignment $v = Au$ yields an operator $A : \mathcal{G}_c(\Omega) \to \mathcal{G}(\Omega)$ which is formally denoted by

$$Au(x) = \int_{\mathbb{R}^d}\int_\Omega \mathrm{e}^{\mathrm{i}(x-y)\xi} a(x,y,\xi)u(y)\,\mathrm{d}y\,đ\xi$$

and called a *generalized pseudo-differential operator*. The operator A is continuous with respect to the sharp topologies. If A is properly supported (this being implied by the existence of amplitudes $a_\varepsilon(x,y,\xi)$ whose support with respect to (x,y) is a proper subset of $\Omega\times\Omega$ independently of ε), then A maps $\mathcal{G}_c(\Omega)$ into itself. The formal transpose of A is also a generalized pseudo-differential operator given by

$$A^{\mathsf{T}}u(x)=\int_{\mathbb{R}^d}\int_\Omega e^{i(x-y)\xi}a(y,x,-\xi)u(y)\,dy\,đ\xi.$$

For proofs of all these statements we refer to [7]. Following [5], the action of a properly supported generalized pseudo-differential operator can be extended to the dual:

$$A:\mathcal{L}(\mathcal{G}_c(\Omega),\widetilde{\mathbb{C}})\to\mathcal{L}(\mathcal{G}_c(\Omega),\widetilde{\mathbb{C}}),\qquad\langle Au,\psi\rangle=\langle u,A^{\mathsf{T}}\psi\rangle.$$

The extended operator is continuous (both for the weak and the strong topologies).

Finally, we recall [7] that the *generalized symbol* of a properly supported, generalized pseudo-differential operator A with a regular amplitude can be defined by

$$\sigma_A(x,\xi)=\int_{\mathbb{R}^d}\int_\Omega e^{i(x-y)\xi}\chi(x,y)a(x,y,\xi+\eta)\,đy\,d\eta$$

with a suitable proper cut-off function χ. Here an amplitude $a\in\widetilde{S}^m(\Omega\times\Omega\times\mathbb{R}^d)$ is called *regular* if it is represented by a family satisfying

$$\forall K\Subset\Omega\times\Omega\,\exists p\geq0\,\forall\alpha,\beta\in\mathbb{N}_0^d:\ |a_\varepsilon|^{(m)}_{K,\alpha,\beta}=\mathcal{O}(\varepsilon^{-p})\text{ as }\varepsilon\to0$$

(with p not depending on α,β). Then σ_A belongs to $\widetilde{S}^m(\Omega\times\mathbb{R}^d)$ and

$$Au(x)=\int_{\mathbb{R}^d}e^{ix\xi}\sigma_A(x,\xi)\mathcal{F}u(\xi)\,đ\xi$$

for all $u\in\mathcal{G}_c(\Omega)$. The symbolic calculus was recently extended to the full class of generalized pseudo-differential operators by [6].

This concludes the brief recall of pseudo-differential operators in the Colombeau setting. Everything that has been said can be (and has been) generalized to operators based on the Hörmander class $S^m_{\rho,\delta}$ with $0\leq\rho<\delta\leq1$.

4. Symmetric hyperbolic systems: Existence theory

In this section we construct solutions in the dual of the Colombeau algebra to the initial value problem

$$\begin{aligned}\partial_t u&=Au+f,\\ u(0,\cdot)&=g.\end{aligned}\tag{4.1}$$

Here

$$g\in\big(\mathcal{L}(\mathcal{G}_c(\mathbb{R}^d),\widetilde{\mathbb{C}})\big)^n,\qquad f\in\big(\mathcal{L}(\mathcal{G}_c([0,\infty)\times\mathbb{R}^d),\widetilde{\mathbb{C}})\big)^n$$

and $A=A(t,x,D_x)$ will be a time-dependent $(n\times n)$-matrix of first-order generalized pseudo-differential operators. We look for solutions

$$u\in\big(\mathcal{L}(\mathcal{G}_c([0,\infty)\times\mathbb{R}^d),\widetilde{\mathbb{C}})\big)^n.$$

To this end, we will first have to clarify the solution concept. The matrix of pseudo-differential operators will be given by representing amplitudes $a_\varepsilon(t,x,y,\xi)$ which are $\mathcal{C}^\infty$-maps of $t \in [0,\infty)$ with values in $S^m(\mathbb{R}^d \times \mathbb{R}^d \times \mathbb{R}^d)$, moderate in the seminorms (3.5), that is,

$$\forall T > 0\, \forall k \in \mathbb{N}_0\, \forall K \Subset \Omega\, \forall \alpha, \beta \in \mathbb{N}_0^d\, \exists p \geq 0 : \ |a_\varepsilon|^{(m)}_{T,k,K,\alpha,\beta} = \mathcal{O}(\varepsilon^{-p}) \text{ as } \varepsilon \to 0.$$

We may identify two such amplitudes if their difference is negligible in the seminorms (3.5). Assuming that the operator $A = A(t,x,\mathrm{D}_x)$ is properly supported at each fixed $t \geq 0$, an inspection of the relevant proofs in [7] shows that

$$A : \big(\mathcal{L}(\mathcal{G}_\mathrm{c}([0,\infty)\times\mathbb{R}^d), \widetilde{\mathbb{C}})\big)^n \to \big(\mathcal{L}(\mathcal{G}_\mathrm{c}([0,\infty)\times\mathbb{R}^d), \widetilde{\mathbb{C}})\big)^n$$

and so does its formal transpose A^T. By a solution to system (4.1) we mean an element $u \in \big(\mathcal{L}(\mathcal{G}_\mathrm{c}([0,\infty)\times\mathbb{R}^d), \widetilde{\mathbb{C}})\big)^n$ such that

$$\langle u, \partial_t \psi\rangle + \langle u, A^\mathsf{T}\psi\rangle + \langle f, \psi\rangle + \langle g, \psi(0,\cdot)\rangle = 0 \tag{4.2}$$

in $\widetilde{\mathbb{C}}$ for all $\psi \in \big(\mathcal{G}_\mathrm{c}([0,\infty)\times\mathbb{R}^d)\big)^n$. Here the first three brackets denote duality on $[0,\infty)\times\mathbb{R}^d$, the fourth on $\mathbb{R}^d$. The concept is motivated by a formal integration by parts.

To arrive at the existence result, a number of hypotheses will be needed.

(H1) (Basic Sobolev structure) The data f, g are basic functionals, given by representing families $(f_\varepsilon)_{\varepsilon\in(0,1]}$, $(g_\varepsilon)_{\varepsilon\in(0,1]}$ such that $f_\varepsilon \in C([0,\infty) : \mathrm{H}^s(\mathbb{R}^d))$ and $g_\varepsilon \in \mathrm{H}^s(\mathbb{R}^d)$ for some $s \in \mathbb{R}$ with the corresponding seminorms moderate, i.e.,

$$\sup_{t\in[0,T]} \|f_\varepsilon(t)\|_{\mathrm{H}^s(\mathbb{R}^d)} = \mathcal{O}(\varepsilon^{-p}), \quad \|g_\varepsilon\|_{\mathrm{H}^s(\mathbb{R}^d)} = \mathcal{O}(\varepsilon^{-p}) \text{ as } \varepsilon \to 0$$

for some $p \geq 0$ and all $T \geq 0$ (p may depend on T).

(H2) (Smooth time-dependence of symbols) A is a matrix of generalized first-order pseudo-differential operators given by families of properly supported amplitudes $a_\varepsilon(t,x,y,\xi)$ which are $\mathcal{C}^\infty$-maps of $t \in [0,\infty)$ with values in $S^1(\mathbb{R}^d \times \mathbb{R}^d \times \mathbb{R}^d)$.

(H3) (Logarithmic estimates) For all $T \geq 0$, $k \in \mathbb{N}_0$, $K \Subset \mathbb{R}^d \times \mathbb{R}^d$, $\alpha, \beta \in \mathbb{N}_0^d$:

$$|a_\varepsilon|^{(1)}_{T,k,K,\alpha,\beta} = \mathcal{O}(|\log\varepsilon|) \text{ as } \varepsilon \to 0.$$

(H4) (Symmetric hyperbolicity) $A + A^\star$ is a matrix of generalized pseudo-differential operators of order 0, where $A^\star = \bar{A}^\mathsf{T}$ denotes the formal adjoint.

(H5) (Compact supports) The symbols $a_\varepsilon(t,x,y,\xi)$ as well as the functions $f_\varepsilon(t,x)$, when restricted to any strip $[0,T]\times\mathbb{R}^d$, have compact support with respect to the variable x, independently of ε; the same holds of the data $g_\varepsilon(x)$.

Remark 4.1. (i) The basic structure required in (H1) is inherent in our method of proof. (ii) What concerns assumption (H2), smooth dependence on t is already required in the classical theory. By what has been said about properly supported, regular pseudo-differential operators in Section 3, the formal transpose $A^\top$ also possesses a family of defining amplitudes smoothly depending on t, as do the symbols of A and $A^\top$. Note that the logarithmic estimate (H3) entails, in particular, that A is a matrix of regular pseudo-differential operators. (iii) The logarithmic estimate (H3) is needed in the energy estimate and cannot be dropped as is seen from simple examples even of ordinary differential equations [18]. However, it is required only up to orders $|\alpha|, |\beta| \le d+1$ and $k = 0$ if moderate estimates of type $\mathcal{O}(\varepsilon^{-p})$ hold for the other values of k, α, β. (iv) The compactness of the supports is convenient; it allows to employ the simple arguments from [19]. In the scalar case, it can be removed by the techniques of [12]. Note that due to the formulas in Section 3, the compactness assumption on the amplitudes of A entails the one on its transpose $A^\top$ and the symbols σ_A and $\sigma_{A^\top}$. (v) The Sobolev index s may well be negative, so data given by distributions or by moderate families of distributions are admitted.

Theorem 4.2. *Under the hypotheses* (H1) *through* (H5), *the initial value problem* (4.1) *has a solution* $u \in \big(\mathcal{L}(\mathcal{G}_c([0,\infty) \times \mathbb{R}^d), \widetilde{\mathbb{C}})\big)^n$.

Proof. Choosing representatives of the data and the symbols in (4.1), it is well known that the system

$$\begin{aligned} \partial_t u_\varepsilon &= A_\varepsilon u_\varepsilon + f_\varepsilon, \\ u_\varepsilon(0,\cdot) &= g_\varepsilon \end{aligned} \tag{4.3}$$

has a unique classical solution

$$u_\varepsilon \in C([0,\infty) : \mathrm{H}^s(\mathbb{R}^d)) \cap C^1([0,\infty) : \mathrm{H}^{s-1}(\mathbb{R}^d));$$

to see this, we may use the compactness of the supports and follow the arguments in [19]. We are going to show that the family $(u_\varepsilon)_{\varepsilon\in(0,1]}$ qualifies as a representative of a solution $u \in \mathcal{L}(\mathcal{G}_c([0,\infty) \times \mathbb{R}^d), \widetilde{\mathbb{C}})$.

First, at fixed $0 \le t \le T$ the operators $A_\varepsilon + A_\varepsilon^\top$ map $\mathrm{L}^2(\mathbb{R}^d)$ continuously into itself with an operator norm C_ε which is an affine function of

$$|\sigma_{A_\varepsilon}|^{(1)}_{T,0,K,\alpha,\beta} \quad \text{and} \quad |\sigma_{A_\varepsilon^\top}|^{(1)}_{T,0,K,\alpha,\beta}$$

for $|\alpha|, |\beta| \le d+1$ where K is a compact set containing the supports of the symbols. Following the arguments in [19], we easily arrive at the energy estimate

$$\|u_\varepsilon(t)\|^2_{\mathrm{H}^s(\mathbb{R}^d)} \le \Big(\|g_\varepsilon\|^2_{\mathrm{H}^s(\mathbb{R}^d)} + \int_0^t \|f_\varepsilon(\tau)\|^2_{\mathrm{H}^s(\mathbb{R}^d)}\,\mathrm{d}\tau\Big) \exp(C_\varepsilon t)$$

for $0 \le t \le T$. By assumption, $C_\varepsilon = \mathcal{O}(|\log\varepsilon|)$; thus the family $(u_\varepsilon)_{\varepsilon\in(0,1]}$ satisfies moderate estimates in the space $C([0,\infty) : \mathrm{H}^s(\mathbb{R}^d))$. In particular, it forms a moderate family of distributions and hence defines a basic functional $u \in \mathcal{L}(\mathcal{G}_c([0,\infty) \times \mathbb{R}^d), \widetilde{\mathbb{C}})$.

Now let $\psi \in \mathcal{G}_c([0,\infty) \times \mathbb{R}^d)$. Its representatives ψ_ε, $\varepsilon \in (0,1]$, belong to $C([0,\infty) : \mathrm{H}_0^{|s|}(\mathbb{R}^d))$, in particular. The action of the functional u on ψ is given in terms of representatives by

$$\int_0^\infty \langle u_\varepsilon(t), \psi_\varepsilon(t,\cdot)\rangle \,\mathrm{d}t$$

where the brackets denote duality of H^s and $\mathrm{H}_0^{|s|}$ (or simply integration if $s \geq 0$). We use that u_ε solves (4.3) and compute

$$\begin{aligned}\int_0^\infty \langle u_\varepsilon(t), (\partial_t + A_\varepsilon^\mathsf{T})\psi_\varepsilon(t,\cdot)\rangle \,\mathrm{d}t + \int_0^\infty \langle f_\varepsilon(t), \psi_\varepsilon(t,\cdot)\rangle \,\mathrm{d}t + \langle g_\varepsilon, \psi_\varepsilon(0,\cdot)\rangle \\ = \int_0^\infty \langle (-\partial_t + A_\varepsilon)u_\varepsilon(t), \psi_\varepsilon(t,\cdot)\rangle \,\mathrm{d}t + \int_0^\infty \langle f_\varepsilon(t), \psi_\varepsilon(t,\cdot)\rangle \,\mathrm{d}t = 0,\end{aligned}$$

thus u is a solution in the sense of (4.2). □

Remark 4.3. Due to the hypotheses on the supports of the data and the operators, the solution u also has compact support on every strip $[0,T] \times \mathbb{R}^d$.

We turn to the question of uniqueness of solutions. We do not know whether the current formulation of our solution concept is strong enough to have uniqueness; but we may enforce uniqueness in the following way. The notion of the Colombeau space $\mathcal{G}_E$ based on a locally convex topological vector space E can be found in [3]. If we take the space $E = C([0,\infty) : \mathrm{H}^s(\mathbb{R}^d)) \cap C^1([0,\infty) : \mathrm{H}^{s-1}(\mathbb{R}^d))$ for such a construction, the arguments above show that problem (4.1) will have a solution in $\mathcal{G}_E$ in the sense that the equations hold in $\mathcal{G}_F$ with $F = C([0,\infty) : \mathrm{H}^{s-1}(\mathbb{R}^d))$. The energy estimates will show that such a solution is unique. This assertion actually encodes the fact that a negligible perturbation (measured in the H^s-norm) of the data leads to a negligible perturbation of the classical solution.

Uniqueness in the sense inherent to the solution concept in $\mathcal{L}(\mathcal{G}_\mathrm{c}([0,\infty) \times \mathbb{R}^d), \widetilde{\mathbb{C}})$ can be inferred for $\mathcal{G}$-regular solutions. A $\widetilde{\mathbb{C}}$-linear functional on $\mathcal{G}_\mathrm{c}([0,\infty) \times \mathbb{R}^d)$ is *$\mathcal{G}$-regular* if it is given by an element of $\mathcal{G}([0,\infty) \times \mathbb{R}^d)$.

Proposition 4.4. *Under the hypotheses* (H1) *through* (H5), *there is at most one $\mathcal{G}$-regular solution with compact support on every strip* $[0,T] \times \mathbb{R}^d$ *to the initial value problem* (4.1).

Proof. The difference $w = u - v$ of two $\mathcal{G}$-regular solutions satisfies

$$\langle w, \partial_t \psi\rangle + \langle w, A^\mathsf{T}\psi\rangle = 0$$

in $\widetilde{\mathbb{C}}$ for all $\psi \in \left(\mathcal{G}_\mathrm{c}([0,\infty) \times \mathbb{R}^d)\right)^n$. Consider the final value problem

$$\begin{aligned}\partial_t \theta &= -A^\mathsf{T}\theta + \bar{w}, \\ \theta(T,\cdot) &= 0.\end{aligned} \tag{4.4}$$

Due to the regularity and support assumption, the representatives w_ε of w belong to $C^k([0,\infty) : \mathrm{H}^s(\mathbb{R}^d))$ for every $k \geq 0$ and $s \in \mathbb{R}$. Due to hyperbolicity, we may mimic the proof of Theorem 4.2 backward in time and obtain solutions θ_ε which also belong to every space $C^k([0,T] : \mathrm{H}^s(\mathbb{R}^d))$ and in addition – by Remark 4.3 – have compact support in the strip $[0,T] \times \mathbb{R}^d$. Since each θ_ε is smooth and vanishes identically at $t = T$, we may extend it by zero for $t > T$ and obtain an element $\theta \in \mathcal{G}_c([0,\infty) \times \mathbb{R}^d)$ in this way which solves (4.4). Therefore,

$$0 = \langle w, \partial_t \theta \rangle + \langle w, A^\mathsf{T} \theta \rangle = \int_0^T \int_{\mathbb{R}^d} |w(t,x)|^2 \, dx \, dt$$

in $\widetilde{\mathbb{C}}$. This implies that $\int_0^T \int_{\mathbb{R}^d} w(t,x)\psi(t,x) \, dx \, dt$ vanishes in $\widetilde{\mathbb{C}}$ for every $\psi \in \big(\mathcal{G}_c([0,T] \times \mathbb{R}^d)\big)^n$. Since T is arbitrary, w is zero as a functional and $u = v$. □

Remark 4.5. $\mathcal{G}$-regular solutions exist: if the data f, g are $\mathcal{G}$-regular with compact support in the appropriate sense as above, then the solution to (4.1) is $\mathcal{G}$-regular.

5. Propagation of singularities: An example

In this section we study an example of a transport equation with discontinuous coefficient and distributional data that demonstrates the advantage of the solution concept in the dual of the Colombeau algebra. The point is that the coefficient is regularized to produce an element of the Colombeau algebra, while the initial data enter in a non-regularized way as members of the dual.

The model equation we study can be written formally as

$$\begin{aligned} &\partial_t u(t,x) + H(t-1)\partial_x u(t,x) = 0, \\ &u(0,x) = \delta(x) \end{aligned} \tag{5.1}$$

where H is the Heaviside function and δ the Dirac measure. For $0 < t < 1$, we have $\partial_t u = 0$, while $(\partial_t + \partial_x)u = 0$ for $t > 1$ – thus heuristically the solution should be of the form

$$u(t,x) = \begin{cases} \delta(x), & 0 \le t < 1, \\ \delta(x - t + 1), & 1 < t. \end{cases} \tag{5.2}$$

Note that the insertion of u in (5.1) a priori has no meaning in the theory of distributions, due to the product of $\partial_x u$ with the jump function at $(t,x) = (1,0)$. We shall construct a solution in the Colombeau dual space $\mathcal{L}(\mathcal{G}_c([0,\infty) \times \mathbb{R}), \widetilde{\mathbb{C}})$, show that its restriction to $\{(t,x) \in [0,\infty) \times \mathbb{R} : t \neq 1\}$ actually is given by (5.2) and compute its microlocal behavior at $(t,x) = (1,0)$.

We begin by regularizing the coefficient. Let $\rho \in \mathcal{D}(\mathbb{R})$, $\operatorname{supp} \rho = [-1,1]$, $\rho > 0$ on $(0,1)$ and symmetric, $\int \rho(x) \, dx = 1$ and put

$$\lambda_\varepsilon(t) = \int_{-\infty}^{(t-1)/\varepsilon} \rho(s) \, ds = \mu\big(\tfrac{t-1}{\varepsilon}\big) \tag{5.3}$$

with $\mu(t) = \int_{-\infty}^t \rho(s) \, ds$. Then $\lambda_\varepsilon(t) = 0$ for $t \le 1 - \varepsilon$, $\lambda_\varepsilon(t) = 1$ for $t \ge 1 + \varepsilon$ and strictly increasing between $1-\varepsilon$ and $1+\varepsilon$. In particular, $\lambda_\varepsilon(t) \to H(t-1)$ as $\varepsilon \to 0$. The family $(\lambda_\varepsilon)_{\varepsilon \in (0,1]}$ is moderate and thus defines an element $\lambda \in \mathcal{G}([0,\infty) \times \mathbb{R})$. We interpret the initial value problem (5.1) as the equation

$$\partial_t u + \lambda \partial_x u = 0, \qquad u(0,\cdot) = \delta \tag{5.4}$$

in $\mathcal{L}(\mathcal{G}_c([0,\infty)\times\mathbb{R}),\widetilde{\mathbb{C}})$ in the sense of formula (4.2). Theorem 4.2 is not required, because we can construct a solution explicitly. Put

$$\Lambda_\varepsilon(t)=\int_0^t \lambda_\varepsilon(s)\,\mathrm{d}s=\begin{cases}0, & 0\le t\le 1-\varepsilon,\\ \int_{1-\varepsilon}^t \lambda_\varepsilon(s)\,\mathrm{d}s, & 1-\varepsilon\le t\le 1+\varepsilon,\\ t-1, & 1+\varepsilon\le t.\end{cases} \tag{5.5}$$

At the level of representatives, equation (5.4) reads

$$\partial_t u_\varepsilon+\lambda_\varepsilon\partial_x u_\varepsilon=0,\qquad u_\varepsilon(0,\cdot)=\delta \tag{5.6}$$

and makes sense in $C^1([0,\infty):\mathcal{D}'(\mathbb{R}))$, say. The unique distributional solution is given by

$$u_\varepsilon(t,x)=\delta(x-\Lambda_\varepsilon(t)).$$

Clearly $(u_\varepsilon)_{\varepsilon\in(0,1]}$ is a moderate family of distributions and thus defines an element of $\mathcal{L}(\mathcal{G}_c([0,\infty)\times\mathbb{R}),\widetilde{\mathbb{C}})$, which solves (5.4).

If $\psi\in\mathcal{G}_c([0,1)\times\mathbb{R})$, then the support of its representatives is contained in $[0,1-\eta]\times\mathbb{R}$ for some $\eta>0$. It follows from (5.5) that $\langle u_\varepsilon,\psi_\varepsilon\rangle=\int_0^1\psi_\varepsilon(t,0)\,\mathrm{d}t$ for sufficiently small ε. This shows that the restriction of u to $[0,1)\times\mathbb{R}$ is equal to the distribution $1\otimes\delta$. Similarly, we can compute the restriction to $(1,\infty)\times\mathbb{R}$ and arrive at the assertion that the restriction of the generalized solution to $\{(t,x)\in[0,\infty)\times\mathbb{R}:t\neq 1\}$ is indeed given by (5.2).

What is more, the distributional limit (in $\mathcal{D}'([0,\infty)\times\mathbb{R})$) of the solutions u_ε as $\varepsilon\to 0$ is the same, more precisely, it is given by the distribution

$$v(t,x)=(H\otimes\delta)(1-t,x)+(H\otimes\delta)(t-1,x-t+1);$$

this follows from the pointwise convergence properties of Λ_ε.

Remark 5.1. (The wave front set of the distributional limit) The classical wave front set of the limiting distribution v is given by $\mathrm{WF}(v)=W_0\cup W_1\cup W_2$ where

$$\begin{aligned}W_0&=\{(t,0,0,\xi):0\le t<1,\xi\neq 0\}\\ W_1&=\{(1,0,\tau,\xi):(\tau,\xi)\neq(0,0)\}\\ W_2&=\{(t,t-1,-\tau,\tau):1<t,\tau\neq 0\}.\end{aligned} \tag{5.7}$$

Indeed, the assertions for $t\neq 1$ are standard. To see that the fiber over the point $(1,0)$ contains no regular direction, we just observe that

$$\partial_t(\partial_t+\partial_x)v(t,x)=-\delta(t-1)\otimes\delta'(x).$$

It is clear that the wave front set of this latter distribution is equal to $\{(1,0,\tau,\xi):(\tau,\xi)\neq(0,0)\}$ and is contained in $\mathrm{WF}(v)$.

Thus the classical wave front set contains no structural information on the singular behavior near the point $(1,0)$. However, the result shows that the wave front sets of the distributions $\partial_x v$ and $H(t-1)$ are not in favorable position; so their product in the sense of Hörmander does not exist.

We are going to calculate the $\mathcal{G}$- and the $\mathcal{G}^\infty$-wave front sets of the solution $u \in \mathcal{L}(\mathcal{G}_c([0,\infty)\times\mathbb{R}),\widetilde{\mathbb{C}})$ and demonstrate that they exhibit a refined structure.

Proposition 5.2. *The $\mathcal{G}$-singular support of the solution $u \in \mathcal{L}(\mathcal{G}_c([0,\infty)\times\mathbb{R}),\widetilde{\mathbb{C}})$ to (5.4) is equal to $S = \{(t,0) : 0 \le t \le 1\} \cup \{(t,t-1) : t \ge 1\}$.*

Proof. It is clear that the support of u equals S, which in turn contains the $\mathcal{G}$-singular support. So let $(t_0,x_0) \in S$ and assume there is a neighborhood V of (t_0,x_0) on which u is given by an element $u_0 \in \mathcal{G}(V)$. Let K be a compact subset of V containing (t_0,x_0) in its interior. By assumption, there is $p \ge 0$ such that

$$\sup_{(t,x)\in K} |u_{0\varepsilon}(t,x)| = \mathcal{O}(\varepsilon^{-p}) \text{ as } \varepsilon \to 0.$$

Choose points $(t_\varepsilon,x_\varepsilon)$ on the support of u_ε that converge to (t_0,x_0) as $\varepsilon \to 0$. Take $\varphi \in \mathcal{D}(\mathbb{R}^2)$ with support in $[-1,1]\times[-1,1]$, $\varphi \ge 0$ and $\varphi \equiv 1$ on $[-1/2,1/2]\times[-1/2,1/2]$ and put

$$\psi_\varepsilon(t,x) = \varphi(\varepsilon^{-p-1}(t-t_\varepsilon),\varepsilon^{-p-1}(x-x_\varepsilon)).$$

Clearly, $(\psi_\varepsilon)_{\varepsilon\in(0,1]}$ is a moderate family of smooth functions with support in K for sufficiently small ε, hence defines an element $\psi \in \mathcal{G}_c([0,\infty)\times\mathbb{R})$. In addition, the support of ψ_ε is contained in the square R_ε with center $(t_\varepsilon,x_\varepsilon)$ and side-length $2\varepsilon^{p+1}$, and $\psi_\varepsilon \equiv 1$ on the sub-square Q_ε with with center $(t_\varepsilon,x_\varepsilon)$ and side-length ε^{p+1}. By (5.5), the slope of the curve $x = \Lambda_\varepsilon(t)$ is bounded by 1. Therefore, the portion of the curve for $t_\varepsilon - \varepsilon^{p+1} \le t \le t_\varepsilon + \varepsilon^{p+1}$ remains inside Q_ε. We conclude that

$$\langle u_\varepsilon,\psi_\varepsilon\rangle = \int_0^\infty \psi_\varepsilon(t,\Lambda_\varepsilon(t))\,\mathrm{d}t \ge \int_{t_\varepsilon-\varepsilon^{p+1}/2}^{t_\varepsilon+\varepsilon^{p+1}/2} \mathrm{d}t = \varepsilon^{p+1}$$

while

$$\iint_V u_{0\varepsilon}(t,x)\psi_\varepsilon(t,x)\,\mathrm{d}t\,\mathrm{d}x \le \iint_{R_\varepsilon} \varepsilon^{-p}\,\mathrm{d}t\,\mathrm{d}x \le 4\varepsilon^{-p}\varepsilon^{2p+2} = 4\varepsilon^{p+2},$$

a contradiction. □

The $\mathcal{G}^\infty$-singular support of u is equal to S as well, since it contains the $\mathcal{G}$-singular support.

We shall now compute the $\mathcal{G}$-wave front set of the generalized solution in $\mathcal{L}(\mathcal{G}_c([0,\infty)\times\mathbb{R}),\widetilde{\mathbb{C}})$. We are going to see that the non-regular directions above the point $(1,0)$ are confined to the cone spanned by the normals to the supporting tangents of S.

Proposition 5.3. *The $\mathcal{G}$-wave front set of the solution $u \in \mathcal{L}(\mathcal{G}_c([0,\infty)\times\mathbb{R}),\widetilde{\mathbb{C}})$ to (5.4) equals*

$$\mathrm{WF}_{\mathcal{G}}(u) \;=\; W_0 \cup W_{1/2} \cup W_2$$

where W_0, W_2 are as in (5.7) and

$$W_{1/2} = \{(1,0,\tau,\xi) : \xi < 0, 0 \le \tau \le -\xi\} \cup \{(1,0,\tau,\xi) : \xi > 0, -\xi \le \tau \le 0\}.$$

Proof. It was shown above that the restriction of u to $\{(t,x) \in [0,\infty) \times \mathbb{R} : t \neq 1\}$ is given by the distribution (5.2). We remarked in Section 3 that the wave front set of a distribution coincides with its $\mathcal{G}$-wave front set. Therefore, the non-regular directions above the points $(t_0, x_0) \neq (1,0)$ are described by the sets W_0 and W_2. It remains to study the fiber above $(1,0)$.

The upper bound. Take $\chi, \phi \in \mathcal{D}(\mathbb{R})$ with $\chi(1) = \phi(0) = 1$. Then

$$\begin{aligned}\mathcal{F}\big(\chi(t)\phi(x)\,\delta(x-\Lambda_\varepsilon(t))\big)(\tau,\xi) &= \big\langle \delta(x-\Lambda_\varepsilon(t)), \chi(t)\phi(x)\,\mathrm{e}^{-it\tau - ix\xi}\big\rangle \\ &= \int_0^\infty \chi(t)\phi(\Lambda_\varepsilon(t))\,\mathrm{e}^{-it\tau - i\Lambda_\varepsilon(t)\xi}\,\mathrm{d}t \\ &= \int_0^\infty \chi(t)\,\mathrm{e}^{-it\tau - i\Lambda_\varepsilon(t)\xi}\,\mathrm{d}t\end{aligned}$$

where we chose ϕ such that $\phi(\Lambda_\varepsilon(t)) = 1$ for $t \in \operatorname{supp}\chi$, which is clearly possible. The prospective regular directions are given by

$$R = \{(\tau,\xi) = (1,\beta) : \ -1 < \beta < \infty\} \ \cup \ \{(\tau,\xi) = (-1,\beta) : \ -\infty < \beta < 1\}.$$

We consider only the first part of R and take any proper sub-cone of the form

$$\Gamma = \{(\tau,\xi) : \tau > 0, \beta_0\tau \leq \xi \leq \beta_1\tau\}$$

with $-1 < \beta_0 < 0$ and $\beta_1 > 0$. Recalling that $0 \leq \lambda_\varepsilon \leq 1$, we have for $(\tau,\xi) \in \Gamma$ that

$$1 + \lambda_\varepsilon(t)\tfrac{\xi}{\tau} \geq 1 + \beta_0 > 0.$$

Thus the equation

$$s = t + \Lambda_\varepsilon(t)\tfrac{\xi}{\tau} \tag{5.8}$$

can be solved for t; denote the solution by $t = M_\varepsilon(s,\tau,\xi)$. Finally, when t varies in the compact support of χ, $t + \Lambda_\varepsilon(t)\frac{\xi}{\tau}$ remains in a compact set K independently of ε for $(\tau,\xi) \in \Gamma$. Thus we can perform the change of coordinates (5.8) in the integral and arrive at

$$\int_0^\infty \chi(t)\,\mathrm{e}^{-it\tau - i\Lambda_\varepsilon(t)\xi}\,\mathrm{d}t = \int_K \chi(M_\varepsilon(s,\tau,\xi))\,\mathrm{e}^{-is\tau}\big(1 + \lambda_\varepsilon(M_\varepsilon(s,\tau,\xi))\tfrac{\xi}{\tau}\big)^{-1}\,\mathrm{d}s.$$

Integrating m-times by parts, we see that the second integral is a finite sum of terms of the form

$$\int_K \chi^{(j)}(M_\varepsilon)\big(1 + \lambda_\varepsilon(M_\varepsilon)\tfrac{\xi}{\tau}\big)^{-k}\Big(\textstyle\prod_l \lambda_\varepsilon^{(l)}(M_\varepsilon)\Big)\big(\tfrac{\xi}{\tau}\big)^{m-1}\big(\tfrac{1}{\tau}\big)^m \mathrm{e}^{-is\tau}\,\mathrm{d}s \tag{5.9}$$

where we wrote $M_\varepsilon = M_\varepsilon(s,\tau,\xi)$ for short and used the observation that the mth derivative of $M_\varepsilon(s,\tau,\xi)$ with respect to s is a sum of products of powers of $(1 + \lambda_\varepsilon(M_\varepsilon)\frac{\xi}{\tau})^{-1}$, derivatives of λ_ε, and the factor $(\xi/\tau)^{m-1}$. We already know that $1 + \lambda_\varepsilon(t)\frac{\xi}{\tau}$ is bounded away from zero by $1 + \beta_0$. Each derivative of λ_ε is globally bounded by some negative power of ε. Further,

$$\tau \geq \min\big(1/\beta_1, 1/|\beta_0|\big)|\xi|$$

for $(\tau,\xi)\in\Gamma$. Thus ξ/τ is bounded and $\tau\geq c(1+|\xi|+|\tau|)$ on Γ for some constant $c>0$ and $\tau\geq 1$, say. In conclusion, given $m\in\mathbb{N}_0$, there is $p\geq 0$, $C>0$ such that

$$\Big|\int_0^\infty \chi(t)\,\mathrm{e}^{-\mathrm{i}t\tau-\mathrm{i}\Lambda_\varepsilon(t)\xi}\,\mathrm{d}t\Big|\leq C\varepsilon^{-p}(1+|\xi|+|\tau|)^{-m}$$

for $(\tau,\xi)\in\Gamma$. Since Γ may be chosen as to contain any given sub-cone of the first component of R, we see that every direction in the first component of R is $\mathcal{G}$-regular. A similar argument applies to the second component. Thus the set $W_{1/2}$ gives an upper bound for the non-regular directions over the point $(1,0)$.

The lower bound. The prospective singular directions are given by

$$D=\{(\tau,\xi)=(-\alpha,1):0\leq\alpha\leq 1\}\cup\{(\tau,\xi)=(\alpha,-1):0\leq\alpha\leq 1\}.$$

We shall consider the first component of D. Given $\alpha\in(0,1)$ and $\varepsilon\in(0,1]$, there is a unique $t_\varepsilon\in[1-\varepsilon,1+\varepsilon]$ such that $\lambda_\varepsilon(t_\varepsilon)=\alpha$. In fact, we compute from (5.3) that

$$t_\varepsilon=1+\varepsilon\mu^{-1}(\alpha)$$

where μ^{-1} is the inverse of μ restricted to $[-1,1]$. We shall use the fact that the Fourier transform of the distribution $\chi(t)\,\delta(x-\Lambda_\varepsilon(t))$ behaves like $\xi^{-1/2}$ in the direction $(-\alpha\xi,\xi)$ as $\xi\to\infty$, with χ as above. This will follow from [11, Theorem 7.7.5] (method of stationary phase). We write

$$t\alpha-\Lambda_\varepsilon(t)=t\lambda_\varepsilon(t_\varepsilon)-\Lambda_\varepsilon(t)=t_\varepsilon\lambda_\varepsilon(t_\varepsilon)-\Lambda_\varepsilon(t_\varepsilon)+f_\varepsilon(t)\tag{5.10}$$

with

$$f_\varepsilon(t)=(t-t_\varepsilon)\lambda_\varepsilon(t_\varepsilon)+\Lambda_\varepsilon(t_\varepsilon)-\Lambda_\varepsilon(t).$$

Then

$$f_\varepsilon'(t)=\lambda_\varepsilon(t_\varepsilon)-\lambda_\varepsilon(t),\quad f_\varepsilon''(t)=-\tfrac{1}{\varepsilon}\rho\big(\tfrac{t-1}{\varepsilon}\big)$$

so

$$f_\varepsilon(t_\varepsilon)=0,\quad f_\varepsilon'(t_\varepsilon)=0,\quad f_\varepsilon''(t_\varepsilon)=-\tfrac{1}{\varepsilon}\rho\big(\mu^{-1}(\alpha)\big)\neq 0$$

and $f_\varepsilon'(t)\neq 0$ except when $t=t_\varepsilon$. Thus the hypotheses of [11, Theorem 7.7.5] are satisfied and we conclude that there is a constant $C_\varepsilon>0$ such that

$$\Big|\int_0^\infty \chi(t)\,\mathrm{e}^{\mathrm{i}f_\varepsilon(t)\xi}\,\mathrm{d}t-\big(2\pi\mathrm{i}/\xi f_\varepsilon''(t_\varepsilon)\big)^{1/2}\chi(t_\varepsilon)\Big|\;\leq\;C_\varepsilon\xi^{-1}\textstyle\sum_{l\leq 2}\sup|\partial_t^l\chi|\tag{5.11}$$

for $\xi>0$. Writing a_ε for the absolute value of $\big(2\pi\mathrm{i}/f_\varepsilon''(t_\varepsilon)\big)^{-1/2}$ and b_ε for $C_\varepsilon\sum\sup|\partial_t^l\chi|$ and observing that the first term on the right-hand side of (5.10) does not depend on t, we infer from inequality (5.11) that

$$\Big|\int_0^\infty \chi(t)\,\mathrm{e}^{\mathrm{i}t\alpha\xi-\mathrm{i}\Lambda_\varepsilon(t)\xi}\,\mathrm{d}t\Big|\;=\;\Big|\int_0^\infty \chi(t)\,\mathrm{e}^{\mathrm{i}tf_\varepsilon(t)\xi}\,\mathrm{d}t\Big|\;\geq\;\frac{a_\varepsilon}{\sqrt{\xi}}-\frac{b_\varepsilon}{\xi},\tag{5.12}$$

valid for each $\varepsilon\in(0,1]$ and $\xi>0$. If $(-\alpha\xi,\xi)$ were a $\mathcal{G}$-regular direction for $\mathcal{F}(\chi u)$, inequality (3.2) should hold, in particular, for $l=1$. This would mean that there are constants $p\geq 0$, $C>0$ and $\varepsilon_0>0$ such that

$$\Big|(1+|\xi|)\int_0^\infty \chi(t)\,\mathrm{e}^{\mathrm{i}t\alpha\xi-\mathrm{i}\Lambda_\varepsilon(t)\xi}\,\mathrm{d}t\Big|\;\leq\;C\varepsilon^{-p}$$

for all $\varepsilon \leq \varepsilon_0$ and all $\xi > 0$. This clearly contradicts the lower estimate (5.12), fixing $\varepsilon = \varepsilon_0$ and letting $\xi \to \infty$. Thus the set of non-regular directions contains all $(-\alpha, 1)$ with $0 < \alpha < 1$. The directions with $\alpha = 0$ and $\alpha = 1$ are also contained in it, sincc this set is closed. The second part of D is treated in the same way, hence the set $W_{1/2}$ provides a lower bound for the non-regular directions over the point $(1, 0)$ as well. □

Proposition 5.4. *The $\mathcal{G}^\infty$-wave front set of the solution $u \in \mathcal{L}(\mathcal{G}_c([0,\infty) \times \mathbb{R}), \widetilde{\mathbb{C}})$ to (5.4) equals*

$$\mathrm{WF}_{\mathcal{G}^\infty}(u) \;=\; W_0 \cup W_1 \cup W_2.$$

Proof. By an argument analogous to the one at the beginning of the proof of Proposition 5.3, the non-$\mathcal{G}^\infty$-regular directions above the points $(t_0, x_0) \neq (1, 0)$ are described by the sets W_0 and W_2. Thus we only have to look at the fiber above the point $(1, 0)$. We will show that $W_1 \subset \mathrm{WF}_{\mathcal{G}^\infty}(\partial_t(\partial_t + \partial_x)u)$; all the more, $W_1 \subset \mathrm{WF}_{\mathcal{G}^\infty}(u)$. Observe that the support of $\partial_t(\partial_t + \partial_x)u$ consists of the single point $(1, 0)$, so the cut-off is superfluous and we simply calculate

$$\begin{aligned} F_\varepsilon(\tau,\xi) &= -\mathcal{F}\big(\partial_t(\partial_t + \partial_x)\delta(x - \Lambda_\varepsilon(t))\big)(\tau,\xi) \\ &= -\big\langle \partial_t(\partial_t + \partial_x)\delta(x - \Lambda_\varepsilon(t)), \mathrm{e}^{-\mathrm{i}t\tau - \mathrm{i}x\xi}\big\rangle \\ &= \tau(\tau+\xi)\int_{1-2\varepsilon}^{1+2\varepsilon} \theta\big(\tfrac{t-1}{\varepsilon}\big)\,\mathrm{e}^{-\mathrm{i}t\tau - \mathrm{i}\Lambda_\varepsilon(t)\xi}\,\mathrm{d}t. \end{aligned}$$

Here we used the fact that, as a distribution, $\partial_t(\partial_t + \partial_x)\delta(x - \Lambda_\varepsilon(t))$ vanishes for $|t-1| > \varepsilon$. The insertion of the cut-off term θ takes care of that, where $\theta \in \mathcal{D}(\mathbb{R})$, $\theta \equiv 1$ on $[-1, 1]$, $\theta \equiv 0$ outside $[-2, 2]$. Next, we insert the explicit expression

$$\Lambda_\varepsilon(t) = \int_{1-\varepsilon}^{t}\int_{-\infty}^{(s-1)/\varepsilon} \rho(r)\,\mathrm{d}r\,\mathrm{d}s = \varepsilon\int_{-1}^{(t-1)/\varepsilon}\int_{-\infty}^{s} \rho(r)\,\mathrm{d}r\,\mathrm{d}s$$

and change coordinates to see that

$$F_\varepsilon(\tau,\xi) = \tau(\tau+\xi)\,\varepsilon\int_{-2}^{2}\theta(t)\,\mathrm{e}^{-\mathrm{i}(1+\varepsilon t)\tau - \mathrm{i}\varepsilon\xi\int_{-1}^{t}\int_{-\infty}^{s}\rho(r)\,\mathrm{d}r\,\mathrm{d}s}\,\mathrm{d}t.$$

Choosing any direction $\tau = \alpha\xi$ with $\alpha \in \mathbb{R}$, $\alpha \neq 0$, we have that

$$F_\varepsilon(\alpha\xi,\xi) = \xi^2\alpha(\alpha+1)\,\varepsilon\,\mathrm{e}^{-\mathrm{i}\alpha\xi}\int_{-2}^{2}\theta(t)\,\mathrm{e}^{-\mathrm{i}\varepsilon\alpha\xi t - \mathrm{i}\varepsilon\xi\int_{-1}^{t}\int_{-\infty}^{s}\rho(r)\,\mathrm{d}r\,\mathrm{d}s}\,\mathrm{d}t.$$

Now suppose there is $p \geq 0$ such that for all $l \in \mathbb{N}_0$

$$\sup_{\xi>0}\langle\xi\rangle^l|F_\varepsilon(\alpha\xi,\xi)| = \mathcal{O}(\varepsilon^{-p}) \text{ as } \varepsilon \to 0. \tag{5.13}$$

Choosing $\xi_\varepsilon = 1/\sqrt{\varepsilon}$, the integral converges to $\int_{-2}^{2}\theta(t)\,\mathrm{d}t$, so

$$\langle\xi_\varepsilon\rangle^l|F_\varepsilon(\alpha\xi_\varepsilon,\xi_\varepsilon)| \approx \varepsilon^{1-(l+2)/2} = \varepsilon^{-l/2}$$

contradicting (5.13) when $l/2 > p$, as long as $\alpha \neq -1$. Thus all directions $\tau = \alpha\xi$ with $\alpha \neq 0$, $\alpha \neq -1$ are not $\mathcal{G}^\infty$-regular. The set of non-regular directions is closed; thus there is no $\mathcal{G}^\infty$-regular direction above $(1,0)$. □

We can extract further information on how the wave front set depends on the regularity of the coefficient λ. Recall that a family $(r_\varepsilon)_{\varepsilon\in(0,1]}$ is termed a *slow scale net* if

$$\forall p > 0 : |r_\varepsilon|^p = \mathcal{O}(\varepsilon^{-1}) \text{ as } \varepsilon \to 0.$$

In place of the definition (5.3) of λ_ε with ε as regularization parameter (i.e., scale $1/\varepsilon$) we could also choose another scale $r_\varepsilon \to \infty$ and put

$$\lambda_\varepsilon(t) = \int_{-\infty}^{(t-1)r_\varepsilon} \rho(s)\,\mathrm{d}s.$$

It is straightforward to see that λ defined in this way belongs to $\mathcal{G}^\infty([0,\infty)\times\mathbb{R})$ if and only if $(r_\varepsilon)_{\varepsilon\in(0,1]}$ is a slow scale net. In fact, one can show [14] that a bounded family of smooth functions defines an element of $\mathcal{G}^\infty$ if and only if all its derivatives satisfy local slow scale bounds.

Proposition 5.5. *If the coefficient λ defined as above belongs to $\mathcal{G}^\infty([0,\infty)\times\mathbb{R})$, then the generalized wave front sets of the solution $u \in \mathcal{L}(\mathcal{G}_c([0,\infty)\times\mathbb{R}),\widetilde{\mathbb{C}})$ to* (5.4) *satisfy*

$$\mathrm{WF}_{\mathcal{G}}(u) \;=\; \mathrm{WF}_{\mathcal{G}^\infty}(u) \;=\; W_0 \cup W_{1/2} \cup W_2.$$

Proof. The proof proceeds along the same lines as the one of Proposition 5.3. When we arrive at the expression (5.9), we observe that all factors are bounded by slow scale nets, hence so is their product. In particular, ε^{-1} gives an asymptotic upper bound, and we conclude that

$$\Big|\int_0^\infty \chi(t)\,\mathrm{e}^{-\mathrm{i}t\tau-\mathrm{i}\Lambda_\varepsilon(t)\xi}\,\mathrm{d}t\Big| \le C\varepsilon^{-1}(1+|\xi|+|\tau|)^{-m}$$

for $(\tau,\xi)\in\Gamma$ and all $m\in\mathbb{N}_0$. Thus every direction in the set R is $\mathcal{G}^\infty$-regular and so

$$\mathrm{WF}_{\mathcal{G}}(u) \;\subset\; \mathrm{WF}_{\mathcal{G}^\infty}(u) \;\subset\; W_0 \cup W_{1/2} \cup W_2.$$

On the other hand, the argument leading to the contradiction with (5.12) does not depend on the choice of scale. Therefore, the directions $(-\alpha\xi,\xi)$ with $0\le\alpha\le 1$ are not $\mathcal{G}$-regular for $\mathcal{F}(\chi u)$, hence

$$W_0 \cup W_{1/2} \cup W_2 \;\subset\; \mathrm{WF}_{\mathcal{G}}(u)$$

as desired. □

Acknowledgments

The author gratefully acknowledges support by the Laboratoire Analyse, Optimisation, Contrôle of the Université des Antilles et de la Guyane during the completion of this work. Thanks are due to an anonymous referee whose careful reading helped eliminating some of the author's lapses.

References

[1] J.F. Colombeau, *New Generalized Functions and Multiplication of Distributions*, North-Holland, Amsterdam, 1984.

[2] J.F. Colombeau, *Elementary Introduction to New Generalized Functions*, North Holland, Amsterdam, 1985.

[3] C. Garetto, Topological structures in Colombeau algebras: topological $\widetilde{\mathbb{C}}$-modules and duality theory, *Acta. Appl. Math.* **88** (2005), 81–123.

[4] C. Garetto, Topological structures in Colombeau algebras: investigation of the duals of $\mathcal{G}_c(\Omega)$, $\mathcal{G}(\Omega)$, and $\mathcal{G}_{\mathcal{S}}(\Omega)$, *Monatsh. Math.* **146** (2005), 203–226.

[5] C. Garetto, Microlocal analysis in the dual of a Colombeau algebra: generalized wave front sets and noncharacteristic regularity, *New York J. Math.* **12** (2006), 275-318.

[6] C. Garetto, Generalized Fourier integral operators on spaces of Colombeau type, in *New Developments in Pseudo-Differential Operators,* Editors: Luigi Rodino and M.W. Wong, Birkhäuser, 2008, 141–188.

[7] C. Garetto, T. Gramchev and M. Oberguggenberger, Pseudodifferential operators with generalized symbols and regularity theory, *Electron. J. Diff. Eqns.* **2005** (116) (2005), 1–43.

[8] C. Garetto and G. Hörmann, Microlocal analysis of generalized functions: pseudodifferential techniques and propagation of singularities, *Proc. Edinb. Math. Soc.* **48** (2005), 603–629.

[9] C. Garetto and G. Hörmann, On duality theory and pseudodifferential techniques for Colombeau algebras: generalized delta functionals, kernels and wave front sets, *Bull. T. CXXXIII Acad. Serbe Sci. Arts, Cl. Sci. Math. Nat., Sci. Math.* **31** (2006), 115–136.

[10] M. Grosser, M. Kunzinger, M. Oberguggenberger and R. Steinbauer. *Geometric Theory of Generalized Functions with Applications to General Relativity*, Kluwer Acad. Publ., Dordrecht, 2001.

[11] L. Hörmander, *The Analysis of Linear Partial Differential Operators I*, 2nd Ed., Springer-Verlag, Berlin, 1990.

[12] G. Hörmann, First-order hyperbolic pseudodifferential equations with generalized symbols, *J. Math. Anal. Appl.* **293** (2004), 40–56.

[13] G. Hörmann and M.V. de Hoop, Microlocal analysis and global solutions of some hyperbolic equations with discontinuous coefficients, *Acta Appl. Math.* **67** (2001), 173–224.

[14] G. Hörmann, M. Oberguggenberger and S. Pilipović, Microlocal hypoellipticity of linear partial differential operators with generalized functions as coefficients, Trans. Amer. Math. Soc. **358** (2006), 3363–3383.

[15] F. Lafon and M. Oberguggenberger, Generalized solutions to symmetric hyperbolic systems with discontinuous coefficients: the multidimensional case, *J. Math. Anal. Appl.* **160** (1991), 93–106.

[16] M. Nedeljkov, S. Pilipović and D. Scarpalézos, *The Linear Theory of Colombeau Generalized Functions*, Longman Scientific & Technical, Harlow, 1998.

[17] M. Oberguggenberger, Hyperbolic systems with discontinuous coefficients: Generalized solutions and a transmission problem in acoustics, *J. Math. Anal. Appl.* **142** (1989), 452–467.

[18] M. Oberguggenberger, *Multiplication of Distributions and Applications to Partial Differential Equations*, Longman Scientific & Technical, Harlow, 1992.

[19] M.E. Taylor, *Pseudodifferential Operators.* Princeton University Press, Princeton, New Jersey 1981.

Michael Oberguggenberger
Arbeitsbereich für Technische Mathematik
Universität Innsbruck
Technikerstraße 13
A-6020 Innsbruck, Austria
e-mail: `michael.oberguggenberger@uibk.ac.at`

Operator Theory:
Advances and Applications, Vol. 189, 137–184

Generalized Fourier Integral Operators on Spaces of Colombeau Type

Claudia Garetto

Abstract. Generalized Fourier integral operators (FIOs) acting on Colombeau algebras are defined. This is based on a theory of generalized oscillatory integrals (OIs) whose phase functions as well as amplitudes may be generalized functions of Colombeau type. The mapping properties of these FIOs are studied as the composition with a generalized pseudo-differential operator. Finally, the microlocal Colombeau regularity for OIs and the influence of the FIO action on generalized wave front sets are investigated. This theory of generalized FIOs is motivated by the need of a general framework for partial differential operators with non-smooth coefficients and distributional data.

Mathematics Subject Classification (2000). Primary 35S30; Secondary 46F30.

Keywords. Fourier integral operators, Colombeau algebras.

1. Introduction

This work is part of a program that aims to solve linear partial differential equations with non-smooth coefficients and highly singular data and investigate the qualitative properties of the solutions. A well-established theory with powerful analytic methods is available in the case of operators with (relatively) smooth coefficients [21], but cannot be applied to many models from physics which involve non-smooth variations of the physical parameters. These models require indeed partial differential operators where the smoothness assumption on the coefficients is dropped. Furthermore, in case of nonlinear operations (cf. [25, 29, 35]), the theory of distribution does not provide a general framework in which solutions exist.

An alternative framework is provided by the theory of Colombeau algebras of generalized functions [4, 19, 35]. We recall that the space of distributions $\mathcal{D}'(\Omega)$ is embedded via convolution with a mollifier in the Colombeau algebra $\mathcal{G}(\Omega)$ of generalized functions on Ω and interpreting the non-smooth coefficients and data as elements of the Colombeau algebra, existence and uniqueness has been established

This work was completed with the support of FWF (Austria), grants T305-N13 and Y237-N13.

for many classes of equations by now [1, 2, 3, 5, 23, 27, 30, 33, 35, 37, 38, 40]. In order to study the regularity of solutions, microlocal techniques have to be introduced into this setting, in particular, pseudo-differential operators with generalized amplitudes and generalized wave front sets. This has been done in the papers [15, 16, 17, 22, 24, 26, 28, 32, 39], with a special attention for elliptic equations and hypoellipticity.

The interest for hyperbolic equations, regularity of solutions and inverse problems (determining the non-smooth coefficients from the data is an important problem in geophysics [8]), leads in the case of differential operators with Colombeau coefficients, to a theory of Fourier integral operators with generalized amplitudes *and* generalized phase functions. This has been initiated in [18] and has provided some first results on propagation of singularities in the dual $\mathcal{L}(\mathcal{G}_c(\Omega), \widetilde{\mathbb{C}})$ of the Colombeau algebra $\mathcal{G}_c(\Omega)$. We recall that within the Colombeau algebra $\mathcal{G}(\Omega)$, regularity theory is based on the subalgebra $\mathcal{G}^\infty(\Omega)$ of regular generalized functions, whose intersection with $\mathcal{D}'(\Omega)$ coincides with $\mathcal{C}^\infty(\Omega)$. Since $\mathcal{G}^\infty(\Omega) \subseteq \mathcal{G}(\Omega) \subseteq \mathcal{L}(\mathcal{G}_c(\Omega), \widetilde{\mathbb{C}})$, two different regularity theories coexist in the dual: one based on $\mathcal{G}(\Omega)$ and one based on $\mathcal{G}^\infty(\Omega)$.

This work can be considered as a compendium of [18], in the sense that collects (without proof) the main results achieved in [18] and studies the composition between a generalized Fourier integral operator and a generalized pseudo-differential operator in addition.

We can now describe the contents in more detail. Section 2 provides the needed background of Colombeau theory. In particular, topological concepts, generalized symbols and the definition of $\mathcal{G}$- and $\mathcal{G}^\infty$-wave front set are recalled. In Subsection 2.5 we elaborate and state in full generality the notion of asymptotic expansion of a generalized symbol introduced for the first time in [15] and we prove a new and technically useful characterization. Section 3 develops the foundations for generalized Fourier integral operators: oscillatory integrals with generalized phase functions and amplitudes. They are then supplemented by an additional parameter in Section 4, leading to the notion of a Fourier integral operator with generalized amplitude and phase function. We study the mapping properties of such operators on Colombeau algebras, the extension to the dual $\mathcal{L}(\mathcal{G}_c(\Omega), \widetilde{\mathbb{C}})$ and we present suitable assumptions on phase function and amplitude which lead to $\mathcal{G}^\infty$-mapping properties. The core of the work is Section 5, where, by making use of some technical preliminaries, we study in Theorem 5.10 the composition $a(x,D)F_\omega(b)$ of a generalized pseudo-differential operator $a(x,D)$ with a generalized Fourier integral operator of the form

$$F_\omega(b)(u)(x) = \int_{\mathbb{R}^n} \mathrm{e}^{i\omega(x,\eta)} b(x,\eta)\widehat{u}(\eta)\, đ\eta.$$

The final Section 6 collects the first results of microlocal analysis for generalized Fourier integral operators obtained in [18, Section 4]. A deeper investigation of the microlocal properties of generalized Fourier integral operators is current topic of research.

2. Basic notions: Colombeau and duality theory

This section gives some background of Colombeau and duality theory for the techniques used in the sequel of the current work. As main sources we refer to [12, 13, 15, 16, 19].

2.1. Nets of complex numbers

Before dealing with the major points of the Colombeau construction we begin by recalling some definitions concerning elements of $\mathbb{C}^{(0,1]}$.

A net $(u_\varepsilon)_\varepsilon$ in $\mathbb{C}^{(0,1]}$ is said to be *strictly nonzero* if there exist $r > 0$ and $\eta \in (0,1]$ such that $|u_\varepsilon| \geq \varepsilon^r$ for all $\varepsilon \in (0,\eta]$.

The regularity issues discussed in Sections 3 and 4 will make use of the following concept of *slow scale net (s.s.n)*. A slow scale net is a net $(r_\varepsilon)_\varepsilon \in \mathbb{C}^{(0,1]}$ such that

$$\forall q \geq 0\, \exists c_q > 0\, \forall \varepsilon \in (0,1] \qquad |r_\varepsilon|^q \leq c_q \varepsilon^{-1}.$$

Throughout this paper we will always consider slow scale nets $(r_\varepsilon)_\varepsilon$ of positive real numbers with $\inf_{\varepsilon\in(0,1]} r_\varepsilon \neq 0$. A net $(u_\varepsilon)_\varepsilon$ in $\mathbb{C}^{(0,1]}$ is said to be *slow scale-strictly nonzero* is there exist a slow scale net $(s_\varepsilon)_\varepsilon$ and $\eta \in (0,1]$ such that $|u_\varepsilon| \geq 1/s_\varepsilon$ for all $\varepsilon \in (0,\eta]$.

2.2. $\widetilde{\mathbb{C}}$-modules of generalized functions based on a locally convex topological vector space E

The most common algebras of generalized functions of Colombeau type as well as the spaces of generalized symbols we deal with are introduced and investigated under a topological point of view by referring to the following models.

Let E be a locally convex topological vector space topologized through the family of seminorms $\{p_i\}_{i\in I}$. The elements of

$$\begin{aligned}
\mathcal{M}_E &:= \{(u_\varepsilon)_\varepsilon \in E^{(0,1]} : \forall i \in I\ \exists N \in \mathbb{N}\quad p_i(u_\varepsilon) = O(\varepsilon^{-N})\ \text{as}\ \varepsilon \to 0\},\\
\mathcal{M}_E^{\mathrm{sc}} &:= \{(u_\varepsilon)_\varepsilon \in E^{(0,1]} : \forall i \in I\ \exists (\omega_\varepsilon)_\varepsilon\ \text{s.s.n.}\quad p_i(u_\varepsilon) = O(\omega_\varepsilon)\ \text{as}\ \varepsilon \to 0\},\\
\mathcal{M}_E^{\infty} &:= \{(u_\varepsilon)_\varepsilon \in E^{(0,1]} : \exists N \in \mathbb{N}\ \forall i \in I\quad p_i(u_\varepsilon) = O(\varepsilon^{-N})\ \text{as}\ \varepsilon \to 0\},\\
\mathcal{N}_E &:= \{(u_\varepsilon)_\varepsilon \in E^{(0,1]} : \forall i \in I\ \forall q \in \mathbb{N}\quad p_i(u_\varepsilon) = O(\varepsilon^{q})\ \text{as}\ \varepsilon \to 0\},
\end{aligned}$$

are called E-moderate, E-moderate of slow scale type, E-regular and E-negligible, respectively. We define the space of *generalized functions based on* E as the factor space $\mathcal{G}_E := \mathcal{M}_E/\mathcal{N}_E$.

The ring of *complex generalized numbers*, denoted by $\widetilde{\mathbb{C}} := \mathcal{E}_M/\mathcal{N}$, is obtained by taking $E = \mathbb{C}$. $\widetilde{\mathbb{C}}$ is not a field since by Theorem 1.2.38 in [19] only the elements which are strictly nonzero (i.e., the elements which have a representative strictly nonzero) are invertible and vice versa. Note that all the representatives of $u \in \widetilde{\mathbb{C}}$ are strictly nonzero once we know that there exists at least one which is strictly nonzero. When u has a representative which is slow scale-strictly nonzero we say that it is *slow scale-invertible*.

For any locally convex topological vector space E the space $\mathcal{G}_E$ has the structure of a $\widetilde{\mathbb{C}}$-module. The $\mathbb{C}$-module $\mathcal{G}_E^{\mathrm{sc}} := \mathcal{M}_E^{\mathrm{sc}}/\mathcal{N}_E$ of *generalized functions of slow scale type* and the $\widetilde{\mathbb{C}}$-module $\mathcal{G}_E^\infty := \mathcal{M}_E^\infty/\mathcal{N}_E$ of *regular generalized functions* are subrings of $\mathcal{G}_E$ with more refined assumptions of moderateness at the level of representatives. We use the notation $u = [(u_\varepsilon)_\varepsilon]$ for the class u of $(u_\varepsilon)_\varepsilon$ in $\mathcal{G}_E$. This is the usual way we adopt to denote an equivalence class.

The family of seminorms $\{p_i\}_{i\in I}$ on E determines a *locally convex $\widetilde{\mathbb{C}}$-linear* topology on $\mathcal{G}_E$ (see [12, Definition 1.6]) by means of the *valuations*

$$\mathrm{v}_{p_i}([(u_\varepsilon)_\varepsilon]) := \mathrm{v}_{p_i}((u_\varepsilon)_\varepsilon) := \sup\{b \in \mathbb{R} : \qquad p_i(u_\varepsilon) = O(\varepsilon^b) \text{ as } \varepsilon \to 0\}$$

and the corresponding *ultra-pseudo-seminorms* $\{\mathcal{P}_i\}_{i\in I}$, where $\mathcal{P}_i(u) = \mathrm{e}^{-\mathrm{v}_{p_i}(u)}$. For the sake of brevity we omit to report definitions and properties of valuations and ultra-pseudo-seminorms in the abstract context of $\widetilde{\mathbb{C}}$-modules. Such a theoretical presentation can be found in [12, Subsections 1.1, 1.2]. We recall that on $\widetilde{\mathbb{C}}$ the valuation and the ultra-pseudo-norm obtained through the absolute value in $\mathbb{C}$ are denoted by $\mathrm{v}_{\widetilde{\mathbb{C}}}$ and $|\cdot|_\mathrm{e}$ respectively. Concerning the space $\mathcal{G}_E^\infty$ of regular generalized functions based on E the moderateness properties of $\mathcal{M}_E^\infty$ allows to define the valuation

$$\mathrm{v}_E^\infty((u_\varepsilon)_\varepsilon) := \sup\{b \in \mathbb{R} : \forall i \in I \qquad p_i(u_\varepsilon) = O(\varepsilon^b) \text{ as } \varepsilon \to 0\}$$

which extends to $\mathcal{G}_E^\infty$ and leads to the ultra-pseudo-norm $\mathcal{P}_E^\infty(u) := \mathrm{e}^{-\mathrm{v}_E^\infty(u)}$.

The Colombeau algebra $\mathcal{G}(\Omega) = \mathcal{E}_M(\Omega)/\mathcal{N}(\Omega)$ can be obtained as a $\widetilde{\mathbb{C}}$-module of $\mathcal{G}_E$-type by choosing $E = \mathcal{E}(\Omega)$. Topologized through the family of seminorms $p_{K,i}(f) = \sup_{x\in K, |\alpha|\le i} |\partial^\alpha f(x)|$ where $K \Subset \Omega$, the space $\mathcal{E}(\Omega)$ induces on $\mathcal{G}(\Omega)$ a metrizable and complete locally convex $\widetilde{\mathbb{C}}$-linear topology which is determined by the ultra-pseudo-seminorms $\mathcal{P}_{K,i}(u) = \mathrm{e}^{-\mathrm{v}_{p_{K,i}}(u)}$. From a structural point of view $\Omega \to \mathcal{G}(\Omega)$ is a fine sheaf of differential algebras on $\mathbb{R}^n$.

The Colombeau algebra $\mathcal{G}_\mathrm{c}(\Omega)$ of generalized functions with compact support is topologized by means of a strict inductive limit procedure. More precisely, setting $\mathcal{G}_K(\Omega) := \{u \in \mathcal{G}_\mathrm{c}(\Omega) : \operatorname{supp} u \subseteq K\}$ for $K \Subset \Omega$, $\mathcal{G}_\mathrm{c}(\Omega)$ is the strict inductive limit of the sequence of locally convex topological $\widetilde{\mathbb{C}}$-modules $(\mathcal{G}_{K_n}(\Omega))_{n\in\mathbb{N}}$, where $(K_n)_{n\in\mathbb{N}}$ is an exhausting sequence of compact subsets of Ω such that $K_n \subseteq K_{n+1}$. We endow $\mathcal{G}_K(\Omega)$ with the topology induced by $\mathcal{G}_{\mathcal{D}_{K'}(\Omega)}$ where K' is a compact subset containing K in its interior. For more details concerning the topological structure of $\mathcal{G}_\mathrm{c}(\Omega)$ see [13, Example 3.7].

Regularity theory in the Colombeau context as initiated in [35] is based on the subalgebra $\mathcal{G}^\infty(\Omega)$ of all elements u of $\mathcal{G}(\Omega)$ having a representative $(u_\varepsilon)_\varepsilon$ belonging to the set

$$\mathcal{E}_M^\infty(\Omega) := \{(u_\varepsilon)_\varepsilon \in \mathcal{E}[\Omega] : \ \forall K \Subset \Omega\, \exists N \in \mathbb{N}\, \forall \alpha \in \mathbb{N}^n \\ \sup_{x\in K} |\partial^\alpha u_\varepsilon(x)| = O(\varepsilon^{-N}) \text{ as } \varepsilon \to 0\}.$$

$\mathcal{G}^\infty(\Omega)$ can be seen as the intersection $\cap_{K\Subset\Omega}\mathcal{G}^\infty(K)$, where $\mathcal{G}^\infty(K)$ is the space of all $u \in \mathcal{G}(\Omega)$ having a representative $(u_\varepsilon)_\varepsilon$ satisfying the condition: $\exists N \in \mathbb{N}$ $\forall\alpha \in \mathbb{N}^n$, $\sup_{x\in K} |\partial^\alpha u_\varepsilon(x)| = O(\varepsilon^{-N})$. The ultra-pseudo-seminorms $\mathcal{P}_{\mathcal{G}^\infty(K)}(u) := \mathrm{e}^{-\mathrm{v}_{\mathcal{G}^\infty(K)}}$, where

$$\mathrm{v}_{\mathcal{G}^\infty(K)} := \sup\{b \in \mathbb{R} : \forall\alpha \in \mathbb{N}^n \quad \sup_{x\in K} |\partial^\alpha u_\varepsilon(x)| = O(\varepsilon^b)\}$$

equip $\mathcal{G}^\infty(\Omega)$ with the topological structure of a *Fréchet* $\widetilde{\mathbb{C}}$*-module*.

Finally, let us consider the algebra $\mathcal{G}^\infty_\mathrm{c}(\Omega) := \mathcal{G}^\infty(\Omega) \cap \mathcal{G}_\mathrm{c}(\Omega)$. On $\mathcal{G}^\infty_K(\Omega) := \{u \in \mathcal{G}^\infty(\Omega) : \operatorname{supp} u \subseteq K\}$ with $K \Subset \Omega$, we define the ultra-pseudo-norm $\mathcal{P}_{\mathcal{G}^\infty_K(\Omega)}(u) = \mathrm{e}^{-\mathrm{v}^\infty_K(u)}$ where $\mathrm{v}^\infty_K(u) := \mathrm{v}^\infty_{\mathcal{D}_{K'}(\Omega)}(u)$ and K' is any compact set containing K in its interior. At this point, given an exhausting sequence $(K_n)_n$ of compact subsets of Ω, the strict inductive limit procedure equips $\mathcal{G}^\infty_\mathrm{c}(\Omega) = \cup_n\mathcal{G}^\infty_{K_n}(\Omega)$ with a complete and separated locally convex $\widetilde{\mathbb{C}}$-linear topology (see [13, Example 3.13]).

2.3. Topological dual of a Colombeau algebra

A duality theory for $\widetilde{\mathbb{C}}$-modules had been developed in [12] in the framework of topological and locally convex topological $\widetilde{\mathbb{C}}$-modules. Starting from an investigation of $\mathcal{L}(\mathcal{G},\widetilde{\mathbb{C}})$, the $\widetilde{\mathbb{C}}$-module of all $\widetilde{\mathbb{C}}$-linear and continuous functionals on $\mathcal{G}$, it provides the theoretical tools for dealing with the topological duals of the Colombeau algebras $\mathcal{G}_\mathrm{c}(\Omega)$ and $\mathcal{G}(\Omega)$. In the paper $\mathcal{L}(\mathcal{G}(\Omega),\widetilde{\mathbb{C}}$ and $\mathcal{L}(\mathcal{G}_\mathrm{c}(\Omega),\widetilde{\mathbb{C}})$ are endowed with the *topology of uniform convergence on bounded subsets*. This is determined by the ultra-pseudo-seminorms

$$\mathcal{P}_{B^\circ}(T) = \sup_{u\in B} |T(u)|_\mathrm{e},$$

where B is varying in the family of all bounded subsets of $\mathcal{G}(\Omega)$ and $\mathcal{G}_\mathrm{c}(\Omega)$ respectively. For general results concerning the relation between boundedness and ultra-pseudo-seminorms in the context of locally convex topological $\widetilde{\mathbb{C}}$-modules we refer to [13, Section 1]. For the choice of topologies illustrated in this section Theorem 3.1 in [13] shows the following chains of continuous embeddings:

$$\mathcal{G}^\infty(\Omega) \subseteq \mathcal{G}(\Omega) \subseteq \mathcal{L}(\mathcal{G}_\mathrm{c}(\Omega),\widetilde{\mathbb{C}}), \tag{2.1}$$

$$\mathcal{G}^\infty_\mathrm{c}(\Omega) \subseteq \mathcal{G}_\mathrm{c}(\Omega) \subseteq \mathcal{L}(\mathcal{G}(\Omega),\widetilde{\mathbb{C}}), \tag{2.2}$$

$$\mathcal{L}(\mathcal{G}(\Omega),\widetilde{\mathbb{C}}) \subseteq \mathcal{L}(\mathcal{G}_\mathrm{c}(\Omega),\widetilde{\mathbb{C}}). \tag{2.3}$$

In (2.1) and (2.2) the inclusion in the dual is given via integration $\big(u \to \big(v \to \int_\Omega u(x)v(x)dx\big)\big)$ (for definitions and properties of the integral of a Colombeau generalized functions see [19]) while the embedding in (2.3) is determined by the inclusion $\mathcal{G}_\mathrm{c}(\Omega) \subseteq \mathcal{G}(\Omega)$. Since $\Omega \to \mathcal{L}(\mathcal{G}_\mathrm{c}(\Omega),\widetilde{\mathbb{C}})$ is a sheaf we can define the *support of a functional* T (denoted by $\operatorname{supp} T$). In analogy with distribution theory, from Theorem 1.2 in [13] we have that $\mathcal{L}(\mathcal{G}(\Omega),\widetilde{\mathbb{C}})$ can be identified with the set of functionals in $\mathcal{L}(\mathcal{G}_\mathrm{c}(\Omega),\widetilde{\mathbb{C}})$ having compact support.

By (2.1) it is meaningful to measure the regularity of a functional in the dual $\mathcal{L}(\mathcal{G}_c(\Omega), \widetilde{\mathbb{C}})$ with respect to the algebras $\mathcal{G}(\Omega)$ and $\mathcal{G}^\infty(\Omega)$. We define the *$\mathcal{G}$-singular support* of T ($\mathrm{singsupp}_\mathcal{G} T$) as the complement of the set of all points $x \in \Omega$ such that the restriction of T to some open neighborhood V of x belongs to $\mathcal{G}(V)$. Analogously replacing $\mathcal{G}$ with $\mathcal{G}^\infty$ we introduce the notion of *$\mathcal{G}^\infty$-singular support* of T denoted by $\mathrm{singsupp}_{\mathcal{G}^\infty} T$. This investigation of regularity is connected with the notions of generalized wave front sets considered in Subsection 2.8 and will be focused on the functionals in $\mathcal{L}(\mathcal{G}_c(\Omega), \widetilde{\mathbb{C}})$ and $\mathcal{L}(\mathcal{G}(\Omega), \widetilde{\mathbb{C}})$ which have a "basic" structure. In detail, we say that $T \in \mathcal{L}(\mathcal{G}_c(\Omega), \widetilde{\mathbb{C}})$ is basic if there exists a net $(T_\varepsilon)_\varepsilon \in \mathcal{D}'(\Omega)^{(0,1]}$ fulfilling the following condition: for all $K \Subset \Omega$ there exist $j \in \mathbb{N}$, $c > 0$, $N \in \mathbb{N}$ and $\eta \in (0, 1]$ such that

$$\forall f \in \mathcal{D}_K(\Omega)\, \forall \varepsilon \in (0, \eta] \qquad |T_\varepsilon(f)| \le c\varepsilon^{-N} \sup_{x \in K, |\alpha| \le j} |\partial^\alpha f(x)|$$

and $Tu = [(T_\varepsilon u_\varepsilon)_\varepsilon]$ for all $u \in \mathcal{G}_c(\Omega)$.
In the same way a functional $T \in \mathcal{L}(\mathcal{G}(\Omega), \widetilde{\mathbb{C}})$ is said to be basic if there exists a net $(T_\varepsilon)_\varepsilon \in \mathcal{E}'(\Omega)^{(0,1]}$ such that there exist $K \Subset \Omega$, $j \in \mathbb{N}$, $c > 0$, $N \in \mathbb{N}$ and $\eta \in (0, 1]$ with the property

$$\forall f \in \mathcal{C}^\infty(\Omega)\, \forall \varepsilon \in (0, \eta] \qquad |T_\varepsilon(f)| \le c\varepsilon^{-N} \sup_{x \in K, |\alpha| \le j} |\partial^\alpha f(x)|$$

and $Tu = [(T_\varepsilon u_\varepsilon)_\varepsilon]$ for all $u \in \mathcal{G}(\Omega)$.
Clearly the sets $\mathcal{L}_\mathrm{b}(\mathcal{G}_c(\Omega), \widetilde{\mathbb{C}})$ and $\mathcal{L}_\mathrm{b}(\mathcal{G}(\Omega), \widetilde{\mathbb{C}})$ of basic functionals are $\widetilde{\mathbb{C}}$-linear subspaces of $\mathcal{L}(\mathcal{G}_c(\Omega), \widetilde{\mathbb{C}})$ and $\mathcal{L}(\mathcal{G}(\Omega), \widetilde{\mathbb{C}})$ respectively. In addition if T is a basic functional in $\mathcal{L}(\mathcal{G}_c(\Omega), \widetilde{\mathbb{C}})$ and $u \in \mathcal{G}_c(\Omega)$ then $uT \in \mathcal{L}(\mathcal{G}(\Omega), \widetilde{\mathbb{C}})$ is basic. We recall that nets $(T_\varepsilon)_\varepsilon$ which define basic maps as above were already considered in [9, 10] with slightly more general notions of moderateness and different choices of notations and language.

2.4. Generalized symbols

For the convenience of the reader we recall a few basic notions concerning the sets of symbols employed in the course of this work. More details can be found in [15, 16] where a theory of generalized pseudo-differential operators acting on Colombeau algebras is developed.

Definitions. Let Ω be an open subset of $\mathbb{R}^n$, $m \in \mathbb{R}$ and $\rho, \delta \in [0, 1]$. $S^m_{\rho,\delta}(\Omega \times \mathbb{R}^p)$ denotes the set of symbols of order m and type (ρ, δ) as introduced by Hörmander in [20]. The subscript (ρ, δ) is omitted when $\rho = 1$ and $\delta = 0$. If V is an open conic set of $\Omega \times \mathbb{R}^p$ we define $S^m_{\rho,\delta}(V)$ as the set of all $a \in \mathcal{C}^\infty(V)$ such that for all $K \Subset V$,

$$\sup_{(x,\xi) \in K^c} \langle \xi \rangle^{-m+\rho|\alpha|-\delta|\beta|} |\partial^\alpha_\xi \partial^\beta_x a(x, \xi)| < \infty,$$

where $K^c := \{(x, t\xi) : (x, \xi) \in K\ t \ge 1\}$. We also make use of the space $S^1_{\mathrm{hg}}(\Omega \times \mathbb{R}^p \setminus 0)$ of all $a \in S^1(\Omega \times \mathbb{R}^p \setminus 0)$ homogeneous of degree 1 in ξ. Note that the assumption of homogeneity allows to state the defining conditions above in terms of the seminorms

$$\sup_{x\in K,\xi\in\mathbb{R}^p\setminus 0} |\xi|^{-1+\alpha}|\partial_\xi^\alpha\partial_x^\beta a(x,\xi)|$$

where K is any compact subset of Ω.

The space of *generalized symbols* $\widetilde{\mathcal{S}}^m_{\rho,\delta}(\Omega\times\mathbb{R}^p)$ is the $\widetilde{\mathbb{C}}$-module of $\mathcal{G}_E$-type obtained by taking $E=S^m_{\rho,\delta}(\Omega\times\mathbb{R}^p)$ equipped with the family of seminorms

$$|a|^{(m)}_{\rho,\delta,K,j}=\sup_{x\in K,\xi\in\mathbb{R}^n}\sup_{|\alpha+\beta|\le j}|\partial_\xi^\alpha\partial_x^\beta a(x,\xi)|\langle\xi\rangle^{-m+\rho|\alpha|-\delta|\beta|},\qquad K\Subset\Omega,\ j\in\mathbb{N}.$$

The valuation corresponding to $|\cdot|^{(m)}_{\rho,\delta,K,j}$ gives the ultra-pseudo-seminorm $\mathcal{P}^{(m)}_{\rho,\delta,K,j}$. $\widetilde{\mathcal{S}}^m_{\rho,\delta}(\Omega\times\mathbb{R}^p)$ topologized through the family $\{\mathcal{P}^{(m)}_{\rho,\delta,K,j}\}_{K\Subset\Omega,j\in\mathbb{N}}$ of ultra-pseudo-seminorms is a Fréchet $\widetilde{\mathbb{C}}$-module. In analogy with $\widetilde{\mathcal{S}}^m_{\rho,\delta}(\Omega\times\mathbb{R}^p)$ we use the notation $\widetilde{\mathcal{S}}^m_{\rho,\delta}(V)$ for the $\widetilde{\mathbb{C}}$-module $\mathcal{G}_{S^m_{\rho,\delta}(V)}$.

$\widetilde{\mathcal{S}}^m_{\rho,\delta}(\Omega_x\times\mathbb{R}^p_\xi)$ has the structure of a sheaf with respect to Ω. So it is meaningful to talk of the support with respect to x of a generalized symbol a ($\mathrm{supp}_x\, a$).

We define the *conic support* of $a\in\widetilde{\mathcal{S}}^m_{\rho,\delta}(\Omega\times\mathbb{R}^p)$ ($\mathrm{cone\,supp}\, a$) as the complement of the set of points $(x_0,\xi_0)\in\Omega\times\mathbb{R}^p$ such that there exists a relatively compact open neighborhood U of x_0, a conic open neighborhood Γ of ξ_0 and a representative $(a_\varepsilon)_\varepsilon$ of a satisfying the condition

$$\forall\alpha\in\mathbb{N}^p\,\forall\beta\in\mathbb{N}^n\,\forall q\in\mathbb{N}\qquad \sup_{x\in U,\xi\in\Gamma}\langle\xi\rangle^{-m+\rho|\alpha|-\delta|\beta|}|\partial_\xi^\alpha\partial_x^\beta a_\varepsilon(x,\xi)|=O(\varepsilon^q)\text{ as }\varepsilon\to 0. \tag{2.4}$$

By definition $\mathrm{cone\,supp}\, a$ is a closed conic subset of $\Omega\times\mathbb{R}^p$. The generalized symbol a is 0 on $\Omega\setminus\pi_x(\mathrm{cone\,supp\,a})$.

Slow scale symbols. In the paper the classes of the factor space $\mathcal{G}^{\mathrm{sc}}_{S^m_{\rho,\delta}(\Omega\times\mathbb{R}^p)}$ are called *generalized symbols of slow scale type.* For simplicity we introduce the notation $\widetilde{S}^{m,\mathrm{sc}}_{\rho,\delta}(\Omega\times\mathbb{R}^p)$. Substituting $S^m_{\rho,\delta}(\Omega\times\mathbb{R}^p)$ with $S^m_{\rho,\delta}(V)$ we obtain the set $\widetilde{S}^{m,\mathrm{sc}}_{\rho,\delta}(V):=\mathcal{G}^{\mathrm{sc}}_{S^m_{\rho,\delta}(V)}$ of slow scale symbols on the open set $V\subseteq\Omega\times(\mathbb{R}^p\setminus 0)$.

Generalized symbols of order $-\infty$. Different notions of regularity are related to the sets $\widetilde{\mathcal{S}}^{-\infty}(\Omega\times\mathbb{R}^p)$ and $\widetilde{\mathcal{S}}^{-\infty,\mathrm{sc}}(\Omega\times\mathbb{R}^p)$ of generalized symbols of order $-\infty$. The space $\widetilde{\mathcal{S}}^{-\infty}(\Omega\times\mathbb{R}^p)$ of generalized symbols of order $-\infty$ is defined as the $\widetilde{\mathbb{C}}$-module $\mathcal{G}_{S^{-\infty}(\Omega\times\mathbb{R}^p)}$. Its elements are equivalence classes a whose representatives $(a_\varepsilon)_\varepsilon$ have the property $|a_\varepsilon|^{(m)}_{K,j}=O(\varepsilon^{-N})$ as $\varepsilon\to 0$, where N depends on the order m of the symbol, on the order j of the derivatives and on the compact set $K\subseteq\Omega$. $\widetilde{\mathcal{S}}^{-\infty,\mathrm{sc}}(\Omega\times\mathbb{R}^p)$ is defined by substituting $O(\varepsilon^{-N})$ with $O(\lambda_\varepsilon)$ in the previous estimate, where $(\lambda_\varepsilon)_\varepsilon$ is a slow scale net depending as above on the order m of the symbol, on the order j of the derivatives and on the compact set $K\subseteq\Omega$. It follows that $(a_\varepsilon)_\varepsilon$ is $\mathcal{G}^\infty$-regular, in the sense that

$$|a_\varepsilon|^{(m)}_{K,j} = O(\varepsilon^{-1})$$

as $\varepsilon \to 0$ for all m, j and $K \Subset \Omega$.

Generalized microsupports. The $\mathcal{G}$- and $\mathcal{G}^\infty$-regularity of generalized symbols on $\Omega \times \mathbb{R}^n$ is measured in conical neighborhoods by means of the following notions of microsupports.

Let $a \in \widetilde{\mathcal{S}}^l_{\rho,\delta}(\Omega \times \mathbb{R}^n)$ and $(x_0, \xi_0) \in T^*(\Omega) \setminus 0$. The symbol a is $\mathcal{G}$-smoothing at (x_0, ξ_0) if there exist a representative $(a_\varepsilon)_\varepsilon$ of a, a relatively compact open neighborhood U of x_0 and a conic neighborhood $\Gamma \subseteq \mathbb{R}^n \setminus 0$ of ξ_0 such that

$$\forall m \in \mathbb{R}\, \forall \alpha, \beta \in \mathbb{N}^n\, \exists N \in \mathbb{N}\, \exists c > 0\, \exists \eta \in (0, 1]\, \forall (x, \xi) \in U \times \Gamma\, \forall \varepsilon \in (0, \eta]$$
$$|\partial^\alpha_\xi \partial^\beta_x a_\varepsilon(x, \xi)| \le c\langle \xi \rangle^m \varepsilon^{-N}. \tag{2.5}$$

The symbol a is $\mathcal{G}^\infty$-smoothing at (x_0, ξ_0) if there exist a representative $(a_\varepsilon)_\varepsilon$ of a, a relatively compact open neighborhood U of x_0, a conic neighborhood $\Gamma \subseteq \mathbb{R}^n \setminus 0$ of ξ_0 and a natural number $N \in \mathbb{N}$ such that

$$\forall m \in \mathbb{R}\, \forall \alpha, \beta \in \mathbb{N}^n\, \exists c > 0\, \exists \eta \in (0, 1]\, \forall (x, \xi) \in U \times \Gamma\, \forall \varepsilon \in (0, \eta]$$
$$|\partial^\alpha_\xi \partial^\beta_x a_\varepsilon(x, \xi)| \le c\langle \xi \rangle^m \varepsilon^{-N}. \tag{2.6}$$

We define the *$\mathcal{G}$-microsupport* of a, denoted by $\mu\,\mathrm{supp}_\mathcal{G}(a)$, as the complement in $T^*(\Omega) \setminus 0$ of the set of points (x_0, ξ_0) where a is $\mathcal{G}$-smoothing and the *$\mathcal{G}^\infty$-microsupport* of a, denoted by $\mu\,\mathrm{supp}_{\mathcal{G}^\infty}(a)$, as the complement in $T^*(\Omega) \setminus 0$ of the set of points (x_0, ξ_0) where a is $\mathcal{G}^\infty$-smoothing.

Continuity Results. By simple reasoning at the level of representatives one proves that the usual operations between generalized symbols, as product and derivation, are continuous. In particular the $\widetilde{\mathbb{C}}$-bilinear map

$$\mathcal{G}_\mathrm{c}(\Omega) \times \widetilde{\mathcal{S}}^m_{\rho,\delta}(\Omega \times \mathbb{R}^p) \to \widetilde{\mathcal{S}}^m_{\rho,\delta}(\Omega \times \mathbb{R}^p) : (u, a) \to a(y, \xi)u(y) \tag{2.7}$$

is continuous. If $l < -p$ each $b \in \widetilde{\mathcal{S}}^l_{\rho,\delta}(\Omega \times \mathbb{R}^p)$ can be integrated on $K \times \mathbb{R}^p$, $K \Subset \Omega$, by setting

$$\int_{K \times \mathbb{R}^p} b(y, \xi)\, dy\, d\xi := \left[\left(\int_{K \times \mathbb{R}^p} b_\varepsilon(y, \xi)\, dy\, d\xi\right)_\varepsilon\right].$$

Moreover if $\mathrm{supp}_y b \Subset \Omega$ we define the integral of b on $\Omega \times \mathbb{R}^p$ as

$$\int_{\Omega \times \mathbb{R}^p} b(y, \xi)\, dy\, d\xi := \int_{K \times \mathbb{R}^p} b(y, \xi)\, dy\, d\xi,$$

where K is any compact set containing $\mathrm{supp}_y b$ in its interior. Integration defines a continuous $\widetilde{\mathbb{C}}$-linear functional on this space of generalized symbols with compact support in y as it is proven in [18, Proposition 1.1, Remark 1.2].

2.5. Asymptotic expansions in $\widetilde{\mathcal{S}}^m_{\rho,\delta}(\Omega\times\mathbb{R}^p)$ and $\widetilde{\mathcal{S}}^{m,\mathrm{sc}}_{\rho,\delta}(\Omega\times\mathbb{R}^p)$

In this subsection we elaborate and state in full generality the notion of asymptotic expansion of a generalized symbol introduced for the first time in [15]. We also provide a technical result which will be useful in Section 5. We begin by working on moderate nets of symbols and we recall that a net $(C_\varepsilon)_\varepsilon \in \mathbb{C}^{(0,1]}$ is said to be of slow scale type if there exists a slow scale net $(\omega_\varepsilon)_\varepsilon$ such that $|C_\varepsilon| = O(\omega_\varepsilon)$.

Definition 2.1. Let $\{m_j\}_{j\in\mathbb{N}}$ be sequences of real numbers with $m_j \searrow -\infty$, $m_0 = m$.

(i) Let $\{(a_{j,\epsilon})_\epsilon\}_{j\in\mathbb{N}}$ be a sequence of elements $(a_{j,\epsilon})_\epsilon \in \mathcal{M}_{S^{m_j}_{\rho,\delta}(\Omega\times\mathbb{R}^p)}$. We say that the formal series $\sum_{j=0}^\infty (a_{j,\epsilon})_\epsilon$ is the asymptotic expansion of $(a_\epsilon)_\epsilon \in \mathcal{E}[\Omega\times\mathbb{R}^n]$, $(a_\epsilon)_\epsilon \sim \sum_j (a_{j,\epsilon})_\epsilon$ for short, iff for all $r \geq 1$

$$\Big(a_\epsilon - \sum_{j=0}^{r-1} a_{j,\epsilon}\Big)_\epsilon \in \mathcal{M}_{S^{m_r}_{\rho,\delta}(\Omega\times\mathbb{R}^p)}.$$

(ii) Let $\{(a_{j,\epsilon})_\epsilon\}_{j\in\mathbb{N}}$ be a sequence of elements $(a_{j,\epsilon})_\epsilon \in \mathcal{M}^{\mathrm{sc}}_{S^{m_j}_{\rho,\delta}(\Omega\times\mathbb{R}^p)}$. We say that the formal series $\sum_{j=0}^\infty (a_{j,\epsilon})_\epsilon$ is the asymptotic expansion of $(a_\epsilon)_\epsilon \in \mathcal{E}[\Omega\times\mathbb{R}^n]$, $(a_\epsilon)_\epsilon \sim_{\mathrm{sc}} \sum_j (a_{j,\epsilon})_\epsilon$ for short, iff for all $r \geq 1$

$$\Big(a_\epsilon - \sum_{j=0}^{r-1} a_{j,\epsilon}\Big)_\epsilon \in \mathcal{M}^{\mathrm{sc}}_{S^{m_r}_{\rho,\delta}(\Omega\times\mathbb{R}^p)}.$$

Theorem 2.2.

(i) *Let $\{(a_{j,\epsilon})_\epsilon\}_{j\in\mathbb{N}}$ be a sequence of elements $(a_{j,\epsilon})_\epsilon \in \mathcal{M}_{S^{m_j}_{\rho,\delta}(\Omega\times\mathbb{R}^p)}$ with $m_j \searrow -\infty$ and $m_0 = m$. Then, there exists $(a_\varepsilon)_\varepsilon \in \mathcal{M}_{S^m_{\rho,\delta}(\Omega\times\mathbb{R}^p)}$ such that $(a_\epsilon)_\epsilon \sim \sum_j (a_{j,\epsilon})_\epsilon$. Moreover, if $(a'_\epsilon)_\epsilon \sim \sum_j (a_{j,\epsilon})_\epsilon$ then $(a_\varepsilon - a'_\varepsilon)_\varepsilon \in \mathcal{M}_{S^{-\infty}(\Omega\times\mathbb{R}^p)}$.*

(ii) *Let $\{(a_{j,\epsilon})_\epsilon\}_{j\in\mathbb{N}}$ be a sequence of elements $(a_{j,\epsilon})_\epsilon \in \mathcal{M}^{\mathrm{sc}}_{S^{m_j}_{\rho,\delta}(\Omega\times\mathbb{R}^p)}$ with $m_j \searrow -\infty$ and $m_0 = m$. Then, there exists $(a_\varepsilon)_\varepsilon \in \mathcal{M}^{\mathrm{sc}}_{S^m_{\rho,\delta}(\Omega\times\mathbb{R}^p)}$ such that $(a_\epsilon)_\epsilon \sim_{\mathrm{sc}} \sum_j (a_{j,\epsilon})_\epsilon$. Moreover, if $(a'_\epsilon)_\epsilon \sim_{\mathrm{sc}} \sum_j (a_{j,\epsilon})_\epsilon$ then $(a_\varepsilon - a'_\varepsilon)_\varepsilon \in \mathcal{M}^{\mathrm{sc}}_{S^{-\infty}(\Omega\times\mathbb{R}^p)}$.*

Proof. The proof follows the classical line of arguments, but we will have to keep track of the ϵ-dependence carefully. We consider a sequence of relatively compact open sets $\{V_l\}$ contained in Ω, such that for all $l \in \mathbb{N}$, $V_l \subset K_l = \overline{V_l} \subset V_{l+1}$ and $\bigcup_{l\in\mathbb{N}} V_l = \Omega$. Let $\psi \in \mathcal{C}^\infty(\mathbb{R}^p)$, $0 \leq \psi(\xi) \leq 1$, such that $\psi(\xi) = 0$ for $|\xi| \leq 1$ and $\psi(\xi) = 1$ for $|\xi| \geq 2$.

(i) We introduce

$$b_{j,\epsilon}(x,\xi) = \psi(\lambda_{j,\varepsilon}\xi)a_{j,\epsilon}(x,\xi),$$

where $\lambda_{j,\varepsilon}$ will be positive constants with $\lambda_{j+1,\varepsilon} < \lambda_{j,\varepsilon} < 1$, $\lambda_{j,\varepsilon} \to 0$ if $j \to \infty$. We can define

$$a_\epsilon(x,\xi) = \sum_{j\in\mathbb{N}} b_{j,\epsilon}(x,\xi). \tag{2.8}$$

This sum is locally finite and therefore $(a_\varepsilon)_\varepsilon \in \mathcal{E}[\Omega \times \mathbb{R}^p]$. We observe that $\partial^\alpha(\psi(\lambda_{j,\varepsilon}\xi)) = \partial^\alpha\psi(\lambda_{j,\varepsilon}\xi)\lambda_{j,\varepsilon}^{|\alpha|}$, $\mathrm{supp}\,(\partial^\alpha\psi(\lambda_{j,\varepsilon}\xi)) \subseteq \{\xi :\ 1/\lambda_{j,\varepsilon} \le |\xi| \le 2/\lambda_{j,\varepsilon}\}$, and that $1/\lambda_{j,\varepsilon} \le |\xi| \le 2/\lambda_{j,\varepsilon}$ implies $\lambda_{j,\varepsilon} \le 2/|\xi| \le 4/(1+|\xi|)$. We first estimate $b_{j,\varepsilon}$. Fixing $K \Subset \Omega$ and $\alpha \in \mathbb{N}^p$, $\beta \in \mathbb{N}^n$, we obtain for $j \in \mathbb{N}$, $\epsilon \in (0,1]$, $x \in K$, $\xi \in \mathbb{R}^p$,

$$\begin{aligned}|\partial_\xi^\alpha\partial_x^\beta b_{j,\epsilon}(x,\xi)| &\le \sum_{\gamma\le\alpha}\binom{\alpha}{\gamma}\lambda_{j,\varepsilon}^{|\alpha-\gamma|}|\partial^{\alpha-\gamma}\psi(\lambda_{j,\varepsilon}\xi)||a_{j,\varepsilon}|^{(m_j)}_{\rho,\delta,K,\gamma,\beta}\langle\xi\rangle^{m_j-\rho|\gamma|+\delta|\beta|}\\ &\le \sum_{\gamma\le\alpha} c(\psi,\gamma)4^{|\alpha-\gamma|}\langle\xi\rangle^{-|\alpha-\gamma|}|a_{j,\varepsilon}|^{(m_j)}_{\rho,\delta,K,\gamma,\beta}\langle\xi\rangle^{m_j-\rho|\gamma|+\delta|\beta|}\\ &\le C_{j,\alpha,\beta,K,\varepsilon}\langle\xi\rangle^{m_j-\rho|\alpha|+\delta|\beta|},\end{aligned}$$

where

$$C_{j,\alpha,\beta,K,\varepsilon} := \sum_{\gamma\le\alpha} c(\psi,\gamma)4^{|\alpha-\gamma|}|a_{j,\varepsilon}|^{(m_j)}_{\rho,\delta,K,\gamma,\beta}.$$

Since $(C_{j,\alpha,\beta,K,\varepsilon})_\varepsilon$ is a moderate net of positive numbers, we have that $(b_{j,\varepsilon})_\varepsilon \in \mathcal{M}_{S^{m_j}_{\rho,\delta}(\Omega\times\mathbb{R}^p)}$. At this point we choose $\lambda_{j,\varepsilon}$ such that for $|\alpha+\beta| \le j$, $l \le j$

$$C_{j,\alpha,\beta,K_l,\varepsilon}\lambda_{j,\varepsilon} \le 2^{-j}. \tag{2.9}$$

Our aim is to prove that $a_\epsilon(x,\xi)$ defined in (2.8) belongs to $\mathcal{M}_{S^m_{\rho,\delta}(\Omega\times\mathbb{R}^p)}$. Since there exists $N_j \in N$ and $\eta_j \in (0,1]$ such that

$$C_{j,\alpha,\beta,K_l,\varepsilon} \le \varepsilon^{-N_j}$$

for $l \le j$ and $|\alpha+\beta| \le j$, we take $\lambda_{j,\varepsilon} = 2^{-j}\varepsilon^{N_j}$ on the interval $(0,\eta_j]$. We observe that

$$\begin{gathered}\forall K \Subset \Omega,\ \exists l \in \mathbb{N}:\ K \subset V_l \subset K_l,\\ \forall\alpha_0 \in \mathbb{N}^p,\ \forall\beta_0 \in \mathbb{N}^n,\ \exists j_0 \in \mathbb{N},\ j_0 \ge l:\ |\alpha_0|+|\beta_0| \le j_0,\quad m_{j_0}+1 \le m,\end{gathered} \tag{2.10}$$

and we write $(a_\epsilon)_\epsilon$ as the sum of the following two terms:

$$\sum_{j=0}^{j_0-1} b_{j,\epsilon}(x,\xi) + \sum_{j=j_0}^{+\infty} b_{j,\epsilon}(x,\xi) = f_\epsilon(x,\xi) + s_\epsilon(x,\xi).$$

For $x \in K$, we have that

$$\begin{aligned}|\partial_\xi^{\alpha_0}\partial_x^{\beta_0} f_\epsilon(x,\xi)| &\le \sum_{j=0}^{j_0-1}|b_{j,\varepsilon}|^{(m_j)}_{\rho,\delta,K,\alpha_0,\beta_0}\langle\xi\rangle^{m_j-\rho|\alpha_0|+\delta|\beta_0|}\\ &\le \left(\sum_{j=0}^{j_0-1}|b_{j,\varepsilon}|^{(m_j)}_{\rho,\delta,K,\alpha_0,\beta_0}\right)\langle\xi\rangle^{m-\rho|\alpha_0|+\delta|\beta_0|},\end{aligned}$$

where

$$\left(\sum_{j=0}^{j_0-1} |b_{j,\varepsilon}|^{(m_j)}_{\rho,\delta,K,\alpha_0,\beta_0}\right)_\varepsilon \in \mathcal{E}_M.$$

We now turn to $s_\epsilon(x,\xi)$. From the estimates on $b_{j,\varepsilon}$ and (2.9), we get for $x \in K$ and $\epsilon \in (0,1]$,

$$|\partial_\xi^{\alpha_0}\partial_x^{\beta_0} s_\epsilon(x,\xi)| \le \sum_{j=j_0}^{+\infty} C_{j,\alpha_0,\beta_0,K_l}\langle\xi\rangle^{m_j-\rho|\alpha_0|+\delta|\beta_0|}$$

$$\le \sum_{j=j_0}^{+\infty} 2^{-j}\lambda_{j,\varepsilon}^{-1}\langle\xi\rangle^{-1}\langle\xi\rangle^{m_j+1-\rho|\alpha_0|+\delta|\beta_0|} \le \sum_{j=j_0}^{+\infty} 2^{-j}\lambda_{j,\varepsilon}^{-1}\langle\xi\rangle^{-1}\langle\xi\rangle^{m-\rho|\alpha_0|+\delta|\beta_0|}.$$

Since $\psi(\xi)$ is identically equal to 0 for $|\xi| \le 1$, we can assume in our estimates that $\langle\xi\rangle^{-1} \le \lambda_{j,\varepsilon}$, and therefore from (2.10), we conclude that

$$|\partial_\xi^{\alpha_0}\partial_x^{\beta_0} s_\epsilon(x,\xi)| \le 2\langle\xi\rangle^{m-\rho|\alpha_0|+\delta|\beta_0|},$$

for all $x \in K$, $\xi \in \mathbb{R}^p$ and $\varepsilon \in (0,1]$.

In order to prove that $(a_\epsilon)_\epsilon \sim \sum_j (a_{j,\epsilon})_\epsilon$ we fix $r \ge 1$ and we write

$$a_\epsilon(x,\xi) - \sum_{j=0}^{r-1} a_{j,\epsilon}(x,\xi) = \sum_{j=0}^{r-1}(\psi(\lambda_{j,\varepsilon}\xi)-1)a_{j,\epsilon}(x,\xi) + \sum_{j=r}^{+\infty}\psi(\lambda_{j,\varepsilon}\xi)a_{j,\epsilon}(x,\xi)$$
$$= g_\epsilon(x,\xi) + t_\epsilon(x,\xi).$$

Recall that $\psi \in \mathcal{C}^\infty(\mathbb{R}^p)$ was chosen such that $\psi - 1 \in \mathcal{C}^\infty_c(\mathbb{R}^p)$ and $\mathrm{supp}(\psi-1) \subseteq \{\xi : |\xi| \le 2\}$. Thus, for $0 \le j \le r-1$,

$$\mathrm{supp}(\psi(\lambda_{j,\varepsilon}\xi)-1) \subseteq \{\xi : |\lambda_{j,\varepsilon}\xi| \le 2\} \subseteq \{\xi : |\xi| \le 2\lambda_{r-1,\varepsilon}^{-1}\}.$$

As a consequence, for fixed $K \Subset \Omega$ and for all $\epsilon \in (0,1]$,

$$|\partial_\xi^\alpha\partial_x^\beta g_\epsilon(x,\xi)| \le \sum_{j=0}^{r-1}\sum_{\alpha'\le\alpha}\binom{\alpha}{\alpha'}\lambda_{j,\varepsilon}^{|\alpha'|}c(\psi,\alpha')\langle 2\lambda_{r-1,\varepsilon}^{-1}\rangle^{m_j-m_r+\rho|\alpha'|}|a_{j,\varepsilon}|^{(m_j)}_{\rho,\delta,K,\alpha-\alpha',\beta}$$
$$\cdot \langle\xi\rangle^{m_r-\rho|\alpha|+\delta|\beta|},$$

where, from our assumptions on $(a_{j,\varepsilon})_\varepsilon$ and $(\lambda_{j,\varepsilon})_\varepsilon$, the nets $(|a_{j,\varepsilon}|^{(m_j)}_{\rho,\delta,K,\alpha-\alpha',\beta})_\varepsilon$ and $(\langle 2\lambda_{r-1,\varepsilon}^{-1}\rangle^{m_j-m_r+\rho|\alpha'|})_\varepsilon$ are both moderate. Repeating the same arguments used in the construction of $(a_\epsilon)_\epsilon$ we have that $(t_\epsilon)_\epsilon$ belongs to $\mathcal{M}_{S^{m_r}_{\rho,\delta}(\Omega\times\mathbb{R}^p)}$. It is clear that $(a_\varepsilon)_\varepsilon$ is uniquely determined by $\sum_j(a_{j,\varepsilon})_\varepsilon$ modulo $\mathcal{M}_{S^{-\infty}(\Omega\times\mathbb{R}^p)}$.

(ii) In the slow scale case one easily sees that $(b_{j,\varepsilon})_\varepsilon \in \mathcal{M}^{\mathrm{sc}}_{S^{m_j}_{\rho,\delta}(\Omega\times\mathbb{R}^p)}$. Moreover, since there exists a slow scale net $\omega_j(\varepsilon)$ and $\eta_j \in (0,1]$ such that

$$C_{j,\alpha,\beta,K_l,\varepsilon} \le \omega_j(\varepsilon)$$

for $l \le j$ and $|\alpha+\beta| \le j$, we can take $\lambda_{j,\varepsilon} = 2^{-j}\omega_j^{-1}(\varepsilon)$ on the interval $(0,\eta_j]$. It follows that $(a_\varepsilon)_\varepsilon \in \mathcal{M}^{\mathrm{sc}}_{S^m_{\rho,\delta}(\Omega\times\mathbb{R}^p)}$ and that both the nets $(g_\varepsilon)_\varepsilon$ and $(t_\varepsilon)_\varepsilon$ belong to $\mathcal{M}^{\mathrm{sc}}_{S^{m_r}_{\rho,\delta}(\Omega\times\mathbb{R}^p)}$. □

Proposition 2.3.

(i) *Let $\{(a_{j,\epsilon})_\epsilon\}_{j\in\mathbb{N}}$ be a sequence of elements $(a_{j,\epsilon})_\epsilon \in \mathcal{M}_{S^{m_j}_{\rho,\delta}(\Omega\times\mathbb{R}^p)}$ with $m_j \searrow -\infty$ and $m_0 = m$. Let $(a_\varepsilon)_\varepsilon \in \mathcal{E}[\Omega\times\mathbb{R}^p]$ such that for all $K \Subset \Omega$, for all α,β there exists $\mu\in\mathbb{R}$ and $(C_\varepsilon)_\varepsilon \in \mathcal{E}_M$ such that*

$$|\partial^\alpha_\xi \partial^\beta_x a_\varepsilon(x,\xi)| \le C_\varepsilon \langle\xi\rangle^\mu, \tag{2.11}$$

for all $x\in K$, $\xi\in\mathbb{R}^p$, $\varepsilon\in(0,1]$. Furthermore, assume that for any $r\ge 1$ and $K\Subset\Omega$ there exists $\mu_r=\mu_r(K)$ and $(C_{r,\varepsilon})_\varepsilon = (C_{r,\varepsilon}(K))_\varepsilon \in \mathcal{E}_M$ such that $\mu_r \to -\infty$ as $r\to+\infty$ and

$$\Big| a_\varepsilon(x,\xi) - \sum_{j=0}^{r-1} a_{j,\varepsilon}(x,\xi)\Big| \le C_{r,\varepsilon}\langle\xi\rangle^{\mu_r} \tag{2.12}$$

for all $x\in K$, $\xi\in\mathbb{R}^p$, $\varepsilon\in(0,1]$. Then, $(a_\varepsilon)_\varepsilon \sim \sum_j (a_{j,\epsilon})_\epsilon$.

(ii) *(i) holds with $(a_{j,\epsilon})_\epsilon \in \mathcal{M}^{\mathrm{sc}}_{S^{m_j}_{\rho,\delta}(\Omega\times\mathbb{R}^p)}$, the nets $(C_\varepsilon)_\varepsilon$ and $(C_{r,\varepsilon})_\varepsilon$ of slow scale type and $(a_\varepsilon)_\varepsilon \sim_{\mathrm{sc}} \sum_j (a_{j,\epsilon})_\epsilon$ in the sense of Definition 2.1(ii).*

The proof of Proposition 2.3 requires the following lemma.

Lemma 2.4. *Let K_1 and K_2 be two compact sets in $\mathbb{R}^p$ such that $K_1 \subset \operatorname{Int} K_2$. Then there exists a constant $C>0$ such that for any smooth function f on a neighborhood of K_2, the following estimate holds:*

$$\Big(\sup_{x\in K_1}\sum_{|\alpha|=1}|D^\alpha f(x)|\Big)^2 \le C \sup_{x\in K_2}|f(x)|\Big(\sup_{x\in K_2}|f(x)| + \sup_{x\in K_2}\sum_{|\alpha|=2}|D^\alpha f(x)|\Big).$$

Proof of Proposition 2.3. (i) By Theorem 2.2 we know that there exists $(b_\varepsilon)_\varepsilon \in \mathcal{M}_{S^m_{\rho,\delta}(\Omega\times\mathbb{R}^p)}$ such that $(b_\varepsilon)_\varepsilon \sim \sum_j (a_{j,\varepsilon})_\varepsilon$.We consider the difference $d_\varepsilon = a_\varepsilon - b_\varepsilon$. From (2.11) and the moderateness of $(b_\varepsilon)_\varepsilon$ we have that for all α,β and $K\Subset\Omega$ there exist $(C'_\varepsilon)_\varepsilon$ and μ' such that

$$|\partial^\alpha_\xi \partial^\beta_x d_\varepsilon(x,\xi)| \le C'_\varepsilon \langle\xi\rangle^{\mu'}, \tag{2.13}$$

for all $x\in K$, $\xi\in\mathbb{R}^p$ and $\varepsilon\in(0,1]$. Combining $(b_\varepsilon)_\varepsilon \sim \sum_j (a_{j,\varepsilon})_\varepsilon$ with (2.12) we obtain that for all $r>0$ and for all $K\Subset\Omega$ there exists $(C_{r,\varepsilon}(K))_\varepsilon \in \mathcal{E}_M$ such that

$$|d_\varepsilon(x,\xi)| \le C_{r,\varepsilon}(K)\langle\xi\rangle^{-r}, \qquad x\in K,\ \xi\in\mathbb{R}^p,\ \varepsilon\in(0,1].$$

Set $d_{\xi,\varepsilon}(x,\theta) = d_\varepsilon(x,\xi+\theta)$. Then, $\partial^\alpha_\theta\partial^\beta_x d_{\xi,\varepsilon}(x,\theta)|_{\theta=0} = \partial^\alpha_\xi\partial^\beta_x d(x,\xi)$, and applying Lemma 2.4 with $K_1 = K\times 0$ and $K_2 = K'\times\{|\theta|\le 1\}$, where $K\subset \mathrm{Int}K' \subset K' \Subset \Omega$, we obtain

$$\Big(\sup_{x\in K}\sum_{|\alpha+\beta|=1}|\partial_\xi^\alpha\partial_x^\beta d_\varepsilon(x,\xi)|\Big)^2 \le C\sup_{x\in K',|\theta|\le 1}|d_\varepsilon(x,\xi+\theta)|\cdot$$
$$\cdot\Big(\sup_{x\in K',|\theta|\le 1}|d_\varepsilon(x,\xi+\theta)|+\sup_{x\in K',|\theta|\le 1}\sum_{|\alpha+\beta|=2}|\partial_\xi^\alpha\partial_x^\beta d_\varepsilon(x,\xi+\theta)|\Big)$$
$$\le CC_{r,\varepsilon}(K')\sup_{|\theta|\le 1}\langle\xi+\theta\rangle^{-r}\Big(C_{r,\varepsilon}(K')\sup_{|\theta|\le 1}\langle\xi+\theta\rangle^{-r}+C'_\varepsilon(K')\sup_{|\theta|\le 1}\langle\xi+\theta\rangle^{\mu'(K,2)}\Big)$$
$$\le C''_{r,\varepsilon}(K)\langle\xi\rangle^{-r}, \tag{2.14}$$

where $C''_{r,\varepsilon}(K)\in\mathcal{E}_M$. By induction one can prove that for all $r>0$, for all $K\Subset\Omega$ and for all $\alpha\in\mathbb{N}^p$, $\beta\in\mathbb{N}^n$, there exists a moderate net $(c_\varepsilon)_\varepsilon$ such that the estimate

$$|\partial_\xi^\alpha\partial_x^\beta d_\varepsilon(x,\xi)|\le c_\varepsilon\langle\xi\rangle^{-r}$$

is valid for all $x\in K$, $\xi\in\mathbb{R}^p$ and $\varepsilon\in(0,1]$. This means that $(d_\varepsilon)_\varepsilon\in\mathcal{M}_{S^{-\infty}(\Omega\times\mathbb{R}^p)}$ and therefore $(a_\varepsilon)_\varepsilon\sim\sum_j(a_{j,\varepsilon})_\varepsilon$.

(ii) It is clear that when we work with nets of slow scale type then $(d_\varepsilon)_\varepsilon\in\mathcal{M}^{\mathrm{sc}}_{S^{-\infty}(\Omega\times\mathbb{R}^p)}$ and $(a_\varepsilon)_\varepsilon\sim_{\mathrm{sc}}\sum_j(a_{j,\epsilon})_\epsilon$. □

Remark 2.5. Proposition 2.3 can be stated for nets of symbols in $\mathcal{M}_{S^m_{\rho,\delta}(\Omega\times\mathbb{R}^p\setminus 0)}$ and $\mathcal{M}^{\mathrm{sc}}_{S^m_{\rho,\delta}(\Omega\times\mathbb{R}^p\setminus 0)}$. The proof make use of (2.13) when $|\xi|\le 1$ and (2.14) when $|\xi|>1$.

Definition 2.6. Let $\{m_j\}_{j\in\mathbb{N}}$ with $m_j\searrow-\infty$ and $m_0=m$.

(i) Let $\{a_j\}_{j\in\mathbb{N}}$ be a sequence of symbols $a_j\in\widetilde{\mathcal{S}}^{m_j}_{\rho,\delta}(\Omega\times\mathbb{R}^p)$. We say that the formal series $\sum_j a_j$ is the asymptotic expansion of $a\in\widetilde{\mathcal{S}}^m_{\rho,\delta}(\Omega\times\mathbb{R}^p)$, $a\sim\sum_j a_j$ for short, iff there exist a representative $(a_\epsilon)_\epsilon$ of a and, for every j, representatives $(a_{j,\epsilon})_\epsilon$ of a_j, such that $(a_\epsilon)_\epsilon\sim\sum_j(a_{j,\epsilon})_\epsilon$.

(ii) Let $\{a_j\}_{j\in\mathbb{N}}$ be a sequence of symbols $a_j\in\widetilde{\mathcal{S}}^{m_j,\mathrm{sc}}_{\rho,\delta}(\Omega\times\mathbb{R}^p)$. We say that the formal series $\sum_j a_j$ is the asymptotic expansion of $a\in\widetilde{\mathcal{S}}^{m,\mathrm{sc}}_{\rho,\delta}(\Omega\times\mathbb{R}^p)$, $a\sim\sum_j a_j$ for short, iff there exist a representative $(a_\epsilon)_\epsilon$ of a and, for every j, representatives $(a_{j,\epsilon})_\epsilon$ of a_j, such that $(a_\epsilon)_\epsilon\sim_{\mathrm{sc}}\sum_j(a_{j,\epsilon})_\epsilon$.

2.6. Generalized pseudo-differential operators

Let Ω be an open subset of $\mathbb{R}^n$ and $a\in\widetilde{\mathcal{S}}^m_{\rho,\delta}(\Omega\times\mathbb{R}^n)$. The generalized oscillatory integral (see [15])

$$\int_{\Omega\times\mathbb{R}^n}\mathrm{e}^{i(x-y)\xi}a(x,\xi)u(y)\,dy\,đ\xi := \Big(\int_{\Omega\times\mathbb{R}^n}\mathrm{e}^{i(x-y)\xi}a_\varepsilon(x,\xi)u_\varepsilon(y)\,dy\,đ\xi\Big)_\varepsilon+\mathcal{N}(\Omega),$$

defines the action of the pseudo-differential operator $a(x,D)$ with generalized symbol $a\in\widetilde{\mathcal{S}}^m_{\rho,\delta}(\Omega\times\mathbb{R}^n)$ on $u\in\mathcal{G}_{\mathrm{c}}(\Omega)$. The operator $a(x,D)$ maps $\mathcal{G}_{\mathrm{c}}(\Omega)$ continuously into $\mathcal{G}(\Omega)$ and can be extended to a continuous $\widetilde{\mathbb{C}}$-linear map from $\mathcal{L}(\mathcal{G}(\Omega),\widetilde{\mathbb{C}})$ to $\mathcal{L}(\mathcal{G}_{\mathrm{c}}(\Omega),\widetilde{\mathbb{C}})$. If a is of slow scale type then $a(x,D)$ maps $\mathcal{G}^\infty_{\mathrm{c}}(\Omega)$ continuously into

$\mathcal{G}^\infty(\Omega)$. Pseudo-differential operators with generalized symbol of order $-\infty$ are regularizing, in the sense that $a(x,D)$ maps $\mathcal{L}_{\rm b}(\mathcal{G}(\Omega),\widetilde{\mathbb{C}})$ to $\mathcal{G}(\Omega)$ if $a\in\widetilde{\mathcal{S}}^{-\infty}(\Omega\times\mathbb{R}^n)$ and $\mathcal{L}_{\rm b}(\mathcal{G}(\Omega),\widetilde{\mathbb{C}})$ to $\mathcal{G}^\infty(\Omega)$ if $a\in\widetilde{\mathcal{S}}^{-\infty,\rm sc}(\Omega\times\mathbb{R}^n)$. Clearly, all the previous results can be stated for pseudo-differential operators given by a generalized amplitude $a(x,y,\xi)\in\widetilde{\mathcal{S}}^m_{\rho,\delta}(\Omega\times\Omega\times\mathbb{R}^n)$. For a complete overview on generalized pseudo-differential operators acting on spaces of Colombeau type we advise the reader to refer to [14, 15, 16]

2.7. Generalized elliptic symbols

One of the main issues in developing a theory of generalized symbols has been the search for a notion of generalized elliptic symbol. This is obviously related to the construction of a generalized pseudo-differential parametrix by means of which to investigate problems of $\mathcal{G}$- and $\mathcal{G}^\infty$-regularity. In the sequel we recall some of the results obtain in this direction in [15, 16], which will be employed in Section 5. We work at the level of representatives and we set $\rho=1$, $\delta=0$. We leave to the reader the proof of the next proposition which is based on [15, Section 6].

Proposition 2.7. *Let* $(a_\varepsilon)_\varepsilon\in\mathcal{M}_{S^m(\Omega\times\mathbb{R}^n\setminus 0)}$ *such that*

(e1) *for all* $K\Subset\Omega$ *there exists* $s\in\mathbb{R}$, $(R_\varepsilon)_\varepsilon\in\mathcal{E}_M$ *strictly nonzero and* $\eta\in(0,1]$ *such that*

$$|a_\varepsilon(x,\xi)|\ge\varepsilon^s\langle\xi\rangle^m,$$

for all $x\in K$, $|\xi|\ge R_\varepsilon$ *and* $\varepsilon\in(0,\eta]$.

Then,

(i) *for all* $K\Subset\Omega$, *for all* $\alpha,\beta\in\mathbb{N}^n$ *there exist* $N\in\mathbb{N}$, $(R_\varepsilon)_\varepsilon\in\mathcal{E}_M$ *strictly nonzero and* $\eta\in(0,1]$ *such that*

$$|\partial^\alpha_\xi\partial^\beta_x a_\varepsilon(x,\xi)|\le\varepsilon^{-N}\langle\xi\rangle^{-|\alpha|}|a_\varepsilon(x,\xi)|$$

for all $x\in K$, $|\xi|\ge R_\varepsilon$ *and* $\varepsilon\in(0,\eta]$;

(ii) (i) *holds for the net* $(a_\varepsilon^{-1})_\varepsilon$;

(iii) *if* $(a'_\varepsilon)_\varepsilon\in\mathcal{M}_{S^{m'}(\Omega\times\mathbb{R}^n\setminus 0)}$ *with* $m'<m$ *then* (e1) *holds for the net* $(a_\varepsilon+a'_\varepsilon)_\varepsilon$.

Let $(a_\varepsilon)_\varepsilon\in\mathcal{M}^{\rm sc}_{S^m(\Omega\times\mathbb{R}^n\setminus 0)}$ *such that*

(e2) *for all* $K\Subset\Omega$ *there exists* $(s_\varepsilon)_\varepsilon$ *with* $(s_\varepsilon^{-1})_\varepsilon$ *s.s.n.*, $(R_\varepsilon)_\varepsilon$ *s.s.n. and* $\eta\in(0,1]$ *such that*

$$|a_\varepsilon(x,\xi)|\ge s_\varepsilon\langle\xi\rangle^m,$$

for all $x\in K$, $|\xi|\ge R_\varepsilon$ *and* $\varepsilon\in(0,\eta]$.

Then,

(iv) *for all* $K\Subset\Omega$, *for all* $\alpha,\beta\in\mathbb{N}^n$ *there exist* $(c_\varepsilon)_\varepsilon$, $(R_\varepsilon)_\varepsilon$ *s.s.n and* $\eta\in(0,1]$ *such that*

$$|\partial^\alpha_\xi\partial^\beta_x a_\varepsilon(x,\xi)|\le c_\varepsilon\langle\xi\rangle^{-|\alpha|}|a_\varepsilon(x,\xi)|$$

for all $x\in K$, $|\xi|\ge R_\varepsilon$ *and* $\varepsilon\in(0,\eta]$;

(v) (i) *holds for the net* $(a_\varepsilon^{-1})_\varepsilon$;

(vi) *if* $(a'_\varepsilon)_\varepsilon\in\mathcal{M}^{\rm sc}_{S^{m'}(\Omega\times\mathbb{R}^n\setminus 0)}$ *with* $m'<m$ *then* (e2) *holds for the net* $(a_\varepsilon+a'_\varepsilon)_\varepsilon$.

Proposition 2.8. *Let $(a_\varepsilon)_\varepsilon$ be a net of elliptic symbols of $S^m(\Omega \times \mathbb{R}^n \setminus 0)$.*

(i) *If $(a_\varepsilon)_\varepsilon \in \mathcal{M}_{S^m(\Omega\times\mathbb{R}^n\setminus 0)}$ fulfills condition* (e1) *then there exists $(p_\varepsilon)_\varepsilon \in \mathcal{M}_{S^{-m}(\Omega\times\mathbb{R}^n\setminus 0)}$ such that for all $\varepsilon \in (0,1]$*

$$p_\varepsilon a_\varepsilon = 1 + r_\varepsilon,$$

where $(r_\varepsilon)_\varepsilon \in \mathcal{M}_{S^{-\infty}(\Omega\times\mathbb{R}^n\setminus 0)}$.

(ii) *If $(a_\varepsilon)_\varepsilon \in \mathcal{M}^{\mathrm{sc}}_{S^m(\Omega\times\mathbb{R}^n\setminus 0)}$ fulfills condition* (e2) *then there exists $(p_\varepsilon)_\varepsilon \in \mathcal{M}^{\mathrm{sc}}_{S^{-m}(\Omega\times\mathbb{R}^n\setminus 0)}$ such that for all $\varepsilon \in (0,1]$*

$$p_\varepsilon a_\varepsilon = 1 + r_\varepsilon,$$

where $(r_\varepsilon)_\varepsilon \in \mathcal{M}^{\mathrm{sc}}_{S^{-\infty}(\Omega\times\mathbb{R}^n\setminus 0)}$.

Proof. As in [15, Proposition 6.4] we define p_ε as

$$\sum_j a_\epsilon^{-1}(x,\xi)\varphi(\frac{\xi}{R_{j,\epsilon}})\psi_j(x),$$

where ψ_j is a partition of unity subordinated to a covering of relatively compact subsets Ω_j of Ω, $(R_{j,\varepsilon})_\varepsilon$ is the radius corresponding to $\overline{\Omega_j}$ and φ is a smooth function on $\mathbb{R}^n$ such that $\varphi(\xi) = 0$ for $|\xi| \le 1$ and $\varphi(\xi) = 1$ for $|\xi| \ge 2$. From Proposition 2.7 we have that (e1) yields $(p_\varepsilon)_\varepsilon \in \mathcal{M}_{S^{-m}(\Omega\times\mathbb{R}^n\setminus 0)}$ and (e2) yields $(p_\varepsilon)_\varepsilon \in \mathcal{M}^{\mathrm{sc}}_{S^{-m}(\Omega\times\mathbb{R}^n\setminus 0)}$. Let $K \Subset \Omega$. By construction, for all $x \in K$,

$$p_\varepsilon(x,\xi)a_\varepsilon(x,\xi) = 1 + \Big(\sum_{j=0}^{j_0}\varphi(\frac{\xi}{R_{j,\epsilon}})\psi_j(x) - 1\Big) = 1 + \sum_{j=0}^{j_0}\Big(\varphi(\frac{\xi}{R_{j,\epsilon}}) - 1\Big)\psi_j(x),$$

and the following estimates hold for all $l > 0$ and $\alpha \in \mathbb{N}^n \setminus 0$:

$$\sup_{\xi\neq 0}\langle\xi\rangle^l|\varphi(\frac{\xi}{R_{j,\epsilon}}) - 1| \le \sup_{|\xi|\le 2R_{j,\epsilon}}\langle\xi\rangle^l|\varphi(\frac{\xi}{R_{j,\epsilon}}) - 1| \le c_\varphi\langle 2R_{j,\varepsilon}\rangle^l,$$

$$\sup_{\xi\neq 0}\langle\xi\rangle^l|\partial_\xi^\alpha\varphi(\frac{\xi}{R_{j,\epsilon}})|(R_{j,\varepsilon})^{-|\alpha|} \le \sup_{R_{j,\varepsilon}\le|\xi|\le 2R_{j,\epsilon}}\langle\xi\rangle^l|\partial_\xi^\alpha\varphi(\frac{\xi}{R_{j,\epsilon}})|(R_{j,\varepsilon})^{-|\alpha|}$$
$$\le c_\varphi\langle 2R_{j,\varepsilon}\rangle^l(R_{j,\varepsilon})^{-|\alpha|}.$$

We deduce that $(p_\varepsilon a_\varepsilon - 1)_\varepsilon$ belongs to $\mathcal{M}_{S^{-\infty}(\Omega\times\mathbb{R}^n\setminus 0)}$ under the hypothesis (e1) on $(a_\varepsilon)_\varepsilon$ and that $(p_\varepsilon a_\varepsilon - 1)_\varepsilon$ belongs to $\mathcal{M}^{\mathrm{sc}}_{S^{-\infty}(\Omega\times\mathbb{R}^n\setminus 0)}$ under the hypothesis (e2) on $(a_\varepsilon)_\varepsilon$. □

2.8. Microlocal analysis in the Colombeau context: Generalized wave front sets in $\mathcal{L}(\mathcal{G}_c(\Omega), \widetilde{\mathbb{C}})$

In this subsection we recall the basic notions of microlocal analysis which involve the duals of the Colombeau algebras $\mathcal{G}_c(\Omega)$ and $\mathcal{G}(\Omega)$ and have been developed in [14]. In this generalized context the role which is classically played by $\mathscr{S}(\mathbb{R}^n)$ is given to the Colombeau algebra $\mathcal{G}_{\mathscr{S}}(\mathbb{R}^n) := \mathcal{G}_{\mathscr{S}(\mathbb{R}^n)}$. $\mathcal{G}_{\mathscr{S}}(\mathbb{R}^n)$ is topologized as in Subsection 2.2 and its dual $\mathcal{L}(\mathcal{G}_{\mathscr{S}}(\mathbb{R}^n), \widetilde{\mathbb{C}})$ is endowed with the topology of uniform convergence on bounded subsets. In the sequel $\mathcal{G}_\tau(\mathbb{R}^n)$ denotes the Colombeau algebra of tempered generalized functions defined as the quotient $\mathcal{E}_\tau(\mathbb{R}^n)/\mathcal{N}_\tau(\mathbb{R}^n)$, where $\mathcal{E}_\tau(\mathbb{R}^n)$ is the algebra of all τ-*moderate* nets $(u_\varepsilon)_\varepsilon \in \mathcal{E}_\tau[\mathbb{R}^n] := \mathcal{O}_M(\mathbb{R}^n)^{(0,1]}$ such that

$$\forall \alpha \in \mathbb{N}^n\, \exists N \in \mathbb{N} \qquad \sup_{x\in\mathbb{R}^n} (1+|x|)^{-N}|\partial^\alpha u_\varepsilon(x)| = O(\varepsilon^{-N}) \qquad \text{as } \varepsilon \to 0$$

and $\mathcal{N}_\tau(\mathbb{R}^n)$ is the ideal of all τ-*negligible* nets $(u_\varepsilon)_\varepsilon \in \mathcal{E}_\tau[\mathbb{R}^n]$ such that

$$\forall \alpha \in \mathbb{N}^n\, \exists N \in \mathbb{N}\, \forall q \in \mathbb{N} \quad \sup_{x\in\mathbb{R}^n} (1+|x|)^{-N}|\partial^\alpha u_\varepsilon(x)| = O(\varepsilon^{q}) \text{ as } \varepsilon \to 0.$$

Theorem 3.8 in [12] shows that we have the chain of continuous embeddings

$$\mathcal{G}_{\mathscr{S}}(\mathbb{R}^n) \subseteq \mathcal{G}_\tau(\mathbb{R}^n) \subseteq \mathcal{L}(\mathcal{G}_{\mathscr{S}}(\mathbb{R}^n), \widetilde{\mathbb{C}}).$$

Moreover, since for any $u \in \mathcal{G}_c(\Omega)$ with $\operatorname{supp} u \subseteq K \Subset \Omega$ and any $K' \Subset \Omega$ with $K \subset \operatorname{Int} K'$ one can find a representative $(u_\varepsilon)_\varepsilon$ with $\operatorname{supp} u_\varepsilon \subseteq K'$ for all $\varepsilon \in (0,1]$, we have that $\mathcal{G}_c(\Omega)$ is continuously embedded into $\mathcal{G}_{\mathscr{S}}(\mathbb{R}^n)$.

The Fourier transform on $\mathcal{G}_{\mathscr{S}}(\mathbb{R}^n)$, $\mathcal{L}(\mathcal{G}_{\mathscr{S}}(\mathbb{R}^n), \widetilde{\mathbb{C}})$ and $\mathcal{L}(\mathcal{G}(\Omega), \widetilde{\mathbb{C}})$. The Fourier transform on $\mathcal{G}_{\mathscr{S}}(\mathbb{R}^n)$ is defined by the corresponding transformation at the level of representatives, as follows:

$$\mathcal{F} : \mathcal{G}_{\mathscr{S}}(\mathbb{R}^n) \to \mathcal{G}_{\mathscr{S}}(\mathbb{R}^n) : u \to [(\widehat{u_\varepsilon})_\varepsilon].$$

$\mathcal{F}$ is a $\widetilde{\mathbb{C}}$-linear continuous map from $\mathcal{G}_{\mathscr{S}}(\mathbb{R}^n)$ into itself which extends to the dual in a natural way. In detail, we define the Fourier transform of $T \in \mathcal{L}(\mathcal{G}_{\mathscr{S}}(\mathbb{R}^n), \widetilde{\mathbb{C}})$ as the functional in $\mathcal{L}(\mathcal{G}_{\mathscr{S}}(\mathbb{R}^n), \widetilde{\mathbb{C}})$ given by

$$\mathcal{F}(T)(u) = T(\mathcal{F}u).$$

As shown in [14, Remark 1.5] $\mathcal{L}(\mathcal{G}(\Omega), \widetilde{\mathbb{C}})$ is embedded in $\mathcal{L}(\mathcal{G}_{\mathscr{S}}(\mathbb{R}^n), \widetilde{\mathbb{C}})$ by means of the map

$$\mathcal{L}(\mathcal{G}(\Omega), \widetilde{\mathbb{C}}) \to \mathcal{L}(\mathcal{G}_{\mathscr{S}}(\mathbb{R}^n), \widetilde{\mathbb{C}}) : T \to \big(u \to T((u_{\varepsilon|\Omega})_\varepsilon + \mathcal{N}(\Omega))\big).$$

In particular, when T is a basic functional in $\mathcal{L}(\mathcal{G}(\Omega), \widetilde{\mathbb{C}})$ we have from [14, Proposition 1.6, Remark 1.7] that the Fourier transform of T is the tempered generalized function obtained as the action of $T(y)$ on $\mathrm{e}^{-iy\xi}$, i.e., $\mathcal{F}(T) = T(\mathrm{e}^{-i\cdot\xi}) = (T_\varepsilon(\mathrm{e}^{-i\cdot\xi}))_\varepsilon + \mathcal{N}_\tau(\mathbb{R}^n)$.

Generalized Wave front sets of a functional in $\mathcal{L}(\mathcal{G}_c(\Omega), \widetilde{\mathbb{C}})$. The notions of $\mathcal{G}$-wave front set and $\mathcal{G}^\infty$-wave front set of a functional in $\mathcal{L}(\mathcal{G}_c(\Omega), \widetilde{\mathbb{C}})$ have been introduced in [14] as direct analogues of the distributional wave front set in [20]. They employ a subset of the space $\mathcal{G}^{\mathrm{sc}}_{S^m(\Omega\times\mathbb{R}^n)}$ of generalized symbols of slow scale type denoted by $\widetilde{\underline{\mathcal{S}}}^m_{\mathrm{sc}}(\Omega \times \mathbb{R}^n)$ (see [16, Definition 1.1]) and a suitable notion of slow scale micro-ellipticity [16, Definition 1.2]. In detail, $(x_0, \xi_0) \notin \mathrm{WF}_{\mathcal{G}}\, T$ (resp. $(x_0, \xi_0) \notin \mathrm{WF}_{\mathcal{G}^\infty}\, T$) if there exists $a(x, D)$ properly supported with $a \in \widetilde{\underline{\mathcal{S}}}^0_{\mathrm{sc}}(\Omega \times$

$\mathbb{R}^n$) such that a is slow scale micro-elliptic at (x_0,ξ_0) and $a(x,D)T \in \mathcal{G}(\Omega)$ (resp. $a(x,D)T \in \mathcal{G}^\infty(\Omega)$).

When T is a basic functional of $\mathcal{L}(\mathcal{G}_c(\Omega),\widetilde{\mathbb{C}})$, Proposition 3.14 in [14] proves that one can limit to classical properly supported pseudo-differential operators in the definition of $\mathrm{WF}_{\mathcal{G}}\,T$ and $\mathrm{WF}_{\mathcal{G}^\infty}\,T$. More precisely,

$$\mathrm{W}_{\mathrm{cl},\mathcal{G}}(T) := \bigcap_{AT\in\mathcal{G}(\Omega)} \mathrm{Char}(A) \tag{2.15}$$

and

$$\mathrm{W}_{\mathrm{cl},\mathcal{G}^\infty}(T) := \bigcap_{AT\in\mathcal{G}^\infty(\Omega)} \mathrm{Char}(A) \tag{2.16}$$

where the intersections are taken over all the classical properly supported operators $A \in \Psi^0(\Omega)$ such that $AT \in \mathcal{G}(\Omega)$ in (2.15) and $AT \in \mathcal{G}^\infty(\Omega)$ in (2.16). $\mathrm{WF}_{\mathcal{G}}T$ and $\mathrm{WF}_{\mathcal{G}^\infty}T$ are both closed conic subsets of $T^*(\Omega)\setminus 0$ and, as proved in [14, Proposition 3.5],

$$\pi_\Omega(\mathrm{WF}_{\mathcal{G}}T) = \mathrm{sing\,supp}_{\mathcal{G}}T$$

and

$$\pi_\Omega(\mathrm{WF}_{\mathcal{G}^\infty}T) = \mathrm{sing\,supp}_{\mathcal{G}^\infty}T.$$

Characterization of $\mathrm{WF}_{\mathcal{G}}T$ and $\mathrm{WF}_{\mathcal{G}^\infty}T$ when T is a basic functional. We will employ a useful characterization of the $\mathcal{G}$-wave front set and the $\mathcal{G}^\infty$-wave front set valid for functionals which are basic. It involves the sets of generalized functions $\mathcal{G}_{\mathscr{S},0}(\Gamma)$ and $\mathcal{G}^\infty_{\mathscr{S},0}(\Gamma)$, defined on the conic subset Γ of $\mathbb{R}^n\setminus 0$, as follows:

$$\mathcal{G}_{\mathscr{S},0}(\Gamma) := \{u \in \mathcal{G}_\tau(\mathbb{R}^n) :\ \exists (u_\varepsilon)_\varepsilon \in u\ \forall l \in \mathbb{R}\ \exists N \in \mathbb{N}\quad \sup_{\xi\in\Gamma}\langle\xi\rangle^l |u_\varepsilon(\xi)| = O(\varepsilon^{-N}) \text{ as } \varepsilon \to 0\},$$

$$\mathcal{G}^\infty_{\mathscr{S},0}(\Gamma) := \{u \in \mathcal{G}_\tau(\mathbb{R}^n) :\ \exists (u_\varepsilon)_\varepsilon \in u\ \exists N \in \mathbb{N}\ \forall l \in \mathbb{R}\quad \sup_{\xi\in\Gamma}\langle\xi\rangle^l |u_\varepsilon(\xi)| = O(\varepsilon^{-N}) \text{ as } \varepsilon \to 0\}.$$

Let $T \in \mathcal{L}(\mathcal{G}_c(\Omega),\widetilde{\mathbb{C}})$. Theorem 3.13 in [14] shows that:

(i) $(x_0,\xi_0) \notin \mathrm{WF}_{\mathcal{G}}T$ if and only if there exists a conic neighborhood Γ of ξ_0 and a cut-off function $\varphi \in \mathcal{C}^\infty_c(\Omega)$ with $\varphi(x_0) = 1$ such that $\mathcal{F}(\varphi T) \in \mathcal{G}_{\mathscr{S},0}(\Gamma)$.

(ii) $(x_0,\xi_0) \notin \mathrm{WF}_{\mathcal{G}^\infty}T$ if and only if there exists a conic neighborhood Γ of ξ_0 and a cut-off function $\varphi \in \mathcal{C}^\infty_c(\Omega)$ with $\varphi(x_0) = 1$ such that $\mathcal{F}(\varphi T) \in \mathcal{G}^\infty_{\mathscr{S},0}(\Gamma)$.

3. Generalized oscillatory integrals: Definition

This section is devoted to a notion of oscillatory integral where both the amplitude and the phase function are generalized objects of Colombeau type.

In the sequel Ω is an arbitrary open subset of $\mathbb{R}^n$. We recall that $\phi(y,\xi)$ is a *phase function* on $\Omega\times\mathbb{R}^p$ if it is a smooth function on $\Omega\times\mathbb{R}^p\setminus 0$, real-valued, positively homogeneous of degree 1 in ξ with $\nabla_{y,\xi}\phi(y,\xi) \neq 0$ for all $y \in \Omega$ and $\xi \in \mathbb{R}^p\setminus 0$. We denote the set of all phase functions on $\Omega\times\mathbb{R}^p$ by $\Phi(\Omega\times\mathbb{R}^p)$ and the set of all nets in $\Phi(\Omega\times\mathbb{R}^p)^{(0,1]}$ by $\Phi[\Omega\times\mathbb{R}^p]$. The notations concerning classes of symbols have been introduced in Subsection 2.4. The proofs of the statements collected in this section can be found in [18]. In the paper [18] the authors deal with generalized symbols in $\widetilde{S}^m_{\rho,\delta}(\Omega\times\mathbb{R}^p)$ as well as with regular generalized symbols. This last class of symbols is modelled on the subalgebra $\mathcal{G}^\infty(\Omega)$ of regular generalized functions and contains the generalized symbols of slow scale type as a submodule. Even though many statements of Section 3, 4 and 6 hold for regular symbols as well, for the sake of simplicity and in order to have uniformity of assumptions between phase functions and symbols, we limit in this work to consider $\widetilde{S}^m_{\rho,\delta}(\Omega\times\mathbb{R}^p)$ and the smaller class $\widetilde{S}^{m,\mathrm{sc}}_{\rho,\delta}(\Omega\times\mathbb{R}^p)$ of generalized symbols of slow scale type.

Definition 3.1. An element of $\mathcal{M}_\Phi(\Omega\times\mathbb{R}^p)$ is a net $(\phi_\varepsilon)_\varepsilon \in \Phi[\Omega\times\mathbb{R}^p]$ satisfying the conditions:

(i) $(\phi_\varepsilon)_\varepsilon \in \mathcal{M}_{S^1_{\mathrm{hg}}(\Omega\times\mathbb{R}^p\setminus 0)}$,

(ii) for all $K \Subset \Omega$ the net

$$\left(\inf_{y\in K,\xi\in\mathbb{R}^p\setminus 0}\left|\nabla\phi_\varepsilon\left(y,\frac{\xi}{|\xi|}\right)\right|^2\right)_\varepsilon$$

is strictly nonzero.

On $\mathcal{M}_\Phi(\Omega\times\mathbb{R}^p)$ we introduce the equivalence relation $\sim$ as follows: $(\phi_\varepsilon)_\varepsilon \sim (\omega_\varepsilon)_\varepsilon$ if and only if $(\phi_\varepsilon - \omega_\varepsilon) \in \mathcal{N}_{S^1_{\mathrm{hg}}(\Omega\times\mathbb{R}^p\setminus 0)}$. The elements of the factor space

$$\widetilde{\Phi}(\Omega\times\mathbb{R}^p) := \mathcal{M}_\Phi(\Omega\times\mathbb{R}^p)/\sim.$$

will be called *generalized phase functions.*

We shall employ the equivalence class notation $[(\phi_\varepsilon)_\varepsilon]$ for $\phi \in \widetilde{\Phi}(\Omega\times\mathbb{R}^p)$. When $(\phi_\varepsilon)_\varepsilon$ is a net of phase functions, i.e., $(\phi_\varepsilon)_\varepsilon \in \Phi[\Omega\times\mathbb{R}^p]$, Lemma 1.2.1 in [20] shows that there exists a family of partial differential operators $(L_{\phi_\varepsilon})_\varepsilon$ such that ${}^tL_{\phi_\varepsilon}\mathrm{e}^{i\phi_\varepsilon} = \mathrm{e}^{i\phi_\varepsilon}$ for all $\varepsilon \in (0,1]$. L_{ϕ_ε} is of the form

$$\sum_{j=1}^p a_{j,\varepsilon}(y,\xi)\frac{\partial}{\partial_{\xi_j}} + \sum_{k=1}^n b_{k,\varepsilon}(y,\xi)\frac{\partial}{\partial_{y_k}} + c_\varepsilon(y,\xi), \tag{3.1}$$

where the coefficients $(a_{j,\varepsilon})_\varepsilon$ belong to $S^0[\Omega\times\mathbb{R}^p]$ and $(b_{k,\varepsilon})_\varepsilon$, $(c_\varepsilon)_\varepsilon$ are elements of $S^{-1}[\Omega\times\mathbb{R}^p]$. The following technical lemma is crucial in proving Proposition 3.3.

Lemma 3.2.

(i) *Let* $\varphi_{\phi_\varepsilon}(y,\xi) := |\nabla\phi_\varepsilon(y,\xi/|\xi|)|^{-2}$. *If* $(\phi_\varepsilon)_\varepsilon \in \mathcal{M}_\Phi(\Omega\times\mathbb{R}^p)$ *then*

$$(\varphi_{\phi_\varepsilon})_\varepsilon \in \mathcal{M}_{S^0_{\mathrm{hg}}(\Omega\times\mathbb{R}^p\setminus 0)}.$$

(ii) *If* $(\phi_\varepsilon)_\varepsilon, (\omega_\varepsilon)_\varepsilon \in \mathcal{M}_\Phi(\Omega\times\mathbb{R}^p)$ *and* $(\phi_\varepsilon)_\varepsilon \sim (\omega_\varepsilon)_\varepsilon$ *then*

$$\big((\partial_{\xi_j}\phi_\varepsilon)\varphi_{\phi_\varepsilon} - (\partial_{\xi_j}\omega_\varepsilon)\varphi_{\omega_\varepsilon}\big)_\varepsilon \in \mathcal{N}_{S^0_{\mathrm{hg}}(\Omega\times\mathbb{R}^p\setminus 0)}$$

for all $j = 1,\dots,p$ *and*

$$\big((\partial_{y_k}\phi_\varepsilon)|\xi|^{-2}\varphi_{\phi_\varepsilon} - (\partial_{y_k}\omega_\varepsilon)|\xi|^{-2}\varphi_{\omega_\varepsilon}\big)_\varepsilon \in \mathcal{N}_{S^{-1}_{\mathrm{hg}}(\Omega\times\mathbb{R}^p\setminus 0)}$$

for all $k = 1,\dots,n$.

Proposition 3.3.

(i) *If* $(\phi_\varepsilon)_\varepsilon \in \mathcal{M}_\Phi(\Omega\times\mathbb{R}^p)$ *then* $(a_{j,\varepsilon})_\varepsilon \in \mathcal{M}_{S^0(\Omega\times\mathbb{R}^p)}$ *for all* $j = 1,\dots,p$, $(b_{k,\varepsilon})_\varepsilon \in \mathcal{M}_{S^{-1}(\Omega\times\mathbb{R}^p)}$ *for all* $k = 1,\dots,n$, *and* $(c_\varepsilon)_\varepsilon \in \mathcal{M}_{S^{-1}(\Omega\times\mathbb{R}^p)}$.

(ii) *If* $(\phi_\varepsilon)_\varepsilon, (\omega_\varepsilon)_\varepsilon \in \mathcal{M}_\Phi(\Omega\times\mathbb{R}^p)$ *and* $(\phi_\varepsilon)_\varepsilon \sim (\omega_\varepsilon)_\varepsilon$ *then*

$$L_{\phi_\varepsilon} - L_{\omega_\varepsilon} = \sum_{j=1}^{p} a'_{j,\varepsilon}(y,\xi)\frac{\partial}{\partial_{\xi_j}} + \sum_{k=1}^{n} b'_{k,\varepsilon}(y,\xi)\frac{\partial}{\partial_{y_k}} + c'_\varepsilon(y,\xi),$$

where $(a'_{j,\varepsilon})_\varepsilon \in \mathcal{N}_{S^0(\Omega\times\mathbb{R}^p)}$, $(b'_{k,\varepsilon})_\varepsilon \in \mathcal{N}_{S^{-1}(\Omega\times\mathbb{R}^p)}$ *and* $(c'_\varepsilon)_\varepsilon \in \mathcal{N}_{S^{-1}(\Omega\times\mathbb{R}^p)}$ *for all* $j = 1,\dots,p$ *and* $k = 1,\dots,n$.

As a consequence of Propositions 3.3 we can claim that any generalized phase function $\phi \in \widetilde{\Phi}(\Omega\times\mathbb{R}^p)$ defines a generalized partial differential operator

$$L_\phi(y,\xi,\partial_y,\partial_\xi) = \sum_{j=1}^{p} a_j(y,\xi)\frac{\partial}{\partial_{\xi_j}} + \sum_{k=1}^{n} b_k(y,\xi)\frac{\partial}{\partial_{y_k}} + c(y,\xi)$$

whose coefficients $\{a_j\}_{j=1}^p$ and $\{b_k\}_{k=1}^n$, c are generalized symbols in $\widetilde{\mathcal{S}}^0(\Omega\times\mathbb{R}^p)$ and $\widetilde{\mathcal{S}}^{-1}(\Omega\times\mathbb{R}^p)$, respectively. By construction, L_ϕ maps $\widetilde{\mathcal{S}}^m_{\rho,\delta}(\Omega\times\mathbb{R}^p)$ continuously into $\widetilde{\mathcal{S}}^{m-s}_{\rho,\delta}(\Omega\times\mathbb{R}^p)$, where $s = \min\{\rho, 1-\delta\}$. Hence L^k_ϕ is continuous from $\widetilde{\mathcal{S}}^m_{\rho,\delta}(\Omega\times\mathbb{R}^p)$ to $\widetilde{\mathcal{S}}^{m-ks}_{\rho,\delta}(\Omega\times\mathbb{R}^p)$.

Before stating the next proposition we recall a classical lemma valid any symbol $\phi \in S^1(\Omega\times\mathbb{R}^p\setminus 0)$.

Lemma 3.4. *For all* $\alpha\in\mathbb{N}^p$ *and* $\beta\in\mathbb{N}^n$,

$$\partial^\alpha_\xi\partial^\beta_y \mathrm{e}^{i\phi(y,\xi)} = \mathrm{e}^{i\phi(y,\xi)} \sum_{\substack{k\le|\alpha+\beta|,\\ \alpha_1+\alpha_2+\dots+\alpha_k=\alpha\\ \beta_1+\beta_2+\dots+\beta_k=\beta}} c_{\alpha_1,\dots,\alpha_k,\beta_1,\dots,\beta_k}\,\partial^{\alpha_1}_\xi\partial^{\beta_1}_y\phi(y,\xi)\dots\partial^{\alpha_k}_\xi\partial^{\beta_k}_y\phi(y,\xi).$$

It follows that

$$\partial^\alpha_\xi\partial^\beta_y \mathrm{e}^{i\phi(y,\xi)} = \mathrm{e}^{i\phi(y,\xi)} a_{\alpha,\beta}(y,\xi),$$

where $a_{\alpha,\beta}\in S^{|\beta|}(\Omega\times\mathbb{R}^p\setminus 0)$ *and*

$$|a_{\alpha,\beta}|^{(|\beta|)}_{K,j} \le c \sup_{y\in K,\xi\neq 0}\ \sup_{|\gamma+\delta|\le|\alpha+\beta|+j} \langle\xi\rangle^{-1+|\gamma|}|\partial^\gamma_\xi\partial^\delta_y\phi(y,\xi)|, \tag{3.2}$$

where the constant c *depends only on* α, β, *and* j.

From (3.2) we have that

$$(\phi_\varepsilon)_\varepsilon \in \mathcal{M}_{S^1(\Omega\times\mathbb{R}^p\setminus 0)} \quad\Rightarrow\quad (a_{\alpha,\beta,\varepsilon})_\varepsilon \in \mathcal{M}_{S^{|\beta|}(\Omega\times\mathbb{R}^p\setminus 0)}$$

or more in general that the net $(a_{\alpha,\beta,\varepsilon})_\varepsilon$ has the "ε-scale properties" of $(\phi_\varepsilon)_\varepsilon$.

Proposition 3.5. *Let $\phi \in \widetilde{\Phi}(\Omega \times \mathbb{R}^p)$. The exponential*

$$e^{i\phi(y,\xi)}$$

is a well-defined element of $\widetilde{\mathcal{S}}^1_{0,1}(\Omega \times \mathbb{R}^p \setminus 0)$.

Proof. From Lemma 3.4 we have that if $(\phi_\varepsilon)_\varepsilon \in \mathcal{M}_\Phi(\Omega \times \mathbb{R}^p)$ then $(e^{i\phi_\varepsilon(y,\xi)})_\varepsilon \in \mathcal{M}_{S^0_{0,1}(\Omega\times\mathbb{R}^p\setminus 0)}$. When $(\phi_\varepsilon)_\varepsilon \sim (\omega_\varepsilon)_\varepsilon$, the equality

$$\begin{aligned} e^{i\omega_\varepsilon(y,\xi)} - e^{i\phi_\varepsilon(y,\xi)} &= e^{i\omega_\varepsilon(y,\xi)}\bigl(1 - e^{i(\phi_\varepsilon - \omega_\varepsilon)(y,\xi)}\bigr) \\ &= e^{i\omega_\varepsilon(y,\xi)} \sum_{j=1}^p e^{i(\phi_\varepsilon-\omega_\varepsilon)(y,\theta\xi)} \partial_{\xi_j}(\phi_\varepsilon - \omega_\varepsilon)(y,\theta\xi) i\xi_j, \end{aligned}$$

with $\theta \in (0,1)$, implies that

$$\sup_{y\in K,\xi\in\mathbb{R}^p\setminus 0} |\xi|^{-1}\bigl|e^{i\omega_\varepsilon(y,\xi)} - e^{i\phi_\varepsilon(y,\xi)}\bigr| = O(\varepsilon^q) \tag{3.3}$$

for all $q \in \mathbb{N}$. At this point writing $\partial^\alpha_\xi\partial^\beta_y(e^{i\omega_\varepsilon(y,\xi)} - e^{i\phi_\varepsilon(y,\xi)})$ as

$$\begin{aligned} &\partial^\alpha_\xi\partial^\beta_y e^{i\omega_\varepsilon(y,\xi)}\bigl(1 - e^{i(\phi_\varepsilon-\omega_\varepsilon)(y,\xi)}\bigr) + \\ &\qquad + \sum_{\alpha'<\alpha,\beta'<\beta} \binom{\alpha}{\alpha'}\binom{\beta}{\beta'} \partial^{\alpha'}_\xi\partial^{\beta'}_y e^{i\omega_\varepsilon(y,\xi)}\bigl(-\partial^{\alpha-\alpha'}_\xi\partial^{\beta-\beta'}_y e^{i(\phi_\varepsilon-\omega_\varepsilon)(y,\xi)}\bigr) \end{aligned}$$

we obtain the characterizing estimate of a net in $\mathcal{N}_{S^1_{0,1}(\Omega\times\mathbb{R}^p\setminus 0)}$, using (3.3) the moderateness of $(e^{i\omega_\varepsilon(y,\xi)})_\varepsilon$ and Lemma 3.4. □

By construction of the operator L_ϕ the equality ${}^tL_\phi e^{i\phi} = e^{i\phi}$ holds in $\widetilde{\mathcal{S}}^1_{0,1}(\Omega \times \mathbb{R}^p \setminus 0)$. In addition, Proposition 3.5 and the properties of L^k_ϕ allow to conclude that

$$e^{i\phi(y,\xi)} L^k_\phi(a(y,\xi)u(y))$$

is a generalized symbol in $\widetilde{\mathcal{S}}^{m-ks+1}_{0,1}(\Omega \times \mathbb{R}^p)$ which is integrable on $\Omega \times \mathbb{R}^p$ in the sense of Section 2 when $m - ks + 1 < -p$. From now on we assume that $\rho > 0$ and $\delta < 1$.

Definition 3.6. Let $\phi \in \widetilde{\Phi}(\Omega\times\mathbb{R}^p)$, $a \in \widetilde{\mathcal{S}}^m_{\rho,\delta}(\Omega\times\mathbb{R}^p)$ and $u \in \mathcal{G}_c(\Omega)$. The *generalized oscillatory integral*

$$\int_{\Omega\times\mathbb{R}^p} e^{i\phi(y,\xi)} a(y,\xi)u(y)\, dy\, đ\xi$$

is defined as

$$\int_{\Omega\times\mathbb{R}^p} e^{i\phi(y,\xi)} L^k_\phi(a(y,\xi)u(y))\, dy\, đ\xi$$

where k is chosen such that $m - ks + 1 < -p$.

The functional

$$I_\phi(a) : \mathcal{G}_c(\Omega) \to \widetilde{\mathbb{C}} : u \to \int_{\Omega\times\mathbb{R}^p} e^{i\phi(y,\xi)} a(y,\xi) u(y)\, dy\, đ\xi$$

belongs to the dual $\mathcal{L}(\mathcal{G}_c(\Omega), \widetilde{\mathbb{C}})$. Indeed, by (2.7), the continuity of L^k_ϕ and of the product between generalized symbols we have that the map

$$\mathcal{G}_c(\Omega) \to \widetilde{\mathcal{S}}^{m-ks+1}_{0,1}(\Omega\times\mathbb{R}^p) : u \to e^{i\phi(y,\xi)} L^k_\phi(a(y,\xi)u(y))$$

is continuous and thus, by an application of the integral on $\Omega\times\mathbb{R}^p$, the resulting functional $I_\phi(a)$ is continuous.

4. Generalized Fourier integral operators

Definition and mapping properties

We now study oscillatory integrals where an additional parameter x, varying in an open subset Ω' of $\mathbb{R}^{n'}$, appears in the phase function ϕ and in the symbol a. The dependence on x is investigated in the Colombeau context. We denote by $\Phi[\Omega';\Omega\times\mathbb{R}^p]$ the set of all nets $(\phi_\varepsilon)_{\varepsilon\in(0,1]}$ of continuous functions on $\Omega'\times\Omega\times\mathbb{R}^p$ which are smooth on $\Omega'\times\Omega\times\mathbb{R}^p\setminus\{0\}$ and such that $(\phi_\varepsilon(x,\cdot,\cdot))_\varepsilon \in \Phi[\Omega\times\mathbb{R}^p]$ for all $x\in\Omega'$.

Definition 4.1. An element of $\mathcal{M}_\Phi(\Omega';\Omega\times\mathbb{R}^p)$ is a net $(\phi_\varepsilon)_\varepsilon \in \Phi[\Omega';\Omega\times\mathbb{R}^p]$ satisfying the conditions:

(i) $(\phi_\varepsilon)_\varepsilon \in \mathcal{M}_{S^1_{\mathrm{hg}}(\Omega'\times\Omega\times\mathbb{R}^p\setminus 0)}$,
(ii) for all $K'\Subset\Omega'$ and $K\Subset\Omega$ the net

$$\left(\inf_{x\in K', y\in K, \xi\in\mathbb{R}^p\setminus 0}\left|\nabla_{y,\xi}\phi_\varepsilon\left(x,y,\frac{\xi}{|\xi|}\right)\right|^2\right)_\varepsilon \tag{4.1}$$

is strictly nonzero.

On $\mathcal{M}_\Phi(\Omega';\Omega\times\mathbb{R}^p)$ we introduce the equivalence relation $\sim$ as follows: $(\phi_\varepsilon)_\varepsilon \sim (\omega_\varepsilon)_\varepsilon$ if and only if $(\phi_\varepsilon-\omega_\varepsilon)_\varepsilon \in \mathcal{N}_{S^1_{\mathrm{hg}}(\Omega'\times\Omega\times\mathbb{R}^p\setminus 0)}$. The elements of the factor space

$$\widetilde{\Phi}(\Omega';\Omega\times\mathbb{R}^p) := \mathcal{M}_\Phi(\Omega';\Omega\times\mathbb{R}^p)/\sim .$$

are called *generalized phase functions with respect to the variables in* $\Omega\times\mathbb{R}^p$.

Lemma 3.2 as well as Proposition 3.3 can be adapted to nets in $\mathcal{M}_\Phi(\Omega';\Omega\times\mathbb{R}^p)$. More precisely, the operator

$$L_{\phi_\varepsilon}(x;y,\xi,\partial_y,\partial_\xi) = \sum_{j=1}^p a_{j,\varepsilon}(x,y,\xi)\frac{\partial}{\partial_{\xi_j}} + \sum_{k=1}^n b_{k,\varepsilon}(x,y,\xi)\frac{\partial}{\partial_{y_k}} + c_\varepsilon(x,y,\xi) \tag{4.2}$$

defined for any value of x by (3.1), has the property ${}^tL_{\phi_\varepsilon(x,\cdot,\cdot)}e^{i\phi_\varepsilon(x,\cdot,\cdot)} = e^{i\phi_\varepsilon(x,\cdot,\cdot)}$ for all $x\in\Omega'$ and $\varepsilon\in(0,1]$ and its coefficients depend smoothly on $x\in\Omega'$.

Lemma 4.2.

(i) *Let*
$$\varphi_{\phi_\varepsilon}(x,y,\xi) := |\nabla_{y,\xi}\phi_\varepsilon(x,y,\xi/|\xi|)|^{-2}. \tag{4.3}$$
If $(\phi_\varepsilon)_\varepsilon \in \mathcal{M}_\Phi(\Omega';\Omega\times\mathbb{R}^p)$ *then* $(\varphi_{\phi_\varepsilon})_\varepsilon \in \mathcal{M}_{S^0_{\rm hg}(\Omega'\times\Omega\times\mathbb{R}^p\setminus 0)}$.

(ii) *If* $(\phi_\varepsilon)_\varepsilon, (\omega_\varepsilon)_\varepsilon \in \mathcal{M}_\Phi(\Omega';\Omega\times\mathbb{R}^p)$ *and* $(\phi_\varepsilon)_\varepsilon \sim (\omega_\varepsilon)_\varepsilon$ *then*
$$\big((\partial_{\xi_j}\phi_\varepsilon)\varphi_{\phi_\varepsilon} - (\partial_{\xi_j}\omega_\varepsilon)\varphi_{\omega_\varepsilon}\big)_\varepsilon \in \mathcal{N}_{S^0_{\rm hg}(\Omega'\times\Omega\times\mathbb{R}^p\setminus 0)}$$
for all $j=1,\dots,p$ *and*
$$\big((\partial_{y_k}\phi_\varepsilon)|\xi|^{-2}\varphi_{\phi_\varepsilon} - (\partial_{y_k}\omega_\varepsilon)|\xi|^{-2}\varphi_{\omega_\varepsilon}\big)_\varepsilon \in \mathcal{N}_{S^{-1}_{\rm hg}(\Omega'\times\Omega\times\mathbb{R}^p\setminus 0)}$$
for all $k=1,\dots,n$.

Proposition 4.3.

(i) *If* $(\phi_\varepsilon)_\varepsilon \in \mathcal{M}_\Phi(\Omega';\Omega\times\mathbb{R}^p)$ *then the coefficients occurring in* (4.2) *satisfy the following:* $(a_{j,\varepsilon})_\varepsilon \in \mathcal{M}_{S^0(\Omega'\times\Omega\times\mathbb{R}^p)}$ *for all* $j=1,\dots,p$, $(b_{k,\varepsilon})_\varepsilon \in \mathcal{M}_{S^{-1}(\Omega'\times\Omega\times\mathbb{R}^p)}$ *for all* $k=1,\dots,n$, *and* $(c_\varepsilon)_\varepsilon \in \mathcal{M}_{S^{-1}(\Omega'\times\Omega\times\mathbb{R}^p)}$.

(ii) *If* $(\phi_\varepsilon)_\varepsilon, (\omega_\varepsilon)_\varepsilon \in \mathcal{M}_\Phi(\Omega';\Omega\times\mathbb{R}^p)$ *and* $(\phi_\varepsilon)_\varepsilon \sim (\omega_\varepsilon)_\varepsilon$ *then*
$$L_{\phi_\varepsilon} - L_{\omega_\varepsilon} = \sum_{j=1}^p a'_{j,\varepsilon}(x,y,\xi)\frac{\partial}{\partial_{\xi_j}} + \sum_{k=1}^n b'_{k,\varepsilon}(x,y,\xi)\frac{\partial}{\partial_{y_k}} + c'_\varepsilon(x,y,\xi), \tag{4.4}$$
where $(a'_{j,\varepsilon})_\varepsilon \in \mathcal{N}_{S^0(\Omega'\times\Omega\times\mathbb{R}^p)}$, $(b'_{k,\varepsilon})_\varepsilon \in \mathcal{N}_{S^{-1}(\Omega'\times\Omega\times\mathbb{R}^p)}$ *and* $(c'_\varepsilon)_\varepsilon \in \mathcal{N}_{S^{-1}(\Omega'\times\Omega\times\mathbb{R}^p)}$ *for all* $j=1,\dots,p$ *and* $k=1,\dots,n$.

Proposition 4.3 yields that any generalized phase function ϕ in $\widetilde{\Phi}(\Omega';\Omega\times\mathbb{R}^p)$ defines a partial differential operator
$$L_\phi(x;y,\xi,\partial_y,\partial_\xi) = \sum_{j=1}^p a_j(x,y,\xi)\frac{\partial}{\partial_{\xi_j}} + \sum_{k=1}^n b_k(x,y,\xi)\frac{\partial}{\partial_{y_k}} + c(x,y,\xi) \tag{4.5}$$
with coefficients $a_j \in \widetilde{\mathcal{S}}^0(\Omega'\times\Omega\times\mathbb{R}^p)$, $b_k, c \in \widetilde{\mathcal{S}}^{-1}(\Omega'\times\Omega\times\mathbb{R}^p)$ such that ${}^tL_\phi e^{i\phi} = e^{i\phi}$ holds in $\widetilde{\mathcal{S}}^1_{0,1}(\Omega'\times\Omega\times\mathbb{R}^p\setminus 0)$. Arguing as in Proposition 3.5 we obtain that $e^{i\phi(x,y,\xi)}$ is a well-defined element of $\widetilde{\mathcal{S}}^1_{0,1}(\Omega'\times\Omega\times\mathbb{R}^p\setminus 0)$. The usual composition argument implies that the map
$$\mathcal{G}_c(\Omega) \to \widetilde{\mathcal{S}}^{m-ks+1}_{0,1}(\Omega'\times\Omega\times\mathbb{R}^p) : u \to e^{i\phi(x,y,\xi)}L^k_\phi(a(x,y,\xi)u(y))$$
is continuous.

The oscillatory integral
$$\begin{aligned} I_\phi(a)(u)(x) &= \int_{\Omega\times\mathbb{R}^p} e^{i\phi(x,y,\xi)}a(x,y,\xi)u(y)\,dy\,đ\xi \\ &:= \int_{\Omega\times\mathbb{R}^p} e^{i\phi(x,y,\xi)}L^k_\phi(a(x,y,\xi)u(y))\,dy\,đ\xi, \end{aligned}$$
where $\phi \in \widetilde{\Phi}(\Omega';\Omega\times\mathbb{R}^p)$ and $a \in \widetilde{\mathcal{S}}^m_{\rho,\delta}(\Omega'\times\Omega\times\mathbb{R}^p)$ is an element of $\widetilde{\mathbb{C}}$ for fixed $x\in\Omega'$. In particular, $I_\phi(a)(u)$ is the integral on $\Omega\times\mathbb{R}^p$ of a generalized amplitude

in $\widetilde{\mathcal{S}}^{l}_{0,1}(\Omega' \times \Omega \times \mathbb{R}^p)$ having compact support in y. The order l can be chosen arbitrarily low.

Theorem 4.4. *Let $\phi \in \widetilde{\Phi}(\Omega'; \Omega \times \mathbb{R}^p)$, $a \in \widetilde{\mathcal{S}}^{m}_{\rho,\delta}(\Omega' \times \Omega \times \mathbb{R}^p)$ and $u \in \mathcal{G}_c(\Omega)$. The generalized oscillatory integral*

$$I_\phi(a)(u)(x) = \int_{\Omega \times \mathbb{R}^p} e^{i\phi(x,y,\xi)} a(x,y,\xi) u(y)\, dy\, d\!\!\!{}^-\xi \tag{4.6}$$

defines a generalized function in $\mathcal{G}(\Omega')$ and the map

$$A : \mathcal{G}_c(\Omega) \to \mathcal{G}(\Omega') : u \to I_\phi(a)(u) \tag{4.7}$$

is continuous.

The operator A defined in (4.7) is called *generalized Fourier integral operator* with amplitude $a \in \widetilde{\mathcal{S}}^{m}_{\rho,\delta}(\Omega' \times \Omega \times \mathbb{R}^p)$ and phase function $\phi \in \widetilde{\Phi}(\Omega'; \Omega \times \mathbb{R}^p)$.

Example. Our outline of a basic theory of Fourier integral operators with Colombeau generalized amplitudes and phase functions is motivated to a large extent by potential applications in regularity theory for generalized solutions to hyperbolic partial (or pseudo-) differential equations with distributional or Colombeau-type coefficients (or symbols) and data (cf. [25, 30, 34]). To illustrate the typical situation we consider here the following simple model: let $u \in \mathcal{G}(\mathbb{R}^2)$ be the solution of the generalized Cauchy-problem

$$\partial_t u + c\,\partial_x u + b\,u = 0 \tag{4.8}$$

$$u\,|_{t=0} = g, \tag{4.9}$$

where g belongs to $\mathcal{G}_c(\mathbb{R})$ and the coefficients b, $c \in \mathcal{G}(\mathbb{R}^2)$. Furthermore, b, c, as well as $\partial_x c$ are assumed to be of local L^∞-log-type (concerning growth with respect to the regularization parameter, cf. [34]), c being generalized real-valued and globally bounded in addition. Let $\gamma \in \mathcal{G}(\mathbb{R}^3)$ be the unique (global) solution of the corresponding generalized characteristic ordinary differential equation

$$\frac{d}{ds}\gamma(x,t;s) = c(\gamma(x,t;s),s)$$
$$\gamma(x,t;t) = x.$$

Then u is given in terms of γ by $u(x,t) = g(\gamma(x,t;0)) \exp(-\int_0^t b(\gamma(x,t;r),r)\,dr)$. Writing g as the inverse of its Fourier transform we obtain the Fourier integral representation

$$u(x,t) = \iint e^{i(\gamma(x,t;0)-y)\xi}\, a(x,t,y,\xi)\, g(y)\, dy\, d\!\!\!{}^-\xi, \tag{4.10}$$

where $a(x,t,y,\xi) := \exp(-\int_0^t b(\gamma(x,t;r),r)\,dr)$ is a generalized amplitude of order 0. The phase function $\phi(x,t,y,\xi) := (\gamma(x,t;0) - y)\xi$ has (full) gradient

$$(\partial_x \gamma(x,t;0), \partial_t \gamma(x,t;0), -\xi, \gamma(x,t;0) - y)$$

and thus defines a generalized phase function ϕ. Therefore (4.10) reads $u = Ag$ where $A : \mathcal{G}_c(\mathbb{R}) \to \mathcal{G}(\mathbb{R}^2)$ is a generalized Fourier integral operator.

Regularity properties

We now investigate the regularity properties of the *generalized Fourier integral operator* A. We will prove that for appropriate generalized phase functions and generalized amplitudes, A maps $\mathcal{G}^\infty_{\rm c}(\Omega)$ into $\mathcal{G}^\infty(\Omega')$. The following example shows that a $\mathcal{G}^\infty$-kind of regularity assumption for the net $(\phi_\varepsilon)_\varepsilon$ with respect to the parameter ε does not entail the desired mapping property.

Example. Let $n = n' = p = 1$ and $\Omega = \Omega' = \mathbb{R}$ and $\phi_\varepsilon(x,y,\xi) = (x - \varepsilon y)\xi$. Then $(\phi_\varepsilon)_\varepsilon \in \mathcal{M}_\Phi(\mathbb{R}; \mathbb{R}\times\mathbb{R})$ and in particular we have $N = 0$ in all moderateness estimates (see Definition 4.1(i))) and $|\nabla_{y,\xi}\phi_\varepsilon(x,y,\xi/|\xi|)|^2 \ge \varepsilon^2$. Choose the amplitude a identically equal to 1. The corresponding generalized operator A does not map $\mathcal{G}^\infty_{\rm c}(\mathbb{R})$ into $\mathcal{G}^\infty(\mathbb{R})$. Indeed, for $0 \neq f \in \mathcal{C}^\infty_{\rm c}(\mathbb{R})$ we have that

$$A[(f)_\varepsilon] = \left[\left(\int_{\mathbb{R}\times\mathbb{R}} \mathrm{e}^{i(x-\varepsilon y)\xi} f(y)\,dy\,đ\xi\right)_\varepsilon\right] = [(\varepsilon^{-1}f(x/\varepsilon))_\varepsilon] \in \mathcal{G}(\mathbb{R}) \setminus \mathcal{G}^\infty(\mathbb{R}).$$

This example suggests that a stronger notion of regularity on generalized phase functions has to be designed. Such is provided by the concept of *slow scale net.*

Definition 4.5. We say that $\phi \in \widetilde{\Phi}(\Omega'; \Omega\times\mathbb{R}^p)$ is a *slow scale generalized phase function in the variables of* $\Omega\times\mathbb{R}^p$ if it has a representative $(\phi_\varepsilon)_\varepsilon$ fulfilling the conditions

(i) $(\phi_\varepsilon)_\varepsilon \in \mathcal{M}^{\rm sc}_{S^1_{\rm hg}(\Omega'\times\Omega\times\mathbb{R}^p\setminus 0)}$,

(ii) for all $K' \Subset \Omega'$ and $K \Subset \Omega$ the net (4.1) is slow scale-strictly nonzero.

In the sequel the set of all $(\phi_\varepsilon)_\varepsilon \in \Phi[\Omega'; \Omega\times\mathbb{R}^p]$ fulfilling (i) and (ii) in Definition 4.5 will be denoted by $\mathcal{M}^{\rm sc}_\Phi(\Omega'; \Omega\times\mathbb{R}^p)$ while we use $\widetilde{\Phi}^{\rm sc}(\Omega'; \Omega\times\mathbb{R}^p)$ for the set of slow scale generalized functions as above. Similarly, using $\nabla_{x,y,\xi}$ in place of $\nabla_{y,\xi}$ in (ii) we define the space $\widetilde{\Phi}^{\rm sc}(\Omega'\times\Omega\times\mathbb{R}^p)$ of slow scale generalized phase functions on $\Omega'\times\Omega\times\mathbb{R}^p$. We refer to [18, Section 3] for the proof of the following theorem.

Theorem 4.6. *Let* $\phi \in \widetilde{\Phi}^{\rm sc}(\Omega'; \Omega\times\mathbb{R}^p)$.

(i) *If* $a \in \widetilde{\mathcal{S}}^{m,\rm sc}_{\rho,\delta}(\Omega'\times\Omega\times\mathbb{R}^p)$ *the corresponding generalized Fourier integral operator*

$$A : u \to \int_{\Omega\times\mathbb{R}^p} \mathrm{e}^{i\phi(x,y,\xi)} a(x,y,\xi) u(y)\,dy\,đ\xi$$

maps $\mathcal{G}^\infty_{\rm c}(\Omega)$ *continuously into* $\mathcal{G}^\infty(\Omega')$.

(ii) *If* $a \in \widetilde{\mathcal{S}}^{-\infty,\rm sc}(\Omega'\times\Omega\times\mathbb{R}^p)$ *then* A *maps* $\mathcal{G}_{\rm c}(\Omega)$ *continuously into* $\mathcal{G}^\infty(\Omega')$.

Extension to the dual

Finally, we prove that under suitable hypotheses on the generalized phase function $\phi \in \widetilde{\Phi}(\Omega'\times\Omega\times\mathbb{R}^p)$, the definition of the generalized Fourier integral operator A can be extended to the dual $\mathcal{L}(\mathcal{G}(\Omega), \widetilde{\mathbb{C}})$.

Definition 4.7. We say that $\phi \in \widetilde{\Phi}(\Omega' \times \Omega \times \mathbb{R}^p)$ is a generalized operator phase function if it has a representative $(\phi_\varepsilon)_\varepsilon$ of operator phase functions satisfying the conditions (i) and (ii) of Definition 4.1 and such that

(iii) for all $K' \Subset \Omega'$ and $K \Subset \Omega$ the net

$$\left(\inf_{x \in K', y \in K, \xi \in \mathbb{R}^p \setminus 0} \left| \nabla_{x,\xi} \phi_\varepsilon \left(x, y, \frac{\xi}{|\xi|} \right) \right|^2 \right)_\varepsilon$$

is strictly nonzero.

It is clear that when ϕ is a generalized operator phase function then by Theorem 4.4 the oscillatory integral

$$\int_{\Omega' \times \mathbb{R}^p} e^{i\phi(x,y,\xi)} a(x,y,\xi) v(x)\, dx\, đ\xi, \tag{4.11}$$

where $a \in \widetilde{\mathcal{S}}^m_{\rho,\delta}(\Omega' \times \Omega \times \mathbb{R}^p)$ and $v \in \mathcal{G}_c(\Omega')$, defines a generalized function in $\mathcal{G}(\Omega)$ and a continuous $\widetilde{\mathbb{C}}$-linear operator from $\mathcal{G}_c(\Omega')$ to $\mathcal{G}(\Omega)$. More precisely, we have the following result.

Proposition 4.8. *Let ϕ be a generalized operator phase function on $\Omega' \times \Omega \times \mathbb{R}^p$, $a \in \widetilde{\mathcal{S}}^m_{\rho,\delta}(\Omega' \times \Omega \times \mathbb{R}^p)$ and $A : \mathcal{G}_c(\Omega) \to \mathcal{G}(\Omega')$ the generalized Fourier integral operator given by* (4.6)–(4.7). *Then,*

(i) *the transposed ${}^tA$ of A is the generalized Fourier integral operator given by* (4.11)*;*
(ii) *the operator A can be extended to a continuous $\widetilde{\mathbb{C}}$-linear map acting from $\mathcal{L}(\mathcal{G}(\Omega), \widetilde{\mathbb{C}})$ to $\mathcal{L}(\mathcal{G}_c(\Omega'), \widetilde{\mathbb{C}})$.*

Proof. Working at the level of representatives, the proof of the first assertion is a simple application of the corresponding classical result. It follows that A can be extended to a $\widetilde{\mathbb{C}}$-linear map from $\mathcal{L}(\mathcal{G}(\Omega), \widetilde{\mathbb{C}})$ to $\mathcal{L}(\mathcal{G}_c(\Omega'), \widetilde{\mathbb{C}})$ by setting

$$A(T)(u) = T({}^tAu),$$

for all $T \in \mathcal{L}(\mathcal{G}(\Omega), \widetilde{\mathbb{C}})$ and $u \in \mathcal{G}_c(\Omega')$. Finally, let B a bounded subset of $\mathcal{G}_c(\Omega')$. From the continuity of ${}^tA$ and T we have that

$$\sup_{u \in B} |A(T)(u)|_e = \sup_{u \in B} |T({}^tAu)| = \sup_{v \in {}^tA(B)} |T(v)|,$$

where ${}^tA(B)$ is a bounded subset of $\mathcal{G}(\Omega)$. This shows that $A : \mathcal{L}(\mathcal{G}(\Omega), \widetilde{\mathbb{C}}) \to \mathcal{L}(\mathcal{G}_c(\Omega'), \widetilde{\mathbb{C}})$ is continuous. $\square$

5. Composition of a generalized Fourier integral operator with a generalized pseudo-differential operator

Generalized Fourier integral operators of the type $F_\omega(b)$

Let Ω and Ω' be open subsets of $\mathbb{R}^n$ and $\mathbb{R}^{n'}$ respectively. We now focus on operators of the form

$$F_\omega(b)(u)(x) = \int_{\mathbb{R}^n} \mathrm{e}^{i\omega(x,\eta)} b(x,\eta)\widehat{u}(\eta)\, đ\eta, \tag{5.1}$$

where $\omega \in \widetilde{\mathcal{S}}^1_{\mathrm{hg}}(\Omega' \times \mathbb{R}^n \setminus 0)$, $b \in \widetilde{\mathcal{S}}^m(\Omega' \times \mathbb{R}^n)$ and $u \in \mathcal{G}_{\mathrm{c}}(\Omega)$.
Note that $\phi(x,y,\eta) := \omega(x,\eta) - y\eta$ is a well-defined generalized phase function belonging to $\widetilde{\Phi}(\Omega';\Omega\times\mathbb{R}^n)$. Indeed, for any $(\omega_\varepsilon)_\varepsilon$ representative of ω we have that $(\omega_\varepsilon(x,\eta)-y\eta)_\varepsilon \in \mathcal{M}_{S^1_{\mathrm{hg}}(\Omega'\times\Omega\times\mathbb{R}^n)}$, if $(\omega_\varepsilon-\omega'_\varepsilon)_\varepsilon \in \mathcal{N}_{S^1_{\mathrm{hg}}(\Omega'\times\mathbb{R}^n)}$ then $(\omega_\varepsilon - y\eta - \omega'_\varepsilon + y\eta)_\varepsilon \in \mathcal{N}_{S^1_{\mathrm{hg}}(\Omega'\times\Omega\times\mathbb{R}^n)}$ and $|\nabla_{y,\eta}\phi(x,y,\eta)| = |(-\eta, \nabla_\eta\omega - y)| \ge |\eta|$. In particular it follows that for any representative $\phi_\varepsilon := \omega_\varepsilon(x,\eta) - y\eta$ and any $K' \Subset \Omega'$, $K \Subset \Omega$, the net $\inf_{x\in K', y\in K, \eta\in\mathbb{R}^n\setminus 0} |\nabla_{y,\eta}\phi_\varepsilon(x,y,\frac{\eta}{|\eta|})|$ is slow scale-strictly non-zero.

We recall that by Lemma 3.4, the estimate (3.2) and Proposition 3.5

- if $\omega \in \widetilde{\mathcal{S}}^1_{\mathrm{hg}}(\Omega' \times \mathbb{R}^n \setminus 0)$ then $\mathrm{e}^{i\omega(x,\eta)} \in \widetilde{\mathcal{S}}^1_{0,1}(\Omega' \times \mathbb{R}^n)$ and

$$\partial^\alpha_\eta \partial^\beta_x \mathrm{e}^{i\omega(x,\eta)} = \mathrm{e}^{i\omega(x,\eta)} a_{\alpha,\beta}(x,\eta),$$

  where $a_{\alpha,\beta} \in \widetilde{\mathcal{S}}^{|\beta|}(\Omega' \times \mathbb{R}^n \setminus 0)$ and the equality is intended in the space $\widetilde{\mathcal{S}}^{1+|\beta|}_{0,1}(\Omega' \times \mathbb{R}^n \setminus 0)$;
- if $\omega \in \widetilde{\mathcal{S}}^{1,\mathrm{sc}}_{\mathrm{hg}}(\Omega' \times \mathbb{R}^n \setminus 0)$ then $a_{\alpha,\beta} \in \widetilde{\mathcal{S}}^{|\beta|,\mathrm{sc}}(\Omega' \times \mathbb{R}^n \setminus 0)$.

An immediate application of Theorem 4.4 and Proposition 4.8 yields the following mapping properties.

Proposition 5.1.

(i) *If $\omega \in \widetilde{\mathcal{S}}^1_{\mathrm{hg}}(\Omega' \times \mathbb{R}^n \setminus 0)$ and $b \in \widetilde{\mathcal{S}}^m(\Omega' \times \mathbb{R}^n)$ then $F_\omega(b)$ maps $\mathcal{G}_{\mathrm{c}}(\Omega)$ continuously into $\mathcal{G}(\Omega')$.*

(ii) *If $\omega \in \widetilde{\mathcal{S}}^1_{\mathrm{hg}}(\Omega' \times \mathbb{R}^n \setminus 0)$ has a representative $(\omega_\varepsilon)_\varepsilon \in \Phi[\Omega' \times \mathbb{R}^n]$ such that for all $K' \Subset \Omega'$*

$$\Big(\inf_{x\in K', \eta\in\mathbb{R}^n\setminus 0} |\nabla_x\omega_\varepsilon(x,\frac{\eta}{|\eta|})|\Big)_\varepsilon$$

is strictly non-zero, then $F_\omega(b)$ can be extended to a continuous $\widetilde{\mathbb{C}}$-linear map from $\mathcal{L}(\mathcal{G}(\Omega),\widetilde{\mathbb{C}})$ to $\mathcal{L}(\mathcal{G}_{\mathrm{c}}(\Omega'),\widetilde{\mathbb{C}})$.

(iii) *If $\omega \in \widetilde{\mathcal{S}}^{1,\mathrm{sc}}_{\mathrm{hg}}(\Omega' \times \mathbb{R}^n \setminus 0)$ and $b \in \widetilde{\mathcal{S}}^{m,\mathrm{sc}}(\Omega' \times \mathbb{R}^n)$ then $F_\omega(b)$ maps $\mathcal{G}^\infty_{\mathrm{c}}(\Omega)$ continuously into $\mathcal{G}^\infty(\Omega')$.*

(iv) *If $\mathrm{supp}_x b \Subset \Omega'$ then $F_\omega(b)$ maps $\mathcal{G}_{\mathrm{c}}(\Omega)$ into $\mathcal{G}_{\mathrm{c}}(\Omega')$ and under the assumptions of* (ii) *maps $\mathcal{L}(\mathcal{G}(\Omega),\widetilde{\mathbb{C}})$ into $\mathcal{L}(\mathcal{G}(\Omega'),\widetilde{\mathbb{C}})$.*

Proof. The first assertion is clear from Theorem 4.4 and the second one from Proposition 4.8(ii).
(iii) Lemma 3.4 and the considerations which precede this proposition entail

$$\partial_x^\beta F_\omega(b)(u)(x) = \sum_{\beta'\le\beta}\binom{\beta}{\beta'}\int_{\mathbb{R}^n} \mathrm{e}^{i\omega(x,\eta)} a_{\beta'}(x,\eta)\partial^{\beta-\beta'}b(x,\eta)\widehat{u}(\eta),\ đ\eta,$$

where $a_{\beta'} \in \widetilde{\mathcal{S}}^{|\beta'|,\mathrm{sc}}(\Omega' \times \mathbb{R}^n \setminus 0)$. Hence, if $b \in \widetilde{\mathcal{S}}^{m,\mathrm{sc}}(\Omega' \times \mathbb{R}^n)$ and $u \in \mathcal{G}_{\mathrm{c}}^\infty(\Omega)$ then $F_\omega(b) \in \mathcal{G}^\infty(\Omega')$. Moreover, since for all $\beta \in \mathbb{N}^{n'}$ and $K' \Subset \Omega'$ there exists $h \in \mathbb{N}$ and $c > 0$ such that for all $g \in \mathcal{C}_K^\infty(\Omega)$ and $\varepsilon \in (0,1]$ the estimate

$$\sup_{x\in K'} |\partial^\beta F_{\omega_\varepsilon}(b_\varepsilon)(g)(x)| \le c \max_{\beta'\le\beta} |a_{\beta',\varepsilon}|_{K',0}^{(|\beta'|)} |b_\varepsilon|_{K',|\beta|}^{(m)} \sup_{y\in K, |\gamma|\le h} |\partial^\gamma g(y)|,$$

holds, we conclude that when $[(a_{\beta',\varepsilon})_\varepsilon]$ and $[(b_\varepsilon)_\varepsilon]$ are symbols of slow scale type then the map $F_\omega(b) : \mathcal{G}_{\mathrm{c}}^\infty(\Omega) \to \mathcal{G}^\infty(\Omega')$ is continuous.
(iv) If $\mathrm{supp}_x b \Subset \Omega'$ from the first assertion we have that $F_\omega(b) \in \mathcal{G}_{\mathrm{c}}(\Omega')$. Under the assumptions of (ii) for the phase function ω we have that ${}^tF_\omega(b)$ maps $\mathcal{G}(\Omega')$ continuously into $\mathcal{G}(\Omega)$ and therefore $F_\omega(b)$ can be extended to a map from $\mathcal{L}(\mathcal{G}(\Omega), \widetilde{\mathbb{C}})$ to $\mathcal{L}(\mathcal{G}(\Omega'), \widetilde{\mathbb{C}})$. □

Remark 5.2. Taking $\Omega = \mathbb{R}^n$ and noting that $\mathcal{G}_{\mathrm{c}}(\Omega') \subseteq \mathcal{G}_{\mathrm{c}}(\mathbb{R}^n)$, it is clear that $F_\omega(b)$ maps $\mathcal{G}_{\mathrm{c}}(\mathbb{R}^n)$ into $\mathcal{G}_{\mathrm{c}}(\mathbb{R}^n)$ when $\mathrm{supp}_x b \Subset \Omega'$. In addition, ${}^tF_\omega(b) : \mathcal{G}(\mathbb{R}^n) \to \mathcal{G}(\mathbb{R}^n)$ and $F_\omega(b) : \mathcal{L}(\mathcal{G}(\mathbb{R}^n), \widetilde{\mathbb{C}}) \to \mathcal{L}(\mathcal{G}(\mathbb{R}^n), \widetilde{\mathbb{C}})$.

In the sequel we assume $\Omega = \Omega' \subseteq \mathbb{R}^n$. Our main purpose is to investigate the composition $a(x,D) \circ F_\omega(b)$, where $a(x,D)$ is a generalized pseudo-differential operator and $F_\omega(b)$ a generalized Fourier integral operator as in (5.1). This requires some technical preliminaries.

Technical preliminaries

The proof of the following lemma can be found in [6, Lemmas A.11, A.12].

Lemma 5.3. *Let $a \in \mathcal{C}^\infty(\Omega \times \mathbb{R}^n \setminus 0)$ and $\omega \in \mathcal{C}^\infty(\Omega \times \mathbb{R}^n \setminus 0)$. Then,*

$$\partial_x^\alpha\partial_\eta^\sigma(a(x,\nabla_x\omega(x,\eta))) = \sum_{\sigma'\le\sigma}\binom{\sigma}{\sigma'} \sum_{|\beta+\gamma|\le|\alpha|}\ \sum_{|\sigma''|\le|\sigma'|} \partial_x^\beta\partial_\eta^{\gamma+\sigma''} a(x,\nabla_x\omega(x,\eta))\cdot$$
$$\cdot P_{\eta,\sigma''}^{\sigma'}(x,\eta)\partial_\eta^{\sigma-\sigma'}P_{x\beta\gamma}^\alpha(x,\eta),$$

where

$$P_{\eta,\sigma''}^{\sigma'}(x,\eta) = 1 \qquad \text{if } \sigma' = 0,$$
$$P_{\eta,\sigma''}^{\sigma'} = \sum_{\substack{\delta_1,\dots,\delta_q\\ s_1,\dots,s_q}} c_{\delta_1,\dots,\delta_q}^{s_1,\dots,s_q}\,\partial_\eta^{\delta_1}\partial_{x_{s_1}}\omega(x,\eta)\dots\partial_\eta^{\delta_q}\partial_{x_{s_q}}\omega(x,\eta) \quad \text{otherwise},$$

with $q = |\sigma''|$, $\sum_{j=1}^{q} |\delta_j| = |\sigma'|$ *and*

$$P^{\alpha}_{x\beta\gamma} = 1 \qquad \textit{if} \quad \gamma = 0,$$
$$P^{\alpha}_{x\beta\gamma} = \sum_{\substack{\delta_1,\dots,\delta_r \\ s_1,\dots,s_r}} d^{s_1,\dots,s_r}_{\delta_1,\dots,\delta_r} \partial^{\delta_1}_x \partial_{x_{s_1}} \omega(x,\eta) \dots \partial^{\delta_r}_x \partial_{x_{s_r}} \omega(x,\eta) \quad \textit{otherwise},$$

with $|\gamma| = r$ *and* $\sum_{j=1}^{r} |\delta_j| + |\beta| = |\alpha|$.

Proposition 5.4.

(h1) *Let* $(\omega_\varepsilon)_\varepsilon \in \mathcal{M}_{S^1_{\mathrm{hg}}}(\Omega \times \mathbb{R}^n \setminus 0)$ *such that* $\nabla_x \omega_\varepsilon \neq 0$ *for all* $\varepsilon \in (0,1]$ *and for all* $K \Subset \Omega$

$$\Big(\inf_{x\in K, \eta \in \mathbb{R}^n \setminus 0} \Big|\nabla_x \omega_\varepsilon(x, \frac{\eta}{|\eta|})\Big|\Big)_\varepsilon$$

is strictly non-zero.

(i) *If* $(a_\varepsilon)_\varepsilon \in \mathcal{M}_{S^m(\Omega\times\mathbb{R}^n\setminus 0)}$ *then* $(a_\varepsilon(x, \nabla_x\omega_\varepsilon(x,\eta)))_\varepsilon \in \mathcal{M}_{S^m(\Omega\times\mathbb{R}^n\setminus 0)}$*;*

(ii) *if* $(a_\varepsilon)_\varepsilon \in \mathcal{N}_{S^m(\Omega\times\mathbb{R}^n\setminus 0)}$ *then* $(a_\varepsilon(x, \nabla_x\omega_\varepsilon(x,\eta)))_\varepsilon \in \mathcal{N}_{S^m(\Omega\times\mathbb{R}^n\setminus 0)}$.

(h2) *Let* $(\omega_\varepsilon)_\varepsilon \in \mathcal{M}^{\mathrm{sc}}_{S^1_{\mathrm{hg}}}(\Omega \times \mathbb{R}^n \setminus 0)$ *such that* $\nabla_x \omega_\varepsilon \neq 0$ *for all* $\varepsilon \in (0,1]$ *and for all* $K \Subset \Omega$

$$\Big(\inf_{x\in K, \eta \in \mathbb{R}^n \setminus 0} \Big|\nabla_x \omega_\varepsilon(x, \frac{\eta}{|\eta|})\Big|\Big)_\varepsilon$$

is slow scale strictly non-zero.

(iii) *If* $(a_\varepsilon)_\varepsilon \in \mathcal{M}^{\mathrm{sc}}_{S^m(\Omega\times\mathbb{R}^n\setminus 0)}$ *then* $(a_\varepsilon(x, \nabla_x\omega_\varepsilon(x,\eta)))_\varepsilon \in \mathcal{M}^{\mathrm{sc}}_{S^m(\Omega\times\mathbb{R}^n\setminus 0)}$.

(h3) *Finally, let* $(\omega_\varepsilon - \omega'_\varepsilon)_\varepsilon \in \mathcal{N}_{S^1_{\mathrm{hg}}}(\Omega \times \mathbb{R}^n \setminus 0)$ *with* $(\omega_\varepsilon)_\varepsilon$ *and* $(\omega'_\varepsilon)_\varepsilon$ *satisfying the hypothesis* (h1) *above.*

(iv) *If* $(a_\varepsilon)_\varepsilon \in \mathcal{M}_{S^m(\Omega\times\mathbb{R}^n\setminus 0)}$ *then*

$$(a_\varepsilon(x, \nabla_x\omega_\varepsilon(x,\eta)) - a_\varepsilon(x, \nabla_x\omega'_\varepsilon(x,\eta)))_\varepsilon \in \mathcal{N}_{S^m(\Omega\times\mathbb{R}^n\setminus 0)}.$$

Proof. From Lemma 5.3 it follows that $\partial^\alpha_x \partial^\sigma_\eta (a_\varepsilon(x, \nabla_x\omega_\varepsilon(x,\eta)))$ is a finite sum of terms of the type

$$\partial^{\alpha'}_x \partial^{\sigma'}_\eta a_\varepsilon(x, \nabla_x\omega_\varepsilon(x,\eta)) g_{\alpha',\sigma',\varepsilon}(x,\eta),$$

where $(g_{\alpha',\sigma',\varepsilon})_\varepsilon$ is a net of symbols in $S^{|\sigma'|-|\sigma|}(\Omega \times \mathbb{R}^n \setminus 0)$. Note that $(g_{\alpha',\sigma',\varepsilon})_\varepsilon$ depends on $(\omega_\varepsilon)_\varepsilon$ and is actually a finite sum of products of derivatives of $(\omega_\varepsilon)_\varepsilon$. One can easily prove that

$$\begin{aligned} (\omega_\varepsilon)_\varepsilon \in \mathcal{M}_{S^1_{\mathrm{hg}}}(\Omega \times \mathbb{R}^n \setminus 0) \quad &\Rightarrow \quad (g_{\alpha',\sigma',\varepsilon})_\varepsilon \in \mathcal{M}_{S^{|\sigma'|-|\sigma|}(\Omega\times\mathbb{R}^n\setminus 0)}, \\ (\omega_\varepsilon)_\varepsilon \in \mathcal{M}^{\mathrm{sc}}_{S^1_{\mathrm{hg}}}(\Omega \times \mathbb{R}^n \setminus 0) \quad &\Rightarrow \quad (g_{\alpha',\sigma',\varepsilon})_\varepsilon \in \mathcal{M}^{\mathrm{sc}}_{S^{|\sigma'|-|\sigma|}(\Omega\times\mathbb{R}^n\setminus 0)} \end{aligned} \tag{5.2}$$

and that the following

$$\forall \alpha', \sigma' \in \mathbb{N}^n\, \forall K \Subset \Omega\, \exists (\lambda_\varepsilon)_\varepsilon \in \mathbb{R}^{(0,1]}\, \forall x \in K\, \forall \eta \in \mathbb{R}^n \setminus 0\, \forall \varepsilon \in (0,1]$$
$$|\partial^{\alpha'}_x \partial^{\sigma'}_\eta a_\varepsilon(x, \nabla_x\omega_\varepsilon(x,\eta))| \le \lambda_\varepsilon \langle \nabla_x \omega_\varepsilon \rangle^{m-|\sigma'|} \tag{5.3}$$

holds, with $(\lambda_\varepsilon)_\varepsilon \in \mathcal{E}_M$ if $(a_\varepsilon)_\varepsilon \in \mathcal{M}_{S^m(\Omega\times\mathbb{R}^n\setminus 0)}$, $(\lambda_\varepsilon)_\varepsilon$ slow scale net if $(a_\varepsilon)_\varepsilon \in \mathcal{M}^{\rm sc}_{S^m(\Omega\times\mathbb{R}^n\setminus 0)}$ and $(\lambda_\varepsilon)_\varepsilon \in \mathcal{N}$ if $(a_\varepsilon)_\varepsilon \in \mathcal{N}_{S^m(\Omega\times\mathbb{R}^n\setminus 0)}$. Now, let us consider $(\nabla_x\omega_\varepsilon(x,\eta))_\varepsilon$. We have that

$$\text{(h1)} \quad \Rightarrow \qquad \forall K \Subset \Omega\, \exists r>0\, \exists c_1,c_2>0\, \exists \eta\in(0,1]\, \forall x\in K\, \forall |\eta|\ge 1\, \forall \varepsilon\in(0,\eta]$$
$$\langle\eta\rangle c_1\varepsilon^r \le |\nabla_x\omega_\varepsilon(x,\eta)| \le c_2\varepsilon^{-r}\langle\eta\rangle,$$

$$\text{(h2)} \quad \Rightarrow \qquad \forall K \Subset \Omega\, \exists(\mu_\varepsilon)_\varepsilon\ \text{s.s.n}\, \exists\eta\in(0,1]\, \forall x\in K\, \forall|\eta|\ge 1\, \forall\varepsilon\in(0,\eta]$$
$$\langle\eta\rangle\mu_\varepsilon^{-1} \le |\nabla_x\omega_\varepsilon(x,\eta)| \le \mu_\varepsilon\langle\eta\rangle.$$

Under the hypothesis (h1), combining (5.2) with (5.3) we obtain the assertions (i) and (ii). Moreover, from the second implications of (5.2) and (5.3) we see that (h2) yields (iii). It remains to prove that if (h3) holds and $(a_\varepsilon)_\varepsilon$ is a moderate net of symbols then $(a_\varepsilon(x,\nabla_x\omega_\varepsilon(x,\eta)) - a_\varepsilon(x,\nabla_x\omega'_\varepsilon(x,\eta)))_\varepsilon \in \mathcal{N}_{S^m(\Omega\times\mathbb{R}^n\setminus 0)}$. If suffices to write $\partial^\alpha_x\partial^\sigma_\eta(a_\varepsilon(x,\nabla_x\omega_\varepsilon(x,\eta)) - a_\varepsilon(x,\nabla_x\omega'_\varepsilon(x,\eta)))$ as the finite sum

$$\sum_{\alpha',\sigma'} \partial^{\alpha'}_x\partial^{\sigma'}_\eta a_\varepsilon(x,\nabla_x\omega_\varepsilon(x,\eta))(g_{\alpha',\sigma'}(\omega_\varepsilon) - g_{\alpha',\sigma'}(\omega'_\varepsilon))$$
$$+ \sum_{\alpha',\sigma'}[\partial^{\alpha'}_x\partial^{\sigma'}_\eta a_\varepsilon(x,\nabla_x\omega_\varepsilon(x,\eta)) - \partial^{\alpha'}_x\partial^{\sigma'}_\eta a_\varepsilon(x,\nabla_x\omega'_\varepsilon(x,\eta))]g_{\alpha',\sigma'}(\omega'_\varepsilon). \quad (5.4)$$

An inspection of Lemma 5.3 shows that the net $(g_{\alpha',\sigma'}(\omega_\varepsilon) - g_{\alpha',\sigma'}(\omega'_\varepsilon))_\varepsilon$ belongs to $\mathcal{N}_{S^{|\sigma'|-|\sigma|}(\Omega\times\mathbb{R}^n\setminus 0)}$ and from the hypothesis (h1) on $(\omega_\varepsilon)_\varepsilon$ it follows that the first summand in (5.4) is an element of $\mathcal{N}_{S^{m-|\sigma|}(\Omega\times\mathbb{R}^n\setminus 0)}$. We use Taylor's formula on the second summand of (5.4). Therefore, for x varying in a compact set K and for ε small enough we can estimate

$$|\partial^{\alpha'}_x\partial^{\sigma'}_\eta a_\varepsilon(x,\nabla_x\omega_\varepsilon(x,\eta)) - \partial^{\alpha'}_x\partial^{\sigma'}_\eta a_\varepsilon(x,\nabla_x\omega'_\varepsilon(x,\eta))|$$

by means of

$$\sum_{j=1}^n \varepsilon^{-N}\langle\nabla_x\omega'_\varepsilon(x,\eta) + \theta(\nabla_x\omega_\varepsilon(x,\eta) - \nabla_x\omega'_\varepsilon(x,\eta))\rangle^{m-|\sigma'|-1}|\partial_{x_j}(\omega_\varepsilon-\omega'_\varepsilon)(x,\eta)|$$
$$\le \varepsilon^q\langle\nabla_x\omega'_\varepsilon(x,\eta) + \theta(\nabla_x\omega_\varepsilon(x,\eta) - \nabla_x\omega'_\varepsilon(x,\eta))\rangle^{m-|\sigma'|-1}\langle\eta\rangle,$$

where $\theta\in[0,1]$. Since, taking ε small and $|\eta|\ge 1$ the following inequalities

$$|\nabla_x\omega'_\varepsilon(x,\eta) + \theta(\nabla_x\omega_\varepsilon(x,\eta) - \nabla_x\omega'_\varepsilon(x,\eta))| \ge \varepsilon^r\langle\eta\rangle - \varepsilon^{r+1}\langle\eta\rangle \ge \frac{\varepsilon^r}{2}\langle\eta\rangle,$$
$$|\nabla_x\omega'_\varepsilon(x,\eta) + \theta(\nabla_x\omega_\varepsilon(x,\eta) - \nabla_x\omega'_\varepsilon(x,\eta))| \le \varepsilon^{-r}\langle\eta\rangle$$

hold for some $r>0$, we conclude that

$$\big(\partial^{\alpha'}_x\partial^{\sigma'}_\eta a_\varepsilon(x,\nabla_x\omega_\varepsilon(x,\eta)) - \partial^{\alpha'}_x\partial^{\sigma'}_\eta a_\varepsilon(x,\nabla_x\omega'_\varepsilon(x,\eta))\big)_\varepsilon \in \mathcal{N}_{S^{m-|\sigma'|}(\Omega\times\mathbb{R}^n\setminus 0)}.$$

Thus, from $(g_{\alpha',\sigma'}(\omega'_\varepsilon))_\varepsilon \in \mathcal{M}_{S^{|\sigma'|-|\sigma|}(\Omega\times\mathbb{R}^n\setminus 0)}$ we have that the second summand of (5.4) belongs to $\mathcal{N}_{S^{m-|\sigma|}(\Omega\times\mathbb{R}^n\setminus 0)}$ and the proof is complete. □

Corollary 5.5. *If $a \in \widetilde{\mathcal{S}}^m(\Omega \times \mathbb{R}^n \setminus 0)$ and $\omega \in \widetilde{\mathcal{S}}^1_{\rm hg}(\Omega \times \mathbb{R}^n \setminus 0)$ has a representative satisfying condition* (h1) *of Proposition* 5.4 *then $a(x, \nabla_x\omega(x, \eta)) \in \widetilde{\mathcal{S}}^m(\Omega \times \mathbb{R}^n \setminus 0)$. If $a \in \widetilde{\mathcal{S}}^{m,\rm sc}(\Omega \times \mathbb{R}^n \setminus 0)$ and $\omega \in \widetilde{\mathcal{S}}^{1,\rm sc}_{\rm hg}(\Omega \times \mathbb{R}^n \setminus 0)$ has a representative satisfying condition* (h2) *of Proposition* 5.4, *then $a(x, \nabla_x\omega(x, \eta)) \in \widetilde{\mathcal{S}}^{m,\rm sc}(\Omega \times \mathbb{R}^n \setminus 0)$.*

Let $\omega \in \widetilde{\mathcal{S}}^1_{\rm hg}(\Omega \times \mathbb{R}^n \setminus 0)$ have a representative satisfying (h1). We want to investigate the properties of

$$D^\beta_z\big(\mathrm{e}^{i\overline{\omega}(z,x,\eta)}\big)|_{z=x}, \tag{5.5}$$

where $\overline{\omega}(z, x, \eta) := \omega(z, \eta) - \omega(x, \eta) - \langle\nabla_x\omega(x, \eta), z - x\rangle$. We make use of the following technical lemma, whose proof can be found in [7, Proposition 15].

Lemma 5.6. *Let $\omega \in \mathcal{C}^\infty(\Omega \times \mathbb{R}^n \setminus 0)$ and $\overline{\omega}(z, x, \eta)$ as above. Then, for $|\beta| \neq 0$, we have*

$$\begin{aligned} D^\beta_z \mathrm{e}^{i\overline{\omega}(z,x,\eta)} = \mathrm{e}^{i\overline{\omega}(z,x,\eta)}\Big[&(\nabla_z\omega(z, \eta) - \nabla_x\omega(x, \eta))^\beta \\ &+ \sum_{j_1} c_{j_1}(\nabla_z\omega(z, \eta) - \nabla_x\omega(x, \eta))^{\theta_{j_1}} \prod_{j_2=1}^{n_{1,j_1}} \partial_z^{\gamma_{j_1,j_2}}\omega(z, \eta) \\ &+ \sum_{j_1} c'_{j_1} \prod_{j_2=1}^{n_{2,j_1}} \partial_z^{\delta_{j_1,j_2}}\omega(z, \eta)\Big], \end{aligned}$$

where c_{j_1}, c'_{j_1} are suitable constants, $|\gamma_{j_1,j_2}| \geq 2$, $|\delta_{j_1,j_2}| \geq 2$ and

$$\theta_{j_1} + \sum_{j_2=1}^{n_{1,j_1}} \gamma_{j_1,j_2} = \sum_{j_2=1}^{n_{2,j_1}} \delta_{j_1,j_2} = \beta.$$

It follows that

$$D^\beta_z\big(\mathrm{e}^{i\overline{\omega}(z,x,\eta)}\big)|_{z=x} = \sum_{j_1} c_{j_1} \prod_{j_2=1}^{n_{1,j_1}} \partial_x^{\gamma_{j_1,j_2}}\omega(x, \eta) + \sum_{j_1} c'_{j_1} \prod_{j_2=1}^{n_{2,j_1}} \partial_x^{\delta_{j_1,j_2}}\omega(x, \eta), \tag{5.6}$$

with

$$\sum_{j_2=1}^{n_{1,j_1}} \gamma_{j_1,j_2} = \sum_{j_2=1}^{n_{2,j_1}} \delta_{j_1,j_2} = \beta.$$

Moreover, from $|\gamma_{j_1,j_2}| \geq 2$, $|\delta_{j_1,j_2}| \geq 2$ we have $|\beta| \geq 2n_{1,j_1}$ and $|\beta| \geq 2n_{2,j_1}$.

Since the constants c_{j_1}, c'_{j_1} do not depend on ω, we can use the formula (5.6) in estimating the net $(D^\beta_z\big(\mathrm{e}^{i\overline{\omega_\varepsilon}(z,x,\eta)}\big)|_{z=x})_\varepsilon$.

Proposition 5.7.

(i) *If $\omega \in \widetilde{\mathcal{S}}^1_{\rm hg}(\Omega \times \mathbb{R}^n \setminus 0)$ then* (5.5) *is a well-defined element of $\widetilde{\mathcal{S}}^{|\beta|/2}(\Omega \times \mathbb{R}^n \setminus 0)$.*

(ii) *If $\omega \in \widetilde{\mathcal{S}}^{1,\rm sc}_{\rm hg}(\Omega \times \mathbb{R}^n \setminus 0)$ then* (5.5) *is a well-defined element of $\widetilde{\mathcal{S}}^{|\beta|/2,\rm sc}(\Omega \times \mathbb{R}^n \setminus 0)$.*

Proof. From (5.6) we have that

$$(\omega_\varepsilon)_\varepsilon \in \mathcal{M}_{S^1_{\rm hg}}(\Omega\times\mathbb{R}^n\setminus 0) \quad\Rightarrow\quad (D^\beta_z(\mathrm{e}^{i\overline{\omega_\varepsilon}(z,x,\eta)})|_{z=x})_\varepsilon \in \mathcal{M}_{S^{|\beta|/2}(\Omega\times\mathbb{R}^n\setminus 0)},$$
$$(\omega_\varepsilon)_\varepsilon \in \mathcal{M}^{\rm sc}_{S^1_{\rm hg}}(\Omega\times\mathbb{R}^n\setminus 0) \quad\Rightarrow\quad (D^\beta_z(\mathrm{e}^{i\overline{\omega_\varepsilon}(z,x,\eta)})|_{z=x})_\varepsilon \in \mathcal{M}^{\rm sc}_{S^{|\beta|/2}(\Omega\times\mathbb{R}^n\setminus 0)}.$$

Noting that $(\omega_\varepsilon-\omega'_\varepsilon)_\varepsilon \in \mathcal{N}_{S^1_{\rm hg}(\Omega\times\mathbb{R}^n\setminus 0)}$ entails

$$\Big(\prod_{j_2=1}^{n_{1,j_1}} \partial_x^{\gamma_{j_1,j_2}}\omega_\varepsilon(x,\eta) - \prod_{j_2=1}^{n_{1,j_1}} \partial_x^{\gamma_{j_1,j_2}}\omega'_\varepsilon(x,\eta)\Big)_\varepsilon \in \mathcal{N}_{S^{|\beta|/2}(\Omega\times\mathbb{R}^n\setminus 0)},$$
$$\Big(\prod_{j_2=1}^{n_{2,j_1}} \partial_x^{\delta_{j_1,j_2}}\omega_\varepsilon(x,\eta) - \prod_{j_2=1}^{n_{2,j_1}} \partial_x^{\delta_{j_1,j_2}}\omega'_\varepsilon(x,\eta)\Big)_\varepsilon \in \mathcal{N}_{S^{|\beta|/2}(\Omega\times\mathbb{R}^n\setminus 0)},$$

we conclude that the net $\big(D^\beta_z(\mathrm{e}^{i\overline{\omega_\varepsilon}(z,x,\eta)})|_{z=x} - D^\beta_z(\mathrm{e}^{i\overline{\omega'_\varepsilon}(z,x,\eta)})|_{z=x}\big)_\varepsilon$ belongs to $\mathcal{N}_{S^{|\beta|/2}(\Omega\times\mathbb{R}^n\setminus 0)}$. □

By combining Corollary 5.5 with Proposition 5.7 we obtain the following statement.

Proposition 5.8. *Let $\alpha\in\mathbb{N}^n$ and*

$$h_\alpha(x,\eta) = \frac{\partial^\alpha_\xi a(x,\nabla_x\omega(x,\eta))}{\alpha!} D^\alpha_z\big(\mathrm{e}^{i\overline{\omega}(z,x,\eta)}b(z,\eta)\big)|_{z=x}. \tag{5.7}$$

(i) *If $a\in\widetilde{\mathcal{S}}^m(\Omega\times\mathbb{R}^n\setminus 0)$, $\omega\in\widetilde{\mathcal{S}}^1_{\rm hg}(\Omega\times\mathbb{R}^n\setminus 0)$ has a representative satisfying condition* (h1) *and $b\in\widetilde{\mathcal{S}}^l(\Omega\times\mathbb{R}^n\setminus 0)$, then $h_\alpha\in\widetilde{\mathcal{S}}^{l+m-|\alpha|/2}(\Omega\times\mathbb{R}^n\setminus 0)$ for all α.*

(ii) *If $a\in\widetilde{\mathcal{S}}^{m,\rm sc}(\Omega\times\mathbb{R}^n\setminus 0)$, $\omega\in\widetilde{\mathcal{S}}^{1,\rm sc}_{\rm hg}(\Omega\times\mathbb{R}^n\setminus 0)$ has a representative satisfying condition* (h2) *and $b\in\widetilde{\mathcal{S}}^{l,\rm sc}(\Omega\times\mathbb{R}^n\setminus 0)$, then $h_\alpha\in\widetilde{\mathcal{S}}^{l+m-|\alpha|/2,\rm sc}(\Omega\times\mathbb{R}^n\setminus 0)$ for all α.*

Our next task is to give a closer look to $\mathrm{e}^{i\omega(x,\eta)}$.

Proposition 5.9. *Let $\omega\in\widetilde{\mathcal{S}}^1_{\rm hg}(\Omega\times\mathbb{R}^n\setminus 0)$ have a representative satisfying condition* (h1)*. Then for any positive integer N there exists $p_N\in\widetilde{\mathcal{S}}^{-2N}(\Omega\times\mathbb{R}^n\setminus 0)$ such that*

$$\mathrm{e}^{i\omega(x,\eta)} = \Big(p_N(x,\eta)\Delta^N_x + r(x,\eta)\Big)\mathrm{e}^{i\omega(x,\eta)}, \tag{5.8}$$

where $r\in\widetilde{\mathcal{S}}^{-\infty}(\Omega\times\mathbb{R}^n\setminus 0)$.
If $\omega\in\widetilde{\mathcal{S}}^1_{\rm hg}(\Omega\times\mathbb{R}^n\setminus 0)$ is of slow scale type and has a representative satisfying condition (h2) *then p_N and r are of slow scale type.*

Proof. Let $(\omega_\varepsilon)_\varepsilon$ be a representative of ω satisfying (h1). We leave to the reader to prove by induction that

$$\Delta^N_x\big(\mathrm{e}^{i\omega_\varepsilon(x,\eta)}\big) = a_\varepsilon(x,\eta)\mathrm{e}^{i\omega_\varepsilon(x,\eta)},$$

where $(a_\varepsilon)_\varepsilon \in \mathcal{M}_{S^{2N}(\Omega\times\mathbb{R}^n\setminus 0)}$ with principal part given by

$$a_{2N,\varepsilon} = (-1)^N |\nabla_x\omega_\varepsilon(x,\eta)|^{2N}.$$

From (h1) we have that $\nabla_x\omega_\varepsilon \neq 0$ for all $\varepsilon \in (0,1]$ and for all $K \Subset \Omega$ there exist $r > 0$ and $\varepsilon_0 \in (0,1]$ such that $|\nabla_x\omega_\varepsilon(x,\eta)| \geq \varepsilon^r|\eta|$ for all $x \in K$, $\eta \neq 0$ and $\varepsilon \in (0,\varepsilon_0]$. Hence,

$$|\nabla_x\omega_\varepsilon(x,\eta)| \geq \frac{\varepsilon^r}{2}\langle\eta\rangle,$$

for $|\eta| \geq 1$, $x \in K$ and $\varepsilon \in (0,\varepsilon_0]$. It follows from Proposition 2.7(iii) that $(a_\varepsilon)_c$ is a net of elliptic symbols of $S^{2N}(\Omega \times \mathbb{R}^n \setminus 0)$ such that for all $K \Subset \Omega$ there exist $s \in \mathbb{R}$, $(R_\varepsilon)_\varepsilon$ strictly nonzero and $\varepsilon_0 \in (0,1]$ such that

$$|a_\varepsilon(x,\eta)| \geq \varepsilon^s\langle\eta\rangle^{2N},$$

for $x \in K$, $|\eta| \geq R_\varepsilon$ and $\varepsilon \in (0,\varepsilon_0]$. By Proposition 2.8(i) we find $(p_{N,\varepsilon})_\varepsilon \in \mathcal{M}_{S^{-2N}(\Omega\times\mathbb{R}^n\setminus 0)}$ and $(r_\varepsilon)_\varepsilon \in \mathcal{M}_{S^{-\infty}(\Omega\times\mathbb{R}^n\setminus 0)}$ such that

$$p_{N,\varepsilon}a_\varepsilon = 1 - r_\varepsilon \tag{5.9}$$

for all ε. Therefore,

$$\mathrm{e}^{i\omega_\varepsilon(x,\eta)} = \Big(p_{N,\varepsilon}(x,\eta)\Delta_x^N + r_\varepsilon(x,\eta)\Big)\mathrm{e}^{i\omega_\varepsilon(x,\eta)}.$$

This equality at the representatives' level implies the equality (5.8) between equivalence classes of $\widetilde{\mathcal{S}}^1_{0,1}(\Omega \times \mathbb{R}^n \setminus 0)$.

Now, let ω be a slow scale symbol with a representative $(\omega_\varepsilon)_\varepsilon$ satisfying condition (h2). From Proposition 2.7(vi) we have that $(a_\varepsilon)_\varepsilon \in \mathcal{M}^{\mathrm{sc}}_{S^{2N}(\Omega\times\mathbb{R}^n\setminus 0)}$ is a net of elliptic symbols such that for some $(s_\varepsilon)_\varepsilon$ inverse of a slow scale net, $(R_\varepsilon)_\varepsilon$ slow scale net and $\varepsilon_0 \in (0,1]$ the inequality

$$|a_\varepsilon(x,\eta)| \geq s_\varepsilon\langle\eta\rangle^{2N},$$

holds for all $x \in K$, $|\eta| \geq R_\varepsilon$ and $\varepsilon \in (0,\varepsilon_0]$. Proposition 2.8(ii) shows that (5.9) is true for some $(p_{N,\varepsilon})_\varepsilon \in \mathcal{M}^{\mathrm{sc}}_{S^{-2N}(\Omega\times\mathbb{R}^n\setminus 0)}$ and $(r_\varepsilon)_\varepsilon \in \mathcal{M}^{\mathrm{sc}}_{S^{-\infty}(\Omega\times\mathbb{R}^n\setminus 0)}$. □

Main Theorems

The make use of the previous propositions in proving the main theorems of this section: Theorems 5.10 and 5.11.

Theorem 5.10. *Let $\omega \in \widetilde{\mathcal{S}}^1_{\mathrm{hg}}(\Omega \times \mathbb{R}^n \setminus 0)$ have a representative satisfying condition* (h1). *Let $a \in \widetilde{\mathcal{S}}^m(\Omega \times \mathbb{R}^n)$ and $b \in \widetilde{\mathcal{S}}^l(\Omega \times \mathbb{R}^n \setminus 0)$ with* $\mathrm{supp}_x\, b \Subset \Omega$. *Then, the operator $a(x,D)F_\omega(b)$ has the following properties:*

(i) *maps $\mathcal{G}_\mathrm{c}(\Omega)$ into $\mathcal{G}(\Omega)$ and $\mathcal{L}(\mathcal{G}(\Omega),\widetilde{\mathbb{C}})$ into $\mathcal{L}(\mathcal{G}_\mathrm{c}(\Omega),\widetilde{\mathbb{C}})$;*
(ii) *is of the form*

$$\int_{\mathbb{R}^n} \mathrm{e}^{i\omega(x,\eta)}h(x,\eta)\widehat{u}(\eta)\,đ\eta + r(x,D)u,$$

where $h \in \widetilde{\mathcal{S}}^{l+m}(\Omega \times \mathbb{R}^n \setminus 0)$ *has asymptotic expansion given by the symbols* h_α *defined in* (5.7) *and* $r \in \widetilde{\mathcal{S}}^{-\infty}(\Omega \times \mathbb{R}^n \setminus 0)$.

Proof. From Proposition 5.1(iv) is clear that $F_\omega(b)$ maps $\mathcal{G}_c(\Omega)$ and $\mathcal{L}(\mathcal{G}(\Omega), \widetilde{\mathbb{C}})$ into themselves respectively. We obtain (i) combining this results with the usual mapping properties of a generalized pseudo-differential operator. We now have to investigate the composition

$$\begin{aligned} a(x,D)F_\omega(b)u(x) &= \int_{\Omega\times\mathbb{R}^n} e^{i(x-z)\theta} a(x,\theta) F_\omega(b)u(z)\, dz\, đ\theta \\ &= \int_{\Omega\times\mathbb{R}^n} e^{i(x-z)\theta} a(x,\theta) \Big(\int_{\mathbb{R}^n} e^{i\omega(z,\eta)} b(z,\eta)\widehat{u}(\eta)\, đ\eta \Big)\, dz\, đ\theta \\ &= \int_{\mathbb{R}^n}\int_{\Omega\times\mathbb{R}^n} e^{i((x-z)\theta+\omega(z,\eta))} a(x,\theta) b(z,\eta)\, dz\, đ\theta\, \widehat{u}(\eta)\, đ\eta, \end{aligned}$$

for $u \in \mathcal{G}_c(\Omega)$. The last integral in dz and $đ\theta$ is regarded as the oscillatory integral

$$\int_{\Omega\times\mathbb{R}^n} e^{i(x-z)\theta} a(x,\theta) b(z,\eta) e^{i\omega(z,\eta)}\, dz\, đ\theta, \tag{5.10}$$

with $b(z,\eta)e^{i\omega(z,\eta)} \in \mathcal{G}_c(\Omega_z)$.

In the sequel we will work at the level of representatives and we will follow the proof of Theorem 4.1.1 in [31].

Step 1. Let$(\sigma_\varepsilon)_\varepsilon$ such that $\sigma_\varepsilon \geq c\varepsilon^s$ for some $c, s > 0$ and for all $\varepsilon \in (0,1]$. We take $\varphi \in \mathcal{C}^\infty(\mathbb{R}^n)$ such that $\varphi(y) = 1$ for $|y| \leq 1/2$ and $\varphi(y) = 0$ for $|y| \geq 1$ and we set

$$b_\varepsilon(z,\eta) = b'_\varepsilon(z,x,\eta) + b''_\varepsilon(z,x,\eta) = \varphi\big(\frac{x-z}{\sigma_\varepsilon}\big) b_\varepsilon(z,\eta) + \big(1 - \varphi\big(\frac{x-z}{\sigma_\varepsilon}\big)\big) b_\varepsilon(z,\eta).$$

We now write the integral in dz and $đ\theta$ of (5.10) as

$$\begin{aligned} &\int_{\Omega\times\mathbb{R}^n} e^{i((x-z)\theta+\omega_\varepsilon(z,\eta))} a_\varepsilon(x,\theta) b'_\varepsilon(z,x,\eta)\, dz\, đ\theta \\ &\quad + \int_{\Omega\times\mathbb{R}^n} e^{i((x-z)\theta+\omega_\varepsilon(z,\eta))} a_\varepsilon(x,\theta) b''_\varepsilon(z,x,\eta)\, dz\, đ\theta := I_{1,\varepsilon}(x,\eta) + I_{2,\varepsilon}(x,\eta) \end{aligned}$$

and we begin to investigate the properties of $(I_{2,\varepsilon})_\varepsilon$. Proposition 5.9 provides the identity

$$e^{i\omega_\varepsilon(z,\eta)} = \Big(p_{N,\varepsilon}(z,\eta)\Delta_z^N + r_\varepsilon(z,\eta) \Big) e^{i\omega_\varepsilon(z,\eta)},$$

where $(p_{N,\varepsilon})_\varepsilon \in \mathcal{M}_{S^{-2N}(\Omega\times\mathbb{R}^n\setminus 0)}$ and $(r_\varepsilon)_\varepsilon \in \mathcal{M}_{S^{-\infty}(\Omega\times\mathbb{R}^n\setminus 0)}$, and allows us to write $(I_{2,\varepsilon})_\varepsilon$ as

$$\begin{aligned} &\int_{\Omega\times\mathbb{R}^n} e^{i\omega_\varepsilon(z,\eta)} \Delta_z^N \Big(e^{i(x-z)\theta} p_{N,\varepsilon}(z,\eta) b''_\varepsilon(z,x,\eta) \Big) a_\varepsilon(x,\theta)\, dz\, đ\theta \\ &+ \int_{\Omega\times\mathbb{R}^n} e^{i(x-z)\theta} a_\varepsilon(x,\theta) b''_\varepsilon(z,x,\eta) r_\varepsilon(z,\eta) e^{i\omega_\varepsilon(z,\eta)}\, dz\, đ\theta := I^1_{2,\varepsilon}(x,\eta) + I^2_{2,\varepsilon}(x,\eta) \end{aligned}$$

The net $(I^2_{2,\varepsilon})_\varepsilon \in \mathcal{M}_{S^{-\infty}}(\Omega\times\mathbb{R}^n\setminus 0)$. Indeed, $I^2_{2,\varepsilon}(x,\eta) = \int_{\mathbb{R}^n}\int_{\mathbb{R}^n} g_\varepsilon(x,\eta,z,\theta)\,dz\,đ\theta$, where

$$g_\varepsilon(x,\eta,z,\theta) = \mathrm{e}^{i(x-z)\theta}a_\varepsilon(x,\theta)b''_\varepsilon(z,x,\eta)r_\varepsilon(z,\eta)\mathrm{e}^{i\omega_\varepsilon(z,\eta)}$$

and the following holds: for all $K \Subset \Omega$, for all $\alpha \in \mathbb{N}^n$ and $d>0$ exist $N \in \mathbb{N}$ and $\varepsilon_0 \in (0,1]$ such that

$$\left|(i\theta)^\alpha \int_{\mathbb{R}^n} g_\varepsilon(x,\eta,z,\theta)\,dz\right| \le \varepsilon^{-N}\langle\theta\rangle^m\langle\eta\rangle^{-d}.$$

This is due to the fact that $\mathrm{supp}_z b_\varepsilon \subseteq K_b \Subset \Omega$ for all ε and $(r_\varepsilon)_\varepsilon \in \mathcal{M}_{S^{-\infty}(\Omega\times\mathbb{R}^n\setminus 0)}$.

Step 2. By construction $b''_\varepsilon(z,x,\eta) = 0$ if $|x-z| \le \sigma_\varepsilon/2$ for all $\varepsilon \in (0,1]$. By making use of the identity

$$\mathrm{e}^{i(x-z)\theta} = |x-z|^{-2k}(-\Delta_\theta)^k\left(\mathrm{e}^{i(x-z)\theta}\right)$$

we have

$$\begin{aligned}
&I^1_{2,\varepsilon}(x,\eta)\\
&= \int_{\Omega\times\mathbb{R}^n} \mathrm{e}^{i\omega_\varepsilon(z,\eta)}\Delta_z^N\left(|x-z|^{-2k}(-\Delta_\theta)^k\left(\mathrm{e}^{i(x-z)\theta}\right)p_{N,\varepsilon}(z,\eta)b''_\varepsilon(z,x,\eta)\right)a_\varepsilon(x,\theta)\,dz\,đ\theta\\
&= \int_{\Omega\times\mathbb{R}^n} \mathrm{e}^{i\omega_\varepsilon(z,\eta)}(-\Delta_\theta)^k a_\varepsilon(x,\theta)\Delta_z^N\left(\mathrm{e}^{i(x-z)\theta}|x-z|^{-2k}p_{N,\varepsilon}(z,\eta)b''_\varepsilon(z,x,\eta)\right)dz\,đ\theta.
\end{aligned}$$

It follows that for $x \in K \Subset \Omega$ and ε small enough

$$\begin{aligned}
|I^1_{2,\varepsilon}(x,\eta)| &\le c\varepsilon^{-N_a-N_p-N_b}(\sigma_\varepsilon)^{-2k}\sum_{|\gamma|\le 2N} c_\gamma\sigma_\varepsilon^{-|\gamma|}\int_{\mathbb{R}^n}\langle\theta\rangle^{m-2k+2N}\,đ\theta\,\langle\eta\rangle^{-2N+l}\\
&\le \varepsilon^{-N'}\int_{\mathbb{R}^n}\langle\theta\rangle^{m-2k+2N}\,đ\theta\,\langle\eta\rangle^{-2N+l}.
\end{aligned}$$

Hence, given $d>0$ and taking N,k such that $-2N+l<-d$ and $m-2k+2N<-n$ we obtain that $(I^1_{2,\varepsilon})_\varepsilon$ is a moderate net of symbols of order $-\infty$ on $\Omega\times\mathbb{R}^n\setminus 0$. Summarizing,

$$\int_{\Omega\times\mathbb{R}^n} \mathrm{e}^{i((x-z)\theta+\omega_\varepsilon(z,\eta))}a_\varepsilon(x,\theta)b_\varepsilon(z,\eta)\,dz\,đ\theta = I_{1,\varepsilon}(x,\eta) + I^1_{2,\varepsilon}(x,\eta) + I^2_{2,\varepsilon}(x,\eta),$$

where $(I^1_{2,\varepsilon})_\varepsilon$ and $(I^2_{2,\varepsilon})_\varepsilon$ belong to $\mathcal{M}_{S^{-\infty}(\Omega\times\mathbb{R}^n\setminus 0)}$.

Step 3. It remains to study

$$I_{1,\varepsilon}(x,\eta) = \int_{\Omega\times\mathbb{R}^n} \mathrm{e}^{i((x-z)\theta+\omega_\varepsilon(z,\eta))}a_\varepsilon(x,\theta)b'_\varepsilon(z,x,\eta)\,dz\,đ\theta.$$

We expand $a_\varepsilon(x,\theta)$ with respect to θ at $\theta = \nabla_x\omega_\varepsilon(x,\eta)$ and we observe that

$$(\theta-\nabla_x\omega_\varepsilon(x,\eta))^\alpha\mathrm{e}^{i(x-z)(\theta-\nabla_x\omega_\varepsilon(x,\eta))} = (-1)^{|\alpha|}D_z^\alpha\mathrm{e}^{i(x-z)(\theta-\nabla_x\omega_\varepsilon(x,\eta))}.$$

By integrating by parts we obtain

$$\begin{aligned}
&\mathrm{e}^{-i\omega_\varepsilon(x,\eta)} I_{1,\varepsilon}(x,\eta)\\
&= \int_{\Omega\times\mathbb{R}^n} \mathrm{e}^{i\overline{\omega}_\varepsilon(z,x,\eta)} \mathrm{e}^{i(x-z)(\theta-\nabla_x\omega_\varepsilon(x,\eta))} b'_\varepsilon(z,x,\eta) a_\varepsilon(x,\theta)\, dz\, đ\theta\\
&= \sum_{|\alpha|<k} \frac{1}{\alpha!} \partial_\xi^\alpha a_\varepsilon(x, \nabla_x\omega_\varepsilon(x,\eta))\cdot\\
&\quad \cdot \int_{\Omega\times\mathbb{R}^n} D_z^\alpha\Big(\mathrm{e}^{i\overline{\omega}_\varepsilon(z,x,\eta)} b'_\varepsilon(z,x,\eta)\Big) \mathrm{e}^{i(x-z)(\theta-\nabla_x\omega_\varepsilon(x,\eta))}\, dz\, đ\theta\\
&\quad + \sum_{|\alpha|=k} \frac{k}{\alpha!} \int_{\Omega\times\mathbb{R}^n} D_z^\alpha\Big(\mathrm{e}^{i\overline{\omega}_\varepsilon(z,x,\eta)} b'_\varepsilon(z,x,\eta)\Big) \mathrm{e}^{-i(x-z)\theta} r_{\alpha,\varepsilon}(x,\eta,\theta)\, dz\, đ\theta,
\end{aligned}$$

where $\overline{\omega}_\varepsilon(z,x,\eta) := \omega_\varepsilon(z,\eta) - \omega_\varepsilon(x,\eta) - (\nabla_x\omega_\varepsilon(x,\eta))(z-x)$ and

$$r_{\alpha,\varepsilon}(x,\eta,\theta) = \int_0^1 (1-t)^{k-1} \partial_\xi^\alpha a_\varepsilon(x, \nabla_x\omega_\varepsilon(x,\eta) - t\theta)\, dt.$$

Since, $b'_\varepsilon(z,x,\eta) = b_\varepsilon(x,\eta)$ if $|x-z| \le \frac{\sigma_\varepsilon}{2}$ we have that

$$\begin{aligned}
&\int_{\Omega\times\mathbb{R}^n} D_z^\alpha\Big(\mathrm{e}^{i\overline{\omega}_\varepsilon(z,x,\eta)} b'_\varepsilon(z,x,\eta)\Big) \mathrm{e}^{i(x-z)(\theta-\nabla_x\omega_\varepsilon(x,\eta))}\, dz\, đ\theta\\
&\qquad = \int_{\Omega\times\mathbb{R}^n} D_z^\alpha\Big(\mathrm{e}^{i\overline{\omega}_\varepsilon(z,x,\eta)} b'_\varepsilon(z,x,\eta)\Big) \mathrm{e}^{i(x-z)\theta}\, dz\, đ\theta\\
&\qquad = D_z^\alpha\Big(\mathrm{e}^{i\overline{\omega}_\varepsilon(z,x,\eta)} b_\varepsilon(z,\eta)\Big)|_{x=z}
\end{aligned}$$

This means that

$$\mathrm{e}^{-i\omega_\varepsilon(x,\eta)} I_{1,\varepsilon}(x,\eta) = \sum_{|\alpha|<k} h_{\alpha,\varepsilon}(x,\eta) + \sum_{|\alpha|=k} \frac{k}{\alpha!} R_{\alpha,\varepsilon}(x,\eta),$$

where $(h_{\alpha,\varepsilon})_\varepsilon$ is defined in (5.7) and

$$R_{\alpha,\varepsilon}(x,\eta) := \int_{\Omega\times\mathbb{R}^n} D_z^\alpha\Big(\mathrm{e}^{i\overline{\omega}_\varepsilon(z,x,\eta)} b'_\varepsilon(z,x,\eta)\Big) \mathrm{e}^{-i(x-z)\theta} r_{\alpha,\varepsilon}(x,\eta,\theta)\, dz\, đ\theta.$$

Step 4. Our next task is to prove moderate symbol estimates for the net $(R_{\alpha,\varepsilon})_\varepsilon$. Let $\chi \in \mathcal{C}_c^\infty(\mathbb{R}^n)$ such that $\chi(\theta) = 1$ for $|\theta| \le 1$ and $\chi(\theta) = 0$ for $|\theta| \ge 3/2$. Let us take a positive net $(\tau_\varepsilon)_\varepsilon$ such that $\tau_\varepsilon \ge c\varepsilon^r$ for some $c > 0$ and $r > 0$. We define the sets

$$W^1_{\tau_\varepsilon,\eta} = \{\theta \in \mathbb{R}^n : |\theta| < \tau_\varepsilon|\eta|\}, \qquad W^2_{\tau_\varepsilon,\eta} = \mathbb{R}^n \setminus W^1_{\tau_\varepsilon,\eta}.$$

Set now $\chi_\varepsilon(\theta) := \chi(\theta/\tau_\varepsilon)$. By construction we have that $\chi_\varepsilon(\theta/|\eta|) = 1$ on $W^1_{\tau_\varepsilon,\eta}$, $\operatorname{supp}\chi_\varepsilon(\cdot/|\eta|) \subseteq W^1_{2\tau_\varepsilon,\eta}$ and $\operatorname{supp}(1-\chi_\varepsilon(\cdot/|\eta|)) \subseteq W^2_{\tau_\varepsilon,\eta}$. We write $R_{\alpha,\varepsilon}(x,\eta)$ as

$$\begin{aligned}\int_{\Omega\times\mathbb{R}^n} & D^\alpha_z\Big(\mathrm{e}^{i\overline{\omega}_\varepsilon(z,x,\eta)}b'_\varepsilon(z,x,\eta)\Big)\mathrm{e}^{-i(x-z)\theta}r_{\alpha,\varepsilon}(x,\eta,\theta)\chi_\varepsilon(\theta/|\eta|)\,dz\,đ\theta\\ &+\int_{\Omega\times\mathbb{R}^n} D^\alpha_z\Big(\mathrm{e}^{i\overline{\omega}_\varepsilon(z,x,\eta)}b'_\varepsilon(z,x,\eta)\Big)\mathrm{e}^{-i(x-z)\theta}r_{\alpha,\varepsilon}(x,\eta,\theta)(1-\chi_\varepsilon(\theta/|\eta|))\,dz\,đ\theta\\ &\qquad := R^1_{\alpha,\varepsilon}(x,\eta)+R^2_{\alpha,\varepsilon}(x,\eta).\end{aligned}$$

We begin by estimating the net $R^1_{\alpha,\varepsilon}$. We make use of the identity

$$\mathrm{e}^{-i(x-z)\theta} = (1+|\eta|^2|x-z|^2)^{-N}(1-|\eta|^2\Delta_\theta)^N\mathrm{e}^{-i(x-z)\theta}$$

which yields

$$\begin{aligned}R^1_{\alpha,\varepsilon}(x,\eta) = \int_{\Omega\times\mathbb{R}^n} & D^\alpha_z\Big(\mathrm{e}^{i\overline{\omega}_\varepsilon(z,x,\eta)}b'_\varepsilon(z,x,\eta)\Big)\mathrm{e}^{-i(x-z)\theta}\cdot\\ &\cdot(1+|\eta|^2|x-z|^2)^{-N}(1-|\eta|^2\Delta_\theta)^N\big(r_{\alpha,\varepsilon}(x,\eta,\theta)\chi_\varepsilon(\theta/|\eta|)\big)\,dz\,đ\theta.\end{aligned}$$

By the moderateness of the net (ω_ε) and Taylor's formula we have the inequality

$$|\nabla_x\omega_\varepsilon(x,\eta)-\nabla_z\omega_\varepsilon(z,\eta)| \le c\varepsilon^{-M}|\eta|^1|x-z|, \tag{5.11}$$

valid for $z\in K_b$, $|x-z|\le\sigma_\varepsilon$ and σ_ε small enough such that $\cup_{\varepsilon\in(0,1]}\{z+\lambda(x-z): z\in K_b,\ |x-z|\le\sigma_\varepsilon,\ \lambda\in[0,1]\}\subseteq K'\Subset\Omega$. Clearly M depends on the compact set K'. By Lemma 5.6 we have that for all x and z as above, $|\eta|\ge 1$ and $\varepsilon\in(0,\varepsilon_0]$, the estimate

$$\Big|D^\beta_z\mathrm{e}^{i\overline{\omega}_\varepsilon(z,x,\eta)}\Big| \le c'\varepsilon^{-M'}\langle\eta\rangle^{\frac{|\beta|}{2}}(1+|\eta|^2|x-z|^2)^{L_\beta}$$

holds for some $L_\beta\in\mathbb{N}$ and $M'\in\mathbb{N}$. Hence, recalling that $\sigma_\varepsilon\ge c\varepsilon^s$ for some $c,s>0$, we are led from the previous considerations to

$$\Big|D^\alpha_z\Big(\mathrm{e}^{i\overline{\omega}_\varepsilon(z,x,\eta)}b'_\varepsilon(z,x,\eta)\Big)\Big| \le C\varepsilon^{-N'}\langle\eta\rangle^{l+\frac{|\alpha|}{2}}(1+|\eta|^2|x-z|^2)^{L_\alpha}, \tag{5.12}$$

valid for $|\eta|\ge 1$, ε small enough and N' depending on K_b, α and the bound $c\varepsilon^s$ of σ_ε. Before considering $(1-|\eta|^2\Delta_\theta)^N\big(r_{\alpha,\varepsilon}(x,\eta,\theta)\chi_\varepsilon(\theta/|\eta|)\big)$ it is useful to investigate the quantity $|\nabla_x\omega_\varepsilon(x,\eta)-t\theta|$ for $x\in K\Subset\Omega$ and $\theta\in W^1_{2\tau_\varepsilon,\eta}$. We recall that there exists $r>0$, $c_0,c_1>0$ and $\varepsilon_0\in(0,1]$ such that

$$c_0\varepsilon^r|\eta| \le |\nabla_x\omega_\varepsilon(x,\eta)| \le c_1\varepsilon^{-r}|\eta|,$$

for all $x\in K$, $\eta\ne 0$ and $\varepsilon\in(0,\varepsilon_0]$. Since, if $\theta\in W^1_{2\tau_\varepsilon,\eta}$ then $|\theta|<2\tau_\varepsilon|\eta|$, we obtain, for all $x\in K$, $\theta\in W^1_{2\tau_\varepsilon,\eta}$ and $t\in[0,1]$, the following estimates:

$$\begin{aligned}|\nabla_x\omega_\varepsilon(x,\eta)-t\theta| &\le |\nabla_x\omega_\varepsilon(x,\eta)|+|\theta| \le (1+2\tau_\varepsilon c_0^{-1}\varepsilon^{-r})|\nabla_x\omega_\varepsilon(x,\eta)|\\ |\nabla_x\omega_\varepsilon(x,\eta)-t\theta| &\ge (1-2\tau_\varepsilon c_0^{-1}\varepsilon^{-r})|\nabla_x\omega_\varepsilon(x,\eta)|.\end{aligned}$$

It follows that assuming $\tau_\varepsilon \le \frac{\varepsilon^r}{4c_0^{-1}}$ the inequality

$$\frac{c_0}{2}\varepsilon^r|\eta| \le \frac{1}{2}|\nabla_x\omega_\varepsilon(x,\eta)| \le |\nabla_x\omega_\varepsilon(x,\eta) - t\theta| \le \frac{3}{2}|\nabla_x\omega_\varepsilon(x,\eta)| \le c_1\frac{3}{2}\varepsilon^{-r}|\eta| \quad (5.13)$$

holds for $x \in K$, $\eta \neq 0$, $\theta \in W^1_{2\tau_\varepsilon,\eta}$, $t \in [0,1]$ and ε small enough. We make use of (5.13) in estimating $(1-|\eta|^2\Delta_\theta)^N\big(r_{\alpha,\varepsilon}(x,\eta,\theta)\chi_\varepsilon(\theta/|\eta|)\big)$ and we conclude that for all $N \in \mathbb{N}$ there exists N'' such that

$$|(1-|\eta|^2\Delta_\theta)^N\big(r_{\alpha,\varepsilon}(x,\eta,\theta)\chi_\varepsilon(\theta/|\eta|)\big)| \le \varepsilon^{-N''}\langle\eta\rangle^{m-|\alpha|}, \quad (5.14)$$

for all $x \in K$, $\eta \neq 0$, $\theta \in W^1_{2\tau_\varepsilon,\eta}$ and $\varepsilon \in (0,\varepsilon_0]$. A combination of (5.12) with (5.14) entails

$$|R^1_{\alpha,\varepsilon}(x,\eta)| \le \varepsilon^{-N'-N''}\langle\eta\rangle^{m+l-\frac{|\alpha|}{2}}\int_{W^1_{2\tau_\varepsilon,\eta}} d\theta \int_{\mathbb{R}^n}(1+|\eta|^2|y|^2)^{L_\alpha-N}\,dy.$$

Therefore, choosing $N \ge L_\alpha + \frac{n+1}{2}$ we obtain

$$\begin{aligned}|R^1_{\alpha,\varepsilon}(x,\eta)| &\le c\varepsilon^{-N'-N''}(\tau_\varepsilon)^n\langle\eta\rangle^{m+l-\frac{|\alpha|}{2}}|\eta|^n\int_{\mathbb{R}^n}\langle z\rangle^{-n-1}\,dz|\eta|^{-n}\\ &\le c_1\varepsilon^{-N_1}\langle\eta\rangle^{m+l-\frac{|\alpha|}{2}},\end{aligned}$$

for $x \in K$ and $|\eta| \ge 1$.

The case $|\eta| \le 1$ requires less precise estimates. More precisely, it is enough to see that from Lemma 5.6 we have that for all α there exists some $d \in \mathbb{R}$ such that

$$\left|D^\alpha_z\Big(\mathrm{e}^{i\overline{\omega}_\varepsilon(z,x,\eta)}b'_\varepsilon(z,x,\eta)\Big)\right| \le C\varepsilon^{-M'}\langle\eta\rangle^d$$

for all $x \in K$, $z \in K_b$ and $|x-z| \le \sigma_\varepsilon$. Thus,

$$|R^1_{\alpha,\varepsilon}(x,\eta)| \le c\varepsilon^{N_2}\langle\eta\rangle^{d-l-\frac{|\alpha|}{2}}\langle\eta\rangle^{m+l-\frac{|\alpha|}{2}} \le c_2\varepsilon^{N_2}\langle\eta\rangle^{m+l-\frac{|\alpha|}{2}},$$

when $|\eta| \le 1$. In conclusion, there exists $(C_{K,\varepsilon})_\varepsilon \in \mathcal{E}_M$ such that

$$|R^1_{\alpha,\varepsilon}(x,\eta)| \le C_{K,\varepsilon}\langle\eta\rangle^{m+l-\frac{|\alpha|}{2}}$$

for all $x \in K$, $\eta \in \mathbb{R}^n \setminus 0$ and $\varepsilon \in (0,1]$.

Step 5. Finally, we consider $R^2_{\alpha,\varepsilon}(x,\eta)$. By Lemma 5.6 we can write

$$D^\alpha_z\Big(\mathrm{e}^{i\overline{\omega}_\varepsilon(z,x,\eta)}b'_\varepsilon(z,x,\eta)\Big)$$

as the finite sum

$$\mathrm{e}^{i\overline{\omega}_\varepsilon(z,x,\eta)}\sum_\beta b_{\alpha,\beta,\varepsilon}(z,x,\eta),$$

where, by making use of the hypotheses on b'_ε and σ_ε, the following holds:

$$\begin{aligned}\forall\beta \in \mathbb{N}^n\,\exists m_\beta \in \mathbb{R}\,\forall\gamma \in \mathbb{N}^n\,\exists(\mu_{\beta,\gamma,\varepsilon})_\varepsilon \in \mathcal{E}_M\,\forall x \in \Omega\,\forall z \in \Omega\,\forall\eta \in \mathbb{R}^n\setminus 0\,\forall\varepsilon \in (0,1]\\ |\partial^\gamma_z b_{\alpha,\beta,\varepsilon}(z,x,\eta)| \le \mu_{\beta,\gamma,\varepsilon}\langle\eta\rangle^{m_\beta},\end{aligned}$$

with $b_{\alpha,\beta,\varepsilon}(z,x,\eta)=0$ for $|x-z|\ge\sigma_\varepsilon$. Hence, we have

$$\begin{aligned}
&R^2_{\alpha,\varepsilon}(x,\eta)\\
&=\sum_\beta\int_{\Omega\times\mathbb{R}^n}\mathrm{e}^{i\overline{\omega}_\varepsilon(z,x,\eta)}b_{\alpha,\beta,\varepsilon}(z,x,\eta)\mathrm{e}^{-i(x-z)\theta}r_{\alpha,\varepsilon}(x,\eta,\theta)(1-\chi_\varepsilon(\theta/|\eta|))\,dz\,đ\theta\\
&=\sum_\beta\int_{\mathbb{R}^n}\mathrm{e}^{-ix\theta}r_{\alpha,\varepsilon}(x,\eta,\theta)(1-\chi_\varepsilon(\theta/|\eta|))\int_\Omega\mathrm{e}^{i\rho_\varepsilon(z,x,\eta,\theta)}b_{\alpha,\beta,\varepsilon}(z,x,\eta)\,dz\,đ\theta,
\end{aligned}$$

where $\rho_\varepsilon(z,x,\eta,\theta)=\overline{\omega}_\varepsilon(z,x,\eta)+z\theta$. Since $\chi_\varepsilon(\theta/|\eta|)=1$ for $\theta\in W^1_{\tau_\varepsilon,\eta}$, we may limit ourselves to consider $\theta\in W^2_{\tau_\varepsilon,\eta}$, i.e., $|\theta|\ge\tau_\varepsilon|\eta|$. We investigate now the properties of the net $(\rho_\varepsilon)_\varepsilon$. We have

$$\nabla_z\rho_\varepsilon(z,x,\eta,\theta)=\theta+\nabla_z\omega_\varepsilon(z,\eta)-\nabla_x\omega_\varepsilon(x,\eta),$$

and therefore (5.11) yields

$$|\theta+\nabla_z\omega_\varepsilon(z,\eta)-\nabla_x\omega_\varepsilon(x,\eta)|\le|\theta|+\varepsilon^{-M}\sigma_\varepsilon|\eta|\le|\theta|(1+\varepsilon^{-M}\sigma_\varepsilon\tau_\varepsilon^{-1})$$

for $\theta\in W^2_{\tau_\varepsilon,\eta}$, $|x-z|<\sigma_\varepsilon$, $z\in K_b$ and ε small enough. We now take σ_ε so small that $\varepsilon^{-M}\sigma_\varepsilon\le\frac{\tau_\varepsilon}{2}$. From (5.11) and the previous assumptions we obtain

$$|\theta+\nabla_z\omega_\varepsilon(z,\eta)-\nabla_x\omega_\varepsilon(x,\eta)|\ge|\theta|-\varepsilon^{-M}\sigma_\varepsilon|\eta|\ge|\theta|-\varepsilon^{-M}\sigma_\varepsilon\tau_\varepsilon^{-1}|\theta|\ge\frac{1}{2}|\theta|.$$

In other words, there exists $(\lambda_{1,\varepsilon})_\varepsilon\in\mathcal{E}_M$ strictly nonzero and $(\lambda_{2,\varepsilon})_\varepsilon\in\mathcal{E}_M$ such that

$$\lambda_{1,\varepsilon}|\theta|\le|\theta+\nabla_z\omega_\varepsilon(z,\eta)-\nabla_x\omega_\varepsilon(x,\eta)|\le\lambda_{2,\varepsilon}|\theta|,$$

for $\theta\in W^2_{\tau_\varepsilon,\eta}$, $|x-z|<\sigma_\varepsilon$, $z\in K_b$ and $\varepsilon\in(0,1]$.
Consider now

$$p_{N,\varepsilon}(z,x,\eta,\theta)=\mathrm{e}^{-i\rho_\varepsilon(z,x,\eta,\theta)}\Delta_z^N\mathrm{e}^{i\rho_\varepsilon(z,x,\eta,\theta)}.$$

Noting that $\partial_z^\gamma\rho_\varepsilon(z,x,\eta,\theta)=\partial_z^\gamma\omega_\varepsilon(z,\eta)$ for $|\gamma|\ge2$, and making use of the previous estimates on $|\nabla_z\rho_\varepsilon(z,x,\eta,\theta)|$, one can prove by induction that

$$\Delta_z^N\mathrm{e}^{i\rho_\varepsilon(z,x,\eta,\theta)}=\mathrm{e}^{i\rho_\varepsilon(z,x,\eta,\theta)}\Big((-1)^N|\nabla_z\rho_\varepsilon(z,x,\eta,\theta)|^{2N}+s_{N,\varepsilon}(z,x,\eta,\theta)\Big),$$

where $(s_{N,\varepsilon})_\varepsilon$ has the following property:

$$\exists l\in[0,2N)\,\forall\gamma\in\mathbb{N}^n\,\exists(s'_{\gamma,N,\varepsilon})_\varepsilon\in\mathcal{E}_M\qquad|\partial_z^\gamma s_{N,\varepsilon}(z,x,\eta,\theta)|\le s'_{\gamma,N,\varepsilon}|\theta|^l,\tag{5.15}$$

for $|\eta|\ge1$, $\theta\in W^2_{\tau_\varepsilon,\eta}$, $|x-z|<\sigma_\varepsilon$ and $z\in K_b$. It follows that

$$\begin{aligned}|p_{N,\varepsilon}(z,x,\eta,\theta)|&\ge\frac{1}{2^{2N}}|\theta|^{2N}-s'_{0,N,\varepsilon}|\theta|^l=|\theta|^{2N}\big(\frac{1}{2^{2N}}-s'_{0,N,\varepsilon}|\theta|^{l-2N}\big)\\
&\ge\frac{1}{2^{2N+1}}|\theta|^{2N},\end{aligned}\tag{5.16}$$

for $\theta\in W^2_{\tau_\varepsilon,\eta}$, $|x-z|<\sigma_\varepsilon$, $z\in K_b$ and $|\eta|\ge\lambda_{N,\varepsilon}:=\tau_\varepsilon^{-1}(2^{2N+1}s'_{0,N,\varepsilon})^{\frac{1}{2N-l}}$. Moreover, we have that for all $\gamma\in\mathbb{N}^n$ there exists $(a_{\gamma,N,\varepsilon})_\varepsilon\in\mathcal{E}_M$ such that

$$|\partial_z^\gamma|\nabla_z\rho_\varepsilon(z,x,\eta,\theta)|^{2N}|\le a_{\gamma,N,\varepsilon}|\theta|^{2N},\tag{5.17}$$

for $|\eta| \ge 1$, $\theta \in W^2_{\tau_\varepsilon,\eta}$, $|x-z| < \sigma_\varepsilon$ and $z \in K_b$. This allows us to prove by induction that

$$\forall \gamma \in \mathbb{N}^n\, \exists (b_{\gamma,N,\varepsilon})_\varepsilon \in \mathcal{E}_M\, \exists (\lambda_{\gamma,N,\varepsilon})_\varepsilon \in \mathcal{E}_M \qquad |\partial_z^\gamma p^{-1}_{N,\varepsilon}(z,x,\eta,\theta)| \le b_{\gamma,N,\varepsilon}|\theta|^{-2N}, \tag{5.18}$$

for $\theta \in W^2_{\tau_\varepsilon,\eta}$, $|x-z| < \sigma_\varepsilon$, $z \in K_b$ and $|\eta| \ge \lambda_{\gamma,N,\varepsilon}$. The assertion (5.18) is clear for $\gamma = 0$ by (5.16). Assume now that (5.18) holds for $|\gamma'| < N$ and take $|\gamma| = N$. From $p^{-1}_{N,\varepsilon}p_{N,\varepsilon} = 1$ we obtain

$$\partial_z^\gamma p^{-1}_{N,\varepsilon}(z,x,\eta,\theta)p_{N,\varepsilon}(z,x,\eta,\theta) = -\sum_{\gamma'<\gamma}\binom{\gamma}{\gamma'}\partial_z^{\gamma'}p^{-1}_{N,\varepsilon}(z,x,\eta,\theta)\partial_z^{\gamma-\gamma'}p_{N,\varepsilon}(z,x,\eta,\theta)$$

and therefore

$$\begin{aligned}|\partial_z^\gamma p^{-1}_{N,\varepsilon}(z,x,\eta,\theta)| &\le \sum_{\gamma'<\gamma} b_{\gamma',N,\varepsilon}|\theta|^{-2N}(a_{\gamma-\gamma',N,\varepsilon}|\theta|^{2N} + s'_{\gamma-\gamma',N,\varepsilon}|\theta|^l)|\theta|^{-2N}\\ &\le b_{\gamma,N,\varepsilon}|\theta|^{-2N},\end{aligned}$$

for $\theta \in W^2_{\tau_\varepsilon,\eta}$, $|x-z| < \sigma_\varepsilon$, $z \in K_b$ and $|\eta| \ge \lambda_{\gamma,N,\varepsilon} := \max_{\gamma'<\gamma}\lambda_{\gamma',N,\varepsilon}$. We make use of the identity

$$\mathrm{e}^{i\rho_\varepsilon(z,x,\eta,\theta)} = \Delta_z^N \mathrm{e}^{i\rho_\varepsilon(z,x,\eta,\theta)}p^{-1}_{N,\varepsilon}(z,x,\eta,\theta)$$

in the integral

$$\int_\Omega \mathrm{e}^{i\rho_\varepsilon(z,x,\eta,\theta)}b_{\alpha,\beta,\varepsilon}(z,x,\eta)\,dz.$$

Since, $\mathrm{supp}_z b_{\alpha,\beta,\varepsilon}(z,x,\eta) \subseteq K_b$, $b_{\alpha,\beta,\varepsilon}(z,x,\eta) = 0$ for $|x-z| \ge \sigma_\varepsilon$ and $1 - \chi_\varepsilon(\theta/|\eta|) = 0$ for $\theta \notin W^2_{\tau_\varepsilon,\eta}$, we can write

$$\begin{aligned}&\int_{\mathbb{R}^n}\mathrm{e}^{-ix\theta}r_{\alpha,\varepsilon}(x,\eta,\theta)(1-\chi_\varepsilon(\theta/|\eta|))\int_\Omega \mathrm{e}^{i\rho_\varepsilon(z,x,\eta,\theta)}b_{\alpha,\beta,\varepsilon}(z,x,\eta)\,dz\,đ\theta\\ &\quad= \int_{\mathbb{R}^n}\mathrm{e}^{-ix\theta}r_{\alpha,\varepsilon}(x,\eta,\theta)(1-\chi_\varepsilon(\theta/|\eta|))\int_\Omega \mathrm{e}^{i\rho_\varepsilon(z,x,\eta,\theta)}\cdot\\ &\qquad\qquad\cdot\Delta_z^N\big(p^{-1}_{N,\varepsilon}(z,x,\eta,\theta)b_{\alpha,\beta,\varepsilon}(z,x,\eta)\big)\,dz\,đ\theta,\end{aligned}$$

where

$$\begin{aligned}&\left|r_{\alpha,\varepsilon}(x,\eta,\theta)(1-\chi_\varepsilon(\theta/|\eta|))\int_\Omega \mathrm{e}^{i\rho_\varepsilon(z,x,\eta,\theta)}\Delta_z^N\big(p^{-1}_{N,\varepsilon}(z,x,\eta,\theta)b_{\alpha,\beta,\varepsilon}(z,x,\eta)\big)\,dz\right|\\ &\qquad\le c|r_{\alpha,\varepsilon}(x,\eta,\theta)||1-\chi_\varepsilon(\theta/|\eta|)|b_{N,\varepsilon}\mu_{\beta,N,\varepsilon}|\theta|^{-2N}\langle\eta\rangle^{m_\beta},\end{aligned}$$

for $|\eta| \ge \lambda_{N,\varepsilon}$ and m_β independent of N. We take $2N = N_1 + N_2$ such that $-N_2 + m_\beta \le 0$. Hence, from $|\theta| \ge \tau_\varepsilon|\eta|$ we have, for some $(c_{N,\varepsilon})_\varepsilon \in \mathcal{E}_M$ and $|\eta| \ge \lambda_{N,\varepsilon}$, the following estimate:

$$\begin{aligned}&\left|r_{\alpha,\varepsilon}(x,\eta,\theta)(1-\chi_\varepsilon(\theta/|\eta|))\int_\Omega \mathrm{e}^{i\rho_\varepsilon(z,x,\eta,\theta)}\Delta_z^N\big(p^{-1}_{N,\varepsilon}(z,x,\eta,\theta)b_{\alpha,\beta,\varepsilon}(z,x,\eta)\big)\,dz\right|\\ &\qquad\le c_{N,\varepsilon}|r_{\alpha,\varepsilon}(x,\eta,\theta)||1-\chi_\varepsilon(\theta/|\eta|)||\theta|^{-N_1}.\end{aligned}$$

By definition of $r_{\alpha,\varepsilon}$ we easily see that for all $K \Subset \Omega$ there exists $(d_\varepsilon)_\varepsilon, (d'_\varepsilon)_\varepsilon \in \mathcal{E}_M$ such that

$$|r_{\alpha,\varepsilon}(x,\eta,\theta)(1-\chi_\varepsilon(\theta/|\eta|))| \le d_\varepsilon \langle\theta\rangle^{m_+}\langle\nabla_x\omega_\varepsilon(x,\eta)\rangle^{m_+} \le d'_\varepsilon\langle\theta\rangle^{2m_+}.$$

Hence for all $h \ge 0$ there exists $2N = N_1 + N_2$ large enough such that, for $x \in K$ and $|\eta| \ge \lambda_{N,\varepsilon}$

$$\left| r_{\alpha,\varepsilon}(x,\eta,\theta)(1-\chi_\varepsilon(\theta/|\eta|)) \int_\Omega e^{i\rho_\varepsilon(z,x,\eta,\theta)}\Delta_z^N\left(p_{N,\varepsilon}^{-1}(z,x,\eta,\theta)b_{\alpha,\beta,\varepsilon}(z,x,\eta)\right)dz\right| \le \nu_{K,\varepsilon}\langle\theta\rangle^{-h} \le \nu_{K,\varepsilon}\tau_\varepsilon^{-h}\langle\eta\rangle^{-h},$$

with $(\nu_{K,\varepsilon})_\varepsilon \in \mathcal{E}_M$. This means that for all $h \ge 0$ there exists $(\lambda_\varepsilon)_\varepsilon \in \mathcal{E}_M$ such that

$$|R^2_{\alpha,\varepsilon}(x,\eta)| \le \nu_{K,\varepsilon}\langle\eta\rangle^{-h}$$

when $x \in K$, $|\eta| \ge \lambda_\varepsilon$ and $\varepsilon \in (0,1]$. A simple investigation of the oscillatory integral which defines $R^2_{\alpha,\varepsilon}(x,\eta)$ shows that there exists some $h' \ge 0$ and some $\nu'_{K,\varepsilon} \in \mathcal{E}_M$ such that the estimate

$$|R^2_{\alpha,\varepsilon}(x,\eta)| \le \nu'_{K,\varepsilon}\langle\eta\rangle^{h'}$$

holds for all $x \in K$, $\eta \in \mathbb{R}^n \setminus 0$ and $\varepsilon \in (0,1]$. This yields for $|\eta| \le \lambda_\varepsilon$

$$|R^2_{\alpha,\varepsilon}(x,\eta)| \le \nu'_{K,\varepsilon}\langle\eta\rangle^{-h}\langle\eta\rangle^{h+h'} \le \nu'_{K,\varepsilon}\langle\lambda_\varepsilon\rangle^{h+h'}\langle\eta\rangle^{-h}.$$

In conclusion, we have that for all $h \ge 0$ there exists $(C_{h,\varepsilon}(K))_\varepsilon \in \mathcal{E}_M$ such that

$$|R^2_{\alpha,\varepsilon}(x,\eta)| \le C_{h,\varepsilon}(K)\langle\eta\rangle^{-h}$$

for all $x \in K$, $\eta \in \mathbb{R}^n \setminus 0$ and $\varepsilon \in (0,1]$.

Step 6. Finally, we combine all the results of the previous steps. We have that

$$a_\varepsilon(x,D)F_{\omega_\varepsilon}(b_\varepsilon)u_\varepsilon(x) = \int_{\mathbb{R}^n} I_{1,\varepsilon}(x,\eta)\widehat{u_\varepsilon}(\eta)\,đ\eta + \int_{\mathbb{R}^n} I_{2,\varepsilon}(x,\eta)\widehat{u_\varepsilon}(\eta)\,đ\eta, \quad (5.19)$$

where $(I_{2,\varepsilon})_\varepsilon \in \mathcal{M}_{S^{-\infty}(\Omega\times\mathbb{R}^n\setminus 0)}$. From Theorem 2.2(i) and Proposition 5.8 there exists $(h_\varepsilon)_\varepsilon \in \mathcal{M}_{S^{m+l}(\Omega\times\mathbb{R}^n\setminus 0)}$ such that $h_\varepsilon(x,\eta) \sim \sum_\alpha h_{\alpha,\varepsilon}(x,\eta)$. We write the first integral in (5.19) as

$$\int_{\mathbb{R}^n} e^{i\omega_\varepsilon(x,\eta)}h_\varepsilon(x,\eta)\widehat{u_\varepsilon}(\eta)\,đ\eta + \int_{\mathbb{R}^n} e^{i\omega_\varepsilon(x,\eta)}\left(e^{-i\omega_\varepsilon(x,\eta)}I_{1,\varepsilon}(x,\eta) - h_\varepsilon(x,\eta)\right)\widehat{u_\varepsilon}(\eta)\,đ\eta$$

and we concentrate on

$$e^{-i\omega_\varepsilon(x,\eta)}I_{1,\varepsilon}(x,\eta) - h_\varepsilon(x,\eta).$$

From the previous computations we have that for all $k \ge 1$ and $K \Subset \Omega$ there exists $(C_{k,\varepsilon}(K))_\varepsilon \in \mathcal{E}_M$ such that

$$\left| e^{-i\omega_\varepsilon(x,\eta)}I_{1,\varepsilon}(x,\eta) - \sum_{|\alpha|<k} h_{\alpha,\varepsilon}(x,\eta)\right| \le C_{k,\varepsilon}(K)\langle\eta\rangle^{m+l-\frac{k}{2}}$$

for all $x \in K$, $\eta \in \mathbb{R}^n \setminus 0$ and $\varepsilon \in (0,1]$. Moreover, regarding $I_{1,\varepsilon}(x,\eta)$ as the oscillatory integral

$$\int_{\Omega\times\mathbb{R}^n} \mathrm{e}^{-iz\theta}\mathrm{e}^{ix\theta+i\omega_\varepsilon(z,\eta)} a_\varepsilon(x,\theta)b'_\varepsilon(z,x,\eta)\,dz\,đ\theta,$$

from Theorem 3.1 in [11], we obtain that for all $\alpha,\beta \in \mathbb{N}^n$ there exists $d \in \mathbb{R}$ and for all $K \Subset \Omega$ there exists $(c_{\alpha,\beta,\varepsilon}(K))_\varepsilon \in \mathcal{E}_M$ such that for all $\eta \in \mathbb{R}^n \setminus 0$ and $\varepsilon \in (0,1]$,

$$\sup_{x\in K} |\partial_\eta^\alpha \partial_x^\beta I_{1,\varepsilon}(x,\eta)| \le c_{\alpha,\beta,\varepsilon}(K)\langle\eta\rangle^d.$$

Recalling that

$$\partial_\eta^\alpha \partial_x^\beta \mathrm{e}^{-i\omega_\varepsilon(x,\eta)} = \mathrm{e}^{-i\omega_\varepsilon(x,\eta)} a_{\alpha,\beta,\varepsilon}(x,\eta),$$

with $(a_{\alpha,\beta,\varepsilon})_\varepsilon \in \mathcal{M}_{S^{|\beta|}(\Omega\times\mathbb{R}^n\setminus 0)}$, we conclude that the net $(\mathrm{e}^{-i\omega_\varepsilon(x,\eta)}I_{1,\varepsilon}(x,\eta))_\varepsilon$ satisfies the hypothesis of Proposition 2.3(i). It follows that

$$(\mathrm{e}^{-i\omega_\varepsilon(x,\eta)}I_{1,\varepsilon}(x,\eta))_\varepsilon \sim \sum_\alpha (h_{\alpha,\varepsilon})_\varepsilon.$$

Hence, by Theorem 2.2(i) we conclude

$$(\mathrm{e}^{-i\omega_\varepsilon(x,\eta)}I_{1,\varepsilon}(x,\eta) - h_\varepsilon(x,\eta))_\varepsilon \in \mathcal{M}_{S^{-\infty}(\Omega\times\mathbb{R}^n\setminus 0)}.$$

Going back to (5.19) we have that there exists $(r_\varepsilon)_\varepsilon \in \mathcal{M}_{S^{-\infty}(\Omega\times\mathbb{R}^n\setminus 0)}$ such that

$$a_\varepsilon(x,D)F_{\omega_\varepsilon}(b_\varepsilon)u_\varepsilon(x) = \int_{\mathbb{R}^n} \mathrm{e}^{i\omega_\varepsilon(x,\eta)}h_\varepsilon(x,\eta)\widehat{u_\varepsilon}(\eta)\,đ\eta + r_\varepsilon(x,D)(u_\varepsilon)(x).$$

□

Theorem 5.11. *Let $\omega \in \widetilde{\mathcal{S}}^{1,\mathrm{sc}}_{\mathrm{hg}}(\Omega\times\mathbb{R}^n\setminus 0)$ have a representative satisfying condition (h2). Let $a \in \widetilde{\mathcal{S}}^{m,\mathrm{sc}}(\Omega\times\mathbb{R}^n)$ and $b \in \widetilde{\mathcal{S}}^{l,\mathrm{sc}}(\Omega\times\mathbb{R}^n\setminus 0)$ with $\mathrm{supp}_x\, b \Subset \Omega$. Then, the operator $a(x,D)F_\omega(b)$ has the following properties:*

(i) *maps $\mathcal{G}^\infty_{\mathrm{c}}(\Omega)$ into $\mathcal{G}^\infty(\Omega)$;*
(ii) *is of the form*

$$\int_{\mathbb{R}^n} \mathrm{e}^{i\omega(x,\eta)}h(x,\eta)\widehat{u}(\eta)\,đ\eta + r(x,D)u,$$

where $h \in \widetilde{\mathcal{S}}^{l+m,\mathrm{sc}}(\Omega\times\mathbb{R}^n\setminus 0)$ has asymptotic expansion given by the symbols h_α defined in (5.7) and $r \in \widetilde{\mathcal{S}}^{-\infty,\mathrm{sc}}(\Omega\times\mathbb{R}^n\setminus 0)$.

Proof. Combining Proposition 5.1(iii) with the usual mapping properties of generalized pseudo-differential operators we have that (i) holds. Concerning assertion (ii), we argue as in the proof of Theorem 5.10 by taking the nets $(\sigma_\varepsilon)_\varepsilon$ and $(\tau_\varepsilon)_\varepsilon$ slow scale strictly nonzero. From the assumptions of slows scale type on ω, a and b we have that all the moderate nets involved are of slow scale type. This leads to the desired conclusion. □

6. Generalized Fourier integral operators and microlocal analysis

Concluding, we present some first results of microlocal analysis for generalized Fourier integral operators provided in [18, Section 4]. A deeper investigation of the microlocal properties of

$$A : \mathcal{G}_c(\Omega) \to \mathcal{G}_c(\Omega') : u \to \int_{\Omega\times\mathbb{R}^p} e^{i\phi(x,y,\xi)} a(x,y,\xi) u(y)\, dy\, đ\xi$$

is current topic of research.

Generalized Singular Supports of the Functional $I_\phi(a)$

We begin with the functional

$$I_\phi(a) : \mathcal{G}_c(\Omega) \to \widetilde{\mathbb{C}} : u \to \int_{\Omega\times\mathbb{R}^p} e^{i\phi(y,\xi)} a(y,\xi) u(y)\, dy\, đ\xi$$

Before defining specific regions depending on the generalized phase function ϕ, we observe that any $\phi \in \widetilde{\Phi}(\Omega \times \mathbb{R}^p)$ can be regarded as an element of $\widetilde{\mathcal{S}}^1_{\mathrm{hg}}(\Omega \times \mathbb{R}^p \setminus 0)$ and consequently $|\nabla_\xi \phi|^2 \in \widetilde{\mathcal{S}}^0_{\mathrm{hg}}(\Omega \times \mathbb{R}^p \setminus 0)$.

Let Ω_1 be an open subset of Ω and $\Gamma \subseteq \mathbb{R}^p \setminus 0$. We say that $b \in \widetilde{\mathcal{S}}^0(\Omega \times \mathbb{R}^p \setminus 0)$ is *invertible on* $\Omega_1 \times \Gamma$ if for all relatively compact subsets U of Ω_1 there exists a representative $(b_\varepsilon)_\varepsilon$ of b, a constant $r \in \mathbb{R}$ and $\eta \in (0,1]$ such that

$$\inf_{y\in U, \xi\in\Gamma} |b_\varepsilon(y,\xi)| \geq \varepsilon^r \tag{6.1}$$

for all $\varepsilon \in (0,\eta]$. In an analogous way we say that $b \in \widetilde{\mathcal{S}}^0(\Omega \times \mathbb{R}^p \setminus 0)$ is *slow scale-invertible* on $\Omega_1 \times \Gamma$ if (6.1) holds with the inverse of some slow scale net $(s_\varepsilon)_\varepsilon$ in place of ε^r. This kind of bounds from below hold for all representatives of the symbol b once they are known to hold for one.

In the sequel π_Ω denotes the projection of $\Omega \times \mathbb{R}^p$ on Ω.

Definition 6.1. Let $\phi \in \widetilde{\Phi}(\Omega \times \mathbb{R}^p)$. We define $C_\phi \subseteq \Omega \times \mathbb{R}^p \setminus 0$ as the complement of the set of all $(x_0,\xi_0) \in \Omega \times \mathbb{R}^p \setminus 0$ with the property that there exist a relatively compact open neighborhood $U(x_0)$ of x_0 and a conic open neighborhood $\Gamma(\xi_0) \subseteq \mathbb{R}^p \setminus 0$ of ξ_0 such that $|\nabla_\xi \phi|^2$ is invertible on $U(x_0) \times \Gamma(\xi_0)$. We set $\pi_\Omega(C_\phi) = S_\phi$ and $R_\phi = (S_\phi)^{\mathrm{c}}$.

By construction C_ϕ is a closed conic subset of $\Omega \times \mathbb{R}^p \setminus 0$ and $R_\phi \subseteq \Omega$ is open. It is routine to check that the region C_ϕ coincides with the classical one when ϕ is classical.

Proposition 6.2. *The generalized symbol $|\nabla_\xi \phi|^2$ is invertible on $R_\phi \times \mathbb{R}^p \setminus 0$.*

The more specific assumption of slow scale-invertibility concerning the generalized symbol $|\nabla_\xi \phi|^2$ is employed in the definition of the following sets.

Definition 6.3. Let $\phi \in \widetilde{\Phi}(\Omega \times \mathbb{R}^p)$. We define $C^{\mathrm{sc}}_\phi \subseteq \Omega \times \mathbb{R}^p \setminus 0$ as the complement of the set of all $(x_0,\xi_0) \in \Omega \times \mathbb{R}^p \setminus 0$ with the property that there exist a relatively compact open neighborhood $U(x_0)$ of x_0 and a conic open neighborhood $\Gamma(\xi_0) \subseteq$

$\mathbb{R}^p \setminus 0$ of ξ_0 such that $|\nabla_\xi \phi|^2$ is on $U(x_0) \times \Gamma(\xi_0)$. We set $\pi_\Omega(C_\phi^{\mathrm{sc}}) = S_\phi^{\mathrm{sc}}$ and $R_\phi^{\mathrm{sc}} = (S_\phi^{\mathrm{sc}})^{\mathrm{c}}$.

By construction C_ϕ^{sc} is a conic closed subset of $\Omega \times \mathbb{R}^p \setminus 0$ and $R_\phi^{\mathrm{sc}} \subseteq R_\phi \subseteq \Omega$ is open. In analogy with Proposition 6.2 we can prove that $|\nabla_\xi \phi|^2$ is slow scale-invertible on $R_\phi^{\mathrm{sc}} \times \mathbb{R}^p \setminus 0$.

Theorem 6.4. *Let $\phi \in \widetilde{\Phi}(\Omega \times \mathbb{R}^p)$ and $a \in \widetilde{\mathcal{S}}^m_{\rho,\delta}(\Omega \times \mathbb{R}^p)$.*

(i) *The restriction $I_\phi(a)|_{R_\phi}$ of the functional $I_\phi(a)$ to the region R_ϕ belongs to $\mathcal{G}(R_\phi)$.*
(ii) *If $\phi \in \widetilde{\Phi}^{\mathrm{sc}}(\Omega \times \mathbb{R}^p)$ and $a \in \widetilde{\mathcal{S}}^{m,\mathrm{sc}}_{\rho,\delta}(\Omega \times \mathbb{R}^p)$ then $I_\phi(a)|_{R_\phi^{\mathrm{sc}}} \in \mathcal{G}^\infty(R_\phi^{\mathrm{sc}})$.*

Theorem 6.4 means that

$$\operatorname{sing\,supp}_{\mathcal{G}} I_\phi(a) \subseteq S_\phi$$

if $\phi \in \widetilde{\Phi}(\Omega \times \mathbb{R}^p)$ and $a \in \widetilde{\mathcal{S}}^m_{\rho,\delta}(\Omega \times \mathbb{R}^p)$ and that

$$\operatorname{sing\,supp}_{\mathcal{G}^\infty} I_\phi(a) \subseteq S_\phi^{\mathrm{sc}}$$

if $\phi \in \widetilde{\Phi}^{\mathrm{sc}}(\Omega \times \mathbb{R}^p)$ and $a \in \widetilde{\mathcal{S}}^{m,\mathrm{sc}}_{\rho,\delta}(\Omega \times \mathbb{R}^p)$.

Example. Returning to the first example in Section 4 we are now in the position to analyze the regularity properties of the generalized kernel functional $I_\phi(a)$ of the solution operator A corresponding to the hyperbolic Cauchy-problem. For any $v \in \mathcal{G}_{\mathrm{c}}(\mathbb{R}^3)$ we have

$$I_\phi(a)(v) = \int \mathrm{e}^{i\phi(x,t,y,\xi)}\, a(x,t,y,\xi)\, v(x,t,y)\, dx\, dt\, dy\, đ\xi, \tag{6.2}$$

where a and ϕ are as in Section 4. Note that in the case of partial differential operators with smooth coefficients and distributional initial values the wave front set of the distributional kernel of A determines the propagation of singularities from the initial data. When the coefficients are non-differentiable functions, or even distributions or generalized functions, matters are not yet understood in sufficient generality. Nevertheless, the above results allow us to identify regions where the generalized kernel functional agrees with a generalized function or is even guaranteed to be a $\mathcal{G}^\infty$-regular generalized function. To identify the set C_ϕ in this situation one simply has to study invertibility of $\partial_\xi \phi(x,t,y,\xi) = \gamma(x,t;0) - y$ as a generalized function in a neighborhood of any given point (x_0, t_0, y_0).

Under the assumptions on c of the example in Section 4, the representing nets $(\gamma_\varepsilon(.,.;0))_{\varepsilon\in(0,1]}$ of γ are uniformly bounded on compact sets (e.g., when c is a bounded generalized constant). For given (x_0, t_0) define the generalized domain of dependence $D(x_0, t_0) \subseteq \mathbb{R}$ to be the set of accumulation points of the net $(\gamma_\varepsilon(x_0, t_0; 0))_{\varepsilon\in(0,1]}$. Then we have that

$$\{(x_0, t_0, y_0) \in \mathbb{R}^3 : y_0 \notin D(x_0, t_0)\} \subseteq R_\phi.$$

When $c \in \widetilde{\mathbb{R}}$ this may be proved by showing that if $(x_0, t_0, y_0) \in C_\phi$ then there exists an accumulation point c' of a representative $(c_\varepsilon)_\varepsilon$ of c such that $y_0 = x_0 - c' t_0$.

Example. As an illustrative example concerning the regions involving the regularity of the functional $I_\phi(a)$ we consider the generalized phase function on $\mathbb{R}^2\times\mathbb{R}^2$ given by $\phi_\varepsilon(y_1,y_2,\xi_1,\xi_2)=-\varepsilon y_1\xi_1-s_\varepsilon y_2\xi_2$ where $(s_\varepsilon)_\varepsilon$ is bounded and $(s_\varepsilon^{-1})_\varepsilon$ is a slow scale net. Clearly $\phi:=[(\phi_\varepsilon)_\varepsilon]\in\widetilde{\Phi}^{\rm sc}(\mathbb{R}^2\times\mathbb{R}^2)$. Simple computations show that $R_\phi=\mathbb{R}^2\setminus(0,0)$ and $R_\phi^{\rm sc}=\mathbb{R}^2\setminus\{y_2=0\}$. We leave it to the reader to check that the oscillatory integral

$$\int_{\mathbb{R}^2}\mathrm{e}^{i\phi(y,\xi)}(1+\xi_1^2+\xi_2^2)^{\frac{1}{2}}\,d\!\!\!{}^-\xi=\Big[\Big(\int_{\mathbb{R}^2}\mathrm{e}^{-i\varepsilon y_1\xi_1-is_\varepsilon y_2\xi_2}(1+\xi_1^2+\xi_2^2)^{\frac{1}{2}}\,d\!\!\!{}^-\xi_1\,d\!\!\!{}^-\xi_2\Big)_\varepsilon\Big]$$

defines a generalized function in $\mathbb{R}^2\setminus(0,0)$ whose restriction to $\mathbb{R}^2\setminus\{y_2=0\}$ is regular.

The Colombeau-regularity of the functional $I_\phi(a)$ is easily proved in the case of generalized symbols of order $-\infty$.

Proposition 6.5.
(i) *If* $\phi\in\widetilde{\Phi}(\Omega\times\mathbb{R}^p)$ *and* $a\in\widetilde{\mathcal{S}}^{-\infty}(\Omega\times\mathbb{R}^p)$ *then* $\mathrm{sing\,supp}_{\mathcal{G}}I_\phi(a)=\emptyset$.
(ii) *If* $\phi\in\widetilde{\Phi}^{\rm sc}(\Omega\times\mathbb{R}^p)$ *and* $a\in\widetilde{\mathcal{S}}^{-\infty,\rm sc}(\Omega\times\mathbb{R}^p)$ *then* $\mathrm{sing\,supp}_{\mathcal{G}^\infty}I_\phi(a)=\emptyset$.

Proposition 6.5 leads to the following result.

Proposition 6.6.
(i) *If* $\phi\in\widetilde{\Phi}(\Omega\times\mathbb{R}^p)$ *and* $a\in\widetilde{\mathcal{S}}^m_{\rho,\delta}(\Omega\times\mathbb{R}^p)$ *then*

$$\mathrm{sing\,supp}_{\mathcal{G}}\,I_\phi(a)\subseteq\pi_\Omega(C_\phi\cap\mathrm{cone\,supp}\,a).$$

(ii) *If* $\phi\in\widetilde{\Phi}^{\rm sc}(\Omega\times\mathbb{R}^p)$ *and* $a\in\widetilde{\mathcal{S}}^{m,\rm sc}_{\rho,\delta}(\Omega\times\mathbb{R}^p)$ *then*

$$\mathrm{sing\,supp}_{\mathcal{G}^\infty}I_\phi(a)\subseteq\pi_\Omega(C^{\rm sc}_\phi\cap\mathrm{cone\,supp}\,a).$$

Generalized wave front sets of the functional $I_\phi(a)$

The next theorem investigates the $\mathcal{G}$-wave front set and the $\mathcal{G}^\infty$-wave front set of the functional $I_\phi(a)$ under suitable assumptions on the generalized symbol a and the phase function ϕ.

Theorem 6.7.

(i) *Let* $\phi\in\widetilde{\Phi}(\Omega\times\mathbb{R}^p)$ *and* $a\in\widetilde{\mathcal{S}}^m_{\rho,\delta}(\Omega\times\mathbb{R}^p)$. *The generalized wave front set* $\mathrm{WF}_{\mathcal{G}}I_\phi(a)$ *is contained in the set* $W_{\phi,a}$ *of all points* $(x_0,\xi_0)\in T^*(\Omega)\setminus 0$ *with the property that for all relatively compact open neighborhoods* $U(x_0)$ *of* x_0, *for all open conic neighborhoods* $\Gamma(\xi_0)\subseteq\mathbb{R}^n\setminus 0$ *of* ξ_0, *for all open conic neighborhoods* V *of* $\mathrm{cone\,supp}\,a\cap C_\phi$ *such that* $V\cap(U(x_0)\times\mathbb{R}^p\setminus 0)\neq\emptyset$ *the generalized number*

$$\operatorname*{Inf}_{\substack{y\in U(x_0),\xi\in\Gamma(\xi_0)\\(y,\theta)\in V\cap(U(x_0)\times\mathbb{R}^p\setminus 0)}}\frac{|\xi-\nabla_y\phi(y,\theta)|}{|\xi|+|\theta|}\tag{6.3}$$

is not invertible.

(ii) *If* $\phi \in \widetilde{\Phi}^{\rm sc}(\Omega \times \mathbb{R}^p)$ *and* $a \in \widetilde{\mathcal{S}}^{m,\rm sc}_{\rho,\delta}(\Omega \times \mathbb{R}^p)$ *then* $\mathrm{WF}_{\mathcal{G}^\infty} I_\phi(a)$ *is contained in the set* $W^{\rm sc}_{\phi,a}$ *of all points* $(x_0, \xi_0) \in T^*(\Omega) \setminus 0$ *with the property that for all relatively compact open neighborhoods* $U(x_0)$ *of* x_0, *for all open conic neighborhoods* $\Gamma(\xi_0) \subseteq \mathbb{R}^n \setminus 0$ *of* ξ_0, *for all open conic neighborhoods* V *of* $\mathrm{cone\,supp}\, a \cap C^{\rm sc}_\phi$ *such that* $V \cap (U(x_0) \times \mathbb{R}^p \setminus 0) \neq \emptyset$ *the generalized number* (6.3) *is not slow scale-invertible.*

Note that when ϕ is a classical phase function the set $W_{\phi,a}$ as well as the set $W^{\rm sc}_{\phi,a}$ coincide with

$$\{(x, \nabla_x \phi(x,\theta)) : (x,\theta) \in \mathrm{cone\,supp}\, a \cap C_\phi\}. \tag{6.4}$$

For more details see [18, Remark 4.13].

Example. Theorem 6.7 can be employed for investigating the generalized wave front sets of the kernel $K_A := I_\phi(a)$ of the Fourier integral operator introduced in the first example of Section 4. For simplicity we assume that c is a bounded generalized constant in $\widetilde{\mathbb{R}}$ and that $a = 1$. Let $((x_0, t_0, y_0), \xi_0) \in \mathrm{WF}_{\mathcal{G}} K_A$. From the first assertion of Theorem 6.7 we know that the generalized number given by

$$\inf_{\substack{(x,t,y) \in U, \xi \in \Gamma \\ ((x,t,y),\theta) \in V \cap (U \times \mathbb{R} \setminus 0)}} \frac{|\xi - (\theta, -c_\varepsilon \theta, -\theta)|}{|\xi| + |\theta|} \tag{6.5}$$

is not invertible, for every choice of neighborhoods U of (x_0, t_0, y_0), Γ of ξ_0 and V of C_ϕ. Note that it is not restrictive to assume that $|\theta| = 1$. We fix some sequences $(U_n)_n$, $(\Gamma_n)_n$ and $(V_n)_n$ of neighborhoods shrinking to (x_0, t_0, y_0), $\{\xi_0 \lambda : \lambda > 0\}$ and C_ϕ respectively. By (6.5) we find a sequence ε_n tending to 0 such that for all $n \in \mathbb{N}$ there exists $\xi_n \in \Gamma_n$, $(x_n, t_n, y_n, \theta_n) \in V_n$ with $|\theta_n| = 1$ and $(x_n, t_n, y_n) \in U_n$ such that

$$|\xi_n - (\theta_n, -c_{\varepsilon_n} \theta_n, -\theta_n)| \le \varepsilon_n (|\xi_n| + 1).$$

In particular, ξ_n remains bounded. Passing to suitable subsequences we obtain that there exist θ such that $(x_0, t_0, y_0, \theta) \in C_\phi$, an accumulation point c' of $(c_\varepsilon)_\varepsilon$ and a multiple ξ' of ξ_0 such that $\xi' = (\theta, -c'\theta, -\theta)$. It follows that

$$\frac{\xi_0}{|\xi_0|} = \frac{\xi'}{|\xi'|} = \frac{1}{\sqrt{2 + (c')^2}|\theta|} (\theta, -c'\theta, -\theta).$$

In other words the $\mathcal{G}$-wave front set of the kernel K_A is contained in the set of points of the form $((x_0, t_0, y_0), (\theta_0, -c'\theta_0, -\theta_0))$ where $(x_0, t_0, y_0, \theta_0) \in C_\phi$ and c' is an accumulation point of a net representing c. Since in the classical case (when $c \in \mathbb{R}$) the distributional wave front set of the corresponding kernel is the set $\{((x_0, t_0, y_0), (\theta_0, -c\theta_0, -\theta_0)) : (x_0, t_0, y_0, \theta_0) \in C_\phi\}$, the result obtained above for $\mathrm{WF}_{\mathcal{G}} K_A$ is a generalization in line with what we deduced about the regions R_ϕ and C_ϕ.

Particular case: Generalized pseudo-differential operators

Finally, we consider a generalized pseudo-differential operator $a(x,D)$ on Ω and its kernel $K_{a(x,D)} \in \mathcal{L}(\mathcal{G}_c(\Omega \times \Omega), \widetilde{\mathbb{C}})$. By Remark 4.15 in [18], we have that $\mathrm{WF}_{\mathcal{G}}(K_{a(x,D)})$ is contained in the normal bundle of the diagonal in $\Omega \times \Omega$ when $a \in \widetilde{\mathcal{S}}^m_{\rho,\delta}(\Omega \times \mathbb{R}^n)$ and that $\mathrm{WF}_{\mathcal{G}^\infty}(K_{a(x,D)})$ is a subset of the normal bundle of the diagonal in $\Omega \times \Omega$ when a is of slow scale type. We define the sets

$$\mathrm{WF}_{\mathcal{G}}(a(x,D)) = \{(x,\xi) \in T^*(\Omega) \setminus 0 : \ (x,x,\xi,-\xi) \in \mathrm{WF}_{\mathcal{G}}(K_{a(x,D)})\}$$

and

$$\mathrm{WF}_{\mathcal{G}^\infty}(a(x,D)) = \{(x,\xi) \in T^*(\Omega) \setminus 0 : \ (x,x,\xi,-\xi) \in \mathrm{WF}_{\mathcal{G}^\infty}(K_{a(x,D)})\}.$$

From Theorem 6.7 one deduces the following.

Proposition 6.8. *Let $a(x,D)$ be a generalized pseudo-differential operator.*

(i) *If $a \in \widetilde{\mathcal{S}}^m_{\rho,\delta}(\Omega \times \mathbb{R}^n)$ then $\mathrm{WF}_{\mathcal{G}}(a(x,D)) \subseteq \mu\,\mathrm{supp}_{\mathcal{G}}(a)$.*

(ii) *If $a \in \widetilde{\mathcal{S}}^{m,\mathrm{sc}}_{\rho,\delta}(\Omega \times \mathbb{R}^n)$ then $\mathrm{WF}_{\mathcal{G}^\infty}(a(x,D)) \subseteq \mu\,\mathrm{supp}_{\mathcal{G}^\infty}(a)$.*

Acknowledgment

The author would like to express her gratitude to Professor L. Rodino and Professor M.W. Wong for the kind invitation to the session on pseudo-differential operators of the 2007 ISAAC Congress at METU in Ankara.

References

[1] H.A. Biagioni, *A Nonlinear Theory of Generalized Functions*, Lecture Notes in Math. **1421**, Springer-Verlag, Berlin, 1990.

[2] H. Biagioni and M. Oberguggenberger, Generalized solutions to the Korteweg–de Vries and the regularized long-wave equations, *SIAM J. Math Anal.* **23**(4) (1992), 923–940.

[3] H. Biagioni and M. Oberguggenberger, Generalized solutions to Burgers' equation, *J. Diff. Eqs.* **97**(2) (1992), 263–287.

[4] J.F. Colombeau, *Elementary Introduction to New Generalized Functions*, North-Holland Mathematics Studies **113**, Elsevier Science Publishers, 1985.

[5] J.F. Colombeau and M. Oberguggenberger, On a hyperbolic system with a compatible quadratic term: generalized solutions, delta waves, and multiplication of distributions, *Comm. Partial Differential Equations* **15**(7) (1990), 905–938.

[6] S. Coriasco, Fourier integral operators in SG classes (I), composition theorems and action on SG Sobolev spaces, *Quad. Dip. di Matematica, Univ. Torino* **11**, 1998.

[7] S. Coriasco, Fourier integral operators in SG classes (I), composition theorems and action on SG Sobolev spaces, *Rend. Sem. Mat. Univ. Pol. Torino* **57**(4) (1999), 249–302.

[8] M.V. de Hoop and C.C. Stolk, Microlocal analysis of seismic inverse scattering in anisotropic, elastic media *Comm. Pure Appl. Math.* **55** (2002), 261–301.

[9] A. Delcroix, Generalized integral operators and Schwartz kernel type theorem, *J. Math. Anal. Appl.* **306**(2) (2005), 481–501.

[10] A. Delcroix and D. Scarpalezos, Topology on asymptotic algebras of generalized functions and applications, *Monatsh. Math.* **129**(1) (2000), 1–14.

[11] C. Garetto, Pseudo-differential operators in algebras of generalized functions and global hypoellipticity, *Acta Appl. Math.* **80**(2) (2004), 123–174.

[12] C. Garetto, Topological structures in Colombeau algebras: investigation of the duals of $\mathcal{G}_c(\Omega)$, $\mathcal{G}(\Omega)$ and $\mathcal{G}_{\mathscr{S}}(\mathbb{R}^n)$, *Monatsh. Math.* **146**(3) (2005), 203–226.

[13] C. Garetto, Topological structures in Colombeau algebras: topological $\widetilde{\mathbb{C}}$-modules and duality theory, *Acta. Appl. Math.* **88**(1) (2005), 81–123.

[14] C. Garetto, Microlocal analysis in the dual of a Colombeau algebra: generalized wave front sets and noncharacteristic regularity, *New York J. Math.* **12** (2006), 275–318.

[15] C. Garetto, T. Gramchev and M. Oberguggenberger, Pseudodifferential operators with generalized symbols and regularity theory, *Electron. J. Diff. Eqns.* **116** (2005), 1–43.

[16] C. Garetto and G. Hörmann, Microlocal analysis of generalized functions: pseudo-differential techniques and propagation of singularities, *Proc. Edinburgh Math. Soc.* **48**(3) (2005), 603–629.

[17] C. Garetto and G. Hörmann, Duality theory and pseudodifferential techniques for Colombeau algebras: generalized kernels and microlocal analysis, *Bull. Cl. Sci. Math. Nat. Sci. Math.* **31** (2006), 115–136.

[18] C. Garetto, G. Hörmann and M. Oberguggenberger, Generalized oscillatory integrals and Fourier integral operators, *Proc. Edinburgh Math. Soc.*, to appear.

[19] M. Grosser, M. Kunzinger, M. Oberguggenberger and R. Steinbauer. *Geometric Theory of Generalized Functions, Mathematics and its Applications* **537**, Kluwer, Dordrecht, 2001.

[20] L. Hörmander. Fourier integral operators I, *Acta Math.*, **127** (1971), 79–183.

[21] L. Hörmander. *The Analysis of Linear Partial Differential Operators*, I–IV. Springer-Verlag, Berlin, 1983–85, 2nd ed. vol. I 1990.

[22] G. Hörmann, Integration and microlocal analysis in Colombeau algebras of generalized functions, *J. Math. Anal. Appl.* **239** (1999), 332–348.

[23] G. Hörmann, First-order hyperbolic pseudodifferential equations with generalized symbols. *J. Math. Anal. Appl.* **293**(1) (2004), 40–56.

[24] G. Hörmann, Hölder-Zygmund regularity in algebras of generalized functions, *Z. Anal. Anwendungen* **23** (2004), 139–165.

[25] G. Hörmann and M.V. de Hoop, Microlocal analysis and global solutions of some hyperbolic equations with discontinuous coefficients, *Acta Appl. Math.* **67** (2001), 173–224.

[26] G. Hörmann and M. Kunzinger, Microlocal analysis of basic operations in Colombeau algebras. *J. Math. Anal. Appl.* **261** (2001), 254–270.

[27] G. Hörmann and M. Oberguggenberger, Elliptic regularity and solvability for partial differential equations with Colombeau coefficients *Electron. J. Diff. Eqns.* **14** (2004), 1–30.

[28] G. Hörmann, M. Oberguggenberger and S. Pilipovic, Microlocal hypoellipticity of linear partial differential operators with generalized functions as coefficients, *Trans. Amer. Math. Soc.* **358** (2006), 3363–3383. 2006.

[29] A.E. Hurd and D.H. Sattinger, Questions of existence and uniqueness for hyperbolic equations with discontinuous coefficients, *Trans. Amer. Math. Soc.* **132** (1968), 159–174.

[30] F. Lafon and M. Oberguggenberger, Generalized solutions to symmetric hyperbolic systems with discontinuous coefficients: the multidimensional case, *J. Math. Anal. Appl.* **160** (1991), 93–106.

[31] M. Mascarello and L. Rodino, *Partial Differential Equations with Multiple Characteristics*, Mathematical Topics **13**, Akademie Verlag, Berlin, 1997.

[32] M. Nedeljkov, S. Pilipović and D. Scarpalézos, *The Linear Theory of Colombeau Generalized Functions*, Pitman Research Notes in Mathematics **385**, Longman, Harlow, 1998.

[33] M. Oberguggenberger, Hyperbolic systems with discontinuous coefficients: examples, in *Generalized Functions, Convergence Structures, and Their Applications*, Editors: B. Stanković, E. Pap, S. Pilipović and V.S. Vladimirov, Plenum Press, New York, 1988, 257–266.

[34] M. Oberguggenberger, Hyperbolic systems with discontinuous coefficients: generalized solutions and a transmission problem in acoustics, *J. Math. Anal. Appl.* **142** (1989), 452–467.

[35] M. Oberguggenberger, *Multiplication of Distributions and Applications to Partial Differential Equations*, Pitman Research Notes in Mathematics **259**, Longman, Harlow, 1992.

[36] M. Oberguggenberger, Hyperbolic systems with discontinuous coefficients: generalized wavefront sets, in *New Developments in Pseudo-Differential Operators*, Editors: Luigi Rodino and M.W. Wong, Birkhäuser, 2008, 121–140.

[37] M. Oberguggenberger and F. Russo, Nonlinear SPDEs: Colombeau solutions and pathwise limits, in *Stochastic Analysis and Related Topics VI*, Editors: L. Decreusefonds, J. Gjerde, B. Øksendal and A.S. Üstünel, Birkhäuser, Boston, 1998, 319–332.

[38] M. Oberguggenberger and F. Russo, Nonlinear stochastic wave equations, *Integral Transforms Spec. Funct.* **6** (1998), 71–83.

[39] S. Pilipović, *Colombeau's Generalized Functions and Pseudo-Differential Operators.* Lect. Math. Sci., Univ. Tokyo, Tokyo, 1994.

[40] R. Steinbauer, Geodesic and geodesic deviation for impulsive gravitational waves, *J. Math. Phys.* **39**(4) (1998), 2201–2212.

Claudia Garetto
Institut für Grundlagen der Bauingenieurwissenschaften
Leopold-Franzens-Universität Innsbruck
Technikerstr. 13
A-6020 Innsbruck, Austria
e-mail: `claudia@mat1.uibk.ac.at`

Operator Theory:
Advances and Applications, Vol. 189, 185–200

On Local and Global Regularity of Fourier Integral Operators

Michael Ruzhansky

Abstract. The aim of this paper is to give a review of local and global properties of Fourier integral operators with real and complex phases, in local L^p, global L^2, and in Colombeau's spaces.

Mathematics Subject Classification (2000). Primary 35S30; Secondary 35L30.

Keywords. Fourier integral operators, hyperbolic partial differential equations.

1. Introduction

Fourier integral operators and their regularity properties have been under study since their appearance in [22]. One of the main motivations comes from the Cauchy problem for hyperbolic partial differential equations. Consider a pseudo-differential operator

$$P(t, x, \partial_t, \partial_x) = \partial_t^m + \sum_{j=1}^{m} P_j(t, x, \partial_x)\partial_t^{m-j}$$

of order m. Here $t \in \mathbb{R}$ and $x \in X$, where $X \subset \mathbb{R}^n$ is an open set. Operators $P_j(t, \cdot, \cdot) \in \Psi^j(X)$ are assumed to be (classical) pseudo-differential operators of order j on X.

Let $\sigma_P(t, \tau, x, \xi)$ be the principal symbol of P, i.e., the top order part of the full symbol of P, positively homogeneous of degree m in ξ. As usual, τ and ξ correspond to $D_t = -i\partial_t$ and $D_x = -i\partial_x$, respectively.

Let P be strictly hyperbolic, i.e., assume that σ_P can be factored as

$$\sigma_P(t, x, \tau, \xi) = \prod_{j=1}^{m} (\tau - \tau_j(t, x, \xi)),$$

where $\tau_j(t, x, \xi)$ are real and distinct for $\xi \neq 0$. Then they are positively homogeneous of degree one in ξ and smooth in (t, x), for $\xi \neq 0$. It is well known that

The author was supported by a Royal Society grant and by the EPSRC Grant EP/E062873/01.

under the strict hyperbolicity condition the Cauchy problem for P is well posed. The Cauchy problem for P is the equation

$$\begin{cases} Pu(t,x) = 0, & t \neq 0, \\ \partial_t^j u|_{t=0} = f_j(x), & 0 \leq j \leq m-1. \end{cases} \tag{1.1}$$

The loss of regularity for solutions $u(t,\cdot)$ compared to the Cauchy data depends on the operator P and on the function spaces. For example, the usual energy conservation that holds in L^2 fails in L^p for $p \neq 2$.

The general question that one is interested in is as follows. Let W_j, W be some function spaces. The questions is whether for (small) t we have the property that

$$f_j \in W_j \text{ (for all } j) \implies u(t,\cdot) \in W \quad ?$$

First, recall the following examples of such mapping properties:

$$f_j \in L^2_{n-j} \implies u(t,\cdot) \in L^2_n \tag{1.2}$$

and

$$f_j \in L^p_{n-j+(n-1)|1/p-1/2|} \implies u(t,\cdot) \in L^p_n. \tag{1.3}$$

Here and in the sequel L^p_s will denote the local Sobolev space over L^p defined by the property that $f \in L^p_s$ if $(1-\Delta)^{s/2} f \in L^p$.

Property (1.2) is the usual conservation of energy for hyperbolic equations in L^2. Property (1.3) shows that the loss of regularity occurs in L^p for $p \neq 2$. In the case of the wave equation this was established in [56], [33], [36]. In general, this property was proved in [54] (see also [55] and [57]). We note that this is quite different from L^p-$L^{p'}$ properties, for some overview of which see, e.g., [47].

The main idea for deriving estimates (1.3) is to use the fact that solution $u(t,x)$ to the Cauchy problem (1.1) can be written as a sum of Fourier integral operators (see, e.g., [30], [15], [24], [16]). Thus, the study of the regularity properties of solutions to the Cauchy problem (1.1) are reduced to the analysis of the regularity properties of Fourier integral operators. A more complicated reduction is also possible for certain classes of hyperbolic systems, even with variable multiplicities, for example for systems with microlocally diagonalisable principal part (see [38], and [26] for generic conditions on characteristics).

2. Fourier integral operators with real phases

Let X, Y be open sets in $\mathbb{R}^n$. One defines the class of Fourier integral operators $T \in I^\mu(X \times Y, C')$ by the (microlocal) formula

$$Tf(x) = \int_Y \int_{\mathbb{R}^n} e^{i\Phi(x,y,\theta)} a(x,y,\theta) f(y) d\theta \, dy, \tag{2.1}$$

where amplitude a is a smooth function satisfying

$$a \in S^\mu : |\partial^\alpha_{x,y} \partial^\beta_\theta a(x,y,\theta)| \leq C_{\alpha\beta}(1+|\theta|)^{\mu-|\beta|},$$

for all multi-indices α, β, and Φ is a phase function. First, we assume that the phase is real-valued and satisfies the following properties:

1. $\Phi(x, y, \lambda\theta) = \lambda\Phi(x, y, \theta)$, for all $\lambda > 0$;
2. $d\Phi \neq 0$;
3. $d_\theta\Phi = 0$ is smooth (e.g., $d_\theta\Phi = 0$ implies $d_{(x,y,\theta)}\frac{\partial\Phi}{\partial\theta_j}$ are independent).

It turns out that Fourier integral operators have a useful invariant quantity, the canonical relation, which is a conic Lagrangian manifold in $(T^*X\backslash 0)\times(T^*Y\backslash 0)$, equipped with the symplectic form $\sigma_X \oplus -\sigma_Y$, where σ_X and σ_Y are canonical symplectic forms in T^*X and T^*Y, respectively. The canonical relation of a Fourier integral operator T is defined by $C' = \{(x, \xi, y, -\eta) : (x, \xi, y, \eta) \in C\}$, where

$$C = C_\Phi = \{(x, d_x\Phi, y, d_y\Phi) : d_\theta\Phi = 0\}$$

is the wave front set of the integral kernel of the operator T. We have the canonical projections:

$$\begin{array}{ccccc} T^*X & \overset{\pi_X}{\longleftarrow} & C \subset T^*X \times T^*Y & \overset{\pi_Y}{\longrightarrow} & T^*Y. \\ & & \Big\downarrow \pi_{X\times Y} & & \\ & & X \times Y & & \end{array}$$

Operators appearing in the analysis of hyperbolic equations satisfy the so-called local graph condition. It means that projections

$$\pi_X : C \to T^*X\backslash 0, \quad \pi_Y : C \to T^*Y\backslash 0$$

are local symplectic diffeomorphisms. In fact, because of the symplectic structure, one of them is diffeomorphic if and only if the other one is. In terms of the phase function Φ, it means that we have

$$\det \partial_y\partial_\theta\Phi \neq 0,$$

or similarly $\det \partial_x\partial_\theta\Phi \neq 0$. Thus, from now on we will assume that Fourier integral operators that we consider are non-degenerate, i.e., that the local graph condition is satisfied. Note that the degenerate case is quite more subtle, see, e.g., [37] for some overview.

The local L^2-boundedness of non-degenerate Fourier integral operators of order zero is well known ([17], [22]). In L^p-spaces with $p \neq 2$ there is a loss of regularity depending on the index p. Indeed, it was shown in [54] that non-degenerate Fourier integral operators of order μ are locally bounded in L^p, $1 < p < \infty$, provided that $\mu \leq -(n-1)|1/p - 1/2|$. This follows by interpolation between the local L^2-boundedness of Fourier integral operators of order zero and local boundedness of Fourier integral operators of order $-(n-1)/2$ from the Hardy space H^1 to L^1, established in [54]. Alternatively, it was also established in [60] that non-degenerate Fourier integral operators of order $-(n-1)/2$ are of weak (1,1)-type. In Section 4 we will discuss the local L^p-properties further.

3. Fourier integral operators with complex phases

Let us now briefly describe Fourier integral operators with complex phase functions. Again, let X, Y be open subsets of $\mathbb{R}^n$. Then a Fourier integral operator of order μ with complex phase is defined by the same formulae as in (2.1), namely

$$Tf(x) = \int_{\mathbb{R}^n} \int_{\mathbb{R}^n} e^{i\Phi(x,y,\theta)} a(x,y,\theta) f(y) d\theta \, dy, \tag{3.1}$$

where amplitude $a \in S^\mu$ is of order μ and the complex-valued phase function $\Phi \in \mathbb{C}$ satisfies conditions:

1. $\Phi(x,y,\lambda\theta) = \lambda\Phi(x,y,\theta)$, for all $\lambda > 0$;
2. $d\Phi \neq 0$;
3. $d_\theta\Phi = 0$ is smooth (e.g., $d_\theta\Phi = 0$ implies $d\frac{\partial\Phi}{\partial\theta_j}$ are independent over $\mathbb{C}$);
4. $\operatorname{Im}\Phi \geq 0$.

The last property is clearly necessary in general, for the integral (3.1) to be well defined. The theory of Fourier integral operators with complex phases was developed by Melin, Sjöstrand in [31] and [32], see also [24] and [61] for slightly alternative descriptions. These operators are also related to the Maslov canonical operator.

In [29], Laptev, Safarov and Vassiliev established global parameterisations of Fourier integral operators with real phases by globally defined complex phases (see also [53]).

Again, the canonical relation of T with complex phase Φ is defined by

$$C = C_\Phi = \{(x, d_x\Phi, y, d_y\Phi) : d_\theta\Phi = 0\}.$$

Now we can identify C with a subset of $\widetilde{T^*(X \times Y)}$, where $\widetilde{T^*(X \times Y)}$ is an almost analytic extension of $T^*(X \times Y)$. Here one needs to work with almost analytic extensions of real sets, for example in order to interpret the eikonal equation $\partial_t\Phi = \tau(t, x, \nabla_x\Phi)$ when Φ is complex-valued.

Local L^2-properties for operators with complex phases are more subtle. For example, in [32], Melin and Sjöstrand showed that operators of order zero are locally bounded in L^2 under a complex-valued version of a local graph condition. This condition and its extension will be discusses in the next section in more detail. A more general result was established by Hörmander in [23]. Assume that for every $\gamma \in C_\mathbb{R}$, the maps $(T_\gamma C)_\mathbb{R}$ to $T_{\pi_X(\gamma)}T^*X$ and to $T_{\pi_Y(\gamma)}T^*Y$ are injective. Then it is shown in [23] that Fourier integral operators of order zero with complex-valued phases are locally bounded in L^2. In fact, one can shows that if the mappings

$$C_\mathbb{R} \to T^*X\backslash 0, \quad C_\mathbb{R} \to T^*Y\backslash 0$$

are injective, then $T \circ T^* \in \Psi^0_{1/2,1/2}$ is a pseudo-differential operator of order zero and type $(1/2, 1/2)$. This is one of the differences with operators with real-valued phases, where $T \circ T^* \in \Psi^0_{1,0}$ is a pseudo-differential operator of order zero and type $(1, 0)$.

4. Local L^p regularity properties of Fourier integral operators

The local L^p-properties of Fourier integral operators can be summarised in the following table:

	$\Phi \in \mathbb{R}$	$\Phi \in \mathbb{C}$
L^2	[17, 22]	[31, 32, 23]
L^p	[54]	[42]

To explain this is more detail, we have the following results for the local L^p-continuity:

- Real L^2 theory: Eskin [17] and Hörmander [22]: operators of order zero with real phases are locally continuous in L^2.
- Real L^p theory: Stein [56], Beals [4], Miyachi [33], Peral [36], Sugimoto [59], and finally Seeger, Sogge and Stein [54]: operators of order μ with real phases are locally continuous in L^p, provided that $\mu \leq -(n-1)|1/p - 1/2|$.
- Complex L^2 theory: Melin and Sjöstrand [31, 32] and Hörmander [23]: operators of order zero with complex phases are locally continuous in L^2.
- Complex L^p theory: Ruzhansky [42]: extension of the properties above to derive the unified local L^p properties for operators with complex phases.

We will now review local L^p results for Fourier integral operators with complex phases in more detail.

Let us introduce the following local graph condition for complex-valued phases. We will assume that there exists some $\tau \in \mathbb{C}$ such that we have

$$\text{(LG)} \qquad \det \partial_\theta \partial_y (\operatorname{Re} \Phi + \tau \operatorname{Im} \Phi) \neq 0.$$

Since the equation of the determinant to zero is polynomial in τ, we can conclude that this condition is equivalent to the existence of $\lambda \in \mathbb{R}$ such that

$$\Psi = \operatorname{Re} \Phi + \lambda \operatorname{Im} \Phi$$

defines a (real) local graph, i.e., that $\det \partial_\theta \partial_y \Psi(x, y, \theta) \neq 0$. Condition (LG) weakens Melin–Sjöstrand's local graph condition, which corresponds to the case $\tau = i$. Following [42], under this condition we have the following results. For the sake of simplicity in notation we will suppress primes in the notation for canonical relations.

Theorem 4.1 ([42]). *Let $C \subset \widetilde{(T^*X\backslash 0) \times (T^*Y\backslash 0)}$ be a smooth complex positive homogeneous canonical relation, closed in $T^*(X \times Y)\backslash 0$. Assume that (LG) holds. Let $\mu = -(n-1)|1/p - 1/2|$, $1 < p < \infty$. Then $T \in I^\mu(X \times Y, C')$ is continuous from $L^p_{\text{comp}}(Y)$ to $L^p_{\text{loc}}(X)$.*

This theorem extends both the result of Melin and Sjöstrand in [31, 32] (L^2, $\Phi \in \mathbb{C}$) and the result of Seeger, Sogge and Stein in [54] (L^p, $\Phi \in \mathbb{R}$). We can also note that Theorem 4.1 can not be obtained from its real-valued version since we only have that there exists a real conic Lagrangian manifold $C_0 \subset T^*(X \times Y)\backslash 0$ such that $C_\mathbb{R} \subset C_0$ and such that the class of operators with complex phase of order μ with canonical relation C is contained in the class of operators of the same order μ with real phases with real canonical relation C_0 of type $(1/2, 1/2)$, i.e., we have $I^\mu_\rho(X,Y;C) \subset I^\mu_{1/2}(X,Y;C_0)$. Thus, we may only conclude that operators with complex phases of order μ are locally bounded in L^p provided that $\mu \le -(n-1/2)|1/p-1/2|$. This order is clearly worse than the order μ given in Theorem 4.1.

5. Cauchy problem for operators with complex characteristics

The results can be applied to establish the local regularity properties to solutions to the Cauchy problem for operators with complex characteristics. Consider a (classical) pseudo-differential operator

$$P(t,x,D_t,D_x) = D_t^m + \sum_{j=1}^m P_j(t,x,D_x) D_t^{m-j}$$

of order m, where $t \in [0,T]$ and $x \in X \subset \mathbb{R}^n$. Here $P_j \in C^\infty([0,T], \Psi^j_{cl}(X))$ are classical pseudo-differential operators of order j.

The principal symbol of operator P is given by

$$\sigma_P(t,x,\tau,\xi) = \tau^m + \sum_{j=1}^m P_j(t,x,\xi)\tau^{m-j} = \prod_{j=1}^m (\tau - \tau_j(t,x,\xi)).$$

The Cauchy problem for operator P is the equation

$$\begin{cases} Pu(t,x) = 0, & t \neq 0, \\ \partial_t^j u|_{t=0} = f_j(x), & 0 \le j \le m-1. \end{cases}$$

It was shown in [61] that the propagator for this Cauchy problem can be expressed in terms of Fourier integral operators with complex phases, under the assumption that characteristic roots $\tau_j \in \mathbb{C}$ are complex-valued and satisfy the following assumptions (A1) and (A2):

$$(A1) \quad \text{Simple characteristics: } \tau_i \neq \tau_j, i \neq j, \xi \neq 0.$$

Therefore, we have that $\tau_j \in C^\infty([0,T] \times (T^*X\backslash 0), \mathbb{C})$ is smooth and positively homogeneous of order one in ξ. We also assume that

$$(A2): \quad \operatorname{Im} \tau_j(t,x,\xi) \ge 0 \text{ in } [0,T] \times (T^*X\backslash 0).$$

Theorem 5.1 ([42]). *Let P satisfy* (A1) *and* (A2)*. Let $1 < p < \infty$. Let f_j be compactly supported. Then we have locally:*

$$f_j \in L^p_{\alpha - j + (n-1)|1/p-1/2|} \; (j = 0, \ldots, m-1) \Longrightarrow u(t,\cdot) \in L^p_\alpha.$$

Also, there exists a constant C_T such that for all $t \in [0,T]$ we have the local estimate

$$||u(t,\cdot)||_{L^p_\alpha} \le C_T \sum_{j=0}^{m-1} ||f_j||_{L^p_{\alpha-j+(n-1)|1/p-1/2|}}.$$

The result is sharp in general, since the loss of $(n-1)|1/p-1/2|$ derivatives is sharp for wave type equations (where $\tau_j \in \mathbb{R}$).

We note that the regularity results for operators with complex-valued functions have other applications, see, e.g., [42] for an application to the oblique derivative problem (see also [45] and [43]).

6. Sharpness of local L^p properties

Let $T \in I^\mu(X,Y;C)$ be a Fourier integral operator with a real-valued phase function $\Phi(x,y,\theta)$. The question for the sharpness of local L^p estimates is to find the largest μ for which operators of order μ are continuous from L^p_{comp} to L^p_{loc}. Alternatively, by using the calculus, one can look for the smallest α such that operators of order zero are continuous from the Sobolev space L^p_α to L^p. In this case we have $\mu = -\alpha$.

Let

$$k := \max_{x,y,\theta} \operatorname{rank} \frac{\partial^2}{\partial\theta^2}\Phi(x,y,\theta).$$

In particular, the meaning of k is that the dimension of the singular support of the integral kernel of T is less or equal to $n+k$ (see [40]). For example, we have $k=n-1$ for the solutions of the wave equation and $k=0$ for pseudo-differential operators. The following result was proved in [39]:

Theorem 6.1 ([39]). *If a Fourier integral operator T of order zero with real phase is locally bounded from L^p_α to L^p, then $\alpha \geq \alpha_p = k|1/p-1/2|$.*

There are the following important cases of this result:

- $p=2$; here $\mu=0$.
- $1<p<\infty$ and $k=n-1$: this is the case of solutions to the wave equations ([36]) and more general hyperbolic equations ([54]); here $\mu=-(n-1)|1/p-1/2|$.
- $k=0$: this is the case of pseudo-differential operators; here $\mu=0$.

The orders in Theorem 6.1 indicate that positive results on the L^p boundedness and the order μ in Theorem 4.1 can be improved if we impose further conditions on the canonical relation C.

Let us first review such positive results for operators with real-valued phases. Define

$$\Sigma = \{(x,y) : \operatorname{rank} \frac{\partial^2}{\partial\theta^2}\Phi(x,y,\theta) = k\}.$$

If $(x,y)\in\Sigma$, the level set of $\nabla_\theta\Phi$ is a linear space of dimension $n-k$ in $C_\Phi \subset T^*(X\times Y)$, corresponding to a linear subspace in the conormal bundle

$$N^*\Sigma = \{(x,y,\xi,\eta)\in T^*(X\times Y) : (x,y)\in\Sigma,\\ \xi(\delta x)+\eta(\delta y)=0, \forall(\delta x,\delta y)\in T_{(x,y)}\Sigma\}.$$

The inclusion $N^*\Sigma \subset C_\Phi$ is dense, and the conormal bundle $N^*\Sigma$ consists of affine fibers.

In [54], the following so-called smooth factorization condition (SF) was introduced. Assume that the mapping

$$(x,y) \mapsto (\theta - \text{ level set of } \nabla_\theta \Phi(x,y,\theta)) : \Sigma \to \mathbb{G}_{n-k}(\mathbb{R}^n)$$

is smoothly extendible from Σ to $\pi_{X\times Y}(C_\Phi)$, the singular support of the integral kernel of operator T with canonical relation C_Φ. Here $\mathbb{G}_{n-k}(\mathbb{R}^n)$ denoted the Grassmanian which is the collection of all $(n-k)$-dimensional linear subspaces of $\mathbb{R}^n$.

We can note that the smooth factorization condition (SF) automatically holds in the case $k=0$ (pseudo-differential operators) and in the case $k=n-1$ (hyperbolic equations).

Theorem 6.2 ([54]). *Under the smooth factorization condition* (SF) *operators Fourier integral operators of order zero with real phases are locally continuous from* $L^p_{\alpha_p}$ *to* L^p, *provided that* $\alpha_p = k|1/p-1/2|$ *and* $1<p<\infty$.

Thus, a question arises when and whether the smooth factorization condition (SF) is satisfied. One answer is given by the following theorem:

Theorem 6.3 ([40]). *For translation invariant Fourier integral operators with analytic phases, the smooth factorization condition* (SF) *is satisfied, provided that* $n \le 4$, *or* $k \le 2$.

As a consequence, one obtains a sharp list of L^p-properties for translation invariant Fourier integral operators in $\mathbb{R}^n$, with $n \le 4$, or $k \le 2$. The analysis is based on the theory of affine fibrations developed in [41].

Theorem 6.3 shows that in order for the smooth factorization condition to break, one needs to have the dimension at least $n \ge 5$. In such case examples for the failure of (SF) have been constructed in [39] and in [41]. A typical example is the phase function defined by

$$\Phi(x,y,\theta) = \langle x-y, \theta\rangle + \theta_1\theta_2^2\theta_5^{-2} + (\theta_3\theta_5 - \theta_2\theta_4)^2\theta_5^{-3}.$$

If the operator is not translation invariant, one can construct simpler examples already in lower dimensions. For example, if $n=3$, the phase function

$$\Phi(x,y,\theta) = \langle x-y,\theta\rangle + \theta_3^{-1}(y_1\theta_1 + y_2\theta_2)^2$$

provides an example of the failure of (SF). However, in [44] it was shown that the loss of regularity for the corresponding Fourier integral operators is the same as for operators satisfying the smooth factorization condition (SF).

An analogue of the smooth factorization condition for complex phases was discussed in [42]. The complex-valued phase function Ψ satisfies the complex smooth factorization condition ($\mathbb{C}$SF) if there exists a real $|\tau| < 1/\sqrt{3}$ such that:

- $\Psi = \operatorname{Re}\Phi + \tau \operatorname{Im}\Phi$ satisfies (LG);
- Ψ satisfies real (SF) for some k.

Theorem 6.4 ([42]). *Let $C \subset \widetilde{(T^*X\backslash 0) \times (T^*Y\backslash 0)}$ be a smooth complex positive homogeneous canonical relation, closed in $T^*\widetilde{(X \times Y)}\backslash 0$. Assume that the complex smooth factorization condition (ℂSF) is satisfied. Let $\mu = -k|1/p - 1/2|$ and $1 < p < \infty$. Then Fourier integral operators of order μ with complex phases and canonical relation C are continuous from $L^p_{\mathrm{comp}}(Y)$ to $L^p_{\mathrm{loc}}(X)$.*

Theorem 6.4 deals with operators with symbols of type $(1,0)$. If the amplitude of a Fourier integral operator with the canonical relation as in Theorem 6.4 is in the class $S^{\mu}_{\rho,1-\rho}$, $\rho \geq 1/2$, then it is continuous from $L^p_{\mathrm{comp}}(Y)$ to $L^p_{\mathrm{loc}}(X)$ provided that $\mu = -(k - (n-k)(1-\rho))|1/p - 1/2|$.

There are close relations between the regularity properties of Fourier integral operators with complex phases to the symplectic and analytic geometries, in particular to singularities of the so-called affine fibrations. We refer to [41] for details.

7. Global L^2 estimates for Fourier integral operators

We now consider globally defined Fourier integral operators on $\mathbb{R}^n$ defined by

$$Tu(x) = \int_{\mathbb{R}^n}\int_{\mathbb{R}^n} e^{i(x\cdot\xi + \phi(y,\xi))} a(x,y,\xi) u(y) d\xi\, dy, \tag{7.1}$$

where real-valued phase function $\phi \in C^\infty(\mathbb{R}^n \times \mathbb{R}^n)$ will satisfy conditions specified below. We note that in the case of pseudo-differential operators we have

$$\phi(y,\xi) = -y\cdot\xi.$$

Similar results will hold for the adjoint operator. In particular, this includes operators of the form

$$Su(x) = \int_{\mathbb{R}^n} e^{i\phi(x,\xi)} a(x,\xi)\widehat{u}(\xi) d\xi,$$

which appear as propagators for some classes of hyperbolic equations.

Global properties on $L^2(\mathbb{R}^n)$ of pseudo-differential operators are well known (see, e.g., Calderon–Vaillancourt [8], Cordes [12], Coifman–Meyer [10], Childs [9], Sugimoto [58], Boulkhemair [6], and many other contributions).

The case of Fourier integral operators was also studied but much less is known (see Asada–Fujiwara [3], Asada [1, 2], Fujiwara [21], Kumano-go [27], Boulkhemair [7]). A certain disadvantage of these results is that in all these papers the authors made an assumption that $\partial_\xi\partial_\xi\phi$ is globally bounded on $\mathbb{R}^n \times \mathbb{R}^n$, which fails in many important situations.

For example, in a typical application to smoothing problems (as in [48]–[52]), the canonical transforms used for the changes of variables on the Fourier transform side have the phase of the form $x\cdot\xi - y\cdot\psi(\xi)$, in which case we have $\phi(y,\xi) = y\cdot\psi(\xi)$, where ψ is positively homogeneous of order one. But then

$$\partial_\xi\partial_\xi\phi(y,\xi) = y\cdot\partial_\xi\partial_\xi\psi(\xi)$$

is unbounded on $\mathbb{R}^n \times \mathbb{R}^n$.

There are several questions that arise. For example, what are the minimal growth conditions on the phase and amplitude for operators (7.1) to be globally bounded on $L^2(\mathbb{R}^n)$. Moreover, if we want to have weighted estimates in $L^2(\mathbb{R}^n)$, or weighted estimates in Sobolev spaces over $L^2(\mathbb{R}^n)$, the question arises of what are the minimal requirements for the global calculus of such operators.

Let us now assume that the phase satisfies the following conditions on supp$a$:

(C1) $|\det \partial_y \partial_\xi \phi(y,\xi)| \geq C > 0, \quad \forall (y,\xi) \in \mathbb{R}^n \times \mathbb{R}^n;$

(C2) $|\partial_y^\alpha \partial_\xi \phi(y,\xi)| \leq C_\alpha,\ |\partial_y \partial_\xi^\beta \phi(y,\xi)| \leq C_\beta$
$\forall (y,\xi) \in \mathbb{R}^n \times \mathbb{R}^n, \quad 1 \leq |\alpha|, |\beta| \leq 2n+2.$

Note that condition (C1) is a global version of the local graph condition, which is necessary even for local L^2-bounds for operators with amplitudes in $S^0_{1,0}$.

The importance of condition (C2) is that now we must take only mixed derivatives with respect to y and ξ, so the phase functions for the canonical transforms in smoothing problems satisfy this condition. Indeed, if $\phi(y,\xi) = y \cdot \psi(\xi)$, where ψ is homogeneous of order one for large ξ and $|\det D\psi(\xi)| \geq C > 0$, then condition (C2) is satisfied for large frequencies. An additional argument is required for small frequencies and it can be found in [50].

It can be noted that conditions (C1) and (C2) are considerably weaker than those appearing in the analysis of SG-pseudo-differential operators (see Cordes [13]), SG-Fourier integral operators, (see Coriasco [14]), or for operator of Shubin type (see Boggiatto, Buzano and Rodino [5]).

We consider first operators of the form

$$Tu(x) = \int_{\mathbb{R}^n} \int_{\mathbb{R}^n} e^{i(x\cdot\xi + \phi(y,\xi))} a(x,\xi) u(y) d\xi \, dy.$$

For such operators we have the following theorem

Theorem 7.1 ([50]). *Let $\phi(y,\xi)$ satisfy conditions* (C1), (C2). *Let $a(x,\xi)$ satisfy one of the following conditions:*

(1) [Calderón-Vaillancourt] $\partial_x^\alpha \partial_\xi^\beta a(x,\xi) \in L^\infty(\mathbb{R}^n_x \times \mathbb{R}^n_\xi)$, $\alpha, \beta \in \{0,1\}^n$.
(2) [Cordes] $\partial_x^\alpha \partial_\xi^\beta a(x,\xi) \in L^\infty(\mathbb{R}^n_x \times \mathbb{R}^n_\xi)$, $|\alpha|, |\beta| \leq [n/2]+1$.
(3) [Cordes] $\exists \lambda, \lambda' > n/2 : (1-\Delta_x)^{\lambda/2}(1-\Delta_\xi)^{\lambda'/2} a(x,\xi) \in L^\infty(\mathbb{R}^n_x \times \mathbb{R}^n_\xi)$.
(4) [Childs] *difference conditions to be found in* [50].
(5) [Coifman-Meyer] $\partial_x^\alpha \partial_\xi^\beta a(x,\xi) \in L^\infty(\mathbb{R}^n_x \times \mathbb{R}^n_\xi)$, $|\alpha| \leq [n/2]+1, \beta \in \{0,1\}^n$.
(6) [Coifman-Meyer] $\exists 2 \leq p < \infty$: $\partial_x^\alpha \partial_\xi^\beta a(x,\xi) \in L^p(\mathbb{R}^n_x \times \mathbb{R}^n_\xi)$, $|\alpha| \leq [n(1/2 - 1/p)]+1, |\beta| \leq 2n$.

Then T is $L^2(\mathbb{R}^n)$-bounded.

In brackets we list the names of the authors of the corresponding results for pseudo-differential operators. This theorem follows from a more general statement for the L^2-boundedness of Fourier integral operators with symbols in Besov spaces that appeared in [50].

There are also global L^2-boundedness theorems for adjoint operators, as well as for Fourier integral operators with general amplitudes. We refer to [50] for details of all such statements, as well as for results in weighted $L^2(\mathbb{R}^n)$ spaces.

The global calculus of operators (7.1) as well as global boundedness theorems in weighted Sobolev spaces can be found in [51].

8. Fourier integral operators in Colombeau's spaces

In this section we will discuss properties of Fourier integral operators in the Colombeau's spaces of new generalised functions. For details on these spaces we refer to, e.g., [11] or [34].

The theory of pseudo-differential operators in Colombeau's spaces has been developed in [18] and [19]. Elements of the corresponding theory of Fourier integral operators has been laid down in [20].

At the same time, hyperbolic partial differential equations in Colombeau's spaces have been studied in [28], [25] by the energy methods, yielding relevant extensions of L^2 type results to the setting of new generalised functions.

In this section we will present corresponding L^p-results in the setting of Colombeau's generalised functions, for Fourier integral operators, with subsequent corresponding implications for solutions to hyperbolic equations.

Let X be a bounded open set. Consider families $(u_\epsilon)_\epsilon$ of functions $u_\epsilon \in L^p_\infty(X)$, $0 < \epsilon \leq 1$, where $L^p_\infty(X)$ stands for the space of function in $L^p(X)$ for which all derivatives also belong to $L^p(X)$. One can single out several important families of such functions dependent on their behaviour with respect to ϵ. Thus, the class of moderate families $\mathcal{E}_{L^p}(X)$ is defined as the collection of families satisfying

$$\forall \alpha \geq 0\ \exists N \geq 0 : \ ||\partial^\alpha u_\epsilon||_{L^p} = O(\epsilon^{\ N}) \text{ as } \epsilon \to 0.$$

The class of null families $\mathcal{N}_{L^p}(X)$ is defined as a subclass of $\mathcal{E}_{L^p}(X)$ satisfying the condition that

$$\forall N \geq 0 : \ ||u_\epsilon||_{L^p} = O(\epsilon^N) \text{ as } \epsilon \to 0.$$

The Colombeau's algebra $\mathcal{G}_{L^p}(X)$ is then defined as

$$\mathcal{G}_{L^p}(X) := \mathcal{E}_{L^p}(X)/\mathcal{N}_{L^p}(X).$$

Distributions $L^p_{-\infty} = \bigcup_{s\in\mathbb{R}} L^p_s$ are embedded in $\mathcal{G}_{L^p}(X)$ by the mapping $\iota(u) = [(u * (\rho_\epsilon)]_\epsilon$, where $\rho_\epsilon(x) = \epsilon^{-n}\rho(x/\epsilon)$ is the standard Friedrichs mollifier.

Subsequently, one can define the corresponding classes of generalized pseudo-differential operators. Thus, the generalized symbol is defined as a family $(a_\epsilon)_\epsilon$ of usual symbols $a_\epsilon \in S^m$ such that

$$\forall k, l\ \exists N : \sup_{|\alpha|\leq k, |\beta|\leq l} \sup_{x,\xi\in\mathbb{R}^n} (1+|\xi|)^{-m+|\alpha|} \left|\partial_\xi^\alpha \partial_x^\beta a_\epsilon(x,\xi)\right| = O(\epsilon^{-N}) \text{ as } \epsilon \to 0.$$

Then $A : \mathcal{G}_{L^p}(X) \to \mathcal{G}_{L^p}(X)$ is a generalised pseudo-differential operator with generalised symbol $(a_\epsilon)_\epsilon$ if, on the representative level, it acts as

$$(u_\epsilon)_\epsilon \mapsto (a_\epsilon(t,x,D_x)u_\epsilon)_\epsilon.$$

To formulate the results, also the notion of the slow scale is required. A generalised symbol $(a_\epsilon)_\epsilon$ is said to be of log-type up to order (k,l) (or to be slow scale) if

$$\sup_{|\alpha|\le k,|\beta|\le l}\ \sup_{x,\xi\in\mathbb{R}^n}(1+|\xi|)^{-m+|\alpha|}\left|\partial_\xi^\alpha\partial_x^\beta a_\epsilon(x,\xi)\right| = O(\log(1/\epsilon)) \text{ as } \epsilon\to 0.$$

In [28], Lafon and Oberguggenberger considered partial differential operators of the form

$$A=\sum_{j=1}^n a_j(t,x)\partial_{x_j}+b(t,x).$$

They investigated the Cauchy problem for operator D_t+A and showed the existence of solutions in $\mathcal{G}_{L^\infty}(X)$ provided that b and $\partial_{x_k}a_j$ are of log-type. Moreover, if a_j and b are constant for large x, then the solution is also unique, and in [35] an example of the non-uniqueness was given if this condition breaks. In [25], Hörmann considered the Cauchy problem for more general pseudo-differential operators A of log-type, for which he showed existence and uniqueness in $\mathcal{G}_{L^2}(X)$ (also giving some estimates on k and l in the log-type assumption). In particular, the non-uniqueness effect disappears in $\mathcal{G}_{L^2}(X)$ compared to $\mathcal{G}_{L^\infty}(X)$.

These results may provide some hints on the behaviour of relevant Fourier integral operators in Colombeau' spaces. Let us consider generalised Fourier integral operators of the form

$$Tu(x)=\int_X\int_{\mathbb{R}^n}e^{i\phi(x,y,\xi)}a(x,y,\xi)u(y)dy\,đ\xi.$$

Let us define the regular Colombeau's algebra. The class of regular families $\mathcal{R}_{L^p}(X)$ is defined as a class of functions satisfying the condition

$$\exists N\ge 0\ \forall\alpha\ge 0:\ ||\partial^\alpha u_\epsilon||_{L^p}=O(\epsilon^{-N}) \text{ as } \epsilon\to 0.$$

Then the regular Colombeau algebra $\mathcal{G}^\infty_{L^p}(X)$ is defined by

$$\mathcal{G}^\infty_{L^p}(X):=\mathcal{R}_{L^p}(X)/\mathcal{N}_{L^p}(X).$$

In [20], Garetto, Hörmann and Oberguggenberger showed that if ϕ is a non-degenerate phase function then T maps $\mathcal{G}_{L^\infty}(X)$ to itself continuously. If the phase function ϕ is a generalised family satisfying the slow scale assumption (e.g., log-type), and $a=(a_\epsilon)_\epsilon$ is a regular family of amplitudes, then T maps locally continuously the space of regular Colombeau's functions $\mathcal{G}^\infty_{L^\infty}(X)$ to itself. The following theorem extends this to Colombeau's algebras over any $L^p(X)$, $1\le p\le\infty$.

Theorem 8.1. *Let $1\le p\le\infty$. If ϕ is a non-degenerate (generalised) phase function then T maps $\mathcal{G}_{L^p}(X)$ to itself continuously. If the phase function ϕ is a generalised family satisfying the slow scale assumption (e.g., log-type), and $a=(a_\epsilon)_\epsilon$ is a regular family of amplitudes, then A maps locally continuously the space of regular Colombeau's generalised functions $\mathcal{G}^\infty_{L^p}(X)$ to itself.*

Consequently, propagators for strictly hyperbolic Cauchy problems are continuous in $\mathcal{G}_{L^p}(X)$, and in $\mathcal{G}^{\infty}_{L^p}(X)$ under the log-type assumption on the phase, for all $1 \le p \le \infty$.

The proof is based on the extension to the Colombeau's setting of the eikonal and transport equations, modulo controllable errors with respect to ϵ, and on estimates for generalised Fourier integral operators in $\mathcal{G}_{L^p}(X)$. Details of proofs and exact losses of regularity in Colombeau's spaces will appear in [46].

References

[1] K. Asada, On the L^2 boundedness of Fourier integral operators in $\mathbb{R}^n$, *Proc. Japan Acad. Ser. A Math. Sci.* **57** (1981), 249–253.

[2] K. Asada, On the L^2 boundedness theorem of nonhomogeneous Fourier integral operators in $\mathbb{R}^n$, *Kodai Math. J.* **7** (1984), 248–272.

[3] K. Asada and D. Fujiwara, On some oscillatory integral transformations in $L^2(\mathbb{R}^n)$. *Japan. J. Math. (N.S.)* **4** (1978), 299–361.

[4] M. Beals, *L^p Boundedness of Fourier Integral Operators,* Mem. Amer. Math. Soc. **264** (1982).

[5] P. Boggiato, E. Buzano and L. Rodino. *Global Hypoellipticity and Spectral Theory,* Akademie Verlag, Berlin, 1996.

[6] A. Boulkhemair, L^2 estimates for pseudodifferential operators. *Ann. Scuola Norm. Sup. Pisa Cl. Sci.* **22** (1995), 155–183.

[7] A. Boulkhemair, Estimations L^2 précisées pour des intégrales oscillantes, *Comm. Partial Differential Equations* **22** (1997), 165–184.

[8] A. P. Calderón and R. Vaillancourt, On the boundedness of pseudo-differential operators, *J. Math. Soc. Japan* **23** (1971), 374–378.

[9] A. G. Childs, On the L^2-boundedness of pseudo-differential operators, *Proc. Amer. Math. Soc.* **61** (1976), 252–254.

[10] R.R. Coifman and Y. Meyer, *Au-delà des opérateurs pseudo-différentiels,* Astérisque **57** (1978).

[11] J.-F. Colombeau, *New Generalized Functions and Multiplication of Distributions,* North-Holland Publishing Co., 1984.

[12] H.O. Cordes, On compactness of commutators of multiplications and convolutions, and boundedness of pseudodifferential operators, *J. Funct. Anal.* **18** (1975), 115–131.

[13] H.O. Cordes, *The Technique of Pseudodifferential Operators*, Cambridge Univ. Press, 1995.

[14] S. Coriasco, Fourier integral operators in SG classes I: composition theorems and action on SG Sobolev spaces, *Rend. Sem. Mat. Univ. Pol. Torino* **57** (1999), 249–302.

[15] J.J. Duistermaat, *Fourier Integral Operators,* Birkhäuser, Boston, 1996.

[16] Yu.V. Egorov, Microlocal analysis, in Encyclopedia Math. Sci., Partial Differential Equations, IV, 33, Springer, 1993, 1–147.

[17] G.I. Eskin, Degenerate elliptic pseudo-differential operators of principal type, *Math. USSR Sbornik,* **11** (1970), 539–585.

[18] C. Garetto, T. Gramchev and M. Oberguggenberger, Pseudodifferential operators with generalized symbols and regularity theory, *Electron. J. Differential Equations* **116** (2005), 1–43.

[19] C. Garetto and G. Hörmann, *Microlocal analysis of generalized functions: pseudo-differential techniques and propagation of singularities, Proc. Edinb. Math. Soc.* **48** (2005), 603–629.

[20] C. Garetto, G. Hörmann and M. Oberguggenberger, Generalized oscillatory integrals and Fourier integral operators, preprint, arXiv:math/0607706.

[21] D. Fujiwara, On the boundedness of integral transformations with highly oscillatory kernels, *Proc. Japan Acad.* **51** (1975), 96–99.

[22] L. Hörmander, Fourier integral operators I, *Acta Math.* **127** (1971), 79–183.

[23] L. Hörmander, L^2 estimates for Fourier integral operators with complex phase, *Arkiv för Matematik* **21** (1983), 283–307.

[24] L. Hörmander, *The Analysis of Linear Partial Differential Operators III–IV*, Springer-Verlag, New York, Berlin, 1985.

[25] G. Hörmann, First-order hyperbolic pseudodifferential equations with generalized symbols, *J. Math. Anal. Appl.* **293** (2004), 40–56.

[26] I. Kamotski and M. Ruzhansky, Regularity properties, representation of solutions and spectral asymptotics of systems with multiplicities, *Comm. Partial Differential Equations* **32** (2007), 1–35.

[27] H. Kumano-go, A calculus of Fourier integral operators on $\mathbb{R}^n$ and the fundamental solution for an operator of hyperbolic type, *Comm. Partial Differential Equations* **1** (1976), 1–44.

[28] F. Lafon and M. Oberguggenberger, Generalized solutions to symmetric hyperbolic systems with discontinuous coefficients: the multidimensional case, *J. Math. Anal. Appl.* **160** (1991), 93–106.

[29] A. Laptev, Yu. Safarov and D. Vassiliev, On global representation of Lagrangian distributions and solutions of hyperbolic equations, *Comm. Pure Appl. Math.* **47** (1994), 1411–1456.

[30] P. Lax, Asymptotic solutions of oscillatory initial value problems, *Duke Math. J.* **24** (1957), 627–646.

[31] A. Melin and J. Sjöstrand, Fourier integral operators with complex-valued phase functions, in *Springer Lecture Notes* **459** (1975), 120–223.

[32] A. Melin and J. Sjöstrand, Fourier integral operators with complex phase functions and parametrix for an interior boundary problem, *Comm. Partial Differential Equations* **1** (1976), 313–400.

[33] A. Miyachi, On some estimates for the wave operator in L^p and H^p. *J. Fac. Sci. Univ. Tokyo Sect. IA Math.* **27** (1980), 331–354.

[34] M. Nedeljkov, S. Pilipović and D. Scarpalézos, *The Linear Theory of Colombeau Generalized Functions, Pitman Research Notes in Mathematics Series*, 1998.

[35] M. Oberguggenberger, Hyperbolic systems with discontinuous coefficients: generalized solutions and a transmission problem in acoustics *J. Math. Anal. Appl.* **142** (1989), 452–467.

[36] J. Peral, L^p estimates for the wave equation, *J. Funct. Anal.* **36** (1980), 114–145.

[37] D.H. Phong, Regularity of Fourier integral operators, in *Proc. Int. Congress Math., Zürich, Switzerland,* 1994, 862–874.

[38] G. Rozenblum, Spectral asymptotic behavior of elliptic systems, *Zap. LOMI*, **96** (1980), 255–271 (in Russian).

[39] M. Ruzhansky, On the sharpness of Seeger-Sogge-Stein orders, *Hokkaido Math. J.* **28** (1999), 357–362.

[40] M. Ruzhansky, Analytic Fourier integral operators, Monge–Ampere equation and holomorphic factorization, *Arch. Math.* **72** (1999), 68–76.

[41] M.V. Ruzhansky, Singularities of affine fibrations in the regularity theory of Fourier integral operators, *Russian Math. Surveys* **55** (2000), 99–170.

[42] M. Ruzhansky, *Regularity Theory of Fourier Integral Operators with Complex Phases and Singularities of Affine Fibrations,* CWI Tract **131**, 2001.

[43] M.V. Ruzhansky, Regularity of the parametrix of a problem with oblique derivative, *Russian Math. Surveys* **56** (2001), 1179–1180.

[44] M. Ruzhansky, On the failure of the factorization condition for non-degenerate Fourier integral operators, *Proc. Amer. Math. Soc.* **130** (2002), 1371–1376.

[45] M. Ruzhansky, Recent progress in the regularity theory of Fourier integrals with real and complex phases and solutions to partial differential equations, in *Banach Center Publications* **60** 2003, 151–160.

[46] M. Ruzhansky, *On the properties of Fourier integral operators in spaces of new generalised functions*, in preparation.

[47] M. Ruzhansky and J. Smith, Global time estimates for higher order hyperbolic equations, in *Journées "Equations aux Dérivées Partielles", Exp. No. XII,* 29 *pp., Ecole Polytech., Palaiseau*, 2005.

[48] M. Ruzhansky and M. Sugimoto, New proof of global smoothing estimates for dispersive equations, in *Advances in Pseudo-Differential Operators*, Editors: R. Ashino, P. Boggiatto and M. W. Wong, Operator Theory: Advances and Applications **155**, Birkhäuser, 2004, 65–75.

[49] M. Ruzhansky and M. Sugimoto, A smoothing property of Schrödinger equations in the critical case, *Math. Ann.* **335** (2006), 645–673.

[50] M. Ruzhansky and M. Sugimoto, Global L^2 boundedness theorems for a class of Fourier integral operators, *Comm. Partial Differential Equations* **31** (2006), 547–569.

[51] M. Ruzhansky and M. Sugimoto, *Global calculus of Fourier integral operators, weighted estimates, and applications to global analysis of hyperbolic equations,* in *Pseudo-Differential Operators and Related Topics*, Editors: P. Boggiatto, L. Rodino, J. Toft and M.W. Wong, Operator Theory: Advances and Applications **164**, Birkhäuser, 2006, 65–78.

[52] M. Ruzhansky and M. Sugimoto, Global boundedness theorems for Fourier integral operators associated with canonical transformations, in *Harmonic Analysis and its Applications*, Yokohama Publ., Yokohama, 2006, 65–75

[53] Yu. Safarov and D. Vassiliev, *The Asymptotic Distribution of Eigenvalues of Partial Differential Operators,* American Mathematics Society, 1996.

[54] A. Seeger, C.D. Sogge and E.M. Stein, Regularity properties of Fourier integral operators, *Ann. Math.* **134** (1991), 231–251.

[55] C.D. Sogge, *Fourier Integrals in Classical Analysis,* Cambridge University Press, 1993.

[56] E.M. Stein, L^p boundedness of certain convolution operators, *Bull. Amer. Math. Soc.* **77** (1971), 404–405.

[57] E.M. Stein, *Harmonic Analysis,* Princeton University Press, Princeton, 1993.

[58] M. Sugimoto, L^2-boundedness of pseudo-differential operators satisfying Besov estimates I, *J. Math. Soc. Japan* **40** (1988), 105–122.

[59] M. Sugimoto, On some L^p-estimates for hyperbolic equations, *Arkiv för Matematik* **30** (1992), 149–162.

[60] T. Tao, The weak-type $(1,1)$ of Fourier integral operators of order $-(n-1)/2$. *J. Aust. Math. Soc.* **76** (2004), 1–21.

[61] F. Treves, *Introduction to Pseudodifferential and Fourier Integral Operators,* Vol. 2, Plenum Press, 1982.

Michael Ruzhansky
Department of Mathematics
Imperial College London
United Kingdom
e-mail: m.ruzhansky@imperial.ac.uk

Operator Theory:
Advances and Applications, Vol. 189, 201–246

Type 1,1-Operators Defined by Vanishing Frequency Modulation

Jon Johnsen

Abstract. This paper presents a general definition of pseudo-differential operators of type $1,1$; the definition is shown to be the largest one that is both compatible with negligible operators and stable under vanishing frequency modulation. Elaborating counter-examples of Ching, Hörmander and Parenti–Rodino, type $1,1$-operators with unclosable graphs are proved to exist; others are shown to lack the microlocal property as they flip the wavefront set of an almost nowhere differentiable function. In contrast the definition is shown to imply the pseudo-local property, so type $1,1$-operators cannot create singularities but only change their nature. The familiar rule that the support of the argument is transported by the support of the distribution kernel is generalised to arbitrary type $1,1$-operators. A similar spectral support rule is also proved. As no restrictions appear for classical type $1,0$-operators, this is a new result which in many cases makes it unnecessary to reduce to elementary symbols. As an important tool, a convergent sequence of distributions is said to converge regularly if it moreover converges as smooth functions outside the singular support of the limit. This notion is shown to allow limit processes in extended versions of the formula relating operators and kernels.

Mathematics Subject Classification (2000). Primary 35S05.

Keywords. Exotic pseudo-differential operators, type $1,1$, pseudo-local, spectral support rule, regular convergence, flipped wavefront sets.

1. Introduction

Pseudo-differential operators are generally well understood as a result of extensive analysis since the mid 1960s; but there is an exception for operators of type $1,1$. These have symbols in the Hörmander class $S^d_{1,1}(\mathbb{R}^n \times \mathbb{R}^n)$, which is sometimes called exotic because of the operators' atypical properties.

Recall that a symbol $a(x,\eta) \in C^\infty(\mathbb{R}^{2n})$ belongs to $S^d_{1,1}(\mathbb{R}^n \times \mathbb{R}^n)$ if it for all multiindices α, β satisfies the estimates

$$|D^\alpha_\eta D^\beta_x a(x,\eta)| \le C_{\alpha,\beta}(1+|\eta|)^{d-|\alpha|+|\beta|}. \tag{1.1}$$

For such a symbol, $a(x,D)u = \mathrm{OP}(a)u = Au$ is defined at least for u in the Schwartz space $\mathcal{S}(\mathbb{R}^n)$ by the usual integral, whereby the Fourier transformation is denoted $\mathcal{F}u(\xi) = \hat{u}(\xi) = \int_{\mathbb{R}^n} e^{-\mathrm{i}\,x\cdot\xi} u(x)\,dx$,

$$a(x,D)u(x) = (2\pi)^{-n} \int_{\mathbb{R}^n} e^{\mathrm{i}\,\xi\cdot\eta} a(x,\eta)\mathcal{F}u(\eta)\,d\eta. \tag{1.2}$$

The purpose of the present article is to suggest a *general* definition of operators with type 1,1-symbols; that is, to define $a(x,D)u$ for u in a maximal subspace $D(A)$ such that

$$\mathcal{S}(\mathbb{R}^n) \subset D(A) \subset \mathcal{S}'(\mathbb{R}^n). \tag{1.3}$$

Seemingly this question has not been addressed directly before. But as a fundamental contribution, L. Hörmander [10, 11] used H^s-estimates to extend type 1,1-operators by continuity from $\mathcal{S}(\mathbb{R}^n)$ and characterised the possible s up to a limit point.

For other questions it seems necessary to have an explicit definition of type 1,1-operators. Consider, e.g., the pseudo-local property,

$$\operatorname{sing\,supp} Au \subset \operatorname{sing\,supp} u \quad \text{for all} \quad u \in D(A). \tag{1.4}$$

In the proof of this, it is of course of little use just to know the action of A on $u \in \mathcal{S}(\mathbb{R}^n)$, as both sets are empty for such u. And to apply the fact that the distribution kernel $K(x,y)$ of A is C^∞ for $x \neq y$ one would have to know more on A and its domain $D(A)$ than just (1.3).

To give a brief account of the present contribution, let $\psi \in C^\infty_0(\mathbb{R}^n)$ denote an auxiliary function for which $\psi = 1$ in a neighbourhood of the origin. Then the frequency modulated versions of $u \in \mathcal{S}'(\mathbb{R}^n)$ and of $a(x,\eta)$ with respect to x are given for $m \in \mathbb{N}$ by

$$u^m = \psi(2^{-m}D)u = \mathcal{F}^{-1}(\psi(2^{-m}\cdot)\mathcal{F}u) \tag{1.5}$$

$$a^m = \psi(2^{-m}D_x)a = \mathcal{F}^{-1}_{\xi\to x}(\psi(2^{-m}\xi)\mathcal{F}_{x\to\xi}a(\xi,\eta)). \tag{1.6}$$

Therefore $a(x,D)$ is said to be *stable* under *vanishing* frequency modulation if for every u in its domain

$$a^m(x,D)u^m \xrightarrow[m\to\infty]{} a(x,D)u \quad \text{in} \quad \mathcal{D}'(\mathbb{R}^n). \tag{1.7}$$

Whilst classical pseudo-differential operators have this property, the purpose is to show that (1.7) can be used as a definition of $a(x,D)u$ when $a \in S^\infty_{1,1}(\mathbb{R}^n \times \mathbb{R}^n)$ is given; hereby $D(a(x,D))$ consists of the $u \in \mathcal{S}'(\mathbb{R}^n)$ for which the limit exists independently of ψ. The limit in (1.7) serves as a substitute of the usual extensions by continuity from $\mathcal{S}(\mathbb{R}^n)$.

In this introduction it is to be understood in (1.7) that, for all $u \in \mathcal{S}'(\mathbb{R}^n)$,

$$a^m(x,D)u^m = \mathrm{OP}(a^m(x,\eta)\psi(2^{-m}\eta))u, \tag{1.8}$$

where the right-hand side is in $\mathrm{OP}(S^{-\infty})$. The expression $a^m(x,D)u^m$ itself is brief, but problematic if taken literally since also $a^m(x,\eta) \in S^\infty_{1,1}$. However, using that $\operatorname{supp}\mathcal{F}(u^m) \Subset \mathbb{R}^n$, it will later be seen that $a^m(x,D)u^m$ can be defined via (1.8) and that this is compatible with (1.7); thenceforth $a^m(x,D)u^m$ will be a short and safe notation.

The definition is discussed in detail below, and shown to imply that type 1, 1-operators are pseudo-local (cf. (1.4) and Theorem 6.4). In comparison they do not in general preserve wavefront sets, for following C. Parenti and L. Rodino [19] a version of a well-known example due to C.H. Ching is shown to flip the wavefront set $\mathrm{WF}(w_\theta) = \mathbb{R}^n \times (\mathbb{R}_+\theta)$ into $\mathbb{R}^n \times (\mathbb{R}_+(-\theta))$ for some w_θ, that when the order $d \in]0,1]$ is an almost nowhere differentiable function.

Moreover the following well-known support rule is extended to arbitrary $a(x,D) \in \mathrm{OP}(S^\infty_{1,1}(\mathbb{R}^n \times \mathbb{R}^n))$ with distribution kernel K (cf. Theorem 8.1),

$$\operatorname{supp} a(x,D)u \subset \overline{\operatorname{supp} K \circ \operatorname{supp} u} \quad \text{for all} \quad u \in D(a(x,D)). \tag{1.9}$$

Here $\operatorname{supp} K \circ \operatorname{supp} u := \{\, x \in \mathbb{R}^n \mid \exists y \in \operatorname{supp} u\colon (x,y) \in \operatorname{supp} K \,\}$, whereby $\operatorname{supp} K$ is thought of as a relation on $\mathbb{R}^n$ that maps, or transports, every set $M \subset \mathbb{R}^n$ to the set $(\operatorname{supp} K) \circ M$ of everything related to an element of M.

There is an analogous result which seems to be new, even for classical symbols $a \in S^\infty_{1,0}$. It gives a *spectral* support rule, relating frequencies $\xi \in \operatorname{supp}\mathcal{F}(Au)$ to those in $\operatorname{supp}\mathcal{F}u$: if only $u \in D(A)$ is such that (1.7) holds in the topology of $\mathcal{S}'(\mathbb{R}^n)$, then (cf. Theorem 8.4)

$$\operatorname{supp}\mathcal{F}(a(x,D)u) \subset \overline{\Xi}, \tag{1.10}$$

$$\Xi = \{\, \xi + \eta \mid (\xi,\eta) \in \operatorname{supp}\mathcal{F}_{x\to\xi} a,\ \eta \in \operatorname{supp}\mathcal{F}u \,\}. \tag{1.11}$$

This is highly analogous to (1.9), for $\Xi = \operatorname{supp}\mathcal{K} \circ \operatorname{supp}\mathcal{F}u$, where $\mathcal{K}$ is the kernel of the conjugated operator $\mathcal{F}a(x,D)\mathcal{F}^{-1}$. There is a forerunner of (1.10)–(1.11) in [15], where it was only possible to cover the case $\mathcal{F}u \in \mathcal{E}'(\mathbb{R}^n)$, as the information on $D(a(x,D))$ was inadequate without the definition in (1.7).

The spectral support rule (1.10) often makes it possible to by-pass a reduction to elementary symbols, that were introduced by R. Coifman and Y. Meyer [5] in order to control spectra like $\operatorname{supp}\mathcal{F}a(x,D)u$ in the L_p-theory of general pseudo-differential operators. Use of (1.10)–(1.11) simplifies the theory, for it would be rather inconvenient to add in (1.7) an extra limit process resulting from approximation of $a(x,\eta)$ by elementary symbols.

Both (1.9) and (1.10) are established as consequences of the formula relating an operator A to its kernel $K \in \mathcal{D}'(\mathbb{R}^n \times \mathbb{R}^n)$,

$$\langle\, Au,\, v \,\rangle = \langle\, K,\, v \otimes u \,\rangle. \tag{1.12}$$

It is shown below (cf. Theorems 7.4 and 8.1) that also the right-hand side makes sense as it stands for $u \in \mathcal{D}'(\mathbb{R}^n)$, although K and $v \otimes u$ are distributions then, as long as v is a test function such that

$$\operatorname{supp} K \bigcap \operatorname{supp} v \otimes u \Subset \mathbb{R}^n \times \mathbb{R}^n, \qquad \operatorname{sing\,supp} K \bigcap \operatorname{sing\,supp} v \otimes u = \emptyset. \tag{1.13}$$

That (1.13) suffices for (1.12) follows from the extendability of the bilinear form $\langle\cdot,\cdot\rangle$ in distribution theory to pairs (u,f) fulfilling analogous conditions. This simple extension of $\langle u, f\rangle$ has the advantage that $\langle u, f^\nu\rangle \to \langle u, f\rangle$ when u or f has compact support and $f^\nu \in C^\infty(\mathbb{R}^n)$ are such that

$$f^\nu \xrightarrow[\nu\to\infty]{} f \quad \text{both in } \mathcal{D}'(\mathbb{R}^n) \text{ and in } C^\infty(\mathbb{R}^n \setminus \operatorname{sing\,supp} f). \tag{1.14}$$

Such sequences (f^ν) are below said to converge *regularly* to f; they are easily obtained by convolution. In these terms, $\langle\cdot,\cdot\rangle$ is stable under regular convergence if one entry is in $\mathcal{E}'$.

This set-up is convenient for the derivation of (1.12)–(1.13) for type 1, 1-operators. Indeed, the kernel K_m of the approximating operator $a^m(x,D)u^m$ equals $K*\mathcal{F}^{-1}(\psi_m\otimes\psi_m)$ conjugated by the coordinate change $(x,y)\mapsto(x,x-y)$, so that K_m converges regularly to K; whence (1.12) results in the limit $m\to\infty$. Based on this the support rules (1.9)–(1.10) follow in a natural way.

However, the simple criterion in (1.13) and its stability under regular convergence, that might be known, could be useful also for other questions.

The main contributions in this paper consist first of all of the definition (1.7) and the spectral support rule (1.10) ff; secondly of the proofs of pseudo-locality (1.4) and the support rule (1.9) as well as the extension of the kernel formula (1.12)–(1.13). Moreover, $a(x,D)u$ is shown to be compatible with the usual pseudo-differential operators (cf. Sections 4–5).

In addition there are various improvements of known results on type 1, 1-operators. This overlap is elucidated (in parenthetic remarks) in the next section.

1.1. On known results for type 1, 1-operators

The pathologies of type 1, 1-operators were revealed around 1972–73. On the one hand, C.H. Ching [4] gave examples of symbols $a\in S^0_{1,1}$ for which the corresponding operators are unbounded from $L^2(\mathbb{R}^n)$ to $L^2(K)$ for every $K\Subset\mathbb{R}^n$ (they can moreover be taken *unclosable* in $\mathcal{S}'(\mathbb{R}^n)$, as shown in Lemma 3.2 below). On the other hand, E.M. Stein (1972-73) showed C^s-boundedness[1] for $s>0$ and orders $d=0$.

Afterwards C. Parenti and L. Rodino [19] discovered that some type 1, 1-operators do not preserve wavefront sets (cf. Section 3.2 where this result of [19] is extended to all $d\in\mathbb{R}$, $n\in\mathbb{N}$). The pseudo-local property of type 1, 1-operators was also claimed in [19], but not backed up by adequate arguments; cf. Remark 6.5 below. (The question is therefore taken up in Theorem 6.4, where the first full proof is given.)

[1]Noted by Y. Meyer [17], with reference to lecture notes at Princeton 1972/73. E.M. Stein stated the C^s-result in [24, VII.1.3]; at the end of Ch. VII its origins were given as "Stein [1973a]" (that is *Singular integrals and estimates for the Cauchy-Riemann equations*, Bull. Amer. Math. Soc. 79 (1973), 440–445) but probably should have been "Stein [1973b]": "*Pseudo-differential operators*, Notes by D.H. Phong for a course given at Princeton University 1972-73".

Around 1980, Y. Meyer [17, 18] obtained the famous property that a composition operator $u \mapsto F(u)$, for a fixed C^∞-function F with $F(0)=0$, acting on $u \in H^s_p(\mathbb{R}^n)$ for $s > n/p$, can be written

$$F(u) = a_u(x,D)u \tag{1.15}$$

for a specific u-dependent symbol $a_u \in S^0_{1,1}$. Namely, when $1 = \sum_{j=0}^\infty \Phi_j$ is a Littlewood–Paley partition of unity,

$$a_u(x,\eta) = \sum_{j=0}^\infty m_j(x)\Phi_j(\eta), \qquad m_j(x) = \int_0^1 F'(\sum_{k<j} \Phi_k(D)u(x) + t\Phi_j(D)u(x))\,dt. \tag{1.16}$$

This gave a convenient proof of the fact that $u \mapsto F(u)$ maps $H^s_p(\mathbb{R}^n)$ into itself for $s > n/p$. Indeed, this follows as Y. Meyer for general $a \in S^d_{1,1}$, using reduction to elementary symbols, established continuity

$$H^{s+d}_p(\mathbb{R}^n) \xrightarrow{a(x,D)} H^s_p(\mathbb{R}^n) \qquad \text{for } s>0,\ 1<p<\infty. \tag{1.17}$$

(In Section 9.2 these results are deduced from the definition in (1.7), and continuity on H^s_p of $u \mapsto F \circ u$ is added in a straightforward way in Theorem 9.4.) It was also realised then that type 1, 1-operators show up in J.-M. Bony's paradifferential calculus [1] of non-linear partial differential equations.

In the wake of this, T. Runst [20] treated the continuity in Besov spaces $B^s_{p,q}$ for $p \in]0,\infty]$ and in Lizorkin–Triebel spaces $F^s_{p,q}$ for $p \in]0,\infty[$, although the necessary control of the frequency changes created by $a(x,D)$ was not quite achieved in [20]. (This flaw was explained and remedied in [15] by means of a less general version of (1.10).)

Around the same time G. Bourdaud proved that a given type 1, 1-operator $a(x,D)\colon C_0^\infty(\mathbb{R}^n) \to \mathcal{D}'(\mathbb{R}^n)$ of order 0 is L_2-bounded if and only if its adjoint $a(x,D)^*\colon C_0^\infty(\mathbb{R}^n) \to \mathcal{D}'(\mathbb{R}^n)$ is also a type 1, 1-operator; cf. [2], [3, Th 3].

Except for a limit point, L. Hörmander characterised the $s \in \mathbb{R}$ for which a given $a \in S^d_{1,1}$ is bounded $H^{s+d} \to H^s$; cf. [10, 11] and also [12] where a few improvements are added. As a novelty in the analysis, an important role was shown to be played by the twisted diagonal

$$\mathcal{T} = \{\, (\xi,\eta) \in \mathbb{R}^n \times \mathbb{R}^n \mid \xi + \eta = 0 \,\}. \tag{1.18}$$

E.g., if the partially Fourier transformed symbol $\hat{a}(\xi,\eta) := \mathcal{F}_{x\to\xi} a(x,\eta)$ vanishes in a conical neighbourhood of a non-compact part of $\mathcal{T}$, i.e., if

$$\exists C \geq 1\colon C(|\xi+\eta|+1) \leq |\eta| \implies \hat{a}(x,\eta) = 0, \tag{1.19}$$

then $a(x,D)\colon H^{s+d} \to H^s$ is continuous for every $s \in \mathbb{R}$. Moreover, continuity for all $s > s_0$ was shown to be equivalent to a specific asymptotic behaviour of $\hat{a}(\xi,\eta)$ at $\mathcal{T}$. For operators with additional properties, a symbolic calculus was also developed together with a sharp Gårding inequality; cf. [10, 11, 12].

For *domains* of type 1,1-operators, the scale $F^s_{p,q}(\mathbb{R}^n)$ of Lizorkin–Triebel spaces was recently shown to play a role, for it was proved in [14, 15] that for all $p \in [1,\infty[$, every $a \in S^d_{1,1}$ gives a bounded linear map

$$F^d_{p,1}(\mathbb{R}^n) \xrightarrow{a(x,D)} L_p(\mathbb{R}^n). \tag{1.20}$$

This is a substitute of boundedness from H^d_p (or of L_p-boundedness for $d = 0$), as $H^s_p = F^s_{p,2} \supsetneq F^s_{p,1}$ for $1 < p < \infty$. Inside the $F^s_{p,q}$ and $B^s_{p,q}$ scales, (1.20) gives maximal domains for $a(x,D)$ in L_p, for it was noted in [15, Lem. 2.3] that already Ching's operator is discontinuous from $F^d_{p,q}$ to $\mathcal{D}'$ and from $B^d_{p,q}$ to $\mathcal{D}'$ for every $q > 1$. Continuity was proved in [15] for $s > \max(0, \frac{n}{p} - n)$, $0 < p < \infty$, as a map

$$F^{s+d}_{p,q}(\mathbb{R}^n) \xrightarrow{a(x,D)} F^s_{p,r}(\mathbb{R}^n) \quad \text{for} \quad r \geq q,\, r > \tfrac{n}{n+s}. \tag{1.21}$$

Moreover, (1.19) was shown to imply (1.21) for every $s \in \mathbb{R}$, $r = q$. Analogously for $B^s_{p,q}$. (In Section 9.1 it is shown how the techniques behind (1.21) apply in the present set-up, and as a special case (1.17) is rederived in this way; cf. Theorem 9.2.)

As indicated, a general definition of $a(x,D)u$ for a given symbol $a \in S^d_{1,1}(\mathbb{R}^n \times \mathbb{R}^n)$ seems to have been unavailable hitherto. L. Hörmander [10, 11] estimated Au for arbitrary $u \in \mathcal{S}(\mathbb{R}^n)$ in the H^s-scale, which of course gives a uniquely defined bounded operator $A\colon H^{s+d} \to H^s$; and an extension of A to $\bigcup_{s>s_0} H^{s+d}(\mathbb{R}^n)$ for some limit s_0 or possibly even $s_0 = -\infty$, depending on a.

R. Torres [25] also estimated Au for $u \in \mathcal{S}(\mathbb{R}^n)$, using the framework of M. Frazier and B. Jawerth [6, 7]. This gave unique extensions by continuity to maps $F^{s+d}_{p,q}(\mathbb{R}^n) \to F^s_{p,q}(\mathbb{R}^n)$ for all s so large that, for all multiindices γ,

$$0 \leq |\gamma| < \max\left(0, \frac{n}{p} - n, \frac{n}{q} - n\right) - s \implies A^*(x^\gamma) = 0. \tag{1.22}$$

(As noted in [25], this is related to the conditions on the twisted diagonal given by L. Hörmander.) This approach will at most define A on $\bigcup F^s_{p,q}(\mathbb{R}^n)$.

In addition it was shown in [15, Prop. 1] that every type 1,1-operator A extends to the space $\mathcal{F}^{-1}\mathcal{E}'(\mathbb{R}^n)$. (Extension to $\mathcal{F}^{-1}\mathcal{E}'$ is also considered in Section 4 in connection with compatibility questions.) Clearly $\mathcal{F}^{-1}\mathcal{E}'$ contains all polynomials $\sum_{|\alpha|\leq k} c_\alpha x^\alpha$, and these do not belong to $\bigcup H^s$, nor to $\bigcup F^s_{p,q}$, so this development only emphasizes the need for a general definition of type 1,1-operators, without reference to spaces other than $\mathcal{S}'(\mathbb{R}^n)$.

1.2. Remarks on the construction

As indicated above, the extension of an operator $a(x,D)$ of type 1,1 from the Schwartz space $\mathcal{S}(\mathbb{R}^n)$ to a larger domain $D(a(x,D))$ in $\mathcal{S}'(\mathbb{R}^n)$ can roughly be made as follows:

Introducing $a^m(x,\eta) = \mathcal{F}^{-1}_{\xi\to x}(\hat{a}(\xi,\eta)\psi_m(\xi))$, $\psi_m = \psi(2^{-m}\cdot)$ for a cut-off function $\psi \in C^\infty_0(\mathbb{R}^n)$ with $\psi = 1$ around the origin, then $a(x,D)u$ is defined when $u \in \mathcal{S}'(\mathbb{R}^n)$ is such that $a_\psi(x,D)u = \lim_{m\to\infty} \mathrm{OP}(a^m(x,\eta)\psi_m(\eta))$ exists in $\mathcal{D}'(\mathbb{R}^n)$ and does not depend on ψ. And in the affirmative case,

$$a(x,D)u := a_\psi(x,D)u = \lim_{m\to\infty} \mathrm{OP}(a^m(x,\eta)\psi_m(\eta))u. \tag{1.23}$$

Fundamentally, the role of $a^m(x,\eta)$ is to make the domain of $a(x,D)$ as large as possible: since $a(x,\eta)$ is less special than $a^m(x,\eta)$, the demands on the pair (a,u) would be stronger if only the $\mathrm{OP}(a(x,\eta)\psi_m(\eta))u$ were required to converge. And the domain of $a(x,D)$ would possibly also be smaller, had not the same ψ been used twice to form $a^m(x,\eta)\psi_m(\eta)$. Finally, to take the limit in $\mathcal{S}'(\mathbb{R}^n)$ instead might also exclude some u from $D(a(x,D))$. (However, the $\mathcal{D}'$-limit makes it more demanding to justify compositions $b(x,D)a(x,D)$ of type 1, 1-operators.)

Although (1.23) is an unconventional definition, it is not as arbitrary as it may seem. In fact, cf. Theorem 5.9 below, the resulting map $a \mapsto a(x,D)$, $a \in S^\infty_{1,1}$, can be characterised as the *largest* extension of (1.2) that both gives operators *stable* under vanishing frequency modulation and is *compatible* with OP on $S^{-\infty}$. For $\delta < \rho$ it is even compatible with the classes $\mathrm{OP}(S^\infty_{\rho,\delta})$ in a certain local sense, termed strong compatibility below.

In addition to this, there are at least three simple indications that the definition is reasonable. First of all, if the symbol $a(x,\eta)$ is classical, say $a \in S^\infty_{1,0}$, then the usual $\mathcal{S}'$-continuous extension of $\mathrm{OP}(a)$ fulfils $\mathrm{OP}(a^m(x,\eta)\psi^m(\eta))u \to \mathrm{OP}(a)u$ as a consequence of standard facts (cf. Proposition 5.4 below).

Secondly, the definition also gives back the usual product au, when $a(x)$ is a symbol in $S^\infty_{1,1}$ independent of η. In fact $a \in C^\infty_{\mathrm{b}}(\mathbb{R}^n)$ then, and since $a^m(x)\psi_m(\eta) \in S^{-\infty}(\mathbb{R}^n \times \mathbb{R}^n)$, every $u \in \mathcal{S}'$ gives the following

$$\mathrm{OP}(a^m(x)\psi_m(\eta))u = a^m \cdot \mathcal{F}^{-1}(\psi_m \hat{u}) = a^m u^m \xrightarrow[m\to\infty]{} au. \tag{1.24}$$

So despite the apparent asymmetry in $\mathrm{OP}(a^m\psi_m)u$, where only the symbol is subjected to frequency modulation, the definition is consistent with the product au. However, the expression $a^m(x,D)u^m$, that enters (1.7), is symmetric in this sense.

Thirdly, continuity properties of $a(x,D)$ can be conveniently analysed using Littlewood–Paley techniques applied to both the symbol a and the distribution u. This is facilitated because the Fourier multiplication by ψ_m occurs in both entries of $a^m(x,D)u^m$. Indeed, one can take ψ_m to be the first $m+1$ terms in a Littlewood–Paley partition of unity $1 = \sum_{j=0}^\infty \Phi_j$; then bilinearity gives a direct transition to the paradifferential splitting that has been used repeatedly for L_p continuity results since the 1980s. The reader is referred to Section 9 for details.

Remark 1.1. Analogously to (1.23), there is an extension of the pointwise product $(u_1,u_2) \mapsto u_1u_2$, where $u_j \in L^{\mathrm{loc}}_{p_j}(\mathbb{R}^n)$ for $j = 1,2$ with $\frac{1}{p_1}, \frac{1}{p_2}, \frac{1}{p_1} + \frac{1}{p_2} \in [0,1]$, to the pairs (u,v) in $\mathcal{S}'(\mathbb{R}^n) \times \mathcal{S}'(\mathbb{R}^n)$ for which there is a ψ-independent limit

$$\pi(u,v) := \lim_{m\to\infty} u^m v^m \tag{1.25}$$

This general product $\pi(u, v)$ was introduced and extensively analysed with paramultiplication in [13]; e.g., the convergence in (1.24) follows directly from [13, Prop. 3.6]. By (1.24) one recovers $\pi(a, u)$ from (1.23) when the symbol $a(x, \eta)$ is independent of η. (An open question for $\pi(\cdot, \cdot)$ is settled in Theorem 6.7, where partial associativity is proved from the fact that multiplication by C^∞-functions commutes with vanishing frequency modulation.)

The definition sketched in (1.7) was used rather implicitly in recent works of the author [14, 15]. In the present article, the purpose is to introduce the definition of $a(x, D)u$ in (1.23) systematically and to show that it is consistent with (1.2).

Section 2 gives a review of notation and some preparations, whereas in Section 3 the special properties of type 1, 1-operators are elaborated. Section 4 deals with preliminary extensions of type 1, 1-operators, using cut-off techniques. The general definition of $a(x, D)$ is given in Section 5, where it is proved to be consistent with the usual one if, say $a(x, \eta)$ coincides (for η running through an open set $\Sigma \subset \mathbb{R}^n$) with an element of the classical symbol class $S^d_{1,0}$, or $S^d_{\rho,\delta}$ with $\rho > \delta$. Section 6 contains the proof of the pseudo-local property. As a preparation, extended action of the bracket $\langle \cdot, \cdot \rangle$ from distribution theory is studied in Section 7, with consequences for distribution kernels. A control of $\operatorname{supp} a(x, D)u$ is proved in Section 8, as is the spectral support rule in a general version. Finally Section 9 deals with continuity in the Sobolev spaces H^s_p and a quick review of the consequences for composite functions.

2. Notation and preparations

The distribution spaces $\mathcal{E}'$, $\mathcal{S}'$ and $\mathcal{D}'$, that are dual to C^∞, $\mathcal{S}$ and C_0^∞ respectively, have the usual meaning as in, e.g., [9]. $\mathcal{O}_M(\mathbb{R}^n)$ stands for the space of slowly increasing functions, i.e., the $f \in C^\infty(\mathbb{R}^n)$ such that to every multiindex α there are $C_\alpha > 0$, $N_\alpha > 0$ such that $|D^\alpha f(x)| \le C_\alpha (1+|x|)^{N_\alpha}$ for all $x \in \mathbb{R}^n$. In addition C^∞_{b} denotes the Frechét space of smooth functions with bounded derivatives of any order. The Sobolev space $H^s_p(\mathbb{R}^n)$ with $s \in \mathbb{R}$ and $1 < p < \infty$ is normed by $\|u\|_{H^s_p} = \|\mathcal{F}^{-1}((1 + |\xi|^2)^{s/2} \mathcal{F}u)\|_p$, whereby $\|u\|_p = (\int_{\mathbb{R}^n} |u|^p \, dx)^{1/p}$ is the norm of $L_p(\mathbb{R}^n)$; similarly $\| \cdot \|_\infty$ denotes that of $L_\infty(\mathbb{R}^n)$. That a subset M of $\mathbb{R}^n$ has compact closure is indicated by $M \Subset \mathbb{R}^n$. As usual c denotes a real constant specific to the place of occurrence.

With the short-hand $\langle \xi \rangle = (1 + |\xi|^2)^{1/2}$, a symbol $a(x, \eta)$ is said to be in $S^d_{\rho,\delta}(\mathbb{R}^n \times \mathbb{R}^n)$ if $a \in C^\infty(\mathbb{R}^{2n})$ and for all multiindices α, β there exists $C_{\alpha,\beta} \ge 0$ such that

$$|D^\alpha_\xi D^\beta_x a(x, \xi)| \le C_{\alpha,\beta} \langle \xi \rangle^{d - \rho|\alpha| + \delta|\beta|}. \tag{2.1}$$

Here it is assumed that the order $d \in \mathbb{R}$ and $0 < \rho \le 1$, $0 \le \delta \le 1$ with $\delta \le \rho$, which is understood throughout unless further restrictions are given.

Along with this there is a pseudo-differential operator $a(x,D)$ defined on every u in the Schwartz space $\mathcal{S}(\mathbb{R}^n)$ by the Lebesgue integral

$$a(x,D)u(x) = \operatorname{OP}(a)u(x) = (2\pi)^{-n}\int_{\mathbb{R}^n} e^{\mathrm{i}\,x\cdot\eta}a(x,\eta)\hat{u}(\eta)\,d\eta. \tag{2.2}$$

Here η is the dual variable to $y\in\mathbb{R}^n$ (u is seen as a function of y), while ξ is used for the dual variable to x. If $\psi\in C_0^\infty(\mathbb{R}^n)$ and $\psi=1$ near 0, then $\psi_m=\psi(2^{-m}\cdot)$ gives:

Lemma 2.1. *$a^m(x,\eta)=\psi_m(D_x)a(x,\eta)$ belongs to $S^d_{\rho,\delta}(\mathbb{R}^n\times\mathbb{R}^n)$ when $a(x,\eta)$ itself does so, and then $a(x,\eta)=\lim_{m\to\infty}a^m(x,\eta)\psi_m(\eta)$ holds in $S^{d'}_{\rho,\delta}(\mathbb{R}^n\times\mathbb{R}^n)$ when $d'\geq d+\delta$ and $d'>d$.*

Proof. Since $a\in S^d_{\rho,\delta}$ is bounded with respect to x, the first part results from

$$|D^\beta_x D^\alpha_\eta a^m(x,\eta)| \leq \int |\check{\psi}(y)||D^\beta_x D^\alpha_\eta a(x-2^{-m}y,\eta)|\,dy \leq C'_{\alpha,\beta}\langle\eta\rangle^{d-\rho|\alpha|+\delta|\beta|}. \tag{2.3}$$

Since $\psi(0)=1$ the mean value theorem gives $a^m\to a$ in $S^{d+\delta}_{\rho,\delta}$; and for any $d'>d$ one has $a^m(x,\eta)\psi_m(\eta)-a^m(x,\eta)\to 0$ in $S^{d'}_{\rho,\delta}$; whence $a^m(x,\eta)\psi_m(\eta)-a\to 0$. □

It is straightforward to show from (2.2) that the bilinear map

$$\operatorname{OP}\colon S^d_{\rho,\delta}(\mathbb{R}^n\times\mathbb{R}^n)\times\mathcal{S}(\mathbb{R}^n)\longrightarrow\mathcal{S}(\mathbb{R}^n) \tag{2.4}$$

is continuous. Hereby $\mathcal{S}(\mathbb{R}^n)$ has a Fréchet space structure with seminorms

$$\|\,\psi\,|\mathcal{S},N\| = \sup\{\,|\langle x\rangle^N D^\beta\psi(x)| \mid x\in\mathbb{R}^n,\ |\beta|\leq N\,\}; \tag{2.5}$$

whilst $S^d_{1,1}(\mathbb{R}^n\times\mathbb{R}^n)$ is a Fréchet space with the least $C_{\alpha,\beta}$ in (2.1) as seminorms.

For fixed symbol a in $S^\infty_{1,1}:=\bigcup_d S^d_{1,1}$ the map (2.4) induces a continuous operator $a(x,D)\colon\mathcal{S}(\mathbb{R}^n)\to\mathcal{S}(\mathbb{R}^n)$, that cannot in general be extended to a continuous map $\mathcal{S}'(\mathbb{R}^n)\to\mathcal{D}'(\mathbb{R}^n)$; this is well known cf. Section 3 below.

The next lemma extends [9, Lem. 8.1.1] from $u\in\mathcal{E}'(\mathbb{R}^n)$ to general $u\in\mathcal{S}'(\mathbb{R}^n)$. The extension is irrelevant for the definition of wavefront sets $\operatorname{WF}(u)$, but useful for calculations. It is hardly a surprising result, but without an adequate reference a proof is given here. Recall that $V\subset\mathbb{R}^n$ is a cone if $\mathbb{R}_+V\subset V$. Throughout $\mathbb{R}_\pm=\{\,t\in\mathbb{R}\mid \pm t>0\,\}$.

First the singular cone $\Sigma(u)$ is defined as the complement in $\mathbb{R}^n\setminus\{0\}$ of those $\xi\neq 0$ contained in an open cone $\Gamma\subset\mathbb{R}^n\setminus\{0\}$ fulfilling that $\hat{u}$ is in L_1^{loc} over Γ and

$$C_N := \sup_\Gamma\langle\eta\rangle^N|\hat{u}(\eta)|<\infty, \qquad N>0. \tag{2.6}$$

Then $\Sigma(u)=\emptyset$ when $u\in\mathcal{S}(\mathbb{R}^n)$, and only then (the unit sphere $\mathbb{S}^{n-1}$ is compact).

Lemma 2.2. *Whenever $u\in\mathcal{S}'(\mathbb{R}^n)$, then $\Sigma(\varphi u)\subset\Sigma(u)$ for all $\varphi\in C_0^\infty(\mathbb{R}^n)$, and*

$$\operatorname{WF}(u)\subset\operatorname{sing\,supp}u\times\Sigma(u). \tag{2.7}$$

Proof. It is well known that $\mathcal{S} * \mathcal{S}' \subset \mathcal{O}_M$, so $\widehat{\varphi u}(\xi) = (2\pi)^{-n}\langle\, \hat{u},\, \hat{\varphi}(\xi - \cdot)\,\rangle$ is C^∞.

Given a cone Γ disjoint from $\Sigma(u)$, it suffices to show that $\sup_{\Gamma_1} \langle\eta\rangle^N |\widehat{\varphi u}(\eta)| < \infty$ on every closed cone $\Gamma_1 \subset \Gamma \cup \{0\}$ with supremum independent of Γ_1. When $\xi \neq 0$ is fixed in Γ_1, then $\frac{\xi}{|\xi|} \in \Gamma_1 \cap \mathbb{S}^{n-1}$ and this set has distance $d > 0$ to $\mathbb{R}^n \setminus \Gamma$, so for $0 < \theta < 1$ one has $\eta \in \Gamma$ in the cone $V_\theta = \{\, \eta \neq 0 \mid |\xi - \eta| < \theta d |\xi|\,\}$.

Supposing $\hat{u} = 0$ in $B(0, \frac{1}{4})$, one can take $\chi_0 + \chi_1 = 1$ on $\mathbb{S}^{n-1}$ such that $\chi_0 \in C^\infty(\mathbb{S}^{n-1})$, $\chi_0(\zeta) = 1$ for $|\zeta - \frac{\xi}{|\xi|}| < \frac{d}{3}$ and $\chi_0(\zeta) = 0$ for $|\zeta - \frac{\xi}{|\xi|}| \geq \frac{d}{2}$. Then $\eta \neq 0$ gives

$$\hat{\varphi}(\xi - \eta) = \varphi_0(\eta) + \varphi_1(\eta), \quad \text{for} \quad \varphi_j(\eta) := \chi_j(\tfrac{\eta}{|\eta|})\hat{\varphi}(\xi - \eta), \tag{2.8}$$

and both terms are in $C^\infty(\mathbb{R}^n \setminus \{0\})$ with respect to η, by stereographic projection and the chain rule. Now $\operatorname{supp} \varphi_0 \subset V_{2/3} \subset \Gamma$ and $\hat{u} \in L_1^{\mathrm{loc}}(\Gamma)$ with rapid decay in Γ so that one can estimate $\langle\, \hat{u},\, \varphi_0\,\rangle$ by means of an integral,

$$\begin{aligned} \langle\xi\rangle^N |\langle\, \hat{u},\, \varphi_0\,\rangle| &\leq 2^N \int_\Gamma \langle\xi - \eta\rangle^N |\chi_0(\tfrac{\eta}{|\eta|})||\hat{\varphi}(\xi - \eta)| \langle\eta\rangle^N |\hat{u}(\eta)|\, d\eta \\ &\leq 2^N C_{N+n+1} \|\hat{\varphi}\langle\cdot\rangle^N\|_\infty \|\chi_0\|_\infty \int \langle\eta\rangle^{-n-1}\, d\eta < \infty. \end{aligned} \tag{2.9}$$

In $\mathbb{R}^n \setminus V_{1/3}$ one has $|\xi| \leq \frac{3}{d}|\xi - \eta|$ and $|\eta| \leq (1 + \frac{3}{d})|\xi - \eta|$, so using the seminorms in (2.5),

$$\begin{aligned} \langle\xi\rangle^N |\langle\, \hat{u},\, \varphi_1\,\rangle| &\leq c \sup_{\eta \notin V_{1/3}, |\alpha| \leq M} \langle\xi\rangle^N \langle\eta\rangle^M |D_\eta^\alpha(\chi_1(\tfrac{\eta}{|\eta|})\hat{\varphi}(\xi - \eta))| \\ &\leq c(1 + \tfrac{3}{d})^{N+M} (\sum_{|\alpha| \leq M} \|D^\alpha \chi_1\|_\infty) \,\|\, \hat{\varphi} \,|\mathcal{S}, N + M\| < \infty. \end{aligned} \tag{2.10}$$

Finally one can take $\tilde{\chi} \in C_0^\infty(\mathbb{R}^n)$ such that $\tilde{\chi} = 1$ for $|\eta| \leq \frac{1}{4}$ with support in $B(0, \frac{1}{2})$, then $\langle\xi\rangle^N |\langle\, \hat{u},\, \tilde{\chi}\,\rangle| \leq c \,\|\, \tilde{\chi} \,|\mathcal{S}, N + M\| \,\|\, \hat{\varphi} \,|\mathcal{S}, N + M\|$ follows as in (2.10).

All bounds are uniform in ξ and in Γ_1, hence $\sup_\Gamma \langle\cdot\rangle^N |\widehat{\varphi u}| < \infty$. This proves $\Sigma(\varphi u) \subset \Sigma(u)$, so (2.7) holds as $\mathrm{WF}(u) = \{\, (x, \xi) \mid \xi \in \bigcap_{\varphi(x) \neq 0,\ \varphi \in C_0^\infty} \Sigma(\varphi u)\,\}$. □

Remark 2.3. In Lemma 2.2 equality obviously holds in (2.7) if the singular cone is a ray, i.e., if $\Sigma(u) = \mathbb{R}_+ \zeta$ for some $\zeta \in \mathbb{R}^n$. In such cases $\mathrm{WF}(u)$ can be easily determined.

3. Special properties of type 1, 1-operators

Many of the pathological properties of type 1,1-operators can be obtained from simple examples of the form

$$a_\theta(x, \eta) = \sum_{j=1}^\infty 2^{jd} e^{-\mathrm{i}\, 2^j x \cdot \theta} \chi(2^{-j}\eta), \tag{3.1}$$

whereby $\chi \in C_0^\infty(\mathbb{R}^n)$ with $\operatorname{supp}\chi \subset \{\,\eta \mid \frac{3}{4} \le |\eta| \le \frac{5}{4}\,\}$; and $\theta \in \mathbb{R}^n$ is fixed. Clearly a_θ is in $S^d_{1,1}$ since the terms are disjoint.

Such symbols were used by C.H. Ching [4] and G. Bourdaud [3] for $d = 0$, $|\theta| = 1$ to show L_2-unboundedness. Refining their counter-examples, L. Hörmander linked continuity from H^s with $s \ge -r$, $r \in \mathbb{N}_0$, to the property that θ is a zero of χ of order r.

This is generalised to $\theta \in \mathbb{R}^n$ here because (3.1) with $|\theta| \ne 1$ enters the proof that type 1, 1-operators do not always preserve wavefront sets. And by consideration of arbitrary orders $d \in \mathbb{R}$ the counter-examples get interesting additional properties; cf. Remark 3.5 ff.

From the definition of $a_\theta(x,\eta)$ in (3.1) it is clear that $\hat{u} \in C_0^\infty(\mathbb{R}^n)$ gives

$$a_\theta(x,D)u = \sum_{j=1}^\infty (2\pi)^{-n} \int e^{\mathrm{i}\,x\cdot\eta} 2^{jd} \chi(2^{-j}\eta + \theta)\hat{u}(\eta + 2^j\theta)\,d\eta. \tag{3.2}$$

Then the adjoint $b_\theta(x,D)\colon \mathcal{S}'(\mathbb{R}^n) \to \mathcal{S}'(\mathbb{R}^n)$ of $a_\theta(x,D)$ fulfils, for all $v \in \mathcal{S}(\mathbb{R}^n)$,

$$\begin{aligned}\langle\, u,\, \overline{b_\theta(x,D)v}\,\rangle &= \langle\, a_\theta(x,D)u,\, \overline{v}\,\rangle \\ &= (2\pi)^{-n} \int \hat{u}(\xi)\Big(\sum_{j=1}^\infty 2^{jd}\bar{\chi}(2^{-j}\xi)\hat{v}(\xi - 2^j\theta)\Big)^{-} d\xi,\end{aligned} \tag{3.3}$$

so $\mathcal{F}b_\theta(x,D)v(\xi) = \sum_{j=1}^\infty 2^{jd}\bar{\chi}(2^{-j}\xi)\hat{v}(\xi - 2^j\theta)$. This gives a convenient way to calculate the H^s-norm of $b_\theta(x,D)v$, for when this is finite the disjoint supports of the $\chi(2^{-j}\cdot)$ imply

$$\begin{aligned}&(2\pi)^n \|b_\theta(x,D)v\|^2_{H^s} \\ &= \sum_{j=1}^\infty \int_{2^{j-1}<|\xi|<2^{j+1}} (1+|\xi|^2)^s 4^{jd} |\chi(2^{-j}\xi)\hat{v}(\xi - 2^j\theta)|^2\,d\xi \\ &= \int \Big(\sum_{\frac{1}{2}<|2^{-j}\xi+\theta|<2} (1+|\xi + 2^j\theta|^2)^s 4^{jd} |\chi(2^{-j}\xi + \theta)|^2 \Big) |\hat{v}(\xi)|^2\,d\xi.\end{aligned} \tag{3.4}$$

Therefore the action of $b_\theta(x,D)$ 'piles up' at $\xi = 0$, say for $s > 0 = d$, $|\theta| = 1$; i.e., the adjoint $b_\theta(x,D)$ is not even of type 1, 1 (cf. [3]). Unless of course $\chi(\theta) = 0$.

This leads to the next result, which for $d = 0$, $|\theta| = 1$ gives back [10, Prop. 3.5]. The identity (3.4) is taken from the proof given there, but (3.4) is used consistently here.

Proposition 3.1. *When $a_\theta(x,\eta)$ is given by (3.1) for $d \in \mathbb{R}$, $\theta \ne 0$, then $a_\theta(x,D)$ extends by continuity to a bounded operator $H^{s+d}(\mathbb{R}^n) \to H^s(\mathbb{R}^n)$ for all*

$$s > -r, \quad r := \inf\{\, |\alpha| \mid D^\alpha\chi(\theta) \ne 0\,\}. \tag{3.5}$$

Conversely, for $|\theta| \in [\frac{3}{4}, \frac{5}{4}]$ existence of a continuous linear extension $H^{s+d}(\mathbb{R}^n) \to \mathcal{D}'(\mathbb{R}^n)$ implies (3.5).

Proof. For sufficiency of (3.5) it is enough to prove the adjoint $b_\theta(x,D)$ continuous

$$b_\theta(x,D)\colon H^t(\mathbb{R}^n)\to H^{t-d}(\mathbb{R}^n)\quad\text{for}\quad t<r. \tag{3.6}$$

This is obtained from (3.4) with $s=t-d$, where the inequalities $\frac{1}{2}<|2^{-j}\xi+\theta|<2$ yield that $(1+4^{jd}|2^{-j}\xi+\theta|^2)^{t-d}=\mathcal{O}(4^{j(t-d)})$ for $j\to\infty$. Moreover $|\chi(\theta+\eta)|\le c|\eta|^r$, so

$$4^{jd}(1+|\xi+2^j\theta|^2)^{t-d}|\chi(2^{-j}\xi+\theta)|^2\le c|\xi|^{2r}4^{j(t-r)}. \tag{3.7}$$

Since $t-r<0$ the geometric series can be estimated by the first term, and using that $|2^{-j}\xi|-|\theta|\le 2$ implies $2^j+1\ge\frac{|\xi|+1}{2+|\theta|}$, and hence $2^j\ge\frac{\langle\xi\rangle}{4+2|\theta|}$, this gives

$$\|b_\theta(x,D)v\|^2_{H^{t-d}}\le\frac{c}{1-4^{t-r}}\int|\xi|^{2r}\left(\frac{\langle\xi\rangle}{4+2|\theta|}\right)^{2(t-r)}|\hat{v}(\xi)|^2\,d\xi\le c\|v\|^2_{H^t}. \tag{3.8}$$

Necessity of (3.5) for $\frac{3}{4}\le|\theta|\le\frac{5}{4}$ follows if $a_\theta(x,D)$ cannot be continuous $H^{-r+d}\to\mathcal{D}'$. So the adjoint is assumed continuous $b_\theta(x,D)\colon C_0^\infty(\mathbb{R}^n)\to H^{r-d}(\mathbb{R}^n)$. Inserting in (3.4) that $\chi(\theta+\eta)=P_r(\eta)+\mathcal{O}(|\eta|^{r+1})$ for a homogeneous polynomial $P_r\not\equiv 0$ of degree r,

$$\begin{aligned}\|b_\theta(x,D)v\|^2_{H^{r-d}}=c\int\Big(\sum_{\frac{1}{2}<|2^{-j}\xi+\theta|<2}&(2^{-j}+|2^{-j}\xi+\theta)|^2)^{r-d}\\ &\times|P_r(\xi)+\mathcal{O}(2^{-j})|^2\Big)|\hat{v}(\xi)|^2\,d\xi.\end{aligned} \tag{3.9}$$

For each ξ the sum runs over all $j\ge J$ for a certain $J\ge 0$, since $\frac{1}{2}<|\theta|<2$. By increasing J if necessary, $|P_r(\eta)+\mathcal{O}(2^{-j})|\ge\frac{1}{2}|P_r(\eta)|$. In addition it holds for all $j\ge 0$ that $(2^{-j}+|2^{-j}\xi+\theta)|^2)^{r-d}\ge\min(5^{r-d},4^{d-r})$. Therefore the series above is estimated from below by $\sum_{j\ge J}\frac{1}{4}|P_r(\xi)|^2=\infty$ for $\xi\in\operatorname{supp}\hat{v}$. This contradicts that $b_\theta(x,D)(C_0^\infty)\subset H^{r-d}$. □

It is with good reason that necessity of (3.5) is obtained only for $\frac{3}{4}\le|\theta|\le\frac{5}{4}$. For if $\theta\notin\operatorname{supp}\chi$, (3.5) would hold with $r=\infty$ and $a_\theta(x,D)$ be continuous from H^s for every $s\in\mathbb{R}$ by the sufficiency (i.e., no necessary condition can be imposed if $|\theta|\notin[\frac{3}{4},\frac{5}{4}]$).

3.1. Unclosed graphs

As an addendum to Proposition 3.1, it is a strengthening fact that Ching's operator $a_\theta(x,D)$ can be taken unclosable in $\mathcal{S}'(\mathbb{R}^n)$. I.e., its graph G, as a subspace of $\mathcal{S}'(\mathbb{R}^n)\times\mathcal{S}'(\mathbb{R}^n)$, can have a closure $\overline{G}$ that is not a graph, for as shown in Lemma 3.2 below $\overline{G}$ will in some cases contain a pair $(0,v)$ for some $v\ne 0$, $v\in\mathcal{S}(\mathbb{R}^n)$.

This is important since it shows that type 1, 1-operators cannot just be defined by closing their graphs; nor can one hope to give a definition by other means, such as (1.7), and reach a closed operator in general.

Lemma 3.2. *Let $a_\theta(x,\eta)$ be given as in* (3.1) *for $d \in \mathbb{R}$ and with $|\theta| = 1$ and $\chi = 1$ on the ball $B(\theta, \frac{1}{10})$. Then $a_\theta(x,D)$ is unclosable since there exist v, $v_N \in \mathcal{S}(\mathbb{R}^n) \setminus \{0\}$ such that*

$$\lim_{N\to\infty} v_N = 0 \text{ in } H^d(\mathbb{R}^n), \qquad \lim_{N\to\infty} a_\theta(x,D)v_N = v \text{ in } \mathcal{S}(\mathbb{R}^n). \tag{3.10}$$

Proof. Take $v \in \mathcal{S}(\mathbb{R}^n) \setminus \{0\}$ with $\operatorname{supp} \hat{v} \subset \{\, |\xi| \le \frac{1}{20} \,\}$ and let

$$\hat{v}_N(\xi) = \tfrac{1}{\log N} \sum_{j=N}^{N^2} \tfrac{2^{-jd}}{j} \hat{v}(\xi - 2^j\theta) = \mathcal{F}\Big(v(x) \sum_{j=N}^{N^2} \frac{2^{-jd} e^{\mathrm{i}\, 2^j x\cdot\theta}}{j \log N} \Big). \tag{3.11}$$

Since the $\operatorname{supp} \hat{v}(\xi - 2^j\theta)$ are disjoint, and $c_1 2^j \le \langle\xi\rangle \le c_2 2^j$ hold on each support,

$$\|v_N\|_{H^d} \le \sum_{j=N}^{N^2} \tfrac{1}{(2\pi)^{n/2}} \Big(\int_{\mathbb{R}^n} \frac{\langle\xi\rangle^{2d}}{j^2 2^{2jd}} |\hat{v}(\xi - 2^j\theta)|^2 \, d\xi \Big)^{1/2} \le c\|v\|_2 \Big(\sum_N^\infty j^{-2} \Big)^{1/2} \searrow 0. \tag{3.12}$$

Because v_N is defined by a finite sum, and $\chi(2^{-j}\cdot) \equiv 1$ on $\operatorname{supp} \hat{v}(\cdot - 2^j\theta)$, a direct computation gives the following limit in $\mathcal{S}(\mathbb{R}^n)$,

$$a_\theta(x,D)v_N(x) = \tfrac{1}{\log N} \big(\tfrac{1}{N} + \tfrac{1}{N+1} + \dots + \tfrac{1}{N^2} \big) v(x) \xrightarrow[N\to\infty]{} v(x). \tag{3.13}$$

Indeed, $1 \le (N^{-1} + \dots + N^{-2}) / \log N \le \log(\frac{N^2}{N-1}) / \log N \searrow 1$. □

The sequence v_N also tends to 0 in the more general Besov and Lizorkin–Triebel spaces $B^d_{p,q}$ and $F^d_{p,q}$ for every $p \in [1,\infty]$ and $q > 1$; cf. [15].

3.2. Violation of the microlocal property

In the proof of Lemma 3.2 the role of the exponential functions in $a_\theta(x,\eta)$ was clearly to move all high frequencies in the spectrum of v_N to a neighbourhood of the origin. So it is perhaps not surprising that another variant of Ching's example will produce frequencies η that are moved to, say $-\eta$.

This indicates that type 1, 1-operators need not have the microlocal property; i.e., the inclusion $\mathrm{WF}(a(x,D)u) \subset \mathrm{WF}(u)$ among wavefront sets is violated for certain symbols $a \in S^\infty_{1,1}$.

This is explicated here, following C. Parenti and L. Rodino [19] who treated $d = 0$ and $n = 1$. Their suggested programme is carried out below with a coverage of all $d \in \mathbb{R}$, $n \in \mathbb{N}$ and arbitrary directions of θ. As a minor improvement, the wavefront sets are explicitly determined here; and due to the uniformly estimated symbols and the fact that v in (3.14) below has compact spectrum, the present proofs are also rather cleaner.

With notation as in the proof of Lemma 3.2, again with $|\theta| = 1$, one can set $w_\theta(x) = w(\theta, d; x) = \sum_{j=1}^\infty 2^{-jd} e^{\mathrm{i}\, 2^j\theta\cdot x} v(x)$ for $v \in \mathcal{S}(\mathbb{R}^n)$ with $\operatorname{supp}\hat{v} \subset B(0, 1/20)$, so that

$$\hat{w}_\theta(\eta) = \sum_{j=1}^\infty 2^{-jd} \hat{v}(\eta - 2^j\theta). \tag{3.14}$$

As shown below, this distribution has the cone $\mathbb{R}^n \times (\mathbb{R}_+\theta)$ as its wavefront set. The counter-example arises by considering w_θ together with the symbol $a_{2\theta} \in S^d_{1,1}(\mathbb{R}^n \times \mathbb{R}^n)$ defined by (3.1) for a χ fulfilling

$$\chi(\eta) = 1 \quad \text{for} \quad \tfrac{9}{10} \le |\eta| \le \tfrac{11}{10}. \tag{3.15}$$

As χ vanishes around 2θ, there are by Proposition 3.1 continuous extensions

$$a_{2\theta}(x, D)\colon H^{s+d}(\mathbb{R}^n) \to H^s(\mathbb{R}^n) \quad \text{for all} \quad s \in \mathbb{R}. \tag{3.16}$$

Moreover, it is easy to see that in this case every (ξ, η) in $\operatorname{supp} \hat{a}_{2\theta}$ lies in the cone $|\eta| \le 2|\xi + \eta|$ so that a fulfils (1.19) for $C = 2$. So neither a large domain, like $\bigcup H^s$ in (3.16), nor the twisted diagonal condition can ensure the microlocal property of a type 1, 1-operator:

Proposition 3.3. *The distributions $w(\theta, d; x)$ are in $H^s(\mathbb{R}^n)$ precisely for $s < d$, and when $a_{2\theta}$ is chosen as in (3.1), (3.15) with $|\theta| = 1$, then*

$$a_{2\theta}(x, D)w(\theta, d; x) = w(-\theta, 0; x). \tag{3.17}$$

Moreover,

$$\mathrm{WF}(w_\theta) = \mathbb{R}^n \times (\mathbb{R}_+\theta), \tag{3.18}$$

$$\mathrm{WF}(a_{2\theta}(x, D)w(\theta, d; x)) = \mathbb{R}^n \times (\mathbb{R}_+(-\theta)), \tag{3.19}$$

so the wavefront sets of w_θ and $a_{2\theta}(x, D)w_\theta$ are disjoint.

Proof. Estimates analogous to (3.12) show that the series for $\hat{w}_\theta$ converges in $L_2(\langle\eta\rangle^{2s} d\eta)$ if and only if $s < d$; hence w_θ is well defined in $\mathcal{S}'$ and belongs to H^s for $s < d$. Since the series for w_θ converges in H^s for $s < d$, the continuity (3.16) and (3.15) imply

$$a_{2\theta}(x, D)w(\theta, d; x) = \sum_{j=1}^{\infty} (2\pi)^{-n} \int e^{\mathrm{i}\, x\cdot(\eta - 2^j 2\theta)} \hat{v}(\eta - 2^j\theta)\, d\eta = w(-\theta, 0; x). \tag{3.20}$$

Therefore $\mathrm{WF}(a_{2\theta}(x, D)w_\theta) \bigcap \mathrm{WF}(w_\theta) = \emptyset$ follows as soon as (3.18) has been proved.

Clearly $\operatorname{supp} \hat{w}_\theta$ intersects each ray $\mathbb{R}_+\zeta$ only in a compact set, except for $\zeta = \theta$ in which case $|\hat{w}_\theta(\eta)| \le 2^{|d|} \|\hat{v}\|_\infty \langle\eta\rangle^{-d}$ is an exact decay rate as $2^{j-1} \le \langle\eta\rangle \le 2^{j+1}$ on $\operatorname{supp} \hat{v}(\eta - 2^j\theta)$, so that $\Sigma(w_\theta) = \mathbb{R}_+\theta$ in the notation of Lemma 2.2. This almost proves (3.18), but a full proof is a little lengthy, because of the overlapping supports in

$$\widehat{\varphi w_\theta} = \sum 2^{-jd} \widehat{\varphi v}(\eta - 2^j\theta), \qquad \varphi \in C_0^\infty(\mathbb{R}^n). \tag{3.21}$$

(This important technicality seems to be overlooked in the sketchy arguments of G. Garello [8], who also dealt with extensions of the results of [19].)

That $\operatorname{sing\,supp} w_\theta = \mathbb{R}^n$ follows if $\varphi \in C_0^\infty(\mathbb{R}^n) \setminus \{0\}$ implies $\varphi w_\theta \notin C_0^\infty(\mathbb{R}^n)$. The last property is invariant under multiplication by a character, so it can be arranged that $|\widehat{\varphi v}|$ attains its maximum at 0. Despite the overlapping supports, $\widehat{\varphi w_\theta}$ can then be seen to decay as $\langle\eta\rangle^{-d}$ along $\mathbb{R}_+\theta$, but not rapidly because $\widehat{\varphi v} \not\equiv 0$.

To carry this out, one can pick $r \in]0, \frac{1}{4}[$ so that

$$|\eta| < r \implies |\widehat{\varphi v}(\eta)| > \tfrac{1}{2}\|\widehat{\varphi v}\|_\infty. \tag{3.22}$$

Since every term in $\widehat{\varphi w_\theta}$ is in $\mathcal{S}(\mathbb{R}^n)$ it is only necessary to estimate those with indices $j > J$, for some J. (The estimates make sense since $\widehat{\varphi w_\theta}$ is a function, in $L_2(\langle\eta\rangle^{2s} d\eta)$.) Using that $c_N := \sup |\eta|^N |\widehat{\varphi v}(\eta)| < \infty$, and $r < 1/4$, one finds with a fixed $N > |d|$ that for $\eta \in B(2^k\theta, r)$, $k > J$,

$$2^{kd} \sum_{J<j<k} 2^{-jd} |\widehat{\varphi v}(\eta - 2^j\theta)| \le \sum_{J<j<k} 2^{(k-j)d} \frac{c_N}{2^{kN}} \left(1 - \frac{2^j}{2^k} - \frac{r}{2^k}\right)^{-N} \le \frac{c_N 4^N}{2^{JN}(1 - 2^{d-N})}. \tag{3.23}$$

Similarly $\sum_{j>k} 2^{(k-j)d} |\widehat{\varphi v}(\eta - 2^j\theta)| \le \frac{c_N 4^N}{2^{JN}(1-2^{-d-N})}$ results by factorising 2^j out of $(\dots)^{-N}$.

It is clear one can take J so large that the right-hand sides are less than $\frac{1}{5}\|\widehat{\varphi v}\|_\infty$. Then (3.22) and the fact that $2^{k-1} \le \langle\eta\rangle \le 2^{k+1}$ on $B(2^k\theta, r)$ give

$$\Big|\sum_{j>J} 2^{-jd} \widehat{\varphi v}(\eta - 2^j\theta)\Big| \ge \left(\tfrac{1}{2} - \tfrac{2}{5}\right) \|\widehat{\varphi v}\|_\infty 2^{|d|} \langle\eta\rangle^{-d} \quad \text{for} \quad \eta \in B(2^k\theta, r). \tag{3.24}$$

This estimate is uniform in $k > J$; hence $\operatorname{sing\,supp} w_\theta = \mathbb{R}^n$. So by Lemma 2.2 and Remark 2.3, $\mathrm{WF}(u) = \operatorname{sing\,supp} w_\theta \times \Sigma(w_\theta) = \mathbb{R}^n \times (\mathbb{R}_+\theta)$, i.e., (3.18) is obtained. □

Remark 3.4. It is clear from (3.20) that $a_{2\theta'}(x,\eta) w(\theta, d; x) = w(\theta'', 0; x)$ for $\theta'' = \theta + 2\theta'$, $|\theta'| = 1 = |\theta|$. But θ'' can point in any direction in $\mathbb{R}^n$, so type 1, 1-operators can make arbitrary directional changes in wavefront sets (as noted in [19]).

Remark 3.5. There is an amusing reason why the counter-example w_θ in Proposition 3.3 is singular on all of $\mathbb{R}^n$, i.e., why $\operatorname{sing\,supp} w_\theta = \mathbb{R}^n$. In fact $w_\theta(x) = v(x) f(x \cdot \theta)$ where $f(t) = \sum_{j=1}^\infty 2^{-jd} e^{\mathrm{i}\, 2^j t}$, and this is for $0 < d \le 1$ a well-known variant of Weierstrass' nowhere differentiable function (a fact that could have substantiated the argument for formula (19) in [19]). That the theory of type 1, 1-operators is linked to this classical construction seems to be previously unobserved.

Remark 3.6. To elucidate Remark 3.5, $f(t) = \sum_{j=1}^\infty 2^{-jd} e^{\mathrm{i}\, 2^j t}$ is investigated here. Clearly $f \in \mathcal{S}'(\mathbb{R})$ for all $d \in \mathbb{R}$, as the Fourier transformed series $2\pi \sum_{j=1}^\infty 2^{-jd} \delta_{2^j}$ converges there.

By uniform convergence f is for $d > 0$ a continuous 2π-periodic and bounded function. Nowhere-differentiability for $0 < d \le 1$ is an easy (maybe not widely known) exercise in distribution theory: $\operatorname{supp} \mathcal{F} f$ is lacunary, so any choice of $\chi \in \mathcal{S}(\mathbb{R})$ such that $\hat{\chi}(1) = 1$ and $\operatorname{supp} \hat{\chi} \subset]\frac{1}{2}, 2[$ will give $\operatorname{supp} \hat{\chi}(2^{-k}) \bigcap \operatorname{supp} \delta_{2^j} \ne \emptyset$ only for $j = k$, which entails

$$2^{-kd} e^{\mathrm{i}\, 2^k t} = 2^k \chi(2^k \cdot) * f(t) = \int_{\mathbb{R}} \chi(z)(f(t - 2^{-k} z) - f(t))\, dz; \tag{3.25}$$

so if f were differentiable at t_0, $G(h) := \frac{1}{h}(f(t_0+h) - f(t_0))$ would be in $C(\mathbb{R}) \cap L_\infty(\mathbb{R})$, and the contradiction $d > 1$ would follow since by majorised convergence

$$2^{(1-d)k} e^{\mathrm{i}\, 2^k t_0} = -\int G(-2^{-k} z) z\chi(z)\, dz \xrightarrow[k\to\infty]{} f'(t_0)\cdot(-D)\hat{\chi}(0) = 0. \quad (3.26)$$

Moreover, if $m < d \leq m+1$ for $m \in \mathbb{N}$ it follows by termwise differentiation that $f \in C^m(\mathbb{R})$, but with $f^{(m)}$ nowhere differentiable; so $\operatorname{sing\,supp} f = \mathbb{R}$ for all $d > 0$.

For $d \leq 0$ one has $f \notin L_1^{\mathrm{loc}}$, for f is invariant in $\mathcal{S}'$ under translation by 2π, so if $f \in L_1^{\mathrm{loc}}$ is assumed, $\langle\, f, \varphi\,\rangle = \int f\varphi\, dt$ holds for φ in $C_0^\infty(\mathbb{R})$ as well as in $\mathcal{S}$, since $|\int f\varphi\, dt| \leq c \sup_{t\in\mathbb{R}}(1+|t|^2)|\varphi(t)|$ follows from the fact that $(1+r^2)^{-1} f(r)$ is in L_1:

$$\int_{\mathbb{R}} \frac{|f(r)|}{1+r^2}\, dr = \sum_{p\in\mathbb{Z}} \int_{2p\pi}^{2(p+1)\pi} \frac{|f(r)|}{1+r^2}\, dr \leq \int_0^{2\pi} |f|\, dr \sum_{p=0}^{\infty} \tfrac{2}{1+(2p\pi)^2} < \infty. \quad (3.27)$$

Therefore the convolution in (3.25) is given by the integral also in this case. By taking t outside a G_δ-set G of measure 0, f is continuous (in $\mathbb{R}\setminus G$) at t, whence

$$2^{-kd} = \Big|\int 2^k \chi(2^k r)(f(t-r) - f(t))\, dr\Big| \xrightarrow[k\to\infty]{} 0. \quad (3.28)$$

In fact, for $\varepsilon > 0$ the part with $|r| < \delta$ is $< \varepsilon$ for some $\delta > 0$, but $\sup_{|r|\geq\delta}(1+|t-r|^2)2^k|\chi(2^k r)| = \mathcal{O}(2^{-k})$, so since $L_1(\mathbb{R})$ by (3.27) contains $r \mapsto (f(t-r) - f(t))/(1+|t-r|^2)$ the limit 0 results. Thence the contradiction $d > 0$.

To complete the picture, Weierstrass' original function $W(t) = \sum_{j=0}^\infty \frac{\cos(b^j t)}{a^j}$, where $b \geq a > 1$, is nowhere differentiable by the same argument. One only has to take $\operatorname{supp}\mathcal{F}\chi \subset \,]\frac{1}{b}, b[$, $\mathcal{F}\chi(1) = 1$, for in $\mathcal{F}\cos(b^j\cdot) = \frac{2\pi}{2}(\delta_{b^j} + \delta_{-b^j})$ the last term is removed by $\hat{\chi}(b^{-j}\cdot)$, so that $\chi(b^{-j}D)$ yields a second microlocalisation of W. As in (3.25)–(3.26) it follows that $(\frac{b}{a})^k e^{\mathrm{i}\, b^k t_0} \to 0$ for $k \to \infty$, contradicting that $b \geq a$. A further study of nowhere differentiable functions by means of microlocalisation can be found in [16].

Remark 3.7. As a precise account of the regularity, $f \in B^d_{\infty,\infty}(\mathbb{R})$; for $0 < d < 1$ this Besov space consists of Hölder continuous functions of order d. Indeed, the norm of f in $B^d_{\infty,\infty}$ is from the left part of (3.25) seen to equal 1, when χ is taken as Φ_1 in a suitable Littlewood–Paley partition of unity $1 = \sum_{j=0}^\infty \Phi_j$. Moreover, $f \in F^d_{p,\infty;\mathrm{loc}}(\mathbb{R})$ when $d \in \mathbb{R}$, $0 < p < \infty$, for the definition in [26] of $F^s_{p,q}$ gives, when $v \in \mathcal{S}(\mathbb{R})$ with $\operatorname{supp}\hat{v} \subset B(0, \frac{1}{20})$,

$$\|vf\|_{F^d_{p,\infty}} = \Big(\int (\sup_j 2^{jd}|\Phi_j(D)(vf)|)^p\, dt\Big)^{1/p} = \big\|\sup_j |v(t)e^{\mathrm{i}\, 2^j t}|\big\|_p = \|v\|_p. \quad (3.29)$$

These are identities, so the $F^d_{p,\infty}$-regularity is sharp. That $\varphi f \in F^s_{p,\infty}(\mathbb{R})$ for $\varphi \in C_0^\infty(\mathbb{R})$ results from $\varphi f = \frac{\varphi}{v} vf \in F^d_{p,\infty}$ when $\varphi \in C_0^\infty(\mathbb{R})$ has support in $\mathbb{R}\setminus\{v = 0\}$; one can reduce to this with a partition of unity on φ and translation of v in each term.

Remark 3.8. Returning to Remark 3.5, note that $w_\theta(x) = v(x)f(x \cdot \theta)$ is almost nowhere differentiable for $0 < d \leq 1$, since v has isolated zeroes. If $d \leq 0$ then $w(\theta, d; \cdot) \notin L_1^{\text{loc}}(\mathbb{R}^n)$ for else one can derive the contradiction $2^{-kd}v(x) = 2^{kn}\chi(2^k\cdot) * [vf(\langle \cdot, \theta \rangle)] \to 0$ by modifying the corresponding part of Remark 3.6. As in Remark 3.7 it follows that

$$w(\theta, d; x) \in F^d_{p,\infty;\text{loc}}(\mathbb{R}^n) \quad \text{for all } p \in]0, \infty[\,,\ d \in \mathbb{R}. \tag{3.30}$$

If $\tilde{w}(\theta, d; x)$ is defined as $w(\theta, d; x)$ except with a further factor $1/j$ in each summand, similar arguments yield that $\tilde{w} \in H^s$ now for $s \leq d$ as well as the other properties in Proposition 3.3, with a nowhere differentiable series for $0 < d < 1$. Moreover, $\tilde{w}(\theta, d; \cdot)$ is in $F^d_{p,q;\text{loc}}(\mathbb{R}^n)$ as soon as $q > 1$ for every $p \in]0, \infty[$. Hence the counter-examples with unclosable graphs and violated microlocal properties are both related to Lizorkin–Triebel spaces $F^d_{p,q}$ with arbitrary $q > 1$; cf. (1.20) and Section 3.1.

4. Preliminary extensions

Throughout $\mathcal{F}_{y\to\eta}$ etc. will denote partial Fourier transformations, that are all homeomorphisms on $\mathcal{S}'(\mathbb{R}^{2n})$. In general the Fourier transformation in all variables is written $\mathcal{F}u$ or $\hat{u}$, except that for a symbol $a \in \mathcal{S}'(\mathbb{R}^{2n})$,

$$\hat{a}(\xi, \eta) := \mathcal{F}_{x\to\xi}a(\xi, \eta). \tag{4.1}$$

Transformation of coordinates via $(x, y) \mapsto (x, x - y)$, that has matrix $M = \left(\begin{smallmatrix} I & 0 \\ I & -I \end{smallmatrix}\right) = M^{-1}$, is indicated by $f \circ M$.

As a preparation some well-known formulae are recalled:

Proposition 4.1. *Let $a \in S^\infty_{1,1}(\mathbb{R}^n \times \mathbb{R}^n)$ and $u, v, f, g \in \mathcal{S}(\mathbb{R}^n)$. Then*

$$\langle a(x, D)u,\, v \rangle = \langle \tfrac{e^{\mathrm{i}\, x\cdot\eta}}{(2\pi)^n} a(x, \eta),\, v(x) \otimes \hat{u}(\eta) \rangle = \langle K,\, v \otimes u \rangle. \tag{4.2}$$

$$K(x, y) = \mathcal{F}^{-1}_{\eta\to y}(a)(x, x - y) = \mathcal{F}^{-1}_{\eta\to y}(a(x, \eta)) \circ \begin{pmatrix} I & 0 \\ I & -I \end{pmatrix}. \tag{4.3}$$

$$\langle \mathcal{F}a(x, D)f,\, \mathcal{F}g \rangle = (2\pi)^{-n} \iint \hat{a}(\xi - \eta, \eta)\hat{f}(\eta)\hat{g}(\xi)\, d\xi d\eta. \tag{4.4}$$

Here $\iint \ldots d\xi d\eta$ is valid as an integral for $a \in \mathcal{S}(\mathbb{R}^{2n})$, but should be read as the scalar product on $\mathcal{S}' \times \mathcal{S}$ for general $a \in S^\infty_{1,1}$.

Proof. By Fubini's theorem, (4.2) holds for $a \in \mathcal{S}(\mathbb{R}^{2n})$, when K is given as in (4.3). The bijection $a \leftrightarrow K$ extends to a homeomorphism on $\mathcal{S}'(\mathbb{R}^{2n})$. So by density of $\mathcal{S}$ in $S^d_{1,1}$, as subsets of $S^{d+1}_{1,1}$ hence of $\mathcal{S}'$, the identities in (4.2) hold for all $a \in S^d_{1,1}$. Formula (4.4) results from (4.2) for $u = f$, $v = \mathcal{F}^2 g$, since $\widehat{(\mathcal{F}^2 g) \otimes \hat{f}} = \mathcal{F}_{\xi\to x}(\hat{g} \otimes \hat{f})$. □

The partially Fourier transformed symbol $\hat{a}(\xi,\eta)$ is closely related to the distribution kernel $K(x,y)$ of $a(x,D)$ as well as to the kernel $\mathcal{K}(\xi,\eta)$ of the conjugation $\mathcal{F}a(x,D)\mathcal{F}^{-1}$ of $a(x,D)$ by the Fourier transformation on $\mathbb{R}^n$. Indeed, modulo simple isomorphisms, $\hat{a}$ gives both $\mathcal{K}$ and the frequencies in K:

Proposition 4.2. *When $a \in S^{\infty}_{1,1}(\mathbb{R}^n \times \mathbb{R}^n)$, and K, $\mathcal{K}$ and M are as above,*

$$\mathcal{F}K(\xi,\eta) = \hat{a}(\xi+\eta,-\eta) = \hat{a} \circ M^t(\xi,\eta) = (2\pi)^n \mathcal{K}(\xi,-\eta). \tag{4.5}$$

Proof. (4.3) implies that $K = \mathcal{F}^{-1}_{\eta\to y}(e^{-\mathrm{i}\,x\cdot\eta}a(x,-\eta))$, since $\mathcal{F}^{-1}$ commutes with reflections in η and y. Then (4.5) follows by application of $\mathcal{F}$ and (4.4). □

The right-hand side of (4.2) is inconvenient for the definition of type 1, 1-operators, as in general both entries of $\langle\, K,\, v \otimes u \,\rangle$ have singularities (in some cases this can be handled, cf. Section 7). However, it is a well-known fact that also in case $\rho = 1 = \delta$ the kernels only have singularities along the diagonal.

Lemma 4.3. *For every $a \in S^d_{1,1}(\mathbb{R}^n \times \mathbb{R}^n)$ the kernel $K(x,y)$ is C^∞ for $x \neq y$.*

Proof. For N so large that $d + |\beta| + |\alpha| - 2N < -n$,

$$|z|^{2N} D^\beta_x D^\alpha_z \mathcal{F}_{\eta\to z}(a(x,\eta)) = \mathcal{F}_{\eta\to z}(\Delta^N_\eta(\eta^\alpha D^\beta_x a(x,\eta))) \tag{4.6}$$

is a continuous function, so any derivative of K is so for $x \neq y$. □

Instead the middle of (4.2) gives a convenient way to prove that every type 1, 1-operator extends to $\mathcal{F}^{-1}\mathcal{E}'(\mathbb{R}^n)$, i.e., to the space of tempered distributions with compact spectrum. This result was first observed in [15], but the following argument should be interesting for its simplicity. When $v \in C^\infty_0(\mathbb{R}^n)$ and $u \in \mathcal{F}^{-1}C^\infty_0(\mathbb{R}^n)$ then (4.2) gives

$$\langle\, a(x,D)u,\, v \,\rangle = \langle\, v(x) \otimes \hat{u}(\eta),\, \tfrac{e^{\mathrm{i}\,x\cdot\eta}}{(2\pi)^n} a(x,\eta) \,\rangle_{\mathcal{E}'\times C^\infty}. \tag{4.7}$$

This suggests to introduce $A\colon \mathcal{F}^{-1}\mathcal{E}'(\mathbb{R}^n) \to C^\infty(\mathbb{R}^n)$ given by

$$Au(x) = \langle\, \hat{u},\, \tfrac{e^{\mathrm{i}\langle\, x,\,\cdot\,\rangle}}{(2\pi)^n} a(x,\cdot) \,\rangle. \tag{4.8}$$

That Au is in C^∞ is a standard fact used, e.g., in the construction of tensor products on $\mathcal{E}'(\mathbb{R}^{n'}) \times \mathcal{E}'(\mathbb{R}^{n''})$; cf. [9, Th. 5.1.1]. By definition of the tensor product of arbitrary $v, \hat{u} \in \mathcal{E}'(\mathbb{R}^n)$ in (4.7), they should act successively on the C^∞-function, so for v in $C^\infty_0(\mathbb{R}^n)$,

$$\langle\, v \otimes \hat{u},\, e^{\mathrm{i}\langle\,\cdot\,,\,\cdot\,\rangle} a \cdot (2\pi)^{-n} \,\rangle = \langle\, v,\, Au \,\rangle = \int_{\mathbb{R}^n} v(x) Au(x)\, dx. \tag{4.9}$$

This and (4.7) gives $Au = a(x,D)u$ for every $u \in \mathcal{F}^{-1}C^\infty_0 = \mathcal{S} \cap \mathcal{F}^{-1}\mathcal{E}'$; hence $a(x,D)$ and A are compatible. Therefore $a(x,D)$ extends to a map

$$a(x,D)\colon \mathcal{S}(\mathbb{R}^n) + \mathcal{F}^{-1}\mathcal{E}'(\mathbb{R}^n) \to C^\infty(\mathbb{R}^n) \tag{4.10}$$

by setting $a(x,D)u = a(x,D)v + Av'$ when $u = v + v'$ for $v \in \mathcal{S}$ and $\mathcal{F}v' \in \mathcal{E}'$ (if $0 = v + v'$, clearly $v = -v'$ is in $\mathcal{F}^{-1}C^\infty_0$, hence gives identical images, i.e., $a(x,D)v + Av' = 0$).

By the duality of $\mathcal{E}'$ and C^∞, the right-hand side of (4.8) should be calculated by multiplying $a(x,\eta)$ by a cut-off function $\chi(\eta)$ equalling 1 on a neighbourhood of $\operatorname{supp}\mathcal{F}u$. The resulting symbol $\chi(\eta)a(x,\eta)$ is clearly in

$$S^{-\infty}(\mathbb{R}^n\times\mathbb{R}^n) := \bigcap_{d,\rho,\delta} S^d_{\rho,\delta}(\mathbb{R}^n\times\mathbb{R}^n). \tag{4.11}$$

A systematic exploitation of localisations $\chi(\eta)a(x,\eta)$ is found in the next section.

4.1. Extension by spectral localisation

For type 1, 1-operators, this section gives a first extension, based on cut-off techniques and arguments from algebra. The latter are trivial, but important for several compatibility questions that are treated here.

Let $\mathcal{S}'_\Sigma(\mathbb{R}^n)$ denote the closed subspace of distributions with spectrum in a given open set $\Sigma\subset\mathbb{R}^n$, i.e.,

$$\mathcal{S}'_\Sigma(\mathbb{R}^n) = \{\, u\in\mathcal{S}'(\mathbb{R}^n) \mid \operatorname{supp}\hat{u}\subset\Sigma\,\}. \tag{4.12}$$

Clearly the intersection $\mathcal{S}_\Sigma(\mathbb{R}^n) := \mathcal{S}(\mathbb{R}^n)\cap\mathcal{S}'_\Sigma(\mathbb{R}^n)$ is dense in $\mathcal{S}'_\Sigma(\mathbb{R}^n)$.

As a basic assumption in this section, $a(x,\eta)\in S^\infty_{1,1}$ should have the properties of a more 'well-behaved' symbol class S as η runs through a given open set $\Sigma\subset\mathbb{R}^n$. It would then be natural, and necessary, to extend $a(x,D)$ to every $u\in\mathcal{S}'_\Sigma(\mathbb{R}^n)$ by letting it act as an operator with symbol in the class S.

To turn this idea into a definition, an arbitrary linear subspace $S\subset\mathcal{S}'(\mathbb{R}^{2n})\cap C^\infty(\mathbb{R}^{2n})$ will in the following be called a *standard* symbol space if, for every $b\in S$, the integral in (2.2) gives an operator $\mathrm{OP}(b)\colon\mathcal{S}\to\mathcal{S}$ which extends to a continuous linear map

$$\mathrm{OP}(b)\colon\mathcal{S}'(\mathbb{R}^n)\to\mathcal{S}'(\mathbb{R}^n). \tag{4.13}$$

(Such an extension is unique, so the notation need not relate $\mathrm{OP}(b)$ to the choice of S. To avoid confusion, the type 1, 1 operator under extension is usually denoted $a(x,D)$.) An example could be $S=S^d_{\rho,\delta}$ with $(\rho,\delta)\neq(1,1)$; whilst $\mathrm{OP}(b)$ could be the extension to $\mathcal{S}'$ of $b(x,D)$ given by the adjoint of $b^*(x,D)\colon\mathcal{S}(\mathbb{R}^n)\to\mathcal{S}(\mathbb{R}^n)$, that in its turn is defined from the adjoint symbol $b^*(x,\xi)=e^{\mathrm{i}\,D_x\cdot D_\xi}\bar{b}(x,\xi)$.

Using this, $a(x,D)$ can be extended if the symbol $a\in S^\infty_{1,1}$ is *locally* in a standard symbol space S in an open set $\Sigma\subset\mathbb{R}^n$. Specifically this means that for every closed set $F\subset\Sigma$ there exists a cut-off function $\chi\in C^\infty_{\mathrm{b}}(\mathbb{R}^n)$, not necessarily supported by Σ, such that

$$\chi\equiv 1 \text{ on a neighbourhood of } F,\qquad \chi(\eta)a(x,\eta)\in S. \tag{4.14}$$

Instead of $a(x,\eta)\chi(\eta)$, the slightly more correct $a(1\otimes\chi)$ is often preferred in the sequel.

Proposition 4.4. *For each symbol $a\in S^\infty_{1,1}(\mathbb{R}^n\times\mathbb{R}^n)$ that is locally in S in an open set $\Sigma\subset\mathbb{R}^n$, there is defined a map $\mathcal{S}'_\Sigma(\mathbb{R}^n)\to\mathcal{S}'(\mathbb{R}^n)$ by*

$$a(x,D)u = \mathrm{OP}(a(1\otimes\chi))(u), \tag{4.15}$$

which has the same value for all $\chi \in C_{\mathrm{b}}^{\infty}(\mathbb{R}^n)$ *satisfying* (4.14) *for* $F = \operatorname{supp}\mathcal{F}u$. *The map is compatible with* $a(x,D)\colon \mathcal{S}(\mathbb{R}^n) \to \mathcal{S}(\mathbb{R}^n)$.

Proof. Let $u \in \mathcal{S}'_{\Sigma}(\mathbb{R}^n)$. By (4.14) and (4.13), $\mathrm{OP}(a(1\otimes\chi))$ is defined on $\mathcal{S}' \ni u$. If χ_1 is another such function, $a(x,\eta)(\chi(\eta)-\chi_1(\eta))$ is in the vector space S and equals 0 for η in some open set $\Sigma_1 \supset \operatorname{supp} u$, so that by density of $\mathcal{S}_{\Sigma_1}$ in $\mathcal{S}'_{\Sigma_1}$,

$$0 = \mathrm{OP}(a(1\otimes(\chi-\chi_1)))u. \tag{4.16}$$

Therefore (4.15) is independent of the choice of χ, so the map $\mathrm{OP}(a(1\otimes\chi))u$ is defined; it equals $a(x,D)u$ for every $u \in \mathcal{S}\cap\mathcal{S}'_{\Sigma}$ by (2.2). □

The compatibility in Proposition 4.4 gives of course a map on the algebraic subspace $\mathcal{S}(\mathbb{R}^n) + \mathcal{S}'_{\Sigma}(\mathbb{R}^n) \subset \mathcal{S}'(\mathbb{R}^n)$; cf. (4.18). But more holds:

Theorem 4.5. *For every* $a \in S^{\infty}_{1,1}(\mathbb{R}^n\times\mathbb{R}^n)$ *that in an open set* $\Sigma \subset \mathbb{R}^n$ *is locally in a standard symbol space* S *(cf.* (4.13)*), the operator* $a(x,D)$ *extends to a linear map*

$$\mathcal{S}(\mathbb{R}^n) + \mathcal{S}'_{\Sigma}(\mathbb{R}^n) \xrightarrow{a(x,D)} \mathcal{S}'(\mathbb{R}^n) \quad \textit{given by} \tag{4.17}$$

$$a(x,D)u = a(x,D)v + \mathrm{OP}(a(1\otimes\chi))v' \tag{4.18}$$

whenever $u = v + v'$ *for* $v \in \mathcal{S}(\mathbb{R}^n)$, $v' \in \mathcal{S}'_{\Sigma}(\mathbb{R}^n)$*; hereby* $\chi \in C_{\mathrm{b}}^{\infty}(\mathbb{R}^n)$ *can be any function fulfilling* (4.14) *for* $F = \operatorname{supp}\mathcal{F}v'$. *The extension is uniquely determined by coinciding with* (4.13) *on* $\mathcal{S}'_{\Sigma}(\mathbb{R}^n)$.

Proof. For uniqueness, let $\widetilde{\mathrm{OP}}(a)$ be any extension agreeing with (4.13) on $\mathcal{S}'_{\Sigma}(\mathbb{R}^n)$. Then linearity gives, for any splitting $u = v + v'$ and χ as in the theorem, that

$$\widetilde{\mathrm{OP}}(a)u = \widetilde{\mathrm{OP}}(a)v + \widetilde{\mathrm{OP}}(a)v' = a(x,D)v + \mathrm{OP}(a)v' = a(x,D)v + \mathrm{OP}(a(1\otimes\chi))v'. \tag{4.19}$$

To show that (4.18) actually defines the desired map, suppose $u = v+v' = w+w'$ for some $v,w \in \mathcal{S}$ and $v',w' \in \mathcal{S}'_{\Sigma}$. Applying $a(x,D)$ to $v-w$ and $\mathrm{OP}(a(1\otimes\chi))$ to $w'-v'$, with χ taken so that $\chi \equiv 1$ on a neighbourhood of $F = \operatorname{supp}\mathcal{F}v' \cup \operatorname{supp}\mathcal{F}w' \supset \operatorname{supp}\mathcal{F}(w'-v')$, it follows from the compatibility in Proposition 4.4 and linearity that, for $\chi = \chi_1 = \chi_2$,

$$a(x,D)v + \mathrm{OP}(a(1\otimes\chi_1))v' = a(x,D)w + \mathrm{OP}(a(1\otimes\chi_2))w'. \tag{4.20}$$

By Proposition 4.4 one can then pass to arbitrary χ_1, χ_2 equalling 1 around $\operatorname{supp}\mathcal{F}v'$, respectively $\operatorname{supp}\mathcal{F}w'$, without changing the left and right-hand sides. This means that (4.18) gives a map, for $a(x,D)v + \mathrm{OP}(a(1\otimes\chi))v'$ is independent of the splitting $u = v + v'$ and of the corresponding choice of χ; thence linear dependence on u follows too. □

Theorem 4.5 gives a basic extension of type 1,1-operators, that could have been a definition (justified by the given arguments). When $a \in S^{\infty}_{1,1}$ happens to be in S too, then $\chi \equiv 1_{\mathbb{R}^n}$ and $v = 0$ yields $a(x,D)u = \mathrm{OP}(a)u$, so the definition (4.18) gives back the $\mathcal{S}'$-continuous operators with symbols in S.

Before these questions are pursued, the construction's dependence on S and Σ is investigated.

Proposition 4.6. *Let S and $\tilde{S}$ be standard symbol spaces, and let $a \in S^{\infty}_{1,1}$ be locally in S in some open set $\Sigma \subset \mathbb{R}^n$ and also locally in $\tilde{S}$ in an open set $\tilde{\Sigma}$. Then the induced maps*

$$a(x,D)\colon \mathcal{S} + \mathcal{S}'_{\Sigma} \to \mathcal{S}', \qquad a(x,D)\colon \mathcal{S} + \mathcal{S}'_{\tilde{\Sigma}} \to \mathcal{S}' \tag{4.21}$$

are compatible when either Σ has the property that χ in (4.14) for every F can be taken with $\operatorname{supp}\chi \subset \Sigma$, *or $\tilde{\Sigma}$ has the analogous property.*

Proof. One can reduce to the case $S = \tilde{S}$ by introducing the subspace $S + \tilde{S} \subset \mathcal{S}'(\mathbb{R}^{2n})$: for every $b \in S$, $\tilde{b} \in \tilde{S}$ the definition by the usual integral shows that

$$\operatorname{OP}(b + \tilde{b}) = \operatorname{OP}(b) + \operatorname{OP}(\tilde{b}) \quad \text{on} \quad \mathcal{S}(\mathbb{R}^n). \tag{4.22}$$

Here $\operatorname{OP}(b + \tilde{b})$ extends to a continuous, linear map $\mathcal{S}'(\mathbb{R}^n) \to \mathcal{S}'(\mathbb{R}^n)$ since the right-hand side does so. If $b + \tilde{b} = b_1 + \tilde{b}_1$ for $b_1 \in S$, $\tilde{b}_1 \in \tilde{S}$, both $\operatorname{OP}(b+\tilde{b})$, $\operatorname{OP}(b_1 + \tilde{b}_1)$ extend to $\mathcal{S}'$, where they coincide as they do so on $\mathcal{S}$. Hence every $b + \tilde{b}$ in $S + \tilde{S}$ gives an unambiguously defined operator on $\mathcal{S}'(\mathbb{R}^n)$, as required in (4.13); i.e., $S + \tilde{S}$ is a standard space.

Let $u = v + w = \tilde{v} + \tilde{w}$ for some $v, \tilde{v} \in \mathcal{S}$, $w \in \mathcal{S}'_{\Sigma}$ and $\tilde{w} \in \mathcal{S}'_{\tilde{\Sigma}}$. By the last assumption there exists, e.g., $\varphi \in C^{\infty}_{\mathrm{b}}(\mathbb{R}^n)$ such that $\operatorname{supp}\varphi \subset \tilde{\Sigma}$, $\varphi \equiv 1$ on a neighbourhood of $\tilde{F}$ and $a(1 \otimes \varphi) \in \tilde{S}$. In particular $1 - \varphi = 0$ around $\tilde{F}$ so

$$u = \varphi(D)(v + w) + (1 - \varphi(D))(\tilde{v} + \tilde{w}) = \tilde{v} + \varphi(D)(v - \tilde{v}) + \varphi(D)w. \tag{4.23}$$

While $v' = \tilde{v} + \varphi(D)(v - \tilde{v})$ is in $\mathcal{S}(\mathbb{R}^n)$, the term $\varphi(D)w$ is in $\mathcal{S}'_{\Sigma\cap\tilde{\Sigma}}(\mathbb{R}^n)$. By taking $\psi = 1$ in a neighbourhood of $\operatorname{supp}\mathcal{F}\varphi(D)w \subset \Sigma \cap \tilde{\Sigma}$, it is clear that one gets

$$a(x,D)u = a(x,D)v' + \operatorname{OP}(a(1 \otimes \psi))\varphi(D)w \tag{4.24}$$

by application of Theorem 4.5 both for $\mathcal{S}'_{\Sigma}$ and $\mathcal{S}'_{\tilde{\Sigma}}$. □

As a simple application for $\Sigma = \mathbb{R}^n$, every $u \in \mathcal{F}^{-1}\mathcal{E}'(\mathbb{R}^n)$ is in $\mathcal{S}'_{\Sigma} = \mathcal{S}'(\mathbb{R}^n)$; and a is locally in $S^{-\infty}$ since $b(x,\eta) = a(x,\eta)\chi(\eta)$ is in $S^{-\infty}$ for every $\chi \in C^{\infty}_0(\mathbb{R}^n)$, in particular when $\chi = 1$ around $\operatorname{supp}\mathcal{F}u$. Therefore Theorem 4.5 yields a unique extension of $a(x,D)$ to a linear map $\mathcal{S}(\mathbb{R}^n) + \mathcal{F}^{-1}\mathcal{E}'(\mathbb{R}^n)$. (Proposition 4.6 shows that one can replace the reference to $S^{-\infty}$ by, e.g., $S^{\infty}_{1,0}$ or let Σ depend on u without changing the image $a(x,D)u$.)

This approach is more elementary than (4.7) ff. In addition it gives that $a(x,D)$ maps $\mathcal{F}^{-1}\mathcal{E}'(\mathbb{R}^n)$ into $\mathcal{O}_M(\mathbb{R}^n)$. Recall that every $a(x,D)$ in $\operatorname{OP}(S^{-\infty}(\mathbb{R}^n \times \mathbb{R}^n))$ is a map $\mathcal{S}' \to \mathcal{O}_M$ (cf. [23, Cor. 3.8]), since if $u \in \mathcal{S}'$, then $(1 + |x|^2)^{-N}u \in H^{-N}$ for some $N > 0$ and every commutator $[D^{\alpha}a(x,D), (1 + |x|^2)^N]$ is by inspection in $\operatorname{OP}(S^{-\infty})$. These facts imply the next result.

Corollary 4.7. *Every operator $a(x,D)$ with symbol in $S^{\infty}_{1,1}$ extends uniquely to a map*

$$a(x,D)\colon \mathcal{S}(\mathbb{R}^n)+\mathcal{F}^{-1}(\mathcal{E}'(\mathbb{R}^n))\to \mathcal{O}_M(\mathbb{R}^n), \tag{4.25}$$

which is given by Theorem 4.5 *with $S=S^{-\infty}$.*

Notice that the corollary's statement is purely algebraic, since continuity properties are not involved in (4.25). Similarly one has another extension result.

Proposition 4.8. *If $a\in S^{\infty}_{1,1}$ is locally in the symbol class $S^{d}_{1,0}(\mathbb{R}^n\times\mathbb{R}^n)$ in an open cone $V\subset\mathbb{R}^n$ (i.e., $t\eta\in V$ for all $t>0$ and $\eta\in V$), then* (4.18) *yields a unique extension*

$$a(x,D)\colon \mathcal{S}+\mathcal{S}'_V\to\mathcal{S}'. \tag{4.26}$$

If some $a\in S^{\infty}_{1,1}$ satisfies the hypotheses of Proposition 4.8, it follows from Proposition 4.6 that the two extensions in (4.25)–(4.26) are compatible with one another.

Example. By Corollary 4.7, the domain of every $a(x,D)$ in $\operatorname{OP}(S^{\infty}_{1,1})$ contains polynomials $\sum_{|\alpha|\le k}c_\alpha x^\alpha$, as their spectra equal $\{0\}$, and, e.g., also the C^∞-functions

$$x^\alpha e^{\mathrm{i}\,x\cdot z}=(2\pi)^n\mathcal{F}^{-1}(\overline{D}^{\alpha}_{\eta}\delta_z(\eta));\qquad \frac{\sin x_1}{x_1}\dots\frac{\sin x_n}{x_n}=\pi^n\mathcal{F}^{-1}1_{[-1;1]^n}. \tag{4.27}$$

5. Definition by vanishing frequency modulation

The full extension of type 1, 1-operators is given here by means of a limiting procedure.

To define $a(x,D)u$ in general for $a\in S^{d}_{1,1}(\mathbb{R}^n\times\mathbb{R}^n)$, $d\in\mathbb{R}$, and suitable $u\in\mathcal{S}'(\mathbb{R}^n)$, it is convenient for an arbitrary $\psi\in C^\infty_0(\mathbb{R}^n)$ with $\psi\equiv 1$ in a neighbourhood of the origin to introduce the following notation, with $\psi_m(\xi):=\psi(2^{-m}\xi)$,

$$u^m=\mathcal{F}^{-1}(\psi_m\hat{u})=\psi_m(D)u, \tag{5.1}$$

$$a^m(x,\eta)=\mathcal{F}^{-1}_{\xi\to x}(\psi_m(\xi)\mathcal{F}_{x\to\xi}a(\xi,\eta))=\psi_m(D_x)a(x,\eta). \tag{5.2}$$

This is referred to as a frequency modulation of u and of $a(x,\eta)$ with respect to x; the full frequency modulation of a will be $a^m(x,\eta)\psi_m(\eta)$, i.e., $a^m(1\otimes\psi_m)$. Since a^m is in $S^{\infty}_{1,1}$ by Lemma 2.1, the compact support of ψ_m gives that

$$a^m(1\otimes\psi_m)\in S^{-\infty}. \tag{5.3}$$

Hence $\operatorname{OP}(a^m(1\otimes\psi_m))$ is defined on $\mathcal{S}'(\mathbb{R}^n)$, and since $\lim_{m\to\infty}a^m(1\otimes\psi_m)=a$ holds in $S^{d+1}_{1,1}$, it should be natural to make a tentative definition of $a(x,D)$ as

$$a(x,D)u=\lim_{m\to\infty}\operatorname{OP}(a^m(1\otimes\psi_m))u. \tag{5.4}$$

Some difficulties that might appear in this connection are dealt with in the formal

Definition 5.1. The pseudo-differential operator $a(x,D)u$ is defined as the limit in (5.4) for those $a \in S^d_{1,1}(\mathbb{R}^n \times \mathbb{R}^n)$ and $u \in \mathcal{S}'(\mathbb{R}^n)$ for which the limit

- exists in the topology of $\mathcal{D}'(\mathbb{R}^n)$ for every $\psi \in C_0^\infty(\mathbb{R}^n)$ equalling 1 in a neighbourhood of the origin, and
- is independent of such ψ.

To show that $a(x,D)$ extends the operator defined on $\mathcal{S}(\mathbb{R}^n)$ by (2.2), it suffices to combine Lemma 2.1 with (2.4). As shown below, Definition 5.1 also gives back both the usual operator $\operatorname{OP}(a)$ if a is, e.g., of type 1, 0 and the extensions in Section 4.

As an elementary observation, by using the definition for a fixed $a \in S^\infty_{1,1}$ and by the calculus of limits, the operator is defined for u in a subspace of $\mathcal{S}'(\mathbb{R}^n)$. This will be denoted by $D(a(x,D))$, or $D(A)$ if $A := a(x,D)$, in the following.

Clearly $D(A) \supset \mathcal{S}(\mathbb{R}^n)$, so A is a densely defined and linear operator from $\mathcal{S}'(\mathbb{R}^n)$ to $\mathcal{D}'(\mathbb{R}^n)$ (borrowing terminology from unbounded operators in Hilbert spaces). This description cannot be improved much in general, for by Lemma 3.2, $a(x,\eta)$ can be chosen so that A with $D(A) = \mathcal{S}(\mathbb{R}^n)$ is unclosable. But one has

Proposition 5.2. *For a, b in $S^\infty_{1,1}(\mathbb{R}^n \times \mathbb{R}^n)$ the following properties are equivalent:*

(i) *$a(x,\eta) = b(x,\eta)$ for all $(x,\eta) \in \mathbb{R}^{2n}$;*
(ii) *$a(x,D) = b(x,D)$ as operators from $\mathcal{S}'(\mathbb{R}^n)$ to $\mathcal{D}'(\mathbb{R}^n)$;*
(iii) *$a(x,D)u = b(x,D)u$ for every $u \in \mathcal{S}(\mathbb{R}^n)$;*
(iv) *the distribution kernels fulfil $K_a = K_b$.*

In particular the map $a \mapsto a(x,D)$ is a bijection $S^d_{1,1} \leftrightarrow \operatorname{OP}(S^d_{1,1})$; and the operator $a(x,D)$ is completely determined by its action on the Schwartz space.

The last property is perhaps not obvious from the outset, because, in general, there is neither density of $\mathcal{S} \subset D(a(x,D))$ nor continuity of $a(x,D)$ to appeal to. However it follows at once, as it is straightforward to see that (i) $\implies$ (ii) $\implies$ (iii) $\implies$ (iv) $\implies$ (i).

The following notion is very convenient for the analysis of $a(x,D)$:

Definition 5.3. A standard symbol space S on $\mathbb{R}^n \times \mathbb{R}^n$ is said to be *stable under vanishing frequency modulation* if in addition to (4.13),

(i) $b^m(1 \otimes \psi_m)(x,\eta) = \psi(2^{-m}D_x)b(x,\eta)\psi_m(\eta)$, is in S for every $b \in S$, $m \in \mathbb{N}$, and every $\psi \in C_0^\infty(\mathbb{R}^n)$ equalling 1 near the origin,
(ii) for every $u \in \mathcal{S}'(\mathbb{R}^n)$, and ψ as above,

$$\operatorname{OP}(b^m(1 \otimes \psi_m))u \to \operatorname{OP}(b)u \quad \text{in } \mathcal{D}'(\mathbb{R}^n) \text{ for } \quad m \to \infty. \tag{5.5}$$

For short S and the operator class $\operatorname{OP}(S)$ are then said to be *stable*.

Note that (i) requires the operator class $\mathrm{OP}(S)$ to be invariant under full frequency modulation; whereas (ii) requires $\mathrm{OP}(S)$ to be invariant under *vanishing* frequency modulation in the sense that the limit gives back the original operator $\mathrm{OP}(b)$.

Although $S^d_{1,1}$ is not a standard space, $\mathrm{OP}(S^\infty_{1,1})$ is also said to be stable, as (5.5) holds by definition for every u in $D(b(x,D))$, $b \in S^\infty_{1,1}$. Other stable spaces exist as well:

Proposition 5.4. *Every* $S^d_{\rho,\delta}(\mathbb{R}^n \times \mathbb{R}^n)$ *with* $\rho > \delta$ *for* $\rho, \delta \in [0,1]$ *is a stable symbol space. Moreover,* (5.5) *holds in the* $\mathcal{S}'$*-topology.*

Proof. By Lemma 2.1 condition (i) is satisfied, and $\lim b^m(1 \otimes \psi_m) = b$ in $S^{d'}_{\rho,\delta}$, $d' > d + \delta$. As $\mathrm{OP}(b)$ is the adjoint of $b^*(x,D) = \mathrm{OP}(\exp(\mathrm{i}\, D_x \cdot D_\eta)\bar{b})$, each $\varphi \in \mathcal{S}(\mathbb{R}^n)$ gives

$$(\,\mathrm{OP}(b^m(1 \otimes \psi_m))u - \mathrm{OP}(b)u \mid \varphi\,) = (\,u \mid \mathrm{OP}(e^{\mathrm{i}\, D_x \cdot D_\eta}(\overline{b^m}(1 \otimes \overline{\psi}_m) - \bar{b}))\varphi\,) \xrightarrow[m\to\infty]{} 0, \tag{5.6}$$

since passage to adjoint symbols $b \mapsto b^*$ is continuous $S^d_{\rho,\delta} \to S^d_{\rho,\delta}$ for $\rho > \delta$. □

Proposition 5.4 makes the definition of $a(x,D)$ by vanishing frequency modulation look natural. To analyse the consistency questions in general, it is recalled that $\mathrm{OP}(a)$ is defined on $\mathcal{S}(\mathbb{R}^n)$ by the integral (2.2) if a is in a standard space S or in $S = S^\infty_{1,1}$. And for a standard space S, $\mathrm{OP}(a)$ extends uniquely to a continuous linear map on $\mathcal{S}'(\mathbb{R}^n)$.

Let now $a \mapsto \widetilde{\mathrm{OP}}(a)$ be an arbitrary assignment such that $\widetilde{\mathrm{OP}}(a)$, for each $a \in S^\infty_{1,1}$, is a linear operator from $\mathcal{S}'(\mathbb{R}^n)$ to $\mathcal{D}'(\mathbb{R}^n)$. Then the maps OP and $\widetilde{\mathrm{OP}}$ are compatible on a standard symbol space S if $D(\widetilde{\mathrm{OP}}(a)) = \mathcal{S}'(\mathbb{R}^n)$ for every $a \in S \cap S^\infty_{1,1}$ and

$$\widetilde{\mathrm{OP}}(a)u = \mathrm{OP}(a)u \quad \text{for all} \quad u \in \mathcal{S}'(\mathbb{R}^n). \tag{5.7}$$

Moreover, $\widetilde{\mathrm{OP}}$ and OP are called *strongly* compatible on S if, whenever a is in $S^\infty_{1,1}$ and belongs to S locally in some open set $\Sigma \subset \mathbb{R}^n$, it will hold that $\mathcal{S}'_\Sigma(\mathbb{R}^n) \subset D(\widetilde{\mathrm{OP}}(a))$ and

$$\widetilde{\mathrm{OP}}(a)u = \mathrm{OP}(a(1 \otimes \chi))u \quad \text{for all} \quad u \in \mathcal{S}'_\Sigma(\mathbb{R}^n). \tag{5.8}$$

Hereby $\chi \in C^\infty_{\mathrm{b}}(\mathbb{R}^n)$ should fulfil (4.14) for $F = \operatorname{supp} \hat{u}$ and $a(1 \otimes \chi) \in S$. (The right-hand side of (5.8) makes sense because of χ, but it does not depend on χ since S is standard.) Taking $\Sigma = \mathbb{R}^n$ and $\chi \equiv 1$, strong compatibility clearly implies compatibility.

As an example Corollary 4.7 shows that, if the preliminary extension of Section 4.1 is written $\widetilde{\mathrm{OP}}$, then $\widetilde{\mathrm{OP}}(a)$ is strongly compatible with OP on $S^{-\infty}$. More generally Theorem 4.5 gives strong compatibility of $\widetilde{\mathrm{OP}}$ with OP on every standard symbol class S.

The following theorem shows that $a(x,D)$ given by Definition 5.1 contains every extension provided by Theorem 4.5 when S is stable.

Theorem 5.5. *Let $a \in S^{\infty}_{1,1}$ and $\Sigma \subset \mathbb{R}^n$ be an open set such that a locally in Σ belongs to a stable symbol class S (such as $S^d_{\rho,\delta}$ for $\rho > \delta$). Then every $u \in \mathcal{S}(\mathbb{R}^n) + \mathcal{S}'_{\Sigma}(\mathbb{R}^n)$ belongs to the domain $D(a(x,D))$ given by Definition* 5.1. *Moreover,*

$$a(x,D)u = \mathrm{OP}(a)v + \mathrm{OP}(a(1 \otimes \chi))v' \tag{5.9}$$

whenever u is split as $u = v + v'$ for $v \in \mathcal{S}(\mathbb{R}^n)$, $v' \in \mathcal{S}'_{\Sigma}(\mathbb{R}^n)$, and $\chi \in C^{\infty}_{\mathrm{b}}(\mathbb{R}^n)$ fulfils (4.14) *for $F = \operatorname{supp} \mathcal{F}v'$.*

Proof. Let $u \in \mathcal{S}'_{\Sigma}$. Since a is locally in S in Σ one can take χ as in the theorem, so that $a(1 \otimes \chi) \in S$. Using that S in particular is a standard space, approximation of $\hat{u}$ from C^{∞}_0 gives $\mathrm{OP}(a^m(x,\eta)(1-\chi(\eta))\psi_m(\eta))u = 0$. Now (5.5) applies, since S is stable; and multiplication by $\chi(\eta)$ and $\psi_m(D_x)$ commute in $\mathcal{S}'(\mathbb{R}^n \times \mathbb{R}^n)$, so

$$\begin{aligned} \mathrm{OP}(a(1 \otimes \chi))u &= \lim_m \mathrm{OP}(\psi_m(\eta)\chi(\eta)\psi_m(D_x)a(x,\eta))u \\ &= \lim_m \mathrm{OP}(a^m(1 \otimes \psi_m))u = a(x,D)u. \end{aligned} \tag{5.10}$$

This shows that $\mathcal{S}'_{\Sigma}(\mathbb{R}^n) \subset D(a(x,D))$. And for $u \in \mathcal{S}(\mathbb{R}^n)$ it is seen already from (2.4) that $a^m(x,D)u^m \to a(x,D)u$ in $\mathcal{S}(\mathbb{R}^n)$ for $m \to \infty$.

Since $a(x,D)$ is linear by Definition 5.1, it follows that every u in $\mathcal{S} + \mathcal{S}'_{\Sigma}$ belongs to $D(a(x,D))$ and that (5.9) holds. In particular the last statement that (5.9) is independent of v, v' and χ is implied by this (and by Theorem 4.5). □

Remark 5.6. It is noteworthy that Theorem 5.5 resolves a dilemma resulting from application of $a(x,D) \in \mathrm{OP}(S^{\infty}_{1,1})$ to $u \in \mathcal{F}^{-1}(\mathcal{E}'(\mathbb{R}^n))$: then $a(x,D)u$ can be calculated by using both Corollary 4.7 and Definition 5.1. But by taking $S = S^{-\infty}$, Theorem 5.5 entails that the two methods give the same result.

It follows from Theorem 5.5 that the assumptions on Σ and $\tilde{\Sigma}$ are unnecessary in Proposition 4.6 in case S is stable (this emphasizes the advantage of using vanishing frequency modulation). As a reformulation of Theorem 5.5 one has

Corollary 5.7. *The operator $a(x,D)$ given for $a \in S^{\infty}_{1,1}(\mathbb{R}^n \times \mathbb{R}^n)$ by Definition* 5.1 *is strongly compatible with* OP *on every stable symbol space S. In particular $a(x,D)u = \mathrm{OP}(a)u$ holds for every $u \in \mathcal{S}'(\mathbb{R}^n)$ when $a \in S^{\infty}_{\rho,\delta}(\mathbb{R}^n \times \mathbb{R}^n)$ for some $\delta < \rho$.*

As a special case $a(x,D)$ gives back $\mathrm{OP}(a)$ on $S^{-\infty}$. This may also be shown by verifying (5.7) directly, but one can only simplify (5.10) slightly by taking $\chi \equiv 1$ on $\mathbb{R}^n$. The various consistency results obtained in this section can be summed up thus:

Corollary 5.8. *Let $a(x,D)$ be given by Definition* 5.1 *for $a \in S^{\infty}_{1,1}$. Then $a(x,D)u$ equals the integral in* (2.2) *for $u \in \mathcal{S}(\mathbb{R}^n)$ or the extension in Corollary* 4.7 *for every $u \in \mathcal{F}^{-1}\mathcal{E}'(\mathbb{R}^n)$; and it coincides with the extension of* $\mathrm{OP}(a)$ *to $\mathcal{S}'(\mathbb{R}^n)$ if a is in $S^d_{\rho,\delta}(\mathbb{R}^n \times \mathbb{R}^n)$ for some $\rho > \delta$.*

To characterise the operators provided by Definition 5.1, it is convenient to ignore that the compatibility of $a(x,D)$ is strong (cf. Corollary 5.7). Indeed, the map $a \mapsto a(x,D)$ is simply the *largest* compatible extension stable under vanishing frequency modulation:

Theorem 5.9. *The operator $a(x,D)$ given by Definition 5.1 is one among the operator assignments $a \mapsto \widetilde{\mathrm{OP}}(a)$, $a \in S^{\infty}_{1,1}(\mathbb{R}^n \times \mathbb{R}^n)$ with the properties that*

(i) *$\widetilde{\mathrm{OP}}(\cdot)$ is compatible with* OP *on $S^{-\infty}$ (cf. (5.7));*

(ii) *each operator $\widetilde{\mathrm{OP}}(b)$ is stable under vanishing frequency modulation, i.e., $\widetilde{\mathrm{OP}}(b)u = \lim_{m\to\infty} \widetilde{\mathrm{OP}}(b^m(1\otimes\psi_m))u$ for every $u \in D(\widetilde{\mathrm{OP}}(b))$, $b \in S^{\infty}_{1,1}$.*

And moreover, whenever $\widetilde{\mathrm{OP}}$ is such a map, then $\widetilde{\mathrm{OP}}(a) \subset a(x,D)$ for every $a \in S^{\infty}_{1,1}$.

Note that (ii) makes sense as $\widetilde{\mathrm{OP}}(b^m(1\otimes\psi_m))$ by (i) is defined on all of $\mathcal{S}'$.

Proof. Let $a \mapsto \widetilde{\mathrm{OP}}(a)$ be any map fulfilling (i) and (ii); such maps exist since $a \mapsto a(x,D)$ was seen above to have these properties. If $u \in D(\widetilde{\mathrm{OP}}(a))$ it follows from (i) that

$$\mathrm{OP}(a^m(1\otimes\psi_m))u = \widetilde{\mathrm{OP}}(a^m(1\otimes\psi_m))u. \tag{5.11}$$

Here the right-hand side converges to $\widetilde{\mathrm{OP}}(a)u$ by (ii); since ψ is arbitrary this means $\widetilde{\mathrm{OP}}(a)u = a(x,D)u$. Hence $\widetilde{\mathrm{OP}}(a) \subset a(x,D)$. □

This section is concluded with a few remarks on the practical aspects of Definition 5.1. From the integral in (2.2), one would at once infer the following *alter egos* for the full frequency modulation of $a(x,D)$: if $\chi \in C_0^{\infty}(\mathbb{R}^n)$ fulfils $\chi = 1$ around $\operatorname{supp}\psi_m$, then

$$a^m(x,D)u^m = \mathrm{OP}(a^m(x,\eta)\chi(\eta))\mathcal{F}^{-1}(\psi_m\hat{u}) = \mathrm{OP}(a^m(x,\eta)\psi_m(\eta))u. \tag{5.12}$$

However, these identities hold also for more general cut-off functions χ.

Lemma 5.10. *For every $a \in S^{\infty}_{1,1}$, $u \in \mathcal{S}'$ and every $\psi \in C_0^{\infty}$ with $\psi = 1$ near the origin, the formula* (5.12) *holds for all m and all $\chi \in C_{\mathrm{b}}^{\infty}$ for which $\chi = 1$ in a neighbourhood of $F = \operatorname{supp}(\psi_m\hat{u})$ and $a(1\otimes\chi) \in S^{-\infty}$.*

Proof. The last part of (5.12) follows from (2.2) if u is a Schwartz function, hence for all $u \in \mathcal{S}'(\mathbb{R}^n)$ since both $a^m(1\otimes\psi_m)$ and $a^m(1\otimes\chi)$ belong to $S^{-\infty}$. Since $u^m \in \mathcal{F}^{-1}\mathcal{E}'$ and $a^m(x,\eta) \in S^{\infty}_{1,1}$, Corollary 5.8 shows that $a^m(x,D)(u^m)$ can be calculated by the extension in Section 4.1; then Theorem 4.5 gives the left-hand side of (5.12). □

In view of (5.12), one could alternatively have defined $a(x,D)u$ as a limit of $a^m(x,D)u^m$. This would be an advantage in as much as the expression $a^m(x,D)u^m$ is a natural point of departure for Littlewood–Paley analysis of $a(x,D)u$ (as explained later, cf. (9.8)); it would also make a and u enter in a more symmetric fashion. But as a drawback the resulting definition of $a(x,D)$ would then have two steps, the first one being an extension to $\mathcal{F}^{-1}\mathcal{E}'$ as in Section 4.1. In comparison the limit in (5.4) only refers to $S^{-\infty}$, cf. (5.3), which made it possible to state Definition 5.1 directly; cf. (1.7).

Formula (5.12) is so self-suggesting that it is convenient to write $a^m(x,D)u^m$ without further explanation, instead of the slightly tedious $\operatorname{OP}(a^m(1\otimes\psi_m))u$, that enters Definition 5.1. (This is permitted as the two expressions are equal for every choice of the auxiliary function ψ, cf. Lemma 5.10.)

Since Definition 5.1 is based on a limit of $a^m(x,D)u^m$, it is useful to relate the distribution kernel $K_m(x,y)$ of $u \mapsto a^m(x,D)u^m$ to the kernel $K(x,y)$ of $a(x,D)$.

The symbol of $a^m(x,D)u^m$ is $a^m(1\otimes\psi_m) \in S^{-\infty}$, cf. (5.12), so (4.3) and the definition of a^m give, for all $u, v \in \mathcal{S}(\mathbb{R}^n)$,

$$\begin{aligned}\langle\, a^m(x,D)u^m,\, v\,\rangle &= \langle\, \mathcal{F}^{-1}_{\eta\to y}(a^m(1\otimes\psi_m))(x,x-y),\, v(x)\otimes u(y)\,\rangle \\ &= \langle\, \mathcal{F}^{-1}(\psi_m(\xi)\psi_m(\eta)\mathcal{F}_{x\to\xi}a(\xi,\eta)))\circ M,\, v\otimes u\,\rangle.\end{aligned} \tag{5.13}$$

Because $\mathcal{F}\mathcal{F}^{-1}_{\eta\to y}\mathcal{F}^{-1}_{\xi\to x} = I$ on $\mathbb{R}^{2n}$, and $M = \left(\begin{smallmatrix} I & 0 \\ I & -I\end{smallmatrix}\right) = M^{-1}$, formula (5.13) shows that

$$K_m(x,x-y) = \mathcal{F}^{-1}((\psi_m\otimes\psi_m)\mathcal{F}(K\circ M))(x,y). \tag{5.14}$$

This can be restated as follows:

Proposition 5.11. *When $a \in S^\infty_{1,1}$ and $\psi \in C_0^\infty(\mathbb{R}^n)$ equals 1 in a neighbourhood of the origin, then the distribution kernel $K_m(x,y)$ of $u \mapsto a^m(x,D)u^m$, cf. (5.12), is the function in $C^\infty(\mathbb{R}^n\times\mathbb{R}^n)$ given by*

$$K_m(x,y) = \mathcal{F}^{-1}(\psi_m\otimes\psi_m) * (K\circ M)(x,x-y), \tag{5.15}$$

which is the conjugation by $\circ M$ of the convolution of $K(x,y)$ by the Schwartz function $4^{nm}\mathcal{F}^{-1}\psi(2^m x)\mathcal{F}^{-1}\psi(2^m y)$.

Naturally, this result will be useful for the discussion in the next section.

6. Preservation of C^∞-smoothness

It is well known that every classical pseudo-differential operator $A = a(x,D)$ is pseudo-local,

$$\operatorname{sing\,supp} Au \subset \operatorname{sing\,supp} u \quad \text{for every} \quad u \in D(A). \tag{6.1}$$

In the context of type 1, 1-operators, the requirement $u \in D(A)$ should be made explicitly as the domain $D(A)$ in many cases will be only a proper subspace of $\mathcal{S}'(\mathbb{R}^n)$.

It could be useful to call $\Omega := \mathbb{R}^n \setminus \operatorname{sing\,supp} u$ the *regular set* of u, for this set has the important property that regularisations of u converge (not just in $\mathcal{S}'(\mathbb{R}^n)$ but also) in the topology of $C^\infty(\Omega)$. This fact could well be folklore, but references seem unavailable, and since it is the crux of the below proof of pseudo-locality, details are given for the reader's convenience.

Lemma 6.1 (The Regular Convergence Lemma). *Let $u \in \mathcal{S}'(\mathbb{R}^n)$ and set $\psi_k(\xi) = \psi(\varepsilon_k \xi)$ for some sequence $\varepsilon_k \searrow 0$ and $\psi \in \mathcal{S}(\mathbb{R}^n)$. Then*

$$\psi_k(D)u \to \psi(0)\cdot u \quad \textit{for} \quad k \to \infty \tag{6.2}$$

in the Fréchet space $C^\infty(\mathbb{R}^n \setminus \operatorname{sing\,supp} u)$. If $\mathcal{F}^{-1}\psi \in C_0^\infty(\mathbb{R}^n)$ the conclusion holds for all $u \in \mathcal{D}'(\mathbb{R}^n)$, if $\psi_k(D)u$ is replaced by $(\mathcal{F}^{-1}\psi_k) * u$.

In the topology of $\mathcal{S}'(\mathbb{R}^n)$ the well-known property (6.2) is easy, for test against $\bar{\varphi} \in \mathcal{S}(\mathbb{R}^n)$ reduces (6.2) to the fact that $\psi(\varepsilon_k\cdot)\hat{\varphi} \to \psi(0)\hat{\varphi}$ in $\mathcal{S}(\mathbb{R}^n)$.

The main case is of course $\psi(0) = 1$. For $\psi(0) = 0$ one obtains the occasionally useful fact that $\psi_k(D)u \to 0$ in C^∞ over the regular set of u.

Proof. Let $K \Subset \mathbb{R}^n \setminus \operatorname{sing\,supp} u =: \Omega$ and take a partition of unity $1 = \varphi + \chi$ with $\varphi \in C^\infty(\mathbb{R}^n)$ such that $\varphi \equiv 1$ on a neighbourhood of K and $\operatorname{supp}\varphi \Subset \Omega$. This gives a splitting $\psi_k(D)u = \psi_k(D)(\varphi u) + \psi_k(D)(\chi u)$ where $\varphi u \in C_0^\infty(\mathbb{R}^n)$. Since $\int \mathcal{F}^{-1}\psi\, dx = \psi(0)$,

$$\begin{aligned} D^\alpha[\check{\psi}_k * (\varphi u) - \psi(0)\varphi u] &= \int \check{\psi}(y)[D^\alpha(\varphi u)(x - \varepsilon_k y) - D^\alpha(\varphi u)(x)]\, dy \\ &= -\sum_{j=1}^n \int\int_0^1 \check{\psi}(y)\partial_{x_j} D^\alpha(\varphi u)(x - \theta\varepsilon_k y)\varepsilon_k y_j\, d\theta dy. \end{aligned} \tag{6.3}$$

Using the seminorms in (2.5) in a crude way,

$$|\partial_j D^\alpha(\varphi u)(x - \theta\varepsilon_k y)\varepsilon_k y_j| \le \varepsilon_k |y|\, \|\, \varphi u \,|\mathcal{S}, |\alpha| + 1\| \searrow 0, \tag{6.4}$$

so consequently $D^\alpha(\check{\psi}_k * (\varphi u)) \to \psi(0)D^\alpha u$ uniformly on K.

For $\psi_k(D)(\chi u)$ it is used that continuity of $\mathcal{S}(\mathbb{R}^n) \xrightarrow{u} \mathbb{C}$ gives c, $N > 0$ such that

$$|\langle\, \chi u,\, D_x^\alpha \check{\psi}_k(x - \cdot)\,\rangle| \le c \sup_{y\in\mathbb{R}^n,\, |\beta|\le N} \frac{\langle y\rangle^N}{\varepsilon_k^{n+|\alpha|}} \big|D_y^\beta(\chi(y) D^\alpha \check{\psi}(\frac{x-y}{\varepsilon_k}))\big|. \tag{6.5}$$

Here $0 < d := \operatorname{dist}(K, \operatorname{supp}\chi) \le |x - y|$ for $x \in K$, $y \in \operatorname{supp}\chi$, so every negative power of ε_k fulfils $\varepsilon_k^{-l} \le (1 + \varepsilon_k^{-1}|x - y|)^l/d^l$. Moreover, $(1 + |y|)^N \le c_K(1 + |x - y|/\varepsilon_k)^N$ for $\varepsilon_k < 1$. So evaluation of an $\mathcal{S}$-seminorm at $\mathcal{F}^{-1}\psi$ yields $\sup_K |D^\alpha(\mathcal{F}^{-1}\psi_k * (\chi u))| \le C\varepsilon_k \searrow 0$.

When $\mathcal{F}^{-1}\psi \in C_0^\infty$, clearly $(\mathcal{F}^{-1}\psi_k) * (\chi u) = 0$ around K eventually. □

In the sequel the main case is the one in which ψ itself has compact support, so the proof above is needed.

6.1. The pseudo-local property

The following sharpening of Lemma 6.1 shows that, in certain situations, one even has convergence $f\psi_k(D)u \to fu$ in $\mathcal{S}$. To obtain this in a general set-up, let $x \in \mathbb{R}^n$ be split in two groups as $x = (x', x'')$ with $x' \in \mathbb{R}^{n'}$, $x'' \in \mathbb{R}^{n''}$.

Proposition 6.2. *Suppose* $u \in \mathcal{S}'(\mathbb{R}^n)$ *has* $\operatorname{sing\,supp} u \subset \{ x = (x', x'') \mid x'' = 0 \}$ *and that* $fu \in \mathcal{S}(\mathbb{R}^n)$ *for every* f *in the subclass* $\mathcal{C} \subset C_{\mathrm{b}}^{\infty}(\mathbb{R}^n)$ *consisting of the* f *for which* $|x'|$ *is bounded on* $\operatorname{supp} f$ *and* $\operatorname{supp} f \cap \{ x \mid x'' = 0 \} = \emptyset$*. Then*

$$f\mathcal{F}^{-1}(\psi_k \hat{u}) \xrightarrow[k\to\infty]{} \psi(0) fu \quad \text{in } \mathcal{S}(\mathbb{R}^n), \tag{6.6}$$

for every sequence $\psi_k = \psi(\varepsilon_k \cdot)$ *given as in Lemma* 6.1.

Proof. For $f \in \mathcal{C}$ it is straightforward to see that there is a δ such that

$$\inf\{ |x''| \mid x \in \operatorname{supp} f \} \geq \delta > 0. \tag{6.7}$$

One can then take $\varphi \in \mathcal{C}$ such that $\varphi \equiv 1$ where $|x''| \geq \delta/2$, hence on $K = \operatorname{supp} f$. Mimicking the proof of Lemma 6.1, compactness of K is not needed since φu is in $\mathcal{S}$ by assumption. Instead of (6.4) one should estimate

$$\langle x \rangle^N |\partial_j D^\alpha(\varphi u)(x - \theta\varepsilon_k y)\varepsilon_k y_j|, \tag{6.8}$$

but $(1+|x|)^N \leq (1+|x-\theta\varepsilon_k y|)^N(1+|y|)^N$ when $\varepsilon_k < 1$, so it follows mutatis mutandis that for an arbitrary seminorm,

$$\| f\check{\psi}_k * (\varphi u) - \psi(0) fu \,|\mathcal{S}, N\| \leq c\varepsilon_k \searrow 0. \tag{6.9}$$

And because $\chi = 1 - \varphi$ fulfils $d = \operatorname{dist}(K, \operatorname{supp}\chi) \geq \frac{\delta}{2} > 0$, one gets as in (6.5),

$$\| f\check{\psi}_k * (\chi u) \,|\mathcal{S}, N\| \leq c\varepsilon_k. \tag{6.10}$$

Indeed, $\langle x \rangle^N \leq (1+|y|)^N(1+|x-y|/\varepsilon_k)^N$ and now factors like $\langle y \rangle^N$ are harmless as

$$(1+|y|)^N \leq (1+|y'|)^N(1+|y''|)^N \leq (1+|x'|)^N(1+|y-x|)^N(1+|y''|)^N, \tag{6.11}$$

where $|y''| < \delta$ on $\operatorname{supp}\chi$ whilst $|x'|$ is bounded on $\operatorname{supp} f$. $\square$

For distribution kernels there is a similar result, but in this case it is well known that one need not assume rapid decay: let $f \in C_{\mathrm{b}}^{\infty}(\mathbb{R}^{2n})$ have its support disjoint from the diagonal $\Delta = \{ (x,x) \mid x \in \mathbb{R}^n \}$ and bounded in the x-direction, that is

$$\begin{gathered} \Delta \cap \operatorname{supp} f = \emptyset \\ \exists R > 0\colon (x,y) \in \operatorname{supp} f \implies |x| \leq R. \end{gathered} \tag{6.12}$$

Then $f(x,y)K(x,y)$ is in $\mathcal{S}(\mathbb{R}^{2n})$ whenever K is the kernel of a type 1, 1-operator. Indeed,

$$(1+|(x,y)|)^N \leq (1+|x|)^N(1+|y|)^N \leq (1+|x|)^{2N}(1+|y-x|)^N \tag{6.13}$$

and here $|x|$ is bounded on $\operatorname{supp} f$, so by setting $z = x - y$ in (4.6) one obtains that $\langle (x,y) \rangle^N D_x^\alpha D_y^\beta (fK)$ is bounded for all N, α, β.

Invoking Proposition 6.2 this gives

Proposition 6.3. *If $a \in S^{\infty}_{1,1}(\mathbb{R}^n \times \mathbb{R}^n)$ has kernel K and K_m is the approximating kernel given by (5.15), then it holds for every $f \in C^{\infty}_{\mathrm{b}}(\mathbb{R}^{2n})$ with the property (6.12) that*

$$fK_m \xrightarrow[m\to\infty]{} fK \quad in \quad \mathcal{S}(\mathbb{R}^{2n}). \tag{6.14}$$

Proof. The class $\mathcal{C}$ of Proposition 6.2 contains $f(x, x-y)$, and Proposition 5.11 gives

$$f(x,x-y)K_m(x,x-y) = f(x,x-y)\mathcal{F}^{-1}(\psi_m \otimes \psi_m) * (K \circ M). \tag{6.15}$$

The right-hand side tends to $(fK) \circ M$ in $\mathcal{S}(\mathbb{R}^{2n})$ according to Proposition 6.2, so it remains to use the continuity of $\circ M$ in $\mathcal{S}(\mathbb{R}^n)$. □

It can now be proved that operators of type 1, 1 are pseudo-local. The argument below is classical up to the appeal to (6.18). In case A is $\mathcal{S}'$-continuous, this formula follows at once from the density of $\mathcal{S}$ in $\mathcal{S}'$. However, in general A is not even closable, but instead the limiting procedure of Definition 5.1 applies via the approximation in Proposition 6.3.

Theorem 6.4. *For every $a \in S^{\infty}_{1,1}(\mathbb{R}^n \times \mathbb{R}^n)$ the operator $A = a(x,D)$ has the pseudo-local property; that is* $\operatorname{sing\,supp} Au \subset \operatorname{sing\,supp} u$ *for every $u \in D(A)$.*

Proof. Let $\psi, \chi \in C^{\infty}_0(\mathbb{R}^n)$ have supports disjoint from $\operatorname{sing\,supp} u$ such that $\chi \equiv 1$ around $\operatorname{supp}\psi$. Then $\chi u \in C^{\infty}_0(\mathbb{R}^n)$ so that also $(1-\chi)u$ is in the subspace $D(A)$ and

$$\psi Au = \psi A(\chi u) + \psi A(1-\chi)u. \tag{6.16}$$

Here $\psi A(\chi u) \in C^{\infty}_0(\mathbb{R}^n)$ since $A\colon \mathcal{S} \to \mathcal{S}$, while $u \mapsto \psi A(1-\chi)u$ is seen at once to have kernel

$$\tilde{K}(x,y) = \psi(x)K(x,y)(1-\chi(y)). \tag{6.17}$$

The function $f(x,y) = \psi(x)(1-\chi(y))$ fulfils (6.12), for Δ contains no contact point of $\{\, f \neq 0 \,\}$ because $\operatorname{dist}(\operatorname{supp}\psi, \operatorname{supp}(1-\chi)) > 0$. Therefore $\tilde{K} \in \mathcal{S}(\mathbb{R}^{2n})$ as seen after (6.12). This strongly suggests that, with $\chi_1 = 1-\chi$,

$$\langle\, \psi A\chi_1 u,\, \varphi \,\rangle = \langle\, \varphi \otimes u,\, \tilde{K} \,\rangle \quad \text{for all } \varphi \in C^{\infty}_0(\mathbb{R}^n). \tag{6.18}$$

And it suffices to prove this identity, for by definition of the tensor product it entails that $\psi A\chi_1 u = \langle\, u,\, \tilde{K}(x,\cdot) \,\rangle$ which is a C^{∞}-function of $x \in \mathbb{R}^n$.

Now if $A_m := \mathrm{OP}(a^m(x,\eta)\psi_m(\eta))$ and K_m is its kernel, one can take $u_l \in C^{\infty}_0(\mathbb{R}^n)$ such that $u_l \to u$ in $\mathcal{S}'$. Applying Definition 5.1 to A, the $\mathcal{S}'$-continuity of A_m gives

$$\langle\, \psi A\chi_1 u,\, \varphi \,\rangle = \lim_{m\to\infty} \langle\, A_m\chi_1 u,\, \psi\varphi \,\rangle = \lim_{m\to\infty}\lim_{l\to\infty} \langle\, K_m,\, (\psi\varphi)\otimes(\chi_1 u_l) \,\rangle. \tag{6.19}$$

Here $K_m \in C^{\infty}(\mathbb{R}^{2n})$ by Lemma 5.11, so for the right-hand side one finds, since $u \mapsto \varphi \otimes u$ is $\mathcal{S}'$-continuous and $fK_m \in \mathcal{S}(\mathbb{R}^{2n})$,

$$\int \psi(x)\varphi(x)\chi_1(y)u_l(y)K_m(x,y)\, d(x,y) \xrightarrow[l\to\infty]{} \langle\, \varphi \otimes u,\, (\psi \otimes \chi_1)K_m \,\rangle. \tag{6.20}$$

As $(\psi \otimes \chi_1)K_m = fK_m \to fK = \tilde{K}$ in $\mathcal{S}(\mathbb{R}^{2n})$ by Proposition 6.3, the proof is complete. □

Remark 6.5. Theorem 6.4 was anticipated by C. Parenti and L. Rodino [19], although they just appealed to the fact that $K(x,y)$ is C^∞ for $x \neq y$. This does not quite suffice as $\psi A \chi_1 u$ should be identified with a C^∞-function, e.g., $\langle\, u,\, \tilde{K}(x,\cdot)\,\rangle$, for $u \in D(A) \setminus \mathcal{S}(\mathbb{R}^n)$; which is non-trivial in the absence of continuity and the usual rules of calculus.

6.2. A digression on products

The opportunity is taken here to settle an open problem for the generalised pointwise product $\pi(u,v)$ mentioned in Remark 1.1.

First the commutation of pointwise multiplication and vanishing frequency modulation is discussed. Let $u \in \mathcal{S}'(\mathbb{R}^n)$ and $f \in \mathcal{O}_M(\mathbb{R}^n)$ be given and $\psi_m = \psi(2^{-m}\cdot)$ for some arbitrary $\psi \in \mathcal{S}(\mathbb{R}^n)$ with $\psi(0) = 1$. Approximating fu in two ways in $\mathcal{S}'$,

$$B_m u := \psi_m(D)(fu) - f\psi_m(D)u \to 0 \quad \text{for} \quad m \to \infty. \tag{6.21}$$

This commutation in the limit is not, however, a direct consequence of pseudo-differential calculus, for the commutator B_m has amplitude $b_m(x,y,\eta) = (f(y) - f(x))\psi(2^{-m}\eta)$, which is in the space of symbols with estimates $|D^\alpha_\eta D^\beta_{x,y} a(x,y,\eta)| \le C_{\alpha,\beta,K,N}\langle\eta\rangle^{-N}$ for all $N > 0$, $K \Subset \mathbb{R}^n \times \mathbb{R}^n$. As such $\operatorname{OP}(b_m(x,y,\eta))$ is only defined on $\mathcal{E}'(\mathbb{R}^n)$.

However, (6.21) is seen at once to hold in $C^\infty(\mathbb{R}^n \setminus \operatorname{sing\,supp} u)$, by using Lemma 6.1 on both terms. The next results confirms that $B_m u \to 0$ even in $C^\infty(\mathbb{R}^n)$, despite the singularities of u. The idea is to use Lemma 6.1 once more to get a reduction to $f \in C_0^\infty(\mathbb{R}^n)$, so that $B_m \to 0$ in the globally estimated class $\operatorname{OP}(S^{-\infty}(\mathbb{R}^n \times \mathbb{R}^n))$:

Proposition 6.6. *When $u \in \mathcal{S}'(\mathbb{R}^n)$, $f \in \mathcal{O}_M(\mathbb{R}^n)$, and $\psi \in \mathcal{S}(\mathbb{R}^n)$ with $\psi(0) = 1$, it holds true that $\lim_{m\to\infty}(\psi_m(D)(fu) - f\psi_m(D)u) = 0$ in the topology of $C^\infty(\mathbb{R}^n)$.*

Proof. When $\chi \in C_0^\infty(\mathbb{R}^n)$ equals 1 on a neighbourhood of a given compact set $K \subset \mathbb{R}^n$, then K is contained in the regular set of $(1-\chi)u$, so it follows as above from Lemma 6.1 that $\sup_{K,|\alpha|\le l} |D^\alpha(\psi_m(D)(f(1-\chi)u) - f\psi_m(D)(1-\chi)u)| \to 0$ for $m \to \infty$.

It now suffices to cover the case in which $K \subset \operatorname{supp} u \subset \operatorname{supp} f \Subset \mathbb{R}^n$. Then B_m has symbol

$$b_m(x,\eta) = (e^{\mathrm{i}\,D_x\cdot D_\eta} - 1)f(x)\psi_m(\eta) \in S^{-\infty}(\mathbb{R}^n \times \mathbb{R}^n). \tag{6.22}$$

However, $u \in H^t$ for some $t < 0$, and $B_m \in \mathbb{B}(H^t, H^s)$ for all $s > 0$, whence

$$\sum_{|\alpha|\le l} \|D^\alpha B_m u\|_\infty \le c\|B_m u\|_{H^s} \le c\|B_m\|\,\|u\|_{H^t} \quad \text{for } s > l + n/p. \tag{6.23}$$

It remains to show that the operator norm $\|B_m\| \to 0$. Using direct estimates as in, e.g., [10, Prop. 2.2], it is enough to show for all $N > 0$, α, β that

$$\lim_{m\to\infty} \sup_{\eta\in\mathbb{R}^n} \langle\eta\rangle^N |D_x^\beta D_\eta^\alpha b_m(x,\eta)| \to 0 \quad \text{for} \quad m \to \infty. \tag{6.24}$$

But D_η^α, D_x^β commute with $e^{\mathrm{i}\,D_x\cdot D_\eta}$, so it suffices to treat $\alpha = 0 = \beta$ for general f and ψ, i.e., to show that uniformly in $x \in \mathbb{R}^n$

$$\begin{aligned} |(e^{\mathrm{i}\,D_x\cdot D_\eta} - 1) f(x)\psi_m(\eta)| &\le c \sum_{|\alpha|+|\beta|\le 2n+2} |D_x^\beta D_\eta^\alpha D_x\cdot D_\eta f(x)\psi_m(\eta)| \\ &= \mathcal{O}(2^{-m(|\alpha|+1)}\langle\eta\rangle^{-N}). \end{aligned} \tag{6.25}$$

The estimate to the left is known, and follows directly from [10, Prop. B.2].

Altogether $\sup_{x\in K, |\alpha|\le l} |D^\alpha B_m u| \to 0$ for $m \to \infty$, as claimed. □

Besides being of interest in its own right, Proposition 6.6 gives at once a natural property of associativity for the product π in Remark 1.1.

Theorem 6.7. *The product $(u,v) \mapsto \pi(u,v)$ is partially associative, i.e., when $(u,v) \in \mathcal{S}'(\mathbb{R}^n) \times \mathcal{S}'(\mathbb{R}^n)$ is in the domain of $\pi(\cdot,\cdot)$ so is (fu,v) and (u,fv) for every $f \in \mathcal{O}_M(\mathbb{R}^n)$ and*

$$f\pi(u,v) = \pi(fu,v) = \pi(u,fv). \tag{6.26}$$

Proof. For every $\varphi \in C^\infty(K) = \{\, \psi \in C_0^\infty(\mathbb{R}^n) \mid \operatorname{supp}\psi \subset K \,\}$, $K \Subset \mathbb{R}^n$

$$\langle (fu)^m v^m, \varphi \rangle - \langle f\cdot u^m v^m, \varphi \rangle = \langle v^m, ((fu)^m - f\cdot u^m)\varphi \rangle \xrightarrow[m\to\infty]{} 0, \tag{6.27}$$

for by Banach–Steinhauss' theorem it suffices that $((fu)^m - f\cdot u^m)\varphi \to 0$ in $C^\infty(K)$, which holds since $(fu)^m - f\cdot u^m \to 0$ in $C^\infty(\mathbb{R}^n)$ according to Proposition 6.6. Hence $f\pi(u,v) = \pi(fu,v)$; the other identity is justified similarly. □

7. Extended action of distributions

To prepare for Section 8 it is exploited that the map $(u,f) \mapsto \langle u, f \rangle$ is defined also for certain u, f in $\mathcal{D}'(\mathbb{R}^n)$ that do not belong to dual spaces. This bilinear form is moreover shown to have a property of stability under regular convergence.

7.1. A review

First it is recalled that the product fu is defined for $f, u \in \mathcal{D}'(\mathbb{R}^n)$ if

$$\operatorname{sing\,supp} f \bigcap \operatorname{sing\,supp} u = \emptyset. \tag{7.1}$$

In fact, $\mathbb{R}^n$ is covered by $Y_1 = \mathbb{R}^n \setminus \operatorname{sing\,supp} u$ and $Y_2 = \mathbb{R}^n \setminus \operatorname{sing\,supp} f$; in Y_1 there is a product $(fu)_{Y_1} \in \mathcal{D}'(Y_1)$ given by $\langle (fu)_{Y_1}, \varphi \rangle = \langle f, u\varphi \rangle$ for $\varphi \in C_0^\infty(Y_1)$, and similarly $\langle u, f\varphi \rangle$, $\varphi \in C_0^\infty(Y_2)$ defines a product $(fu)_{Y_2} \in \mathcal{D}'(Y_2)$; and for $\varphi \in C_0^\infty(Y_1 \cap Y_2)$ both products are given by the C^∞-function $f(x)u(x)$ so they coincide on $Y_1 \cap Y_2$; hence fu is well defined in $\mathcal{D}'(\mathbb{R}^n)$ and given on $\varphi \in C_0^\infty(\mathbb{R}^n)$

by the following expression, where the splitting $\varphi = \varphi_1 + \varphi_2$ for $\varphi_j \in C_0^\infty(Y_j)$ is obtained from a partition of unity,

$$\langle fu, \varphi \rangle = \langle f, u\varphi_1 \rangle + \langle u, f\varphi_2 \rangle. \tag{7.2}$$

This follows from the *recollement de morceaux* theorem, cf. [21, Thm. I.IV] or [9, Thm. 2.2.4]; by the proof of this, (7.2) does not depend on how the partition is chosen.

Remark 7.1. Therefore, when F_1, F_2 are closed sets in $\mathbb{R}^n$ given with the properties $\operatorname{sing\,supp} u \subset F_1$, $\operatorname{sing\,supp} f \subset F_2$ and $F_1 \cap F_2 = \emptyset$ (so that $\mathbb{R}^n$ is covered by their complements) one can always take the splitting in (7.2) such that $\varphi_1 \in C_0^\infty(\mathbb{R}^n \setminus F_1)$, $\varphi_2 \in C_0^\infty(\mathbb{R}^n \setminus F_2)$.

Secondly $f \mapsto \langle u, f \rangle$ for $u \in \mathcal{D}'(\mathbb{R}^n)$ is a well-defined linear map on the subspace of $f \in \mathcal{D}'(\mathbb{R}^n)$ such that (7.1) holds together with

$$\operatorname{supp} u \cap \operatorname{supp} f \Subset \mathbb{R}^n. \tag{7.3}$$

In fact $\langle u, f \rangle := \langle fu, 1 \rangle$ is possible: fu is defined by (7.1) and is in $\mathcal{E}'$ by (7.3), so by [9, Th 2.2.5] the map $\psi \mapsto \langle fu, \psi \rangle$ extends from $C_0^\infty(\mathbb{R}^n)$ to all $\psi \in C^\infty(\mathbb{R}^n)$, uniquely among the extensions that vanish when $\operatorname{supp} \psi \cap \operatorname{supp} fu = \emptyset$; hence it is defined on the canonical choice $\psi \equiv 1$, and for all $\varphi \in C_0^\infty(\mathbb{R}^n)$ equal to 1 around $\operatorname{supp} fu$,

$$\langle u, f \rangle = \langle fu, 1 \rangle = \langle fu, \varphi \rangle. \tag{7.4}$$

These constructions have been quoted in a slightly modified form from [9, Sect. 3.1]. The definition implies that $(f, u) \mapsto fu$ is bilinear; it is clearly commutative and is partially associative in the sense that $\psi(fu) = (\psi f)u = f(\psi u)$ when $\psi \in C^\infty(\mathbb{R}^n)$ while f, u fulfill (7.1). This also yields

$$\operatorname{supp} fu \subset \operatorname{supp} f \cap \operatorname{supp} u. \tag{7.5}$$

When applying cut-off functions, partial associativity entails $(\chi f)(\varphi u) = fu$ when χ, φ equal 1 around $\operatorname{supp} f \cap \operatorname{supp} u$. Therefore test against 1 gives that $\langle \varphi u, \chi f \rangle = \langle u, f \rangle$.

7.2. Stability under regular convergence

The product fu is not continuous, for $f = 0$ is the limit in $\mathcal{D}'$ of $f^\nu = e^{-\nu|x|^2} \in C^\infty$ and for $u = \delta_0$ it is clear that $f^\nu u = \delta_0 \not\to 0 = fu$. As a remedy it is noted that fu is separately stable under regular convergence; cf. Lemma 6.1. This carries over to the extended bilinear form $\langle \cdot, \cdot \rangle$ under a compactness condition:

Theorem 7.2. *Let u, $f \in \mathcal{D}'(\mathbb{R}^n)$, $f^\nu \in C^\infty(\mathbb{R}^n)$ fulfil $\lim_\nu f^\nu = f$ in both $\mathcal{D}'(\mathbb{R}^n)$ and $C^\infty(\mathbb{R}^n \setminus F)$ for $F = \operatorname{sing\,supp} f$. When u, f have disjoint singular supports, cf. (7.1), then*

$$f^\nu u \to fu \quad \text{in } \mathcal{D}'(\mathbb{R}^n) \text{ for } \quad \nu \to \infty. \tag{7.6}$$

If moreover $\operatorname{supp} u \bigcap \operatorname{supp} f$ *is compact and* $\chi \in C_0^\infty(\mathbb{R}^n)$ *equals* 1 *around this set,*

$$\lim_{\nu\to\infty} \langle \chi u, f^\nu \rangle = \lim_{\nu\to\infty} \langle u, \chi f^\nu \rangle = \langle u, f \rangle. \tag{7.7}$$

Here one can take $\chi \equiv 1$ *on* $\mathbb{R}^n$ *if a compact set contains* $\operatorname{supp} u$ *or* $\bigcap_\nu \operatorname{supp}(f^\nu u)$. *The conclusions hold verbatim when* $F \subset \mathbb{R}^n$ *is closed and* $\operatorname{sing\,supp} f \subset F \subset (\mathbb{R}^n \setminus \operatorname{sing\,supp} u)$.

Proof. To show (7.6) for a general F, note that (7.2) applies to the product $f^\nu u$ of $f^\nu \in C^\infty$ and $u \in \mathcal{D}'$. Using Remark 7.1 and that $f^\nu \to f$ in $C^\infty(\mathbb{R}^n \setminus F)$, one has $f^\nu \varphi_2 \to f\varphi_2$ in $C_0^\infty(\mathbb{R}^n \setminus F)$; the other term on the right-hand side of (7.2) converges by the $\mathcal{D}'$-convergence of the f^ν. Therefore $\langle f^\nu u, \varphi \rangle \to \langle f, u\varphi_1 \rangle + \langle u, f\varphi_2 \rangle = \langle fu, \varphi \rangle$.

By the definition of $\langle u, f \rangle$ above, when χ is as in the theorem, then the just proved fact that $f^\nu u \to fu$ in $\mathcal{D}'$ leads to (7.7) since

$$\langle u, f \rangle = \langle fu, 1 \rangle = \langle fu, \chi \rangle = \lim_{\nu \to \infty} \langle f^\nu u, \chi \rangle. \tag{7.8}$$

When $\bigcup_\nu \operatorname{supp}(f^\nu u)$ is precompact and $\chi = 1$ on a neighbourhood, then $0 = \langle f^\nu u, 1 - \chi \rangle$ can be added to (7.8), which yields $\lim_\nu \langle u, f^\nu \rangle$ by the extended definition of $\langle \cdot, \cdot \rangle$. □

Remark 7.3. In general (7.7) cannot hold without the cut-off function χ. E.g., for $n \ge 2$ and $x = (x', x_n)$ one may take $f = 1_{\{x_n \le 0\}}$ and $u = 1_{\{x_n \ge 1/|x'|^2\}}$, so that $\langle u, f \rangle = 0$. Setting $f^\nu = 2^{n\nu}\varphi(2^\nu \cdot) * f$ for $\varphi \in C_0^\infty$ with $\varphi \ge 0$, $\int \varphi = 1$, and $\operatorname{supp}\varphi \subset \{ (y', y_n) \mid 1 \le y_n \le 2,\ |y'| \le 1 \}$, it holds for $x \in \Sigma_\nu = \{ 0 \le x_n \le 2^{-\nu} \}$ that $x_n - 2^{-\nu} y_n \le 0$ on $\operatorname{supp}\varphi$ so that

$$f^\nu(x) = \int \varphi(y) f(x - 2^{-\nu} y)\, dy = \int \varphi\, dy = 1. \tag{7.9}$$

Hence $\operatorname{supp} u \cap \operatorname{supp} f^\nu$ is unbounded, so $\langle u, f^\nu \rangle$ is undefined (hardly just a technical obstacle as $\langle u, f^\nu \rangle = \int u f^\nu\, dx = \infty$ would be the value).

7.3. Consequences for kernels

Although it is on the borderline of the present subject, it would not be natural to omit that Theorem 7.2 gives an easy way to extend the link between an operator and its kernel:

Theorem 7.4. *Let* $A\colon \mathcal{S}'(\mathbb{R}^n) \to \mathcal{S}'(\mathbb{R}^n)$ *be a continuous linear map with distribution kernel* $K(x, y) \in \mathcal{S}'(\mathbb{R}^n \times \mathbb{R}^n)$. *Suppose that* $u \in \mathcal{S}'(\mathbb{R}^n)$ *and* $v \in C_0^\infty(\mathbb{R}^n)$ *satisfy*

$$\operatorname{supp} K \bigcap \operatorname{supp} v \otimes u \Subset \mathbb{R}^n \times \mathbb{R}^n, \tag{7.10}$$

$$\operatorname{sing\,supp} K \bigcap \operatorname{sing\,supp} v \otimes u = \emptyset. \tag{7.11}$$

Then $\langle Au, v \rangle = \langle K, v \otimes u \rangle$, *with extended action of* $\langle \cdot, \cdot \rangle$. *When* A *is a continuous linear map* $\mathcal{D}'(\mathbb{R}^n) \to \mathcal{D}'(\mathbb{R}^n)$, *this is valid for* $u \in \mathcal{D}'(\mathbb{R}^n)$, $v \in C_0^\infty(\mathbb{R}^n)$ *fulfilling* (7.10)–(7.11).

Proof. By the conditions on u and v, the expression $\langle K, v \otimes u \rangle$ is well defined. By mollification there is regular convergence to u of a sequence $u_\nu \in C^\infty(\mathbb{R}^n)$; this gives

$$v \otimes u_\nu \xrightarrow[\nu\to\infty]{} v \otimes u \quad \text{in} \quad \mathcal{S}'(\mathbb{R}^n \times \mathbb{R}^n) \quad \text{and} \quad C^\infty(\Omega) \tag{7.12}$$

when $\Omega = (\mathbb{R}^n \times \mathbb{R}^n) \setminus (\operatorname{supp} v \times \operatorname{sing\,supp} u) = \mathbb{R}^{2n} \setminus \operatorname{sing\,supp}(v \otimes u)$. Applying Theorem 7.2 on $\mathbb{R}^{2n}$, the cut-off function may be taken as $\kappa(x)\chi(y)$ for some $\kappa, \chi \in C_0^\infty(\mathbb{R}^n)$ such that κ equals 1 on $\operatorname{supp} v$ and $\kappa \otimes \chi = 1$ on the compact set $\operatorname{supp} K \cap \operatorname{supp}(v \otimes u)$. This gives

$$\langle K, v \otimes u \rangle = \lim_{\nu\to\infty} \langle (\kappa \otimes \chi) K, v \otimes u_\nu \rangle = \lim_{\nu\to\infty} \langle A(\chi u_\nu), v \rangle = \langle A(\chi u), v \rangle. \tag{7.13}$$

For $\chi = \psi(2^{-m}\cdot)$ and $\psi = 1$ near 0, the conclusion follows from the continuity of A since $\psi(2^{-m}\cdot)u \to u$ in $\mathcal{S}'$. The $\mathcal{D}'$-case is similar. $\square$

Remark 7.5. The conditions (7.10)–(7.11) are far from optimal, for $(v \otimes u)K$ acts on 1 if its just an integrable distribution, that is if $(v \otimes u)K$ belongs to $\mathcal{D}'_{L^1} = \bigcup_m W_1^m$ on $\mathbb{R}^{2n}$. Similarly (7.11) is not necessary for $(v \otimes u) \cdot K$ to make sense; e.g., it suffices that $(x, \xi) \notin \mathrm{WF}(K) \cap (-\mathrm{WF}(v \otimes u))$ whenever $(x, \xi) \in \mathbb{R}^{2n}$. More generally the existence of the product $\pi(K, v \otimes u)$ would suffice; cf. Remark 1.1.

The above result applies in particular to the pseudo-differential operators A corresponding to a standard symbol space S, such as $S^d_{1,0}(\mathbb{R}^n \times \mathbb{R}^n)$. So does the next consequence.

Corollary 7.6. *When A is as in Theorem 7.4, it holds for every $u \in \mathcal{S}'(\mathbb{R}^n)$ that*

$$\operatorname{supp} Au \subset \overline{\operatorname{supp} K \circ \operatorname{supp} u}. \tag{7.14}$$

Hereby $\operatorname{supp} K \circ \operatorname{supp} u = \{\, x \in \mathbb{R}^n \mid \exists y \in \operatorname{supp} u \colon (x, y) \in \operatorname{supp} K \,\}$, *which is a closed set if* $\operatorname{supp} u \Subset \mathbb{R}^n$. *The result extends to $u \in \mathcal{D}'(\mathbb{R}^n)$ when A is $\mathcal{D}'$-continuous.*

Proof. Whenever $v \in C_0^\infty(\mathbb{R}^n)$ fulfils $\operatorname{supp} v \Subset \mathbb{R}^n \setminus \overline{\operatorname{supp} K \circ \operatorname{supp} u}$, then

$$\operatorname{supp} K \bigcap \operatorname{supp}(v \otimes u) = \emptyset. \tag{7.15}$$

For else some $(x, y) \in \operatorname{supp} K$ would fulfill $y \in \operatorname{supp} u$ and $x \in \operatorname{supp} v$, in contradiction with the support condition on v. By (7.15) the assumptions of Theorem 7.4 are satisfied, so $\langle Au, v \rangle = \langle K, v \otimes u \rangle = 0$. Hence $Au = 0$ holds outside the closure of $\operatorname{supp} K \circ \operatorname{supp} u$. $\square$

Remark 7.7. The argument of Corollary 7.6 is completely standard for $u \in C_0^\infty$, cf. [9, Thm 5.2.4] or [22, Prop 3.1]; a limiting argument then implies (7.14) for general u. However, the proof above is a direct generalisation of the C_0^∞-case, made possible by the extended action of $\langle \cdot, \cdot \rangle$ in Theorem 7.4. This method may be interesting in its own right; e.g., it extends to type 1, 1-operators also when these are not $\mathcal{S}'$-continuous, cf. Section 8.

8. Kernels and transport of support

Using the preceding section, the well-known support rule is here extended to operators of type 1, 1. As a novelty also a spectral support rule is deduced.

8.1. The support rule for type 1,1-operators

As analogues of Theorem 7.4 and Corollary 7.6 one has:

Theorem 8.1. *If $a \in S^{\infty}_{1,1}(\mathbb{R}^n \times \mathbb{R}^n)$ has kernel K, then $\langle\, a(x, D)u,\, v\,\rangle = \langle\, K,\, v \otimes u\,\rangle$ whenever $u \in D(a(x, D))$, $v \in C^{\infty}_0(\mathbb{R}^n)$ fulfill (7.10)–(7.11). And for all $u \in D(a(x, D))$ the support rule holds, i.e.,* $\operatorname{supp} Au \subset \overline{\operatorname{supp} K \circ \operatorname{supp} u}$.

Proof. $a(x, D)u = \lim_{m\to\infty} A_m u$ where $A_m = \operatorname{OP}(a^m(1 \otimes \psi_m)) \in \operatorname{OP}(S^{-\infty})$; its kernel K_m is given by Proposition 5.11. However, K_m need not fulfil (7.10) together with u, v, but by use of convolutions and cut-off functions one can find u_ν in $C^{\infty}_0(\mathbb{R}^n)$ such that $u_\nu \to u$ in $\mathcal{S}'(\mathbb{R}^n)$ and in $C^{\infty}(\mathbb{R}^n \setminus \operatorname{sing\,supp} u)$ for $\nu \to \infty$. Then Theorem 7.4 gives

$$\langle\, Au,\, v\,\rangle = \lim_{m\to\infty} \lim_{\nu\to\infty} \langle\, A_m u_\nu,\, v\,\rangle = \lim_{m\to\infty} \lim_{\nu\to\infty} \langle\, K_m,\, v \otimes u_\nu\,\rangle. \tag{8.1}$$

To control the supports, one can take a function f fulfilling (6.12) by setting $f(x, y) = g(x)h(x-y)$ for some $g \in C^{\infty}_0(\mathbb{R}^n)$ with $g = 1$ on $\operatorname{supp} v$ and $h \in C^{\infty}(\mathbb{R}^n)$ such that $h(y) = 0$ for $|y| < 1$ while $h(y) = 1$ for $|y| > 2$. Then $K_m = fK_m + (1 - f)K_m$, where the fK_m tend to fK in $\mathcal{S}$ according to Proposition 6.3. The supports of $(1 - f)K_m(v \otimes u_\nu)$, $m, \nu \in \mathbb{N}$, all lie in the precompact set $B(0, R) \times B(0, R+2)$ when $B(0, R) \supset \operatorname{supp} v$, so since u, v are assumed to fulfil (7.10)–(7.11), Theorem 7.2 gives

$$\begin{aligned}\langle\, Au,\, v\,\rangle &= \lim_{m\to\infty} \lim_{\nu\to\infty} \langle\, fK_m,\, v \otimes u_\nu\,\rangle + \lim_{m\to\infty} \lim_{\nu\to\infty} \langle\, (1 - f)K_m,\, v \otimes u_\nu\,\rangle \\ &= \langle\, fK,\, v \otimes u\,\rangle + \langle\, (1 - f)K,\, v \otimes u\,\rangle = \langle\, K,\, v \otimes u\,\rangle.\end{aligned} \tag{8.2}$$

Now the support rule follows by repeating the proof of Corollary 7.6. □

8.2. The spectral support rule

Although it has not attracted much attention, it is a natural and useful task to determine the frequencies entering $x \mapsto a(x, D)u(x)$. But since

$$\mathcal{F}a(x, D)u = \mathcal{F}a(x, D)\mathcal{F}^{-1}(\hat{u}) \tag{8.3}$$

the task is rather to control how the support of $\hat{u}$ is changed by $\mathcal{F}a(x, D)\mathcal{F}^{-1}$, i.e., by the conjugation of $a(x, D)$ by the Fourier transformation.

Even for $A \in \operatorname{OP}(S^{\infty}_{1,0})$ this has seemingly not been carried out before. However, since $\mathcal{F}A\mathcal{F}^{-1}\colon \mathcal{S}'(\mathbb{R}^n) \to \mathcal{S}'(\mathbb{R}^n)$ is continuous for such A, it is straightforward to apply Theorem 7.4 and Corollary 7.6 to the distribution kernel

$$\mathcal{K}(\xi, \eta) = (2\pi)^{-n}\hat{a}(\xi - \eta, \eta) \tag{8.4}$$

of $\mathcal{F}A\mathcal{F}^{-1}$; cf. Proposition 4.2.

This yields at once the following general result:

Theorem 8.2. *If $a \in S^{\infty}_{1,0}(\mathbb{R}^n \times \mathbb{R}^n)$ and $\mathcal{K}$ is as above, then*

$$\operatorname{supp} \mathcal{F}a(x,D)u \subset \overline{\operatorname{supp} \mathcal{K} \circ \operatorname{supp} \mathcal{F}u} \quad \text{for every} \quad u \in \mathcal{S}'(\mathbb{R}^n). \tag{8.5}$$

Here the right-hand side is closed if $\operatorname{supp} \mathcal{F}u \Subset \mathbb{R}^n$.

Here (8.5) may also be written explicitly as in (1.10)–(1.11). It is easily generalised to standard symbol spaces S such as $S^{\infty}_{\rho,\delta}(\mathbb{R}^n \times \mathbb{R}^n)$ with $\delta < \rho$. For elementary symbols in the sense of [5] the spectral support rule (8.5) follows at once, but as it stands Theorem 8.2 seems to be a new result even for classical type 1, 0-operators. The reader is referred to [15, Sect. 1.2] for more remarks on Theorem 8.2, in particular that it makes it unnecessary to reduce to elementary symbols in the L_p-theory (which is implicitly sketched in Section 9 below).

To extend the above to type 1, 1-operators, the next result applies to the conjugated operator $\mathcal{F}a(x,D)\mathcal{F}^{-1}$ instead of Theorem 7.4.

Theorem 8.3. *Let $a \in S^{\infty}_{1,1}(\mathbb{R}^n \times \mathbb{R}^n)$ and denote by $\mathcal{K}(\xi,\eta) = (2\pi)^{-n}\hat{a}(\xi-\eta,\eta)$ the distribution kernel of $\mathcal{F}a(x,D)\mathcal{F}^{-1}$; and suppose $u \in D(a(x,D)) \subset \mathcal{S}'(\mathbb{R}^n)$ is such that, for some ψ as in Definition* 5.1,

$$a(x,D)u = \lim_{m\to\infty} a^m(x,D)u^m \quad \text{holds in} \quad \mathcal{S}'(\mathbb{R}^n), \tag{8.6}$$

and that $\hat{v} \in C_0^{\infty}(\mathbb{R}^n)$ satisfies

$$\operatorname{supp} \mathcal{K} \bigcap \operatorname{supp} \hat{v} \otimes \hat{u} \Subset \mathbb{R}^n \times \mathbb{R}^n, \qquad \operatorname{sing\,supp} \mathcal{K} \bigcap \operatorname{sing\,supp} \hat{v} \otimes \hat{u} = \emptyset. \tag{8.7}$$

Then it holds, with extended action of $\langle \cdot, \cdot \rangle$,

$$\langle \mathcal{F}a(x,D)\mathcal{F}^{-1}(\hat{u}), \hat{v} \rangle = \langle \mathcal{K}, \hat{v} \otimes \hat{u} \rangle. \tag{8.8}$$

Proof. For $u \in D(a(x,D))$ the left-hand side of (8.8) makes sense by (8.6); and the right-hand side does so by (8.7), cf. Section 7. The equality follows from (8.6):

Letting $\psi_m = \psi(2^{-m}\cdot)$ there is some ν such that $\psi_\nu = 1$ on a neighbourhood of $\operatorname{supp}\psi$, so $\psi_{m+\nu}\psi_m = \psi_m$ for all m. Then $1 \otimes \psi_m$ and $\psi_m(\xi-\eta)\psi_m(\eta)$ equal 1 on the intersection of the supports in (8.7) for all sufficiently large m, so

$$\langle \mathcal{K}, \hat{v} \otimes \hat{u} \rangle = \langle \psi_m(\xi-\eta)\psi_{m+\nu}(\eta)\mathcal{K}(\xi,\eta), \hat{v}(\xi)\psi_m(\eta)\hat{u}(\eta) \rangle. \tag{8.9}$$

Here $\mathcal{K}_m = \psi_m(\xi-\eta)\psi_{m+\nu}(\eta)\mathcal{K}(\xi,\eta)$ is the kernel of $\mathcal{F}A_m\mathcal{F}^{-1}$, when $A_m = \operatorname{OP}(a^m(1 \otimes \psi_{m+\nu}))$; cf. Proposition 4.2. Clearly A_m has symbol in $S^{-\infty}$.

Moreover, mollification of $\psi_m\hat{u} = \mathcal{F}u^m$ gives a sequence $(\mathcal{F}u^m)_k$ of functions in $C_0^{\infty}(\mathbb{R}^n)$, that all have their supports in a fixed compact set M. Invoking regular convergence, cf. Lemma 6.1, it follows that

$$\hat{v} \otimes (\mathcal{F}u^m)_k \xrightarrow[k\to\infty]{} \hat{v} \otimes \mathcal{F}u^m \quad \text{in } \mathcal{S}'(\mathbb{R}^{2n}) \text{ and } C^{\infty}(\Omega) \tag{8.10}$$

$$\mathbb{R}^{2n} \setminus \Omega = \operatorname{supp} \hat{v} \times \operatorname{sing\,supp} \mathcal{F}u^m = \operatorname{sing\,supp}(\hat{v} \otimes \mathcal{F}u^m). \tag{8.11}$$

Since all supports are contained in $\operatorname{supp}\hat{v} \times M$, Theorem 7.2 applied on $\mathbb{R}^{2n}$ and the $\mathcal{S}'$-continuity of A_m imply

$$\langle \mathcal{K}_m, \hat{v} \otimes (\mathcal{F}u^m) \rangle = \lim_{k\to\infty} \langle \mathcal{K}_m, \hat{v} \otimes (\mathcal{F}u^m)_k \rangle = \langle \mathcal{F}A_m\mathcal{F}^{-1}(\mathcal{F}u^m), \hat{v} \rangle. \quad (8.12)$$

According to Lemma 5.10 the factor $\psi_{m+\nu}$ can here be removed from the symbol of A_m, so it is implied by (8.9), (8.12) and the explicit assumption of $\mathcal{S}'$-convergence in (8.6) that

$$\begin{aligned} \langle \mathcal{K}, \hat{v} \otimes \hat{u} \rangle &= \langle \mathcal{F}A_m u^m, \hat{v} \rangle \\ &= \lim_{m\to\infty} \langle a^m(x,D)u^m, \mathcal{F}^2 v \rangle = \langle a(x,D)u, \mathcal{F}^2 v \rangle, \end{aligned} \quad (8.13)$$

since $\mathcal{F}^2 v \in \mathcal{S}$. Transposing $\mathcal{F}$, formula (8.8) results. □

The assumption of $\mathcal{S}'$-convergence in (8.6) cannot be omitted from the above proof, although in the last line $\langle a^m(x,D)u^m, \mathcal{F}^2 v \rangle$ is independent of m. E.g., $\cos x$ is in $\mathcal{S}'(\mathbb{R})$, but since $\mathcal{F}(\sum_{j=0}^m \frac{(-1)^j}{(2j)} x^{2j}) = 2\pi\delta_0 + c_2\delta_0'' + \dots + c_m\delta_0^{(m)}$ the power series converges to $\cos x$ in $\mathcal{D}'$ but not in $\mathcal{S}'$ as cosine isn't a polynomial. Moreover, if $\mathcal{F}v = 1$ around 0 for some $\mathcal{F}v \in C_0^\infty([-1,1])$, one clearly has $2\pi = \langle \sum_{j=0}^m \frac{(-1)^j}{(2j)} x^{2j}, \mathcal{F}^2 v \rangle$ for every m as the derivatives of $\mathcal{F}v$ vanish at the origin. And yet $\langle \cos, \mathcal{F}^2 v \rangle = 2\pi \langle \frac{1}{2}(\delta_1 + \delta_{-1}), \mathcal{F}v \rangle = 0 \neq 2\pi$.

The next result extends [15, Prop. 1.4] from the case of $u \in C^\infty(\mathbb{R}^n)$ with $\operatorname{supp}\hat{u} \Subset \mathbb{R}^n$ to almost arbitrary distributions $u \in D(a(x,D))$; but the proof is significantly simpler here. Instead of (8.5), the explicit form given in (1.10)–(1.11) is preferred for practical purposes.

Theorem 8.4 (The Spectral Support Rule). *Let $a \in S^\infty_{1,1}(\mathbb{R}^n \times \mathbb{R}^n)$ and suppose $u \in D(a(x,D))$ is such that, for some $\psi \in C_0^\infty(\mathbb{R}^n)$ equalling 1 around the origin, the convergence of Definition 5.1 holds in the topology of $\mathcal{S}'(\mathbb{R}^n)$, i.e.,*

$$a(x,D)u = \lim_{m\to\infty} a^m(x,D)u^m \quad \text{in} \quad \mathcal{S}'(\mathbb{R}^n). \quad (8.14)$$

Then (8.5) holds, that is with $\Xi = \operatorname{supp}\mathcal{K} \circ \operatorname{supp}\hat{u}$ one has

$$\operatorname{supp}\mathcal{F}(a(x,D)u) \subset \overline{\Xi}, \quad (8.15)$$

$$\Xi = \{\, \xi + \eta \mid (\xi,\eta) \in \operatorname{supp}\hat{a},\ \eta \in \operatorname{supp}\hat{u} \,\}. \quad (8.16)$$

When $u \in \mathcal{F}^{-1}\mathcal{E}'(\mathbb{R}^n)$ then (8.14) holds automatically and Ξ is closed for such u.

Proof. That $\Xi = \operatorname{supp}\mathcal{K} \circ \operatorname{supp}\hat{u}$ has the form in (8.16) follows by substituting $\zeta = \xi + \eta$. Using Theorem 8.3 instead of Theorem 7.4, the proof of Corollary 7.6 can now be repeated mutatis mutandis; which gives the inclusion in question.

The redundancy of (8.14) for $u \in \mathcal{F}^{-1}\mathcal{E}'$ follows since, by Lemma 5.10, one can for large m write $a^m(x,D)u^m = \operatorname{OP}(a^m(1\otimes\chi)\psi_m)u$ for a fixed cut-off function χ. Then Proposition 5.4 gives $\mathcal{S}'$-convergence, for multiplication by $1\otimes\chi$ commutes with $\psi_m(D_x)$ and $a(1\otimes\chi) \in S^{-\infty}$. That Ξ is closed then is straightforward to verify. □

Remark 8.5. The set Ξ in (8.16) need not be closed if $\operatorname{supp}\mathcal{F}u$ is non-compact, for $\operatorname{supp}\hat{a}$ may contain points arbitrarily close to the twisted diagonal. E.g., if $n=1$ and $\operatorname{supp}\hat{u}=[2,\infty[$ whilst $\operatorname{supp}\hat{a}$ consists of the (ξ,η) such that $\eta\ge-\xi-\frac{1}{\xi}$ for $\xi\le-1$, and $\eta\ge2$ for $\xi\ge-1$, then $(\xi_k,\eta_k)=(-k,k+1/k)$ fulfils $\Xi\ni\xi_k+\eta_k=\frac{1}{k}\searrow0$, although $0\notin\Xi$.

That $a(x,D)u$ should be in $\mathcal{S}'$ in (8.14) is natural in order that $\mathcal{F}a(x,D)u$ makes sense before its support is investigated. One could conjecture that the condition of convergence in $\mathcal{S}'$ is redundant, so that it would suffice to assume $a(x,D)u$ is an element of $\mathcal{S}'$. But it is not clear (whether and) how this can be proved.

9. Continuity in Sobolev spaces

As a last justification of Definition 5.1 its close connection to estimates in Sobolev spaces will be indicated.

9.1. Littlewood–Paley decompositions

For the purposes of this section, one may for $a\in S^d_{1,1}(\mathbb{R}^n\times\mathbb{R}^n)$ consider the limit

$$a_\psi(x,D)u=\lim_{m\to\infty}\operatorname{OP}(\psi(2^{-m}D_x)a(x,\eta)\psi(2^{-m}\eta))u. \tag{9.1}$$

By the definition, u is in $D(a(x,D))$ if $a_\psi(x,D)u$ exists for all ψ and is independent of ψ, as ψ runs through $C_0^\infty(\mathbb{R}^n)$ with $\psi=1$ around the origin.

From $a_\psi(x,D)$ there is a particularly easy passage to the paradifferential decomposition used by J.-M. Bony [1]. For this purpose, note that to each fixed ψ there exist $R>r>0$ satisfying

$$\psi(\xi)=1\quad\text{for}\quad|\xi|\le r;\qquad\psi(\xi)=0\quad\text{for}\quad|\xi|\ge R\ge1. \tag{9.2}$$

Moreover it is convenient to let h stand for an integer such that $R\le r2^{h-2}$.

To obtain a Littlewood–Paley decomposition from ψ, define $\varphi=\psi-\psi(2\cdot)$. Then it is clear that $\varphi(2^{-k}\cdot)$ is supported in a corona,

$$\operatorname{supp}\varphi(2^{-k}\cdot)\subset\{\,\xi\mid r2^{k-1}\le|\xi|\le R2^k\,\},\qquad\text{for }k\ge1. \tag{9.3}$$

The identity $1=\psi(\xi)+\sum_{k=1}^\infty\varphi(2^{-k}\xi)$ follows by letting $m\to\infty$ in the telescopic sum,

$$\psi(2^{-m}\xi)=\psi(\xi)+\varphi(\xi/2)+\cdots+\varphi(\xi/2^m). \tag{9.4}$$

Using this one can localise functions $u(x)$ and symbols $a(x,\eta)$ to frequencies $|\eta|\approx 2^j$ for $j\ge1$ by setting

$$u_j=\varphi(2^{-j}D)u,\qquad a_j(x,\eta)=\varphi(2^{-j}D_x)a(x,\eta)=\mathcal{F}^{-1}_{\xi\to x}(\varphi(2^{-j}\xi)\hat{a}(\xi,\eta)). \tag{9.5}$$

$$u_0=\psi(D)u,\qquad a_0(x,\eta)=\psi(D_x)a(x,\eta) \tag{9.6}$$

Similarly localisation to balls $|\eta|\le R2^j$ are written, now with upper indices, as

$$u^j=\psi(2^{-j}D)u,\qquad a^j(x,\eta)=\psi(2^{-j}D_x)a(x,\eta)=\mathcal{F}^{-1}_{\xi\to x}(\psi(2^{-j}\xi)\mathcal{F}_{x\to\xi}a(\xi,\eta)). \tag{9.7}$$

Moreover, $u^0 = u_0$ and $a^0 = a_0$. In addition both, e.g., $a_j = 0$ and $a^j = 0$ should be understood when $j < 0$. (In order not to have two different meanings of sub- and superscripts on functions, the dilations $\psi(2^{-j}\cdot)$ are simply written as such; and the corresponding Fourier multiplier as $\psi(2^{-j}D)$.) Note that $a^k(x,D) = \operatorname{OP}(\psi(2^{-k}D_x)a(x,\eta))$ etc.

However, returning to (9.1), the relation (9.4) applies twice, whence bilinearity gives

$$\begin{aligned} a^m(x,D)u^m &= \operatorname{OP}([a_0(x,\eta)+\dots+a_m(x,\eta)][\psi(\eta)+\dots+\varphi(2^{-m}\eta)])u \\ &= \sum_{j,k=0}^{m} a_j(x,D)u_k. \end{aligned} \tag{9.8}$$

Of course the sum may be split in three groups in which $j \le k-h$, $|j-k| < h$ and $k \le j-h$, respectively. In the limit $m \to \infty$ this gives the decomposition

$$a_\psi(x,D)u = a_\psi^{(1)}(x,D)u + a_\psi^{(2)}(x,D)u + a_\psi^{(3)}(x,D)u, \tag{9.9}$$

whenever a and u fit together such that the three series below converge in $\mathcal{D}'(\mathbb{R}^n)$:

$$a_\psi^{(1)}(x,D)u = \sum_{k=h}^{\infty}\sum_{j\le k-h} a_j(x,D)u_k = \sum_{k=h}^{\infty} a^{k-h}(x,D)u_k \tag{9.10}$$

$$\begin{aligned} a_\psi^{(2)}(x,D)u = \sum_{k=0}^{\infty}\big(&a_{k-h+1}(x,D)u_k+\dots+a_{k-1}(x,D)u_k+a_k(x,D)u_k \\ &+a_k(x,D)u_{k-1}+\dots+a_k(x,D)u_{k-h+1}\big) \end{aligned} \tag{9.11}$$

$$a_\psi^{(3)}(x,D)u = \sum_{j=h}^{\infty}\sum_{k\le j-h} a_j(x,D)u_k = \sum_{j=h}^{\infty} a_j(x,D)u^{j-h}. \tag{9.12}$$

Also (9.11) has a brief form, namely

$$a_\psi^{(2)}(x,D)u = \sum_{k=0}^{\infty}\big((a^k-a^{k-h})(x,D)u_k + a_k(x,D)(u^{k-1}-u^{k-h})\big). \tag{9.13}$$

One advantage of the decomposition is that the terms of the first and last series fulfil a dyadic corona condition; whereas the in second the spectra are in general only restricted to balls:

Proposition 9.1. *If $a \in S^d_{1,1}(\mathbb{R}^n\times\mathbb{R}^n)$ and $u \in \mathcal{S}'(\mathbb{R}^n)$, and r, R are chosen as in (9.2) for each auxiliary function ψ, then every $h \in \mathbb{N}$ such that $R \le r2^{h-2}$ gives*

$$\operatorname{supp}\mathcal{F}(a^{k-h}(x,D)u_k) \subset \Big\{\,\xi \mid \frac{r}{4}2^k \le |\xi| \le \frac{5R}{4}2^k\,\Big\} \tag{9.14}$$

$$\operatorname{supp}\mathcal{F}(a_k(x,D)u^{k-h}) \subset \Big\{\,\xi \mid \frac{r}{4}2^k \le |\xi| \le \frac{5R}{4}2^k\,\Big\}. \tag{9.15}$$

Moreover, for $a^{(2)}_\psi(x,D)$,

$$\operatorname{supp}\mathcal{F}\big(a_k(x,D)(u^{k-1}-u^{k-h})+(a^k-a^{k-h})(x,D)u_k\big)\subset\{\,\xi\mid|\xi|\le 2R2^k\,\}\tag{9.16}$$

If a *satisfies* (1.19) *this support is contained in*

$$\Big\{\,\xi\mid\frac{r}{2^{h+1}C}\le|\xi|\le 2R2^k\,\Big\}\tag{9.17}$$

for all $k\ge h+1+\log_2(C/r)$.

This proposition follows straightforwardly from the spectral support rule in Theorem 8.4, with a special case explained in [15], so further details should hardly be needed here. (The constants are different in [15], since h only fulfils $R<r2^{h-1}$, that suffices in general.)

In addition one can estimate each series, using the Hardy–Littlewood maximal operator $Mu_k(x)$. This gives, e.g., for $\nu=1$, when the Fefferman–Stein inequality is used in the last step,

$$\begin{aligned}\Big(\int_{\mathbb{R}^n}\Big(\sum_{k=h}^{\infty}|2^{sk}a^{k-h}(x,D)u_k(x)|^2\Big)^{\frac{p}{2}}\,dx\Big)^{\frac1p}&\le c(a)\Big\|\Big(\sum_{k=0}^{\infty}|2^{(s+d)k}Mu_k(\cdot)|^2\Big)^{\frac12}\Big\|_p\\&\le c'c(a)\|u\|_{H^{s+d}_p};\end{aligned}\tag{9.18}$$

here $c(a)$ is a continuous seminorm on $a\in S^d_{1,1}$. Similar estimates are obtained for $\nu=2$ and $\nu=3$. The reader is referred to [15] for brevity here. (Although the set-up was more general there with Besov spaces $B^s_{p,q}$ and Triebel–Lizorkin spaces $F^s_{p,q}$, it is easy to specialise to the present H^s_p-framework, mainly by setting $q=2$ in the treatment of the $F^s_{p,q}$. One difference in the framework of [15] is that certain functions $\tilde\Phi_j$ enter the expressions $a_{j,k}(x,D)u_k$ there, but the $\tilde\Phi_j$ amount to special choices of χ in the above formula (5.12), hence may be removed when convenient.)

Combining such estimates with Proposition 9.1 it follows in a well-known way that for $\nu=1$ and $\nu=3$,

$$\|a^{(\nu)}_\psi(x,D)u\|_{H^s_p}\le c\|u\|_{H^{s+d}_p}\quad\text{for}\quad s\in\mathbb{R},\ 1<p<\infty.\tag{9.19}$$

For $\nu=2$ this holds for $s>0$, because the coronas are replaced by balls.

Consequently $u\mapsto a_\psi(x,D)u=a^{(1)}_\psi(x,D)u+a^{(2)}_\psi(x,D)u+a^{(3)}_\psi(x,D)u$ is a bounded linear operator $H^{s+d}_p(\mathbb{R}^n)\to H^s_p(\mathbb{R}^n)$ for $s>0$ and

$$\|a_\psi(x,D)u\|_{H^s_p}\le C(a)\|u\|_{H^{s+d}_p},\tag{9.20}$$

where $C(a)$ is a continuous seminorm on $a\in S^d_{1,1}$. Moreover, if a fulfils (1.19), then the last part of Proposition 9.1 leads to continuity for all $s\in\mathbb{R}$.

Moreover, density of the Schwartz space $\mathcal{S}(\mathbb{R}^n)$ in $H^{s+d}_p(\mathbb{R}^n)$ yields that $a_\psi(x,D)$ is independent of ψ, for they all agree with $\operatorname{OP}(a)u$ whenever $u\in\mathcal{S}(\mathbb{R}^n)$; cf. (9.9), (9.8) and (2.4). So by Definition 5.1 it follows that $a(x,D)u$ is defined on every $u\in H^{s+d}_p$ with $s>0$.

More precisely one has

Theorem 9.2. *Let $a(x,\eta)$ be a symbol in $S^d_{1,1}(\mathbb{R}^n\times\mathbb{R}^n)$. Then for every $s>0$, $1<p<\infty$ the type $1,1$-operator $a(x,D)$ has $H^{s+d}_p(\mathbb{R}^n)$ in its domain and it is a continuous linear map*

$$a(x,D)\colon H^{s+d}_p(\mathbb{R}^n)\to H^s_p(\mathbb{R}^n). \tag{9.21}$$

This extends to all $s\in\mathbb{R}$ when a fulfils the twisted diagonal condition (1.19).

In [15] a similar proof was given for Besov and Lizorkin–Triebel spaces, i.e., for $B^s_{p,q}$ and $F^s_{p,q}$. But this contains the above Theorem 9.2 in view of the well-known identification $H^s_p(\mathbb{R}^n)=F^s_{p,2}(\mathbb{R}^n)$ for $1<p<\infty$, which through a reduction to $s=0$ results from the Littlewood–Paley inequality. The reader is referred to the more general continuity results in [15], which also cover the Hölder–Zygmund classes because of the identification $C^s=B^s_{\infty,\infty}$.

However, a little precaution is needed because $\mathcal{S}(\mathbb{R}^n)$ is not dense in $B^s_{\infty,q}$. Even so $a(x,D)$ is defined on and bounded from $B^s_{\infty,q}$ for $s>d$ (and $s=d$, $q=1$ cf. (1.20) ff and [15, (1.6)]), which may be seen from the Besov space estimates of [15] and the argument preceding Theorem 9.2 as follows. By lowering s one can arrange that $q<\infty$, in which case it is well known that $B^s_{\infty,q}\bigcap\mathcal{F}^{-1}\mathcal{E}'$ is dense; whence $a_\psi(x,D)u$ is independent of ψ for all $u\in B^s_{\infty,q}$ if it is so for all $u\in\mathcal{F}^{-1}\mathcal{E}'$. This last property is a consequence of the fact that $\mathcal{F}^{-1}\mathcal{E}'$ is in the domain of $a(x,D)$; cf. Theorem 5.5 and Remark 5.6.

Remark 9.3. It is evident that the counter-example in Proposition 3.3 relied on an extension of continuity of $a_{2\theta}(x,D)$ to a bounded operator $H^{s+d}\to H^s$ for arbitrary $s<d$. Moreover, this extension has not previously been identified with the definition of $a_{2\theta}(x,D)$ by vanishing frequency modulation. However, by the density of $\mathcal{S}$, it follows from the last part of Theorem 9.2 that these two extensions are identical on H^s for all $s\in\mathbb{R}$, whence the operators in Definition 5.1 lack the microlocal property in the treated cases.

9.2. Composite functions

Finally it is verified that the formal definition of type 1, 1-operators by vanishing frequency modulation also plays well together with Y. Meyer's formula for composite functions.

Consider the map $u\mapsto F\circ u$ given by $F(u(x))$ for a fixed $F\in C^\infty(\mathbb{R})$ and a real-valued $u\in H^{s_0}_{p_0}(\mathbb{R}^n)$ for $s_0>n/p_0$, $1<p_0<\infty$. Then u is uniformly continuous and bounded on $\mathbb{R}^n$ as well as in $L_p(\mathbb{R}^n)$ for $p_0\le p\le\infty$. Note that with the notation of the previous section, and in particular (9.4), one has in $L_p(\mathbb{R}^n)$

$$u_0+u_1+\dots+u_m=u^m=2^{mn}\mathcal{F}^{-1}\psi(2^m\cdot)*u\xrightarrow[m\to\infty]{}u. \tag{9.22}$$

Assuming that $F(0) = 0$ when $p < \infty$, then $v \mapsto F \circ v$ is Lipschitz continuous on the metric subspace $L_p(\mathbb{R}^n, B)$ for every ball $B \Subset \mathbb{R}$,

$$F(w(x)) - F(v(x)) = \int_0^1 F'(v(x) + t(w(x) - v(x)))\,dt \cdot (w(x) - v(x)) \quad (9.23)$$

$$\|F \circ w - F \circ v\|_p \le \sup_B |F'| \cdot \|w - v\|_p. \quad (9.24)$$

Since $\|u^m\|_\infty \le \|\mathcal{F}^{-1}\psi\|_1 \|u\|_\infty$, one can take B so large that it contains $u(\mathbb{R}^n)$ and $u^m(\mathbb{R}^n)$ for every m, so since $u^k - u^{k-1} = u_k$ it follows from the Lipschitz continuity that with limits in L_p, $p_0 \le p \le \infty$,

$$\begin{aligned} F(u(x)) &= \lim_{m\to\infty} F(u^m(x)) = F(0) + \lim_{m\to\infty} \sum_{k=0}^{m} (F(u^k(x)) - F(u^{k-1}(x))) \\ &= F(0) + \sum_{k=0}^{\infty} \int_0^1 F'(u^{k-1}(x) + tu_k(x))\,dt \cdot \varphi(2^{-k}D)u(x). \end{aligned} \quad (9.25)$$

Setting $m_k(x) = \int_0^1 F'(u^{k-1}(x) + tu_k(x))\,dt$ it is not difficult to see that $m_k \in C^\infty(\mathbb{R}^n)$ with bounded derivatives of any order because $D^\beta u_k = 2^{k(n+|\beta|)} \check{\varphi}(2^k\cdot) * u$; and since $2^k \approx |\eta|$ on $\operatorname{supp} \varphi(2^{-k}\cdot)$ that

$$a_u(x,\eta) := \sum_{k=0}^{\infty} m_k(x)\varphi(2^{-k}\eta) \in S^0_{1,1}(\mathbb{R}^n \times \mathbb{R}^n). \quad (9.26)$$

That the corresponding type 1,1-operator $a_u(x,D)$ gives important information on the function $F(u(x))$ was the idea of Y. Meyer [17, 18], and it is now confirmed that his results remain valid when the operators are based on Definition 5.1:

Theorem 9.4. *When $u \in H^{s_0}_{p_0}(\mathbb{R}^n)$ for $s_0 > n/p_0$, $1 < p_0 < \infty$ is real valued, and $F \in C^\infty(\mathbb{R})$ with $F(0) = 0$, then the type 1,1-operator $a_u(x,D)$ is a bounded linear operator*

$$a_u(x,D)\colon H^s_p(\mathbb{R}^n) \to H^s_p(\mathbb{R}^n) \quad \text{for every} \quad s > 0,\ 1 < p < \infty. \quad (9.27)$$

Taking $s = s_0$, $p = p_0$ one has $a_u(x,D)u(x) = F(u(x))$, and the map $u \mapsto F \circ u$ is continuous on $H^{s_0}_{p_0}(\mathbb{R}^n, \mathbb{R})$.

Proof. The continuity on H^s_p follows from Theorem 9.2 since $a_u \in S^0_{1,1}$. As the proof of this theorem shows, the operator norm $\|b(x,D)\|$ in $\mathbb{B}(H^s_p)$ is estimated by a seminorm $c(b)$ on $b \in S^0_{1,1} \subset S^1_{1,1}$, and (9.26) converges in $S^1_{1,1}$, so one has in $\mathbb{B}(H^s_p)$ that

$$\operatorname{OP}\Big(\sum_{k=0}^{m} m_k(x)\varphi(2^{-k}\eta)\Big) \xrightarrow[m\to\infty]{} a_u(x,D). \quad (9.28)$$

By (9.25) this implies that in the space L_{p_0}

$$\begin{aligned} a_u(x,D)u &= \lim_{m\to\infty} \operatorname{OP}\Big(\sum_{k=0}^{m} m_k(x)\varphi(2^{-k}\eta)\Big)u \\ &= \sum_{k=0}^{\infty} m_k(x)\varphi(2^{-k}D)u = F\circ u. \end{aligned} \tag{9.29}$$

Hence $u\mapsto F\circ u$ is a map $H^{s_0}_{p_0}\to H^{s_0}_{p_0}$, which is continuous since for $v\to u$

$$\begin{aligned} F(v(x)) - F(u(x)) = a_u(x,D)(v-u) &+ [a_v(x,D)-a_u(x,D)]u \\ &+ [a_v(x,D)-a_u(x,D)](v-u)\to 0. \end{aligned} \tag{9.30}$$

Indeed, by continuity of $a_u(x,D)$ the first term tends to 0, and by the Banach–Steinhauss theorem the two other terms do so if only $a_v(x,D)\to a_u(x,D)$ in $\mathbb{B}(H^{s_0}_{p_0})$, i.e., if $a_v\to a_u$ in $S^0_{1,1}$. However, the non-linear map $u\mapsto a_u$ is continuous from $H^{s_0}_{p_0}$ to $S^0_{1,1}$, for $(1+|\eta|)^{|\alpha|-|\beta|}|D^\alpha_\eta D^\beta_x(a_v(x,\eta)-a_u(x,\eta))|$ is at each η estimated uniformly by terms that may have $\|v-u\|_\infty$ as a factor or contains $\sup_{x\in\mathbb{R}^n}\int_0^1 |F^{(l)}(v^{k-1}(x)+tv_k(x)) - F^{(l)}(u^{k-1}(x)+tu_k(x))|\,dt$, which tends to 0 by the uniform continuity of $F^{(l)}$ on a sufficiently large ball. □

Among the merits of the theorem, note that for non-integer s it is non-trivial to prove that $F(u(x))$ is in H^s_p when u is so. When needed the reader may derive similar results for the $B^s_{p,q}$ and $F^s_{p,q}$ from the estimates in [15]. Moreover, continuity of $u\mapsto F\circ u$ is as shown a straightforward consequence of the factorisation $a_u(x,D)u$, but this was not mentioned in [17, 18, 12].

Remark 9.5. As a small extension of the above, it may be noted that when F' is bounded on $\mathbb{R}$, then the assumption on u can be relaxed to $u\in L_{p_0}$ for $1\le p_0\le\infty$, for $F(u(x))$ is defined, and the linearisation formula (9.25) still holds as $u\mapsto F\circ u$ is Lipschitz continuous on $L_{p_0}(\mathbb{R}^n,\mathbb{R})$ in this case (however, the symbol $a_u(x,\eta)$ has much weaker properties).

Acknowledgement

My thanks are due to the anonymous referee for requesting a more explicit comparison with the existing literature.

References

[1] J.-M. Bony, Calcul symbolique et propagations des singularités pour les équations aux dérivées partielles non linéaires, *Ann. scient. Éc. Norm. Sup.* **14** (1981), 209–246.

[2] G. Bourdaud, *Sur les opérateurs pseudo-différentiels à coefficients peu reguliers*, Thèse, Univ. de Paris-Sud, 1983.

[3] ———, Une algèbre maximale d'opérateurs pseudo-différentiels, *Comm. Partial Differential Equations* **13** (1988), no. 9, 1059–1083.

[4] C.H. Ching, Pseudo-differential operators with nonregular symbols, *J. Differential Equations* **11** (1972), 436–447.

[5] R.R. Coifman and Y. Meyer, *Au delà des opérateurs pseudo-différentiels*, Astérisque, vol. 57, Société Mathématique de France, Paris, 1978.

[6] M. Frazier and B. Jawerth, Decomposition of Besov spaces, *Indiana Univ. Math. J.* **34** (1985), 777–799.

[7] M. Frazier and B. Jawerth, A discrete transform and decomposition of distribution spaces, *J. Funct. Anal.* **93** (1990), 34–170.

[8] G. Garello, Microlocal properties for pseudodifferential operators of type 1, 1, *Comm. Partial Differential Equations* **19** (1994), 791–801.

[9] L. Hörmander, *The Analysis of Linear Partial Differential Operators*, Grundlehren der mathematischen Wissenschaften, Springer-Verlag, Berlin, 1983, 1985.

[10] ______, Pseudo-differential operators of type 1, 1, *Comm. Partial Differential Equations* **13** (1988), no. 9, 1085–1111.

[11] ______, Continuity of pseudo-differential operators of type 1, 1, *Comm. Partial Differential Equations* **14** (1989), no. 2, 231–243.

[12] ______, *Lectures on Nonlinear Differential Equations*, Mathématiques & applications, vol. 26, Springer-Verlag, Berlin, 1997.

[13] J. Johnsen, Pointwise multiplication of Besov and Triebel–Lizorkin spaces, *Math. Nachr.* **175** (1995), 85–133.

[14] ______, Domains of type 1, 1 operators: a case for Triebel–Lizorkin spaces, *C. R. Acad. Sci. Paris Sér. I Math.* **339** (2004), no. 2, 115–118.

[15] ______, Domains of pseudo-differential operators: a case for the Triebel–Lizorkin spaces, *J. Function Spaces Appl.* **3** (2005), 263–286.

[16] ______, *Simple proofs of nowhere-differentiability for Weierstrass' function and cases of slow growth*, Preprint R-2008-02, Aalborg University, 2008.

[17] Y. Meyer, Régularité des solutions des équations aux dérivées partielles non linéaires (d'après J.-M. Bony), Bourbaki Seminar, Vol. 1979/80, *Lecture Notes in Math.* **842**, Springer, Berlin, 1981, 293–302.

[18] ______, Remarques sur un théorème de J.-M. Bony, in *Proceedings of the Seminar on Harmonic Analysis (Pisa, 1980)*, 1981, 1–20.

[19] C. Parenti and L. Rodino, A pseudo differential operator which shifts the wave front set, *Proc. Amer. Math. Soc.* **72** (1978), 251–257.

[20] T. Runst, Pseudodifferential operators of the "exotic" class $L^0_{1,1}$ in spaces of Besov and Triebel-Lizorkin type, *Ann. Global Anal. Geom.* **3** (1985), no. 1, 13–28.

[21] L. Schwartz, *Théorie des Distributions*, Revised and Enlarged ed., Hermann, Paris, 1966.

[22] M.A. Shubin, *Pseudodifferential Operators and Spectral Theory*, Springer-Verlag, Berlin, 1987, Translated from the 1978 Russian original by Stig I. Andersson.

[23] X. Saint Raymond, *Elementary Introduction to the Theory of Pseudodifferential Operators*, Studies in Advanced Mathematics, CRC Press, Boca Raton, FL, 1991.

[24] E.M. Stein, *Harmonic Analysis: Real-Variable Methods, Orthogonality, and Oscillatory Integrals*, Princeton Mathematical Series, vol. 43, Princeton University Press, Princeton, NJ, 1993.

[25] R.H. Torres, Continuity properties of pseudodifferential operators of type $1,1$, *Comm. Partial Differential Equations* **15** (1990), 1313–1328.

[26] H. Triebel, *Theory of Function Spaces*, Monographs in mathematics, **78**, Birkhäuser Verlag, Basel, 1983.

Jon Johnsen
Department of Mathematical Sciences
Aalborg University
Fredrik Bajers Vej 7G
DK-9220 Aalborg Øst, Denmark
e-mail: jjohnsen@math.aau.dk

Operator Theory:
Advances and Applications, Vol. 189, 247–264

Regularity for Quasi-Elliptic Pseudo-Differential Operators with Symbols in Hölder Classes

Gianluca Garello and Alessandro Morando

Abstract. After proving a result of continuity for pseudo-differential operators whose symbols belong to Hölder spaces and satisfy a decay of quasi-homogeneous type, the authors construct a suitable symbolic calculus and a parametrix for quasi-elliptic operators; these tools are applied to the study of quasi-elliptic linear partial differential equations with Hölder coefficients.

Mathematics Subject Classification (2000). Primary 35S05; Secondary 35A17.

Keywords. Non regular pseudo-differential operators, Besov spaces, Sobolev spaces.

1. Introduction

It is well known that pseudo-differential operators of classical type, defined by

$$a(x,D)u = (2\pi)^{-n}\int e^{i\langle x,\xi\rangle}a(x,\xi)\hat{u}(\xi)\,d\xi, \quad u\in\mathcal{S}(\mathbb{R}^n), \tag{1.1}$$

where $\langle x,\xi\rangle=\sum_{j=1}^n x_j\xi_j$ and $\hat{u}=\mathcal{F}u$ is the Fourier transform of u, with symbols in the Hörmander classes $S^m_{\rho,\delta}$:

$$|\partial_\xi^\alpha\partial_x^\beta a(x,\xi)| < C_{\alpha,\beta}(1+|\xi|)^{m-\rho|\alpha|+\delta|\beta|}, \quad C_{\alpha,\beta}>0,\ x,\xi\in\mathbb{R}^n, \tag{1.2}$$

are not in general L^p bounded, for $p\neq 2$, when $m=0$, $0\le\delta<\rho<1$; see, e.g., Feffermann [1].

In the present paper we deal with symbols $a(x,\xi)$ which take values in Hölder spaces of quasi-homogeneous type in the x variable, and whose derivatives satisfy a decay of type (1.2), with respect to a quasi-homogeneous weight $\langle\xi\rangle_M$ precisely defined in the next (2.1). The pseudo-differential operators thus obtained satisfy in natural way the Marcinkiewicz-Lizorkin Lemma for Fourier multipliers, see [9, §IV.6], thanks to the quasi-homogeneous structure of their symbols; hence their L^p continuity follows, when $m=0$.

L^p continuity for pseudo-differential operators with symbols in Besov, Hölder and Sobolev weighted classes are considered by the authors themselves also for more general weight functions, but always in the case $\delta = 0$, see, e.g., [4], [5].

About this subject let us notice that also the smooth quasi-homogeneous symbol classes are defined by Lascar [6] with $\rho = 1$ and $\delta = 0$, precisely only a "logarithmic" increase of $\partial_x a(x, \xi)$ is considered there. We then need also a variant of Lascar calculus for pseudo-differential operators. Namely we obtain asymptotic expansion for adjoint and compositions of operators and we construct a parametrix in the *quasi-elliptic* case.

Applications to quasi elliptic partial differential equations with weighted Hölder coefficients may be then obtained.

The plan of the paper is the following: In §2 the quasi-homogeneous weights $\langle\xi\rangle_M$ are introduced and the related quasi-homogeneous Sobolev and Hölder classes, $H^{s,p}_M$ and $B^{s,M}_{\infty,\infty}$ defined, by means of a generalization of the dyadic partition of unity, see [12]. In the same section we introduce the quasi-homogeneous symbol classes and prove the continuity results. In §3 we improve the Lascar's pseudo-differential calculus in [6], for adjoint and composition.

We then define the quasi-elliptic symbols and provide results of regularity for quasi-elliptic linear partial differential operators with Hölder coefficients.

A number of applications of pseudo-differential operators with non regular symbols in the study of regularity to non linear partial differential equations are given in the literature of the last twenty years, see, e.g., [10], [11] and the references given there. We hope in the next future in similar applications of our techniques.

2. Symbol classes and continuity for pseudo-differential operators

In the following $M = (m_1, \dots, m_n)$ is a weight vector with positive integer components, such that $\min\limits_{1\le j\le n} m_j = 1$ and

$$|\xi|_M := \left(\sum_{j=1}^{n} \xi_j^{2m_j}\right)^{\frac{1}{2}}, \quad \xi \in \mathbb{R}^n \tag{2.1}$$

is called *quasi-homogeneous weight function on* $\mathbb{R}^n$.

We set $m^* := \max\limits_{1\le j\le n} m_j$, $\frac{1}{M} := \left(\frac{1}{m_1}, \dots, \frac{1}{m_n}\right)$ and $\langle\xi\rangle_M^2 := (1+|\xi|_M^2)$. Clearly the usual Euclidean norm $|\xi|$ corresponds to the quasi-homogeneous weight in the case $M = (1, \dots, 1)$ (in this case, we also set $\langle\xi\rangle^2 := 1 + |\xi|^2$).

By easy computations, see, e.g., [3] we obtain the following

Proposition 2.1. *For any weight vector M there exists a suitable positive constant C such that*

i) $\frac{1}{C}\langle\xi\rangle \le \langle\xi\rangle_M \le C\langle\xi\rangle^{m^*}, \quad \xi \in \mathbb{R}^n$,
ii) $|\xi + \eta|_M \le C(|\xi|_M + |\eta|_M), \quad \xi, \eta \in \mathbb{R}^n$;

iii) (*quasi-homogeneity*) *for any* $t > 0$, $|t^{1/M}\xi|_M = t|\xi|_M$, *where* $t^{1/M}\xi = (t^{1/m_1}\xi_1, \dots, t^{1/m_n}\xi_n)$;

iv) $\xi^\gamma \partial^{\alpha+\gamma}|\xi|_M \le C_{\alpha,\gamma}\langle\xi\rangle_M^{1-\langle\frac{1}{M},\alpha\rangle}$, *for any* $\alpha, \gamma \in \mathbb{Z}_+^n$ *and* $\xi \neq 0$.

The detailed proof, that we omit in the following, can be found in [5] and in the references given there.

Proposition 2.2 (Quasi-Homogeneous Dyadic Decomposition). *For some* $K > 1$ *let us consider the cut-off function* $\phi(t) \in C_0^\infty([0, +\infty])$ *such that* $0 \le \phi(t) \le 1, \phi(t) = 1$ *for* $0 \le t \le \frac{1}{2K}$, $\phi(t) = 0$, *when* $t > K$. *Set now* $\varphi_0(\xi) = \phi\left(\left|2^{-1/M}\xi\right|_M\right) - \phi(|\xi|_M)$ *and*

$$\varphi_{-1}(\xi) = \phi(|\xi|_M), \quad \varphi_h(\xi) = \varphi_0\left(2^{-h/M}\xi\right). \tag{2.2}$$

Then for any $\alpha, \gamma \in \mathbb{Z}_+^n$ *a positive constant* $C_{\alpha,\gamma,K}$ *exists such that:*

$$\begin{aligned} &\operatorname{supp}\varphi_{-1} \subset C_{-1}^{K,M} := \left\{\xi \in \mathbb{R}^n \,,\ |\xi|_M \le K\right\} \\ &\operatorname{supp}\varphi_h \subset C_h^{K,M} := \left\{\xi \in \mathbb{R}^n \,;\ \frac{1}{K}2^{h-1} \le |\xi|_M \le K2^{h+1}\right\},\ h \ge 0; \end{aligned} \tag{2.3}$$

$$\sum_{h=-1}^{\infty} \varphi_h(\xi) = 1, \textit{for all } \xi \in \mathbb{R}^n; \tag{2.4}$$

$$\sum_{h=-1}^{\infty} \varphi_h(D)u = u, \quad \textit{with convergence in } \mathcal{S}'(\mathbb{R}^n); \tag{2.5}$$

$$\left|\xi^\gamma \partial^{\alpha+\gamma}\varphi_h(\xi)\right| \le C_{\alpha,\gamma,K}2^{-\langle\frac{1}{M},\alpha\rangle h}, \quad \xi \in \mathbb{R}^n, h = -1, 0, \dots. \tag{2.6}$$

Moreover for any fixed $\xi \in \mathbb{R}^n$ *the sum in* (2.4) *reduces to a finite number of terms, independent of the choice of* ξ *itself.*

We say the sequence $\varphi := \{\varphi_h\}_{h=-1}^\infty$, defined in (2.2), *quasi-homogeneous partition of unity.* In the following for brevity we set $u_h = \varphi_h(D)u$ for $u \in \mathcal{S}'(\mathbb{R}^n)$ and $h = -1, 0, \dots$. For each $l \ge -1$ we also set $\phi_l := \sum_{h \le l} \varphi_h$. Moreover, we will use the short notation $\langle\xi\rangle_M \sim 2^h$ to mean that ξ belongs to the quasi-homogeneous annulus $C_h^{K,M} := \{\xi \in \mathbb{R}^n : \frac{1}{K}2^{h-1} \le \langle\xi\rangle_M \le K2^{h+1}\}$.

Following the arguments in [12, §10.1] we can introduce now the classes of quasi-homogeneous Sobolev and Hölder functions and state their properties in suitable way.

For any $s \in \mathbb{R}$ and $u \in \mathcal{S}'(\mathbb{R}^n)$ we say that:

- u belongs to the *quasi-homogeneous Hölder space* $B_{\infty,\infty}^{s,M}$ if

$$\|u\|_{B_{\infty,\infty}^{s,M}}^{\varphi} := \sup_{h=-1,\dots} 2^{sh}\|u_h\|_{L^\infty} < \infty \tag{2.7}$$

  is satisfied for some quasi-homogeneous partition of unity φ;
- u belongs to the *quasi-homogeneous Sobolev space* $H_M^{s,p}$, $1 < p < \infty$ if

$$\|u\|_{H_M^{s,p}} := \|\langle D\rangle_M^s u\|_{L^p} := \|\mathcal{F}^{-1}\langle\cdot\rangle_M^s \mathcal{F}u\|_{L^p} < \infty. \tag{2.8}$$

Different choices of the partition of unity φ in (2.7) give raise to equivalent norms, noted by $\|\cdot\|_{B^{s,M}_{\infty,\infty}}$. When $M=(1,\dots,1)$ and $s>0$, $B^{s,M}_{\infty,\infty}$ is the usual Hölder-Zygmund space.

Both the spaces have a Banach structure and $H^{s,p}_M$ admits $\mathcal{S}(\mathbb{R}^n)$ as dense subspace.

Proposition 2.3. *Let us consider $s\in\mathbb{R}$ and $p\in]1,+\infty[$; then, for every quasi-homogeneous dyadic partition of unity $\varphi:=\{\varphi_h\}_{h=-1}^{\infty}$, there exists a constant $C>1$ such that the inequalities*

$$\frac{1}{C}\left\|\left(\sum_{h=-1}^{\infty}4^{sh}|\varphi_h(D)u|^2\right)^{\frac{1}{2}}\right\|_{L^p}\le\|u\|_{H^{s,p}_M}\le C\left\|\left(\sum_{h=-1}^{\infty}4^{sh}|\varphi_h(D)u|^2\right)^{\frac{1}{2}}\right\|_{L^p}\tag{2.9}$$

are verified for any $u\in\mathcal{S}'(\mathbb{R}^n)$.

Proposition 2.3 is a straightforward generalization of [12, Theorem 2.5.6], where characterizations of classical Sobolev spaces are considered.
The main difference consists of using a vector-valued version of the Marcinkiewicz-Lizorkin Lemma for L^p continuity of Fourier Multipliers [9, §IV.6], instead of the Michlin-Hörmander Lemma used in [12, Theorem 2.5.6].

The following Proposition gives a dyadic characterization of quasi-homogeneous Sobolev and Hölder spaces

Proposition 2.4. *Let us consider a sequence of Schwartz distributions $\{u_h\}_{h=-1}^{\infty}\subset\mathcal{S}'(\mathbb{R}^n)$ and a constant $K>1$ such that $\operatorname{supp}\hat{u}_h\subset C_h^{K,M}$ for any $h\ge-1$. Set now $u:=\sum_{h=-1}^{\infty}u_h$.*

The following properties are satisfied:

$$\begin{aligned}&\sup_{h\ge-1}\left(2^{rh}\|u_h\|_{L^\infty}\right)<\infty\Rightarrow u\in B^{r,M}_{\infty,\infty},\quad r\in\mathbb{R},\\ \text{and}\qquad&\|u\|_{B^{r,M}_{\infty,\infty}}\le C_1\sup_{h\ge-1}\left(2^{rh}\|u_h\|_{L^\infty}\right).\end{aligned}\tag{2.10}$$

$$\begin{aligned}&\left\{4^{sh}u_h\right\}_{h=-1}^{\infty}\in L^p\left(\mathbb{R}^n;\ell^2\right)\Rightarrow u\in H^{s,p}_M,\quad s\in\mathbb{R}\\ \text{and}\qquad&\|u\|_{H^{s,p}_M}\le C_2\left\|\left(\textstyle\sum_{h=-1}^{\infty}4^{sh}|u_h|^2\right)^{1/2}\right\|_{L^p}.\end{aligned}\tag{2.11}$$

Here $p\in]1,\infty[$ and the constants C_1,C_2 are independent of the sequence $\{u_h\}_{h=-1}^{\infty}$.

Proof. The proof of (2.11) follows plainly from [8, Propositions 1.1, 1.2], where the result is stated for classical Sobolev spaces. To show that with the assumption in (2.10) $u\in B^{r,M}_{\infty,\infty}$ let us take a smooth partition of unity $\varphi=\{\varphi_k\}_{k=-1}^{+\infty}$; since $\operatorname{supp}\hat{u}_h\subset C_h^{K,M}$ we can find a positive integer N_0 such that for all $k\ge-1$:

$$\|\varphi_k(D)u\|_{L^\infty}\le\sum_{h=\max\{-1,k-N_0\}}^{k+N_0}\|\varphi_k(D)u_h\|_{L^\infty}.\tag{2.12}$$

Moreover we can estimate as follows:

$$\begin{aligned}\|\varphi_k(D)u_h\|_{L^\infty} &\le \|\mathcal{F}^{-1}\varphi_k\|_{L^1}\|u_h\|_{L^\infty}\\ &\le \|\mathcal{F}^{-1}\varphi_k\|_{L^1}2^{-rh}\sup_{q=-1,0,\dots}2^{rq}\|u_q\|_{L^\infty}.\end{aligned} \tag{2.13}$$

It can be proved that the sequence $\left\{\|\mathcal{F}^{-1}\varphi_k\|_{L^1}\right\}_{k=-1}^{+\infty}$ is upper bounded; hence, from inequalities (2.12), (2.13) we obtain

$$\begin{aligned}\|\varphi_k(D)u\|_{L^\infty} &\le C\sup_{q=-1,0,\dots}2^{rq}\|u_q\|_{L^\infty}\sum_{h=\max\{-1,k-N_0\}}^{k+N_0}2^{-rh}\\ &\le C_1 2^{-kr}\sup_{q=-1,0,\dots}2^{rq}\|u_q\|_{L^\infty}, \quad k=-1,0,\dots,\end{aligned}$$

where C, C_1 are positive constants independent of u and k. This proves that $u\in B^{r,M}_{\infty,\infty}$ and gives the estimate in (2.10). □

With straightforward computations we obtain the following

Proposition 2.5. *When $s>0, r>0$ and $p\in]1,+\infty[$ there exist positive constants C_1, C_2 such that the properties* (2.10), (2.11) *are true for all the sequences of Schwartz distributions $\{u_h\}_{h=-1}^{\infty}\subset\mathcal{S}'(\mathbb{R}^n)$ with $\operatorname{supp}\hat{u}_h\subseteq B_h^{K,M}:=\{\xi\in\mathbb{R}^n : |\xi|_M\le K2^{h+1}\}$, $h=-1,0,\dots$.*

Definition 2.6. Given $m\in\mathbb{R}$ and $\delta\in[0,1]$, $S^m_{M,\delta}$ will be the class of functions $a(x,\xi)\in C^\infty(\mathbb{R}^n\times\mathbb{R}^n)$ such that for all multi-indices $\alpha,\beta\in\mathbb{Z}^n_+$ the following holds true:

$$|D_x^\beta D_\xi^\alpha a(x,\xi)|\le C_{\alpha,\beta}\langle\xi\rangle_M^{m-\langle\alpha,\frac{1}{M}\rangle+\delta\langle\beta,\frac{1}{M}\rangle}, \quad \forall x,\,\xi\in\mathbb{R}^n, \tag{2.14}$$

with some positive constant $C_{\alpha,\beta}$.

Exploiting the estimates in Proposition 2.1 (i) one checks that:

$$S^m_{M,\delta}\subset S^{\max\{mm^*,m\}}_{1/m^*,\delta m^*}. \tag{2.15}$$

The above inclusion establishes a relation between quasi-homogeneous classes and the usual Hörmander's symbol classes $S^m_{\rho,\delta}$, with $\rho\le\frac{1}{m^*}<1$.

Definition 2.7. For $r>0$, $m\in\mathbb{R}$ and $\delta\in[0,1]$, $B^{r,M}_{\infty,\infty}S^m_{M,\delta}$ is defined to be the set of measurable functions $a(x,\xi)$ such that for every $\alpha\in\mathbb{Z}^n_+$, the following inequalities hold true

$$|D_\xi^\alpha a(x,\xi)|\le C_\alpha\langle\xi\rangle_M^{m-\langle\alpha,\frac{1}{M}\rangle}, \quad \forall x,\,\xi\in\mathbb{R}^n; \tag{2.16}$$

$$\|D_\xi^\alpha a(\cdot,\xi)\|_{B^{r,M}_{\infty,\infty}}\le C_\alpha\langle\xi\rangle_M^{m-\langle\alpha,\frac{1}{M}\rangle+\delta r}, \quad \forall\xi\in\mathbb{R}^n. \tag{2.17}$$

Later on, we will set $S^m_M:=S^m_{M,0}$, $S^m_{M,M}:=S^m_{M,1}$ (same notations for non regular classes $B^{r,M}_{\infty,\infty}S^m_{M,\delta}$).

Repeating the arguments of [4] (cf. also [10]), we are lead to the following:

Lemma 2.8. *Let $r > 0$ and $\delta \in [0,1]$ be fixed. There exists a sequence of positive numbers $\{c_k\}_{k\in\mathbb{Z}^n}$, satisfying $\sum_{k\in\mathbb{Z}^n} c_k < \infty$, such that for every symbol $a(x,\xi) \in B^{r,M}_{\infty,\infty}S^0_{M,\delta}$ the following decomposition*

$$a(x,\xi) = \sum_{k\in\mathbb{Z}^n} c_k a_k(x,\xi), \tag{2.18}$$

holds true, with absolute convergence in $L^\infty(\mathbb{R}^{2n})$. Moreover, for each $k \in \mathbb{Z}^n$ we have:

$$a_k(x,\xi) = \sum_{h=-1}^{+\infty} d^k_h(x)\psi^k_h(\xi), \tag{2.19}$$

where $\{d^k_h\}_{h=-1}^\infty$ and $\{\psi^k_h\}_{h=-1}^\infty$ are sequences in $B^{r,M}_{\infty,\infty}$ and $C^\infty_0(\mathbb{R}^n)$ respectively, satisfying the following assumptions:

i) *There exists a positive constant $L > 0$ such that:*

$$\begin{gathered}|d^k_h(x)| \le L, \quad \|d^k_h\|_{B^{r,M}_{\infty,\infty}} \le L2^{hr\delta}, \\ \forall x \in \mathbb{R}^n,\ h = -1,0,\dots,\ \forall k \in \mathbb{Z}^n;\end{gathered} \tag{2.20}$$

ii) *There exist positive constants C and $K > 1$ such that for all $k \in \mathbb{Z}^n$ and $h = -1,0,\dots$:*

$$\begin{aligned} supp\,\psi^k_h &\subseteq C^{K,M}_h; \\ |\xi^\gamma D^{\alpha+\gamma}\psi^k_h(\xi)| &\le C2^{-\langle\alpha,\frac{1}{M}\rangle h}, \quad \forall\xi \in \mathbb{R}^n. \end{aligned}$$

Starting from the previous lemma and following the arguments of Taylor [10, Thm. 2.1.A], we obtain the following

Proposition 2.9. *If $r > 0$, $m \in \mathbb{R}$, $\delta \in [0,1]$ and $a(x,\xi) \in B^{r,M}_{\infty,\infty}S^m_{M,\delta}$, then for all $s \in](\delta-1)r, r[$ and $p \in]1,+\infty[$ the following*

$$a(x,D) : H^{s+m,p}_M \to H^{s,p}_M \tag{2.21}$$

$$a(x,D) : B^{s+m,M}_{\infty,\infty} \to B^{s,M}_{\infty,\infty} \tag{2.22}$$

are linear continuous operators.

If $\delta < 1$ the mapping property (2.22) is still true for $s = r$

Proof. We can reduce to prove the result for $m = 0$; moreover, in view of Lemma 2.8 it will be enough proving the statement for an elementary symbol. Thus, let us consider a zeroth order symbol of the form

$$a(x,\xi) = \sum_{h=-1}^{\infty} d_h(x)\psi_h(\xi),$$

where $\{d_h\}$ and $\{\psi_h\}$ are sequences in $B^{r,M}_{\infty,\infty}$ and $C_0^\infty(\mathbb{R}^n)$ respectively, fulfilling the assumptions of Lemma 2.8. Then, using a quasi-homogeneous partition of unity $\{\varphi_l\}$, the corresponding operator $a(x,D)$ can be decomposed as

$$a(x,D)u = T_1^{(N_0)}u + T_2^{(N_0)}u + T_3^{(N_0)}u$$

where

$$T_1^{(N_0)}u := \sum_{h=N_0-1}^{\infty} \sum_{l=-1}^{h-N_0} d_{h,l}u_h\,, \tag{2.23}$$

$$T_2^{(N_0)}u := \sum_{h=-1}^{\infty} \sum_{l=\max\{-1,h-N_0+1\}}^{h+N_0-1} d_{h,l}u_h\,, \tag{2.24}$$

$$T_3^{(N_0)}u := \sum_{l=N_0-1}^{\infty} \sum_{h=-1}^{l-N_0} d_{h,l}u_h\,, \tag{2.25}$$

and N_0 is a sufficiently large positive integer; for all $h,l=-1,0,\dots,$ we have set $u_h=\psi_h(D)u$ and $d_{h,l}=\varphi_l(D)d_h$.

Therefore, Proposition 2.9 becomes a consequence of Lemmas 2.10–2.12 below; in all of them, we make the same hypotheses of Proposition 2.9. □

Lemma 2.10. *For all $s\in\mathbb{R}$ and $p\in]1,+\infty[$ the following*

$$T_1^{(N_0)} : H^{s,p}_M \to H^{s,p}_M\,, \tag{2.26}$$

$$T_1^{(N_0)} : B^{s,M}_{\infty,\infty} \to B^{s,M}_{\infty,\infty} \tag{2.27}$$

are linear continuous operators.

Proof. Since $\widehat{d_{h,l}u_h}$ is supported on $\langle\xi\rangle_M \sim 2^h$, for $-1\le l\le h-N_0$, by Proposition 2.4, we compute:

$$\|T_1^{(N_0)}u\|_{H^{s,p}_M} \le C\left\|\left(\sum_{h=N_0-1}^{\infty} 4^{sh}\left|\sum_{l=-1}^{h-N_0} d_{h,l}u_h\right|^2\right)^{\frac{1}{2}}\right\|_{L^p}. \tag{2.28}$$

On the other hand,

$$\sum_{l=-1}^{h-N_0} d_{h,l}(x) = \phi_{h-N_0}(D)d_h(x) = (\mathcal{F}^{-1}\phi_{h-N_0} * d_h)(x),$$

hence

$$\left|\sum_{l=-1}^{h-N_0} d_{h,l}(x)\right| \le \|\mathcal{F}^{-1}\phi_{h-N_0}\|_{L^1}\|d_h\|_{L^\infty} \le C\|d_h\|_{L^\infty}, \quad \forall x\in\mathbb{R}^n.$$

Using the previous inequality to estimate the right-hand side of (2.28) and accounting the boundedness of $\{d_h\}$ in L^∞, we find

$$\|T_1^{(N_0)}u\|_{H^{s,p}_M} \leq C \left\| \left(\sum_{h=-1}^{\infty} 4^{sh}|u_h|^2 \right)^{\frac{1}{2}} \right\|_{L^p} \leq C\|u\|_{H^{s,p}_M}.$$

This shows the mapping property (2.26). To prove property (2.27), we use estimate (2.10) in Proposition 2.4, together with the boundedness of $\|d_{h,l}\|_{L^\infty}$, to get

$$\begin{aligned}\|T_1^{(N_0)}u\|_{B^{s,M}_{\infty,\infty}} &\leq C \sup_{h=-1,0,\dots} 2^{sh} \left\| \sum_{l=-1}^{h-N_0} d_{h,l}u_h \right\|_{L^\infty} \\ &\leq C \sup_{h=-1,0,\dots} 2^{sh}\|u_h\|_{L^\infty} \leq C\|u\|_{B^{s,M}_{\infty,\infty}}. \qquad \square\end{aligned}$$

Lemma 2.11. *For all $s > (\delta-1)r$ and $p \in]1,+\infty[$ the following*

$$T_2^{(N_0)} : H^{s,p}_M \to H^{s+r(1-\delta),p}_M \tag{2.29}$$

$$T_2^{(N_0)} : B^{s,M}_{\infty,\infty} \to B^{s+r(1-\delta),M}_{\infty,\infty} \tag{2.30}$$

are linear continuous operators.

Proof. Since $\widehat{d_{h,l}u_h}$ is supported in a quasi-homogeneous ball $\{\langle\xi\rangle_M \leq K2^{h+1}\}$, as long as $\max\{-1, h-N_0+1\} \leq l \leq h+N_0-1$ and $s+(1-\delta)r > 0$, we can estimate $\|T_2^{(N_0)}u\|_{H^{s+(1-\delta)r,p}_M}$ by

$$C\left\| \left(\sum_{h=-1}^{\infty} 4^{s+(1-\delta)r}h \left| \sum_{l=\max\{-1,h-N_0+1\}}^{h+N_0-1} d_{h,l}u_h \right|^2 \right)^{\frac{1}{2}} \right\|_{L^p}. \tag{2.31}$$

On the other hand, the inequalities $\|d_{h,l}\|_{L^\infty} \leq 2^{-rl}\|d_h\|_{B^{r,M}_{\infty,\infty}} \leq 2^{hr\delta - rl}$ yield

$$\begin{aligned}\left| \sum_{l=\max\{-1,h-N_0+1\}}^{h+N_0-1} d_{h,l}(x)u_h(x) \right| &\leq \sum_{l=\max\{-1,h-N_0+1\}}^{h+N_0-1} 2^{r(\delta h - l)}|u_h(x)| \\ &\leq C_{N_0}2^{r(\delta-1)h}|u_h(x)|, \quad \forall x \in \mathbb{R}^n,\end{aligned} \tag{2.32}$$

where $C_{N_0} > 0$ depends only on N_0, r and δ. Using the latter inequalities to majorize (2.31), we find the mapping property (2.29).

To prove (2.30), we only need to use the dyadic characterization of positive order Hölder spaces in Proposition 2.5, instead of the same for Sobolev spaces; then using again the estimates $\|d_{h,l}\|_{L^\infty} \leq 2^{hr\delta - rl}$ and (2.32), we manage as before to find

$$\|T_2^{(N_0)}u\|_{B^{s+(1-\delta)r,M}_{\infty,\infty}} \le C \sup_{h=-1,0,\dots} 2^{(s+(1-\delta)r)h} \left\| \sum_{l=\max\{-1,h-N_0+1\}}^{h+N_0-1} d_{h,l}u_h \right\|_{L^\infty}$$
$$\le C \sup_{h=-1,0,\dots} 2^{sh}\|u_h\|_{L^\infty} \le C\|u\|_{B^{s,M}_{\infty,\infty}},$$

with positive constants C depending only on r, s, δ and N_0. □

Lemma 2.12. *For all $s < r$ and $p \in]1, +\infty[$ the following*

$$T_3^{(N_0)} : H_M^{s+(\delta-1)r,p} \to H_M^{s,p} \tag{2.33}$$
$$T_3^{(N_0)} : B^{s+(\delta-1)r,M}_{\infty,\infty} \to B^{s,M}_{\infty,\infty} \tag{2.34}$$

are linear continuous operators. If $\delta < 1$ then

$$T_3^{(N_0)} : B^{r\varepsilon,M}_{\infty,\infty} \to B^{r,M}_{\infty,\infty} \tag{2.35}$$

is also a continuous operator for any $0 \le \delta < \varepsilon$.

Proof. Since for each index $l \ge N_0-1$, $\widehat{d_{h,l}u_h}$ is supported on an annulus $\langle\xi\rangle_M \sim 2^l$, as long as $-1 \le h \le l - N_0$, applying again the dyadic characterization of Sobolev spaces gives

$$\|T_3^{(N_0)}u\|_{H_M^{s,p}} \le C \left\| \left(\sum_{l=N_0-1}^{\infty} 4^{sl} \left| \sum_{h=-1}^{l-N_0} d_{h,l}u_h \right|^2 \right)^{\frac{1}{2}} \right\|_{L^p}$$
$$\le C \left\| \left\{ \sum_{l=N_0-1}^{\infty} 4^{sl} \left(\sum_{h=-1}^{l-N_0} |d_{h,l}||u_h| \right)^2 \right\}^{\frac{1}{2}} \right\|_{L^p}.$$

Then, using once more the estimates $\|d_{h,l}\|_{L^\infty} \le 2^{r\delta h - rl}$, the argument of the last L^p-norm above can be bounded by

$$\left\{ \sum_{l=N_0-1}^{\infty} \left(\sum_{h=-1}^{l-N_0} 2^{(s-r)(l-h)} 2^{(s+(\delta-1)r)h} |u_h| \right)^2 \right\}^{\frac{1}{2}}.$$

Provided that $s - r < 0$, by Young's inequality, the latter quantity is estimated in its turn by

$$C \left(\sum_{h=N_0-1}^{\infty} 4^{(s+r(\delta-1))h} |u_h|^2 \right)^{\frac{1}{2}},$$

with $C := \sum_{l=N_0-1}^{\infty} 2^{(s-r)l}$.

From previous calculations, we derive

$$\|T_3^{(N_0)}u\|_{H^{s,p}_M} \le C\left\|\left(\sum_{h=-1}^{\infty} 4^{(s+(\delta-1)r)h}|u_h|^2\right)^{\frac{1}{2}}\right\|_{L^p} \le C\|u\|_{H^{s+(\delta-1)r,p}_M},$$

which shows the mapping property (2.33).

To establish (2.34) it is now enough to use the first part of Proposition 2.4, and repeat the previous calculations to find

$$\begin{aligned}\|T_3^{(N_0)}u\|_{B^{s,M}_{\infty,\infty}} &\le C \sup_{l\ge N_0-1} 2^{sl} \sum_{h=-1}^{l-N_0} \|d_{h,l}\|_{L^\infty}\|u_h\|_{L^\infty}\\ &\le C \sup_{l\ge N_0-1} \sum_{h=-1}^{l-N_0} 2^{(s-r)(l-h)}2^{(s+(\delta-1)r)h}\|u_h\|_{L^\infty}\\ &\le C \sup_{h=-1,0,\dots} 2^{(s+(\delta-1)r)h}\|u_h\|_{L^\infty} \le C\|u\|_{B^{s+(\delta-1)r,M}_{\infty,\infty}},\end{aligned}$$

since $\sup_{l\ge N_0-1}\sum_{h=-1}^{l-N_0} 2^{(s-r)(l-h)}$ is finite, as long as $s-r<0$.

To prove (2.35), let $\varepsilon>\delta$ be arbitrarily fixed; using (2.10) we find:

$$\begin{aligned}\|T_3^{(N_0)}u\|_{B^{r,m}_{\infty,\infty}} &\le C \sup_{l\ge N_0-1} 2^{rl} \sum_{h=-1}^{l-N_0} \|d_{h,l}\|_{L^\infty}\|u_h\|_{L^\infty}\\ &\le C \sup_{l\ge N_0-1} \sum_{h=-1}^{l-N_0} 2^{hr\delta}\|u_h\|_{L^\infty} \qquad (2.36)\\ &\le C \sup_{h\ge -1} 2^{hr\varepsilon}\|u_h\|_{L^\infty} \sup_{l\ge N_0-1}\sum_{h=-1}^{l-N_0} 2^{hr(\delta-\varepsilon)} \le C\|u\|_{B^{r\varepsilon,M}{\infty,\infty}},\end{aligned}$$

since $\sup_{l\ge N_0-1}\sum_{h=-1}^{l-N_0} 2^{hr(\delta-\varepsilon)}$ is finite, as $\delta-\varepsilon<0$; in (2.36) C stands for suitable different positive constants, depending only on r,ε and δ. □

Remark 2.13. For $\delta=0$, the statement of Proposition 2.9 provides the already known Sobolev and Hölder continuity properties for operators with non regular symbols in $B^{r,M}_{\infty,\infty}S^m_M$, see [5].

On the extremely opposite border case $\delta=1$, Proposition 2.9 tells that operators with non regular symbols in $B^{r,M}_{\infty,\infty}S^m_{M,1}$, $r>0$, possess the mapping properties (2.21), (2.22) for every $s\in]0,r[$.

Since, of course, the inclusions $S^m_{M,\delta}\subset B^{r,M}_{\infty,\infty}S^m_{M,\delta}$ are true for all positive r, as a consequence of Proposition 2.9 we have the following

Corollary 2.14. *If* $a(x,\xi)\in S^m_{M,\delta}$, *for* $0\le\delta<1$, *then the mapping properties* (2.21), (2.22) *are true for all* $s\in\mathbb{R}$.

3. Symbolic calculus and parametrix for quasi-elliptic pseudo-differential operators

Proposition 3.1. Symbolic calculus. *For $m, m' \in \mathbb{R}$, $\delta, \delta' \in [0,1]$, $\mu, \nu \in \mathbb{Z}_+^n$ the following are true:*

$$\sigma \in S^m_{M,\delta} \Rightarrow \sigma^{(\mu)}_{(\nu)} := D^\mu_\xi D^\nu_x \sigma \in S^{m-\langle \mu, \frac{1}{M}\rangle + \delta\langle \nu, \frac{1}{M}\rangle}_{M,\delta}; \tag{3.1}$$

$$\sigma \in S^m_{M,\delta}\,, \quad \tau \in S^{m'}_{M,\delta'} \Rightarrow \sigma \cdot \tau \in S^{m+m'}_{M,\max\{\delta,\delta'\}}. \tag{3.2}$$

Proof. Let $\alpha, \beta \in \mathbb{Z}^n_+$ be arbitrarily fixed. Then

$$|D^\alpha_\xi D^\beta_x \sigma^{(\mu)}_{(\nu)}(x,\xi)| = |D^{\alpha+\mu}_\xi D^{\beta+\nu}_x \sigma(x,\xi)| \le C\langle \xi \rangle_M^{m-\langle \alpha+\mu, \frac{1}{M}\rangle + \delta\langle \beta+\nu, \frac{1}{M}\rangle},$$

with positive constant $C = C_{\alpha,\beta,\mu,\nu}$, shows (3.1). Leibnitz's rule also implies:

$$\begin{aligned}|D^\alpha_\xi D^\beta_x(\sigma\cdot\tau)(x,\xi)| &\le \sum_{0\le\mu\le\alpha\,,\ 0\le\nu\le\beta} \binom{\alpha}{\mu}\binom{\beta}{\nu}|D^{\alpha-\mu}_\xi D^{\beta-\nu}_x\sigma(x,\xi)||D^\mu_\xi D^\nu_x \tau(x,\xi)| \\ &\le \sum_{0\le\mu\le\alpha\,,\ 0\le\nu\le\beta} C_{\alpha,\beta,\mu,\nu}\langle\xi\rangle_M^{m-\langle\alpha-\mu,\frac{1}{M}\rangle+\delta\langle\beta-\nu,\frac{1}{M}\rangle}\langle\xi\rangle_M^{m'-\langle\mu,\frac{1}{M}\rangle+\delta'\langle\nu,\frac{1}{M}\rangle} \\ &\le C\langle\xi\rangle_M^{m+m'-\langle\alpha,\frac{1}{M}\rangle+\max\{\delta,\delta'\}\langle\beta,\frac{1}{M}\rangle}\end{aligned}$$

which shows (3.2). □

Definition 3.2. Asymptotic expansion. Let $\sigma \in S^m_{M,\delta}$. We say that a sequence $\{\sigma_j\}_{j\ge 0}$ of symbols $\sigma_j \in S^{m_j}_{M,\delta}$ is an *asymptotic expansion* of σ if $\{m_j\}_{j\ge0}$ is a decreasing sequence with $m_0 = m$, $\lim\limits_{j\to+\infty} m_j = -\infty$, and for any positive integer N:

$$\sigma - \sum_{j<N} \sigma_j \in S^{m_N}_{M,\delta}.$$

In this case, we write $\sigma \sim \sum\limits_j \sigma_j$.

Henceforth, we set $S^{-\infty} := \bigcap_{m\in\mathbb{R}} S^m_{1,0}$. The use of inequalities in Proposition 2.1 yields that for all weight vector M and $\delta \in [0,1]$ there holds $S^{-\infty} = \bigcap_{m\in\mathbb{R}} S^m_{M,\delta}$.

Lemma 3.3. *Let $\{m_j\}_{j\ge0}$ be a real decreasing sequence with $\lim\limits_{j\to+\infty} m_j = -\infty$. For every $j \ge 0$ let σ_j be a symbol in $S^{m_j}_{M,\delta}$. Then there exists a symbol $\sigma \in S^{m_0}_{M,\delta}$ such that*

$$\sigma \sim \sum_j \sigma_j.$$

Moreover, if τ is any other symbol for which the sequence $\{\sigma_j\}_{j\ge0}$ is an asymptotic expansion, then $\sigma - \tau \in S^{-\infty}$.

Proof. Let $\psi(\xi)$ be a function in $C^\infty(\mathbb{R}^n)$ satisfying

$$0 \le \psi(\xi) \le 1\,,\ \forall \xi \in \mathbb{R}^n\,,\ \psi(\xi) = 0\,,\ \text{for } |\xi|_M \le 1\,,\ \psi(\xi) = 1\,,\ \text{for } |\xi|_M \ge 2.$$

Take a positive sequence $\{\varepsilon_j\}_{j\ge0}$ such that

$$1 > \varepsilon_0 > \varepsilon_1 > \cdots > \varepsilon_j \to 0\,,\ \text{as } j \to +\infty.$$

We define the symbol $\sigma(x,\xi)$ by

$$\sigma(x,\xi) = \sum_{j=0}^{+\infty} \psi(\varepsilon_j^{\frac{1}{M}}\xi)\sigma_j(x,\xi).$$

Then one can repeat the arguments of [13] to prove that $\sigma \in S^{m_0}_{M,\delta}$ and $\sigma \sim \sum_j \sigma_j$. □

In order to develop a functional calculus for pseudo-differential operators with symbols in the quasi-homogeneous classes $S^m_{M,\delta}$ it is convenient to introduce a class of apparently more general operators involving multiple symbols.

Definition 3.4. For $m \in \mathbb{R}$ and $\delta_1, \delta_2 \in [0,1]$, $S^m_{M,\delta_1,\delta_2}$ will be the class of functions $a(x,y,\xi) \in C^\infty(\mathbb{R}^n \times \mathbb{R}^n \times \mathbb{R}^n)$ such that for all multi-indices $\alpha, \beta, \gamma \in \mathbb{Z}^n_+$ there exists a positive constant $C_{\alpha,\beta,\gamma}$ such that

$$|D^\alpha_\xi D^\beta_x D^\gamma_y a(x,y,\xi)| \le C_{\alpha,\beta,\gamma} \langle\xi\rangle_M^{m-\langle\alpha,\frac{1}{M}\rangle+\delta_1\langle\beta,\frac{1}{M}\rangle+\delta_2\langle\gamma,\frac{1}{M}\rangle}, \quad x,y,\xi \in \mathbb{R}^n. \tag{3.3}$$

When $a(x,y,\xi) \in S^m_{M,\delta_1,\delta_2}$ we define the operator $A = a(x,D,y)$ by

$$Au(x) = (2\pi)^{-n} \iint e^{i\langle x-y,\xi\rangle} a(x,y,\xi)u(y)\,dy\,d\xi\,, \quad u \in \mathcal{S}(\mathbb{R}^n). \tag{3.4}$$

Under the above conditions, for $v \in \mathcal{S}(\mathbb{R}^n)$

$$\langle Au\,,\, v\rangle = (2\pi)^{-n} \iiint e^{i\langle x-y,\xi\rangle} a(x,y,\xi)u(y)v(x)\,dx\,dy\,d\xi \tag{3.5}$$

is defined as an ordinary oscillatory integral. It is easily verified that, for fixed u, the expression (3.5), viewed as a functional of v, defines a distribution $Au \in \mathcal{S}'(\mathbb{R}^n)$. Therefore a linear operator

$$A : \mathcal{S}(\mathbb{R}^n) \to \mathcal{S}'(\mathbb{R}^n)$$

is defined; we will formally write it as the integral (3.4).

Proposition 3.5. *If $a(x,y,\xi) \in S^m_{M,\delta_1,\delta_2}$ with $0 \le \delta_2 < \frac{1}{m^*}$ then the operator (3.4) belongs to $\mathrm{Op}S^m_{M,\delta}$ with $\delta = \max\{\delta_1,\delta_2\}$; in fact $A = q(x,D)$, where $q \in S^m_{M,\delta}$ has the asymptotic expansion*

$$q(x,\xi) \sim \sum_{\alpha \ge 0} \frac{i^{|\alpha|}}{\alpha!} D^\alpha_\xi D^\alpha_y\, a(x,y,\xi)_{|\,y=x}. \tag{3.6}$$

Proof. A formal calculation gives

$$Au(x) = (2\pi)^{-n} \int e^{i\langle x,\xi\rangle} q(x,\xi)\widehat{u}(\xi)d\xi\,,$$

with

$$q(x,\xi) = (2\pi)^{-n} \int e^{i\langle x-y,\xi-\eta\rangle} a(x,y,\xi)d\xi dy = \int \widehat{b}(x,\eta,\xi+\eta)d\eta\,, \tag{3.7}$$

and

$$\hat{b}(x,\eta,\xi) := (2\pi)^{-n} \int e^{-i\langle y,\eta\rangle} a(x,x+y,\xi)dy.$$

The hypotheses on $a(x,y,\xi)$ imply

$$|D_\xi^\alpha D_x^\beta \hat{b}(x,\eta,\xi)| \le C_{\nu,\alpha,\beta}\langle\xi\rangle_M^{m-\langle\alpha,\frac{1}{M}\rangle+\delta\langle\beta,\frac{1}{M}\rangle+\delta_2\nu}\langle\eta\rangle_M^{-\nu}$$

for all $\alpha,\beta\in\mathbb{Z}_+^n$, $\nu\in\mathbb{Z}_+$ and suitable $C_{\alpha,\beta,\nu}>0$. A Taylor's expansion of $\hat{b}(x,\eta,xi+\eta)$ in the last argument about ξ gives

$$\Big|\hat{b}(x,\eta,\xi+\eta)-\sum_{|\alpha|<N}\tfrac{1}{\alpha!}(iD_\xi)^\alpha\hat{b}(x,\eta,\xi)\eta^\alpha\Big| \le C\langle\eta\rangle_M^N\langle\eta\rangle_M^{-\nu}\sup_{0\le t\le 1}\langle\xi+t\eta\rangle_M^{m-\frac{N}{m^*}+\delta_2\nu}$$

Integrating with respect to η and observing that

$$(iD_\xi)^\alpha D_y^\alpha a(x,y,\xi)_{|\,y=x}=\int(iD_\xi)^\alpha\hat{b}(x,\eta,\xi)\eta^\alpha d\eta$$

we obtain (3.6) by repeating the arguments in [10, Ch. 0]. □

Proposition 3.6. Adjoint and product.

1. *If $\sigma(x,D)\in \mathrm{Op}S^m_{M,\delta}$ with $0\le\delta<\frac{1}{m^*}$, the adjoint operator $\sigma(x,D)^*\in \mathrm{Op}S^m_{M,\delta}$ and its symbol $\sigma^*(x,\xi)$ satisfies*

$$\sigma^*(x,\xi)\sim\textstyle\sum_{\alpha\ge 0}\frac{i^{|\alpha|}}{\alpha!}D_\xi^\alpha D_x^\alpha\overline{\sigma(x,\xi)}.$$

2. *If $\sigma_1(x,D)\in\mathrm{Op}S^{m_1}_{M,\delta_1}$, $\sigma_2(x,D)\in\mathrm{Op}S^{m_2}_{M,\delta_2}$ with $0\le\delta_2<\frac{1}{m^*}$, then*

$$\sigma_1(x,D)\sigma_2(x,D)\in\mathrm{Op}S^{m_1+m_2}_{M,\delta},$$

with $\delta:=\max\{\delta_1,\delta_2\}$, and the symbol $\tau(x,\xi)$ of the product satisfies

$$q(x,\xi)\sim\textstyle\sum_{\alpha\ge 0}\frac{i^{|\alpha|}}{\alpha!}D_\xi^\alpha\sigma_1(x,\xi)D_x^\alpha\sigma_2(x,\xi).$$

Proof. To prove the statement 1, we apply the result of Proposition 3.5 to the operator

$$\sigma(x,D)^*u(x)=(2\pi)^{-n}\iint e^{i\langle x-y,\xi\rangle}\overline{\sigma(y,\xi)}u(y)dyd\xi,$$

observing that under the hypotheses for σ, one has that $\overline{\sigma(y,\xi)}\in S^m_{M,0,\delta}$. For the statement 2, we first notice that $\sigma_1(x,D)\sigma_2(x,D)=\sigma_1(x,D)\sigma^*(x,D)^*$ and then apply Proposition 3.5 to the operator A as in (3.4) with

$$a(x,y,\xi)=\sigma_1(x,\xi)\overline{\sigma_2^*(y,\xi)}.$$

Under the hypotheses about σ_1,σ_2, the latter belongs to $S^{m_1+m_2}_{M,\delta_1,\delta_2}$. □

We say that a symbol $p(x,\xi)$ in $S^m_{M,\delta}$ or respectively in $B^{r,M}_{\infty,\infty}S^m_{M,\delta}$, $m\in\mathbb{R}$, $r>0$, $0\le\delta<1$ is *M-quasi-elliptic* if there exist two positive constants C,K such that

$$|p(x,\xi)|>C\langle\xi\rangle_M^m,\quad\text{when } |\xi|_M>K. \tag{3.8}$$

Proposition 3.7. *Let $p(x,\xi) \in S^m_{M,\delta}$ be M-quasi-elliptic, for $m \in \mathbb{R}$ and $0 \le \delta < \frac{1}{m^*}$; then we can find $q(x,\xi) \in S^{-m}_{M,\delta}$ such that*

$$\begin{aligned} q(x,D)p(x,D) &= I + R_1(x,D), \quad R_1(x,\xi) \in S^{-\infty} \\ p(x,D)q(x,D) &= I + R_2(x,D), \quad R_2(x,\xi) \in S^{-\infty}. \end{aligned} \tag{3.9}$$

Proof. Let K be the constant in (3.8) and choose $\psi(\xi) \in C^\infty(\mathbb{R}^n)$ be equal to 0 for $|\xi|_M \le K$ and equal to 1 for $|\xi|_M \ge 2K$. Set now

$$q_0(x,\xi) = \frac{\psi(\xi)}{p(x,\xi)}. \tag{3.10}$$

Since the derivatives of $\psi(\xi)$ vanish for great ξ, using the chain rule and observing that $|\partial_x^\alpha \partial_\xi^\beta (p(x,\xi)^{-1})| < C_{\alpha,\beta} \langle\xi\rangle_M^{-m-\langle\alpha,\frac{1}{M}\rangle+\delta\langle\beta,\frac{1}{M}\rangle}$, for $|\xi|_M > K$, we obtain that $q_0(x,\xi) \in S^{-m}_{M,\delta}$. As long as $0 \le \delta < \frac{1}{m^*} < 1$, Proposition 3.6 applies and thus

$$\begin{aligned} q_0(x,D)p(x,D) &= I + r_0(x,D), \\ p(x,D)q_0(x,D) &= I + \tilde{r}_0(x,D); \end{aligned} \qquad r_0(x,\xi), \tilde{r}_0(x,\xi) \in S^{\frac{1}{m^*}}_{M,\delta}. \tag{3.11}$$

Let now $s(x,D)$ be the formal expansion $r_0(x,D) - r_0(x,D)^2 + \cdots$ and set $q(x,D) = (I - s(x,D))q_0(x,D)$. Since $s(x,D) \in \mathrm{Op}S^0_{M,\delta}$, we obtain

$$q(x,D)p(x,D) = I + R(x,D), \quad R(x,\xi) \in S^{-\infty}. \tag{3.12}$$

In the same way we can build $\tilde{q}(x,D)$ and $\tilde{R}(x,D)$ such that $p(x,D)\tilde{q}(x,D) = I + \tilde{R}(x,D)$. Since $(q(x,D)p(x,D))\tilde{q}(x,D) = q(x,D)(p(x,D)\tilde{q}(x,D))$, $q(x,D) = \tilde{q}(x,D) \mathrm{mod}\, S^{-\infty}$ follows, then (3.9) holds. □

In the lack of a suitable parametrix of M-quasi-elliptic operators with non smooth symbols, for studying the regularity of solutions to M-quasi-elliptic equations with non smooth coefficients, following the approach in Taylor [10, §1.3], we introduce the next symbol decomposition.

Let ϕ be a fixed C^∞ function such that $\phi(\xi) = 1$ for $\langle\xi\rangle_M \le 1$ and $\phi(\xi) = 0$ for $\langle\xi\rangle_M > 2$. For a given $\varepsilon > 0$ we set $\phi(\varepsilon^{\frac{1}{M}}\xi) := \phi(\varepsilon^{\frac{1}{m_1}}\xi_1, \ldots, \varepsilon^{\frac{1}{m_n}}\xi_n)$. The following quasi-homogeneous version of [10, Lemma 1.3.A] can be readily proved.

Lemma 3.8. *For all $r > t \ge 0$ and $\varepsilon \in]0,1]$ the following estimates hold true*

$$\|D_x^\beta \phi(\varepsilon^{\frac{1}{M}} D)u\|_{B^{r,M}_{\infty,\infty}} \le C_\beta \varepsilon^{-\langle\beta,\frac{1}{M}\rangle} \|u\|_{B^{r,M}_{\infty,\infty}}, \tag{3.13}$$

$$\|u - \phi(\varepsilon^{\frac{1}{M}} D)u\|_{B^{r-t,M}_{\infty,\infty}} \le C\varepsilon^t \|u\|_{B^{r,M}_{\infty,\infty}}, \tag{3.14}$$

$$\|u - \phi(\varepsilon^{\frac{1}{M}} D)u\|_{L^\infty} \le C\varepsilon^r \|u\|_{B^{r,M}_{\infty,\infty}}. \tag{3.15}$$

with positive constants C_β, C independent of ε.

Proof. To prove (3.13), we show that $\{\varepsilon^{\langle \frac{1}{M},\beta\rangle}\xi^\beta\phi(\varepsilon^{\frac{1}{M}}\xi)\}_{0<\varepsilon\le 1}$ is a bounded set in S^0_M; indeed for any $\alpha\in\mathbb{Z}^n_+$ one computes:

$$\begin{aligned} D^\alpha\xi^\beta\phi(\varepsilon^{\frac{1}{M}}\xi) &= \sum_{\nu\le\alpha,\,\nu\le\beta} C_{\alpha,\beta,\nu}\xi^{\beta-\nu}\varepsilon^{\langle\frac{1}{M},\alpha-\nu\rangle}(D^{\alpha-\nu}\phi)(\varepsilon^{\frac{1}{M}}\xi)\\ &\le C_\beta\langle\xi\rangle_M^{\langle\frac{1}{M},\beta-\nu\rangle}\varepsilon^{\langle\frac{1}{M},\alpha-\nu\rangle}\le C_{\alpha,\beta}\varepsilon^{-\langle\beta,\frac{1}{M}\rangle}, \end{aligned}$$

since $\langle\xi\rangle_M\le\frac{2}{\varepsilon}$ within $\operatorname{supp}\phi(\varepsilon^{\frac{1}{M}}\cdot)$. This proves the assertion.

To establish (3.14), we prove that $\{\varepsilon^{-t}\langle\xi\rangle^t_M(1-\phi(\varepsilon^{\frac{1}{M}}\xi))\}_{0<\varepsilon\le1}$ is a bounded set in S^0_M; for any non zero $\alpha\in\mathbb{Z}^n_+$ we compute:

$$\begin{aligned} D^\alpha(\langle\xi\rangle_M^{-t}(1-\phi(\varepsilon\xi)) &= -\sum_{\nu\le\alpha} C_{\alpha,\nu}D^\nu\langle\xi\rangle_M^{-t}\varepsilon^{\langle\frac{1}{M},\alpha-\nu\rangle}(D^{\alpha-\nu}\phi)(\varepsilon^{\frac{1}{M}}\xi)\\ &\le \sum_{\nu\le\alpha} C_{\alpha,\nu}\langle\xi\rangle_M^{-t-\langle\frac{1}{M},\nu\rangle}\langle\xi\rangle_M^{-\langle\frac{1}{M},\alpha-\nu\rangle}\le\varepsilon^t C_\alpha\langle\xi\rangle_M^{-\langle\alpha,\frac{1}{M}\rangle}, \end{aligned}$$

since $\langle\xi\rangle_M\sim\frac{1}{\varepsilon}$ within $\operatorname{supp}(1-\phi(\varepsilon^{\frac{1}{M}}\cdot))$.

For the proof of (3.15), the same arguments used in [10, Lemma 1.3.C] can be plainly repeated. □

Let $a(x,\xi)$ belong to $B^{r,M}_{\infty,\infty}S^m_M$ and take $\delta\in]0,1]$; we define $a^\#(x,\xi)$ by the following

$$a^\#(x,\xi):=\sum_{h=-1}^{\infty}\phi(2^{-\frac{h\delta}{M}}D_x)a(x,\xi)\varphi_h(\xi).$$

We also set $a^b(x,\xi):=a(x,\xi)-a^\#(x,\xi)$. As a consequence of Lemma 3.8, one can prove the following

Proposition 3.9. *If $a(x,\xi)\in B^{r,M}_{\infty,\infty}S^m_M$, with $r>0$, $m\in\mathbb{R}$, and $\delta\in]0,1]$, then*

$$a^\#(x,\xi)\in S^m_{M,\delta};$$

moreover $a^b(x,\xi)\in B^{r,M}_{\infty,\infty}S^{m-r\delta}_{M,\delta}$.

Proof. For arbitrary $\alpha,\beta\in\mathbb{Z}^n_+$, we compute:

$$\|D^\beta_xD^\alpha_\xi a^\#(\cdot,\xi)\|_{B^{r,M}_{\infty,\infty}}\le\sum_{\nu\le\alpha}\binom{\alpha}{\nu}\sum_{h=-1}^{+\infty}\|D^\beta_x\phi(2^{-\frac{\delta h}{M}}D)D^{\alpha-\nu}_\xi a(\cdot,\xi)\|_{B^{r,M}_{\infty,\infty}}|D^\nu_\xi\varphi_h(\xi)|.$$

By (3.13) of Lemma 3.8 (with $\varepsilon=2^{-h\delta}$) we estimate

$$\begin{aligned} \|D^\beta_x\phi(2^{-\frac{\delta h}{M}}D)D^{\alpha-\nu}_\xi a(\cdot,\xi)\|_{B^{r,M}_{\infty,\infty}} &\le C_\beta 2^{h\delta\langle\beta,\frac{1}{M}\rangle}\|D^{\alpha-\nu}_\xi a(\cdot,\xi)\|_{B^{r,M}_{\infty,\infty}}\\ \le C_\beta 2^{h\delta\langle\beta,\frac{1}{M}\rangle}\langle\xi\rangle_M^{m-\langle\alpha-\nu,\frac{1}{M}\rangle} &\le C_\beta\langle\xi\rangle_M^{m-\langle\alpha-\nu,\frac{1}{M}\rangle+\delta\langle\beta,\frac{1}{M}\rangle},\ \text{for } \langle\xi\rangle_M\sim 2^h; \end{aligned}$$

moreover

$$|D^\nu_\xi\varphi_h(\xi)|\le C_\nu 2^{-h\langle\nu,\frac{1}{M}\rangle}\le C_\nu\langle\xi\rangle_M^{-\langle\nu,\frac{1}{M}\rangle}.$$

Gathering the preceding inequalities, we obtain:

$$\|D^\beta_xD^\alpha_\xi a^\#(\cdot,\xi)\|_{B^{r,M}_{\infty,\infty}}\le C_{\alpha,\beta}\langle\xi\rangle_M^{m-\langle\alpha,\frac{1}{M}\rangle+\delta\langle\beta,\frac{1}{M}\rangle},\quad\forall\xi\in\mathbb{R}^n.$$

Since $B^{r,M}_{\infty,\infty}$ is continuously embedded in $L^\infty\cap C^0$, being $r>0$, the latter inequalities prove that $a^\#(x,\xi)\in S^m_{M,\delta}$ (cf. Definition 2.6).

Since $\sum_h \varphi_h = 1$, we can write

$$a^b(x,\xi) = \sum_{h=-1}^{\infty} (I - \phi(2^{-\frac{h\delta}{M}} D))a(x,\xi)\varphi_h(\xi).$$

For an arbitrary multi-index $\alpha \neq 0$, we find:

$$D_\xi^\alpha a^b(x,\xi) = \sum_{\nu\le\alpha} \binom{\alpha}{\nu} \sum_{h=-1}^{\infty} (I - \phi(2^{-\frac{h\delta}{M}} D))D_\xi^{\alpha-\nu} a(x,\xi) D_\xi^\nu \varphi_h(\xi).$$

Using now (3.15) to estimate the L^∞ norm of $a^b(x,\xi)$, we find:

$$\begin{aligned}\|D_\xi^\alpha a(\cdot,\xi)\|_{L^\infty} &\le \sum_{\nu\le\alpha} \binom{\alpha}{\nu} \sum_{h=-1}^{\infty} \|(Id - \phi(2^{-\frac{h\delta}{M}} D))D_\xi^{\alpha-\nu} a(\cdot,\xi)\|_{L^\infty} |D_\xi^\nu \varphi_h(\xi)| \\ &\le \sum_{\nu\le\alpha} \binom{\alpha}{\nu} \sum_{h=-1}^{\infty} 2^{-h\delta r} \|D^{\alpha-\nu} a(\cdot,\xi)\|_{B^{r,M}_{\infty,\infty}} 2^{-h\langle \nu, \frac{1}{M}\rangle} \le C\langle\xi\rangle_M^{m-\delta r-\langle\alpha,\frac{1}{M}\rangle}.\end{aligned}$$

The latter provides an estimate of type (2.16) with $m - \delta r$ instead of m. To check the validity of an estimate of type (2.17), with $m - \delta r$ instead of m, we use (3.14) with $t = 0$ to estimate the $B^{r,M}_{\infty,\infty}$ norm of $D_\xi^\alpha a^b(x,\xi)$. □

Proposition 3.10. *If the symbol $a(x,\xi) \in B^{r,M}_{\infty,\infty} S^m_M$, $m \in \mathbb{R}$, is M-quasi-elliptic, then for any $0 < \delta \le 1$, $a^\sharp(x,\xi) \in S^m_{M,\delta}$ satisfies the same condition.*

Proof. By the M-quasi-ellipticity there exist positive constants C_1, R_1 such that $|a(x,\xi)| > C_1\langle\xi\rangle_M^m$ when $|\xi|_M > R_1$. From the embedding $B^{r,M}_{\infty,\infty} \hookrightarrow L^\infty$ the same estimate follows for $\|a(\cdot,\xi)\|_{B^{r,M}_{\infty,\infty}}$. For any $R_0 > 0$ we can find a positive integer h_0, which increases together with R_0, such that $\psi_h(\xi) = 0$ as long as $|\xi|_M > R_0$ and $h = -1, \dots, h_0 - 1$. We can then write:

$$a^\sharp(x,\xi) = \sum_{h=h_0}^{\infty} \phi\left(2^{-h\frac{\delta}{M}} D_x\right) a(x,\xi)\varphi_h(\xi), \quad |\xi|_M > R_0. \tag{3.16}$$

Set for brevity $\phi\left(2^{-h\frac{\delta}{M}}\cdot\right) = \phi_h(\cdot)$

By means of (3.16), the Cauchy-Schwarz inequality and (3.15) in Lemma 3.8, when $|\xi|_M > R_0$ we can estimate

$$\begin{aligned}&|a^\sharp(x,\xi) - a(x,\xi)|^2 \\ &= \left|\sum_{h=h_0}^{\infty} (\phi_h(D_x) - I)\, a(x,\xi)\varphi_h(\xi)\right|^2 \\ &= \sum_{h=h_0}^{\infty} \sum_{k=h-N_0}^{h+N_0} \langle (\phi_h(D_x)) - I)\, a(x,\xi)\varphi_h(\xi), (\phi_k(D_x) - I)\, a(x,\xi)\varphi_k(\xi)\rangle \\ &= \sum_{t=-N_0}^{N_0} \sum_{h=h_0}^{\infty} \langle (\phi_h(D_x) - I)\, a(x,\xi)\varphi_h(\xi), (\phi_{h+t}(D_x) - I)\, a(x,\xi)\varphi_{h+t}(\xi)\rangle \\ &\le \sum_{t=-N_0}^{N_0} \sum_{h=h_0}^{\infty} \|\, (\phi_h(D_x)) - I)\, a(\cdot,\xi)\|_\infty |\varphi_h(\xi)| \|\, (\phi_{h+t}(D_x) - I)\, a(\cdot,\xi)\|_\infty |\varphi_{h+t}(\xi)|\end{aligned}$$

$$\leq C^2 \sum_{t=-N_0}^{N_0} \sum_{h=h_0}^{\infty} 2^{-h\delta r} 2^{-(h+t)\delta r} \|a(\cdot,\xi)\|^2_{B^{r,M}_{\infty,\infty}}$$
$$\leq C^2 \sum_{h=h_0}^{\infty} 2^{-2h\delta r} \|a(\cdot,\xi)\|^2_{B^{r,M}_{\infty,\infty}} \leq C^2 2^{-2h_0\delta r} \|a(\cdot,\xi)\|^2_{B^{r,M}_{\infty,\infty}},$$

where C denotes different positive constants depending only on δ, N_0 and r.

Since $\|a(\cdot,\xi)\|_{B^{r,M}_{\infty,\infty}} < c\langle\xi\rangle_M^m$, let us fix R_0 large enough to have: $C2^{-h_0\delta r} < \frac{C_1}{2c}$. Then for $|\xi|_M > \max\{R_0, R_1\}$

$$\begin{aligned} |a^\sharp(x,\xi)| &\geq |a(x,\xi)| - |a^\sharp(x,\xi) - a(x,\xi)| \\ &> C_1\langle\xi\rangle_M^m - \tfrac{C_1}{2}\langle\xi\rangle_M^m = \tfrac{C_1}{2}\langle\xi\rangle_M^m \end{aligned} \tag{3.17}$$

follows and the proof is concluded. □

As an application of the previous considerations on smooth and non-smooth pseudo-differential symbol classes with quasi-homogeneous decay let us consider the linear partial differential operator of quasi-homogeneous order $m \in \mathbb{Z}_+$

$$A(x,D)u := \sum_{\langle\alpha,\frac{1}{M}\rangle\leq m} a_\alpha(x)D^\alpha, \quad a_\alpha \in B^{r,M}_{\infty,\infty}, \quad r > 0. \tag{3.18}$$

Since the symbol $A(x,\xi)$ belongs to $B^{r,M}_{\infty,\infty}S^m_M$, for given $0 < \delta < 1$ we can split it

$$\begin{aligned} &A(x,\xi) = A^\sharp(x,\xi) + A^b(x,\xi), \\ &A^\sharp(x,\xi) \in S^m_{M,\delta}, \ A^b(x,\xi) \in B^{r,M}_{\infty,\infty}S^{m-\delta r}_{M,\delta}. \end{aligned} \tag{3.19}$$

Consider now the linear partial differential equation

$$A(x,D)u = f \tag{3.20}$$

and split it by means of (3.19). Assuming now that $A(x,D)$ is M-quasi-elliptic, then $A^\sharp(x,D)$ admits a left parametrix $B(x,\xi) \in S^{-m}_{M,\delta}$, when $0 < \delta < \frac{1}{m^*}$, see Propositions 3.10, 3.7. Applying now $B(x,D)$ to both sides of (3.20) and using Proposition 3.7 we obtain:

$$u = B(x,D)f - R(x,D)u - B(x,D)A^b(x,D)u, \quad R(x,\xi) \in S^{-\infty} \tag{3.21}$$

Assume now that $f \in H_M^{s-m,p}$ (resp. $f \in B^{s-m,M}_{\infty,\infty}$) and consider u in $H_M^{s-\delta r,p}$ (resp. in $B^{s-\delta r,M}_{\infty,\infty}$), then by means of the continuity properties in Proposition 2.9 it follows that $u \in H_M^{s,p}$ (resp. $u \in B^{s,M}_{\infty,\infty}$), when $(\delta-1)r + m < s < r + m$ (resp. $(\delta-1)r + m < s \leq r + m$), $0 < \delta < \frac{1}{m^*}$.

We have then proved the following

Theorem 3.11. *Let $A(x,D)u = f$ be a linear partial differential equation with Hölder coefficients introduced by means of (3.18). Assume that $A(x,D)$ is M-multi-quasi-elliptic. Then it follows for any solution u:*

$$\begin{aligned} &u \in H_M^{s-\delta r,p}, \quad f \in H_M^{s-m,p} \quad \Rightarrow \quad u \in H_M^{s,p}, \quad \text{if} \quad (\delta-1)r + m < s < r + m; \\ &u \in B^{s-\delta r,M}_{\infty,\infty}, \quad f \in B^{s-m,M}_{\infty,\infty} \quad \Rightarrow \quad u \in B^{s,M}_{\infty,\infty}, \quad \text{if} \quad (\delta-1)r + m < s \leq r + m. \end{aligned}$$

for any $0 < \delta < \frac{1}{m^*}$, $1 < p < \infty$.

References

[1] C. Feffermann, L^p bounds for pseudo-differential operators, *Israel J. Math.* **14** (1973), 413–417.

[2] G. Garello, Pseudodifferential operators with symbols in weighted Sobolev spaces and regularity for non linear partial differential equations, *Math. Nachr.* **239-240** (2002), 62–79.

[3] G. Garello and A. Morando, L^p-bounded pseudodifferential operators and regularity for multi-quasi-elliptic equations, *Integral Equations Operator Theory* **51** (2005), 501–517.

[4] G. Garello and A. Morando, L^p boundedness for pseudodifferential operators with non smooth symbols and applications, *Boll. Unione Mat. Ital. Sez. B Artic. Ric. Mat.* **8**(8) (2005), 461–503.

[5] G. Garello and A. Morando, Continuity in quasi-homogeneous Sobolev spaces for pseudo-differential operators with Besov symbols, in *Modern Trends in Pseudo-Differential Operators*, Editors: J. Toft, M.W. Wong and H. Zhu, Operator Theory: Advances and Applications **172**, Birkhäuser, 2007, 161–172.

[6] R. Lascar, Propagation des singularités des solutions d'équations pseudodifférentielles quasi-homogènes, *Ann. Inst. Fourier Grenoble* **27** (1977), 79–123.

[7] P.I. Lizorkin, (L_p, L_q)-multipliers of Fourier integrals, *Dokl. Akad. Nauk SSSR* **152** (1963), 808–811.

[8] J. Marschall, Pseudo-differential operators with coefficients in Sobolev spaces, *Trans. Amer. Math. Soc.* **307**(1) (1988), 335–361.

[9] E.M. Stein, *Singular Integrals and Differentiability Properties of Functions*, Princeton Mathematical Series **30**, Princeton University Press, Princeton, N.J., 1970.

[10] M.E. Taylor, *Pseudodifferential Operators and Nonlinear PDE*, Birkhäuser, Boston, Basel, Berlin, 1991.

[11] M.E. Taylor, *Tools for PDE: Pseudodifferential Operators, Paradifferential Operators and Layer Potentials*, American Mathematical Society, 2000.

[12] H. Triebel, *Theory of Function Spaces*, Birkhäuser Verlag, Basel, Boston, Stuttgart, 1983.

[13] M.W. Wong, *An Introduction to Pseudo-Differential Operators*, Second Edition, World Scientific Publishing Co., Inc., Singapore, 1999.

Gianluca Garello
Department of Mathematics, University of Torino
Via Carlo Alberto 10
I-10123 Torino, Italy
e-mail: `gianluca.garello@unito.it`

Alessandro Morando
Department of Mathematics, University of Brescia
Via Valotti 9
I-25133, Brescia, Italy
e-mail: `morando@ing.unibs.it`

Operator Theory:
Advances and Applications, Vol. 189, 265–273

Multi-Anisotropic Gevrey Regularity of Hypoelliptic Operators

C. Bouzar and A. Dali

Abstract. We show a multi-anisotropic Gevrey regularity of solutions of hypoelliptic equations. This result is a precision of a classical result of Hörmander

Mathematics Subject Classification (2000). Primary 35H10; Secondary 35D10, 35H30.

Keywords. Hypoelliptic operators, Gevrey regularity, Multi-anisotropic Gevrey spaces, Newton polyhedron, Multi-quasiellipticity, Gevrey vectors.

1. Introduction

An important problem among others of linear partial differential equations is the $\mathcal{C}^\infty$ or Gevrey regularity of solutions of these equations. L. Hörmander has completely characterized the $\mathcal{C}^\infty$ regularity (hypoellipticity) of linear partial differential operators with complex constant coefficients, see [6]. An another fundamental result obtained by L. Hörmander says that every hypoelliptic differential operator $P(D)$ is anisotropic Gevrey hypoelliptic, i.e., $\exists \varrho = (\varrho_1, \dots, \varrho_n) \in \mathbb{R}_+^n$ such that

$$u \in \mathcal{D}'(\Omega) \quad \text{and} \quad P(D)\,u = 0 \Longrightarrow u \in G^{\varrho}(\Omega),$$

where $G^{\varrho}(\Omega)$ is an anisotropic Gevrey space associated with $P(D)$.

A large class of hypoelliptic differential operators is the class of multi-quasi-elliptic differential operators, see V.P. Mikhaïlov [8], J. Friberg [3] and S.G. Gindikin, L.R. Volevich [4].

L. Zanghirati [10], proved that multi-quasielliptic differential operators are multi-anisotropic Gevrey hypoelliptic, i.e.,

$$u \in \mathcal{D}'(\Omega) \quad \text{and} \quad P(D)\,u \in G^{s,\Gamma}(\Omega) \Longrightarrow u \in G^{s,\Gamma}(\Omega),$$

where $G^{s,\,\Gamma}(\Omega)$ is a Gevrey multi-anisotropic space associated with $P(D)$. This result clarifies the classical result of Hörmander in the case of multi-quasielliptic operators. The result of L. Zanghirati has been extended by C. Bouzar and R. Chaïli in [1] to multi-quasielliptic systems of differential operators.

The aim of this paper is to prove the multi-anisotropic Gevrey regularity of hypoelliptic linear differential operators with complex constant coefficients, and consequently we precise the result of Hörmander and extend the result of Zanghirati.

2. Multi-quasiellipticity

Let Ω be an open subset of $\mathbb{R}^n$, if $\alpha = (\alpha_1, \dots, \alpha_n) \in \mathbb{Z}_+^n$, $q = (q_1, \dots, q_n) \in \mathbb{R}_+^n$ and $\xi = (\xi_1, \dots, \xi_n) \in \mathbb{R}^n$, we set

$$|\alpha| = \alpha_1 + \cdots + \alpha_n, \qquad \langle \alpha, q\rangle = \sum\nolimits_{j=1}^n \alpha_j q_j, \qquad \xi^\alpha = \xi_1^{\alpha_1} \dots \xi_n^{\alpha_n},$$

$$D^\alpha = D_1^{\alpha_1} \dots D_n^{\alpha_n}, D_j = \frac{1}{i}\frac{\partial}{\partial \xi_j}, \; j = 1, \dots, n,$$

$$\mathbb{R}_+^n = \{\xi \in \mathbb{R}^n : \xi_j > 0, \; j = 1, \dots, n\}.$$

The space $\mathcal{C}_0^\infty(\Omega)$ is the space of functions $u \in \mathcal{C}^\infty$ with compact support in Ω. The space of distributions on Ω is denoted $\mathcal{D}'(\Omega)$.

Definition 2.1. Let A be a finite subset of $\overline{\mathbb{R}_+^n}$, the Newton's polyhedron of A, denoted $\Gamma(A)$, is the convex hull of $\{0\} \cup A$.

A Newton's polyhedron Γ is always characterized by

$$\Gamma = \bigcap_{q \in \mathcal{A}(\Gamma)} \left\{\alpha \in \overline{\mathbb{R}_+^n}, \; \langle q, \alpha\rangle \le 1\right\},$$

where $\mathcal{A}(\Gamma)$ is a finite subset of $\mathbb{R}^n \; \{0\}$.

Definition 2.2. Let $\Gamma = \bigcap_{q \in \mathcal{A}(\Gamma)} \left\{\alpha \in \overline{\mathbb{R}_+^n}, \; \langle q, \alpha\rangle \le 1\right\}$ be a Newton's polyhedron, Γ is said to be regular, if

$$q_j > 0, \; \forall j = 1, \dots, n; \; \forall q = (q_1, \dots, q_n) \in \mathcal{A}(\Gamma)$$

We associate with a regular Newton's polyhedron Γ the following elements

$$\mathcal{V}(\Gamma) = \left\{s^0 = 0, \; s^1, \dots, s^{m(\Gamma)}\right\} \text{ the set of vertices of } \Gamma$$

$$|\xi|_\Gamma = \sum_{\nu \in \mathcal{V}(\Gamma)} |\xi|^\nu, \; \xi \in \mathbb{R}^n, \text{ where } |\xi|^\nu = |\xi_1|^{\nu_1} \dots |\xi_n|^{\nu_n}$$

$$k(\alpha, \Gamma) = \inf\left\{t > 0, \; t^{-1}\alpha \in \Gamma\right\} = \max_{q \in \mathcal{A}(\Gamma)} \langle \alpha, q\rangle, \; \alpha \in \mathbb{R}_+^n$$

$$\mu(\Gamma) = \max_{\substack{q \in \mathcal{A}(\Gamma) \\ 1 \le j \le n}} q_j^{-1} \text{ called the formal order of } \Gamma$$

A differential operators with complex constant coefficients $P(D) = \sum_\alpha a_\alpha D^\alpha$ has its complete symbol

$$P(\xi) = \sum_\alpha a_\alpha \xi^\alpha.$$

Definition 2.3. The Newton's polyhedron of P, denoted $\Gamma(P)$, is the convex hull of the set $\{0\} \cup \{\alpha \in \mathbb{Z}_+^n : a_\alpha \neq 0\}$.

Define the weight function

$$|\xi|_{\mathbb{P}} = \sum_{\alpha \in \mathcal{V}(P)} |\xi^\alpha|, \ \forall \xi \in \mathbb{R}^n,$$

where $\mathcal{V}(P) = \mathcal{V}(\Gamma(P))$ is the set of vertices of $\Gamma(P)$. Recall

$$d(\xi) := \operatorname{dist}(\xi, N(P)), \text{ where } N(P) := \{\zeta \in \mathbb{C}^n : P(\zeta) = 0\}$$

Definition 2.4. The differential operator $P(D)$ is said hypoelliptic in Ω, if

$$\operatorname{singsupp} P(D)u = \operatorname{singsupp} u, \ \forall u \in \mathcal{D}'(\Omega)$$

The characterization of hypoelliptic differential operators with constant coefficients is du to L. Hörmander. The following result, see Theorem 4.1.3 of [6], gives some characterizations of the hypoellipticity.

Theorem 2.5. *Let $P(D)$ be a differential operator with constant coefficients, the following properties are equivalent:*

i) *The operator $P(D)$ is hypoelliptic.*
ii) $\exists C > 0$, $\exists d > 0$, $|\xi|^d \leq C d(\xi)$, $\forall \xi \in \mathbb{R}^n$, $|\xi|$ *large.*
iii) *If* $\xi \in \mathbb{R}^n$, $|\xi| \to +\infty$, *then* $\dfrac{|D^\alpha P(\xi)|}{|P(\xi)|} \to 0$, $\forall \alpha \neq 0$.
iv) $\exists C > 0$, $\exists \rho > 0$, $\dfrac{|D^\alpha P(\xi)|}{|P(\xi)|} \leq C |\xi|^{-\rho|\alpha|}$, $\forall \xi \in \mathbb{R}^n$, $|\xi|$ *large.*

The connection between an hypoelliptic operator and its Newton's polyhedron is given by the following proposition.

Proposition 2.6. *The Newton's polyhedron of an hypoelliptic differential operator is regular.*

Proof. See [3]. □

Remark 2.7. The converse is not true, $\square = D_x^2 - D_y^2$ has a regular Newton's polyhedron with vertices $\{(0,0), (2,0), (0,2)\}$, but the operator $\square$ is not hypoelliptic.

We introduce multi-quasielliptic polynomials which are a natural generalization of the classical quasi-elliptic operators. These operators were characterized first by V.P. Mikhaïlov [8], then studied by J. Friberg [3] and finally far developed by S.G. Gindikin and L.R. Volevich [4].

Definition 2.8. The polynomial $P(\xi) = \sum_\alpha a_\alpha \xi^\alpha$ is said to be multi-quasielliptic, if

i) its Newton's polyhedron $\Gamma(P)$ is regular.
ii) $\exists C > 0$ such that

$$|\xi|_{\mathbb{P}} \leq C(1 + |P(\xi)|), \ \forall \xi \in \mathbb{R}^n$$

Proposition 2.9. *A multi-quasielliptic operator $P(D)$ is hypoelliptic.*

Proof. See [3] or [4]. □

Remark 2.10. The converse is not true. Indeed, consider the following polynomial

$$P(\xi,\eta) = i\xi^5 + i\xi\eta^4 - 4i\xi^4\eta - 4i\xi^2\eta^3 + 6i\xi^3\eta^2 + i\xi^3 + i\xi\eta^2 + \xi^4\eta^2 + \eta^6 - 4\xi^3\eta^3 - 4\xi\eta^5 + 6\xi^2\eta^4 + \eta^2\xi^2 + \eta^4,$$

which is hypoelliptic thanks to Theorem 4.1.9 of [6]. We have

$$P_{(1,1)}(\xi,\eta) = \eta^2\left(\xi^4 + \eta^4 - 4\xi^3\eta - 4\xi\eta^3 + 6\xi^2\eta^2\right) = \eta^2(\xi-\eta)^4$$

The $q = (1,1)$-quasiprincipal part of $P(\xi,\eta)$ degenerates on the straight $\xi = \eta$, hence the polynomial $P(\xi,\eta)$ is not multi-quasielliptic, see [4].

3. Multi-anisotropic Gevrey vectors

The multi-anisotropic Gevrey spaces were explicitly defined by L. Zanghirati in [10] for studying the multi-anistropic Gevrey regularity of multi-quasielliptic differential operators by the method of elliptic iterates.

Definition 3.1. Let Ω be an open subset of $\mathbb{R}^n$, Γ a regular Newton's polyhedron and $s \geq 1$. Denote $G^{s,\,\Gamma}(\Omega)$ the space of functions $u \in \mathcal{C}^\infty(\Omega)$ such that $\forall K \subset \Omega, \exists C > 0, \forall \alpha \in \mathbb{Z}_+^n$,

$$\sup_{x\in K} |D^\alpha u(x)| \leq C^{|\alpha|+1} k(\alpha,\Gamma)^{sk(\alpha,\Gamma)} \tag{3.1}$$

Example. If Γ is the regular Newton's polyhedron defined by

$$\Gamma = \left\{\alpha \in \overline{\mathbb{R}_+^n} : \sum_{j=1}^n m_j^{-1}\alpha_j \leq 1, m_j \in \mathbb{R}_+\right\},$$

then

$$G^{s,\,\Gamma}(\Omega) = \left\{ \begin{array}{c} u \in \mathcal{C}^\infty(\Omega), \forall K \subset \Omega, \exists C > 0, \forall \alpha \in \mathbb{Z}_+^n \\ |D^\alpha u(x)| \leq C^{|\alpha|+1} \langle \alpha, q\rangle^{s\langle\alpha,q\rangle} \end{array} \right\},$$

where $q := \left(\dfrac{m}{m_1}, \ldots, \dfrac{m}{m_n}\right)$ and $m := \max\limits_j m_j$, i.e., $G^{s,\,\Gamma}(\Omega)$ is the classical anisotropic Gevrey space $G^{s,\,q}(\Omega)$. If $m_1 = m_2 = \cdots = m_n$, we obtain the classical isotropic Gevrey space $G^s(\Omega)$.

Definition 3.2. Let Γ be the regular Newton's polyhedron of $P(D)$ and $s \geq 1$, the space of Gevrey vectors of $P(D)$, denoted $G^s(\Omega, P)$, is the space of $u \in \mathcal{C}^\infty(\Omega)$ such that, $\forall K$ compact of Ω, $\exists C > 0, \forall l \in \mathbb{N}$,

$$\left\|P^l u\right\|_{L^\infty(K)} \leq C^{l+1}(l!)^{s\mu(\Gamma)}$$

Remark 3.3. We can take $l^{sl\mu(\Gamma)}$ instead of $(l!)^{s\mu(\Gamma)}$.

We recall a result of L. Zanghirati [10] and C. Bouzar and R. Chaïli [1] which gives the multi-anisotropic Gevrey regularity of Gevrey vectors of multi-quasielliptic operators.

Theorem 3.4. *Let Ω be an open subset of $\mathbb{R}^n$, $s > 1$ and P a linear differential operator with complex constant coefficients with regular Newton's polyhedron Γ. Then the following assertions are equivalent:*

i) *P is multi-quasielliptic in Ω*
ii) $G^s(\Omega, P) = G^{s,\,\Gamma}(\Omega)$

4. Multi-anisotropic Gevrey hypoellipticity of hypoelliptic operators

In this section, $P = \sum\limits_{\alpha} a_\alpha D^\alpha$ is an hypoelliptic differential operator with complex constant coefficients.

Definition 4.1. A finite set $\mathcal{H} \subset \overline{\mathbb{R}^n_+}$ is said a polyhedron of hypoellipticity of P, if

1. $\forall \nu \in \mathcal{H}, \exists C > 0, \forall \xi \in \mathbb{R}^n,\ |\xi^\nu| \le C\,(1 + d\,(\xi))$
2. $\mathcal{H}$ has vertices with rational components.
3. $\mathcal{H}$ is regular.

Remark 4.2. If ν belongs to the convex hull of $\mathcal{H}$, i.e., $\nu = \sum\limits_{\substack{i\in I\\ I\ fini}} \lambda_i\beta_i$, where $\beta_i \in \mathcal{H}$ and $\sum\limits_{i\in I} \lambda_i = 1, \lambda_i \ge 0$, then $|\xi|^\nu \le C\,(1 + d(\xi))\,, \forall \xi \in \mathbb{R}^n$, therefor it is natural to assume that $\mathcal{H}$ is convex.

Definition 4.3. Denote σ be the smallest natural integer such that

$$\sigma\mathcal{V}\,(\mathcal{H}) \subset 2\mathbb{N}_0^n,$$

and define the differential operator $Q_{\mathcal{H}}\,(D)$, by

$$Q_{\mathcal{H}}\,(D) = \sum_{\alpha\in\mathcal{V}(\mathcal{H})} D^{\sigma\alpha}$$

Proposition 4.4. *The operator $Q_{\mathcal{H}}\,(D)$ is multi-quasielliptic.*

Proof. The Newton's polyhedron of the differential operator $Q_{\mathcal{H}}$ has vertices with even positive integer components. Then

$$|Q_{\mathcal{H}}\,(\xi)| = \sum_{\alpha\in\mathcal{V}(\mathcal{H})} |\xi^{\sigma\alpha}| = |\xi|_{Q_{\mathcal{H}}}\,,$$

hence

$$1 + |\xi|_{Q_{\mathcal{H}}} \le (1 + |Q_{\mathcal{H}}\,(\xi)|)\,,\ \forall \xi \in \mathbb{R}^n \qquad \square$$

Let $v \in \mathcal{C}_0^\infty(\mathbb{R}^n)$, $s \in \mathbb{Z}_+$ and $\varepsilon > 0$, then

$$|||v|||_{s,\varepsilon}^2 := \int_{\mathbb{R}^n} (1 + \varepsilon d(\xi))^s |\widehat{v}(\xi)|^2 \, d\xi$$

The following result is Lemma 4.4.3 of [6].

Lemma 4.5. *Let u be a solution of the equation $Pu = 0$ defined in the ball $B_\varepsilon = \{x \in \mathbb{R}^n : |x| < \varepsilon\}$, and let $\varphi \in \mathcal{C}_0^\infty(B_1)$ and the integer $s \geq 1$. Then*

$$\sum_{\alpha \neq 0} \varepsilon^{-2|\alpha|} \left|\left|\left| P^{(\alpha)}(D)(\varphi^\varepsilon u) \right|\right|\right|_{s,\varepsilon}^2 \leq C \sum_{\alpha \neq 0} \varepsilon^{-2|\alpha|} \int_{B_\varepsilon} \left| P^{(\alpha)}(D) u \right|^2 dx, \tag{4.1}$$

where C is independent of ε and u.

Remark 4.6. In the lemma φ^ε denotes $\varphi^\varepsilon(x) := \varphi\left(\frac{x}{\varepsilon}\right)$.

Thanks to this lemma, we obtain the following result.

Lemma 4.7. *Let $\beta \in \mathbb{Z}_+^n \cap \sigma\mathcal{H}$, then there exists a constant $C > 0$, such that for every solution u of $Pu = 0$ in B_ε and $\varepsilon \in \,]0, 1[$, we have*

$$\varepsilon^{2\sigma} \sum_{\alpha \neq 0} \varepsilon^{-2|\alpha|} \int_{B_{\frac{\varepsilon}{2}}} \left| P^{(\alpha)}(D) D^\beta u \right|^2 dx \leq C \sum_{\alpha \neq 0} \varepsilon^{-2|\alpha|} \int_{B_\varepsilon} \left| P^{(\alpha)}(D) u \right|^2 dx$$

Proof. Let $\beta \in \mathbb{Z}_+^n \cap \sigma\mathcal{H}$, from (1) of definition 4.1, we have

$$\left|\xi^\beta\right| \leq C^\sigma (1 + d(\xi))^\sigma,$$

hence $\exists C > 0$, $\forall \varepsilon \in \,]0, 1[$, $\forall \xi \in \mathbb{R}^n$,

$$\varepsilon^\sigma \left|\xi^\beta\right| \leq C^\sigma d_{\sigma,\varepsilon}(\xi) \tag{4.2}$$

Multiplying (4.2) by $(2\pi)^{-n} |\widehat{v}(\xi)|$ and integrating with respect to ξ, we obtain

$$\varepsilon^{2\sigma} \int \left| D^\beta v \right|^2 dx \leq C^2 |||v|||_{\sigma,\varepsilon}^2 \tag{4.3}$$

Let $\varphi \in C_0^\infty(B_1)$ equals 1 in $B_{\frac{1}{2}}$ and apply the estimate (4.3) to $v = P^{(\alpha)}(D)(\varphi^\varepsilon u)$, then

$$\varepsilon^{2\sigma} \int \left| P^{(\alpha)}(D) D^\beta (\varphi^\varepsilon u) \right|^2 dx \leq C^2 \left|\left|\left| P^{(\alpha)}(D)(\varphi^\varepsilon u) \right|\right|\right|_{\sigma,\varepsilon}^2$$

$$\varepsilon^{2\sigma} \sum_{\alpha \neq 0} \varepsilon^{-2|\alpha|} \int \left| P^{(\alpha)}(D) D^\beta (\varphi^\varepsilon u) \right|^2 dx \leq C^2 \sum_{\alpha \neq 0} \varepsilon^{-2|\alpha|} \left|\left|\left| P^{(\alpha)}(D)(\varphi^\varepsilon u) \right|\right|\right|_{\sigma,\varepsilon}^2,$$

consequently Lemma 4.6 gives

$$\varepsilon^{2\sigma} \sum_{\alpha \neq 0} \varepsilon^{-2|\alpha|} \int \left| P^{(\alpha)}(D) D^\beta (\varphi^\varepsilon u) \right|^2 dx \leq C \sum_{\alpha \neq 0} \varepsilon^{-2|\alpha|} \int_{B_\varepsilon} \left| P^{(\alpha)}(D)(u) \right|^2 dx$$

As $\varphi^{\varepsilon}(x) = \varphi\left(\frac{x}{\varepsilon}\right) = 1$ in $B_{\frac{\varepsilon}{2}}$, then

$$\varepsilon^{2\sigma}\sum_{\alpha\neq 0}\varepsilon^{-2|\alpha|}\int_{B_{\frac{\varepsilon}{2}}}\left|P^{(\alpha)}(D)\,D^{\beta}u\right|^{2}dx \leq C\sum_{\alpha\neq 0}\varepsilon^{-2|\alpha|}\int_{B_{\varepsilon}}\left|P^{(\alpha)}(D)\,(u)\right|^{2}dx \qquad \square$$

Proposition 4.8. *Let Ω be a bounded open set in $\mathbb{R}^n$ and $\beta \in \mathbb{Z}_{+}^{n} \cap \sigma\mathcal{H}$, then there exists a constant $C > 0$, such that for every u solution of $Pu = 0$ in Ω and $\delta \in \left]0,1\right[$, we have*

$$\sum_{\alpha\neq 0}\delta^{-2|\alpha|}\int_{\Omega_{\delta}}\left|P^{(\alpha)}(D)\,D^{\beta}u\right|^{2}dx \leq C\delta^{-2\sigma}\sum_{\alpha\neq 0}\delta^{-2|\alpha|}\int_{\Omega}\left|P^{(\alpha)}(D)\,u\right|^{2}dx,$$

where

$$\Omega_{\delta} = \{x \in \Omega : \operatorname{dist}(x, \partial\Omega) > \delta\}$$

Proof. The proof is obtained from the precedent lemma and follows the same reasoning as the proof of Theorem 4.4.2 of [6]. $\square$

Corollary 4.9. *Let $P(D)$ an hypoelliptic operator, then $\exists C > 0$ such that for every solution of $P(D)\,u = 0$ in Ω,$\forall\varepsilon \in \left]0,1\right[$, $\forall j = 1, 2, \ldots$, we have*

$$\begin{aligned}\varepsilon^{2\sigma}\sum_{0\neq\alpha\in\mathbb{N}_0^n}\varepsilon^{-2|\alpha|}\left\|Q_{\mathcal{H}}(D)\,P^{(\alpha)}(D)\,u\right\|^{2}_{L^2(\Omega_{\varepsilon j})}\\ \leq C\sum_{0\neq\alpha\in\mathbb{N}_0^n}\varepsilon^{-2|\alpha|}\left\|P^{(\alpha)}(D)\,u\right\|^{2}_{L^2\left(\Omega_{\varepsilon(j-1)}\right)}\end{aligned} \tag{4.4}$$

The principal result of this section is the following theorem.

Theorem 4.10. *Let u be a solution of the hypoelliptic equation $P(D)\,u = 0$ in Ω, then for every $\omega \subset\subset \Omega$, there is a constant $C > 0$, such that $\forall j \in \mathbb{N}$, we have*

$$\left\|Q^{j}_{\mathcal{H}}(D)\,u\right\|_{L^2(\omega)} \leq C^{(j+1)}j^{\sigma j} \tag{4.5}$$

Proof. Since $\rho = \rho(\omega, \partial\Omega) > 0$, then there exists $\delta \in \left]0,\rho\right[$ such that $\omega \subset \Omega_{\delta} \subset \Omega$. Take $\varepsilon = \dfrac{\delta}{j}$, $j \in \mathbb{N}$, and let us show by induction on j the following estimate

$$\varepsilon^{2j\sigma+2m}\sum_{0\neq\alpha\in\mathbb{N}_0^n}\varepsilon^{-2|\alpha|}\left\|\left(Q^{j}_{\mathcal{H}}(D)\,P^{(\alpha)}(D)\,u\right)\right\|^{2}_{L^2(\Omega_{j\varepsilon})} < C^{2(j+1)}, \tag{4.6}$$

where m is the order of P.

As every solution u of an hypoelliptic equation is $\mathcal{C}^{\infty}$, then there exists $C > 0$ such that (4.6) is satisfied for $j = 0$. Suppose that (4.6) is true for $j \leq l$ $(l \geq 0)$, we have to prove that it remains true for $j = l + 1$. Since $v = Q^{l}_{\mathcal{H}}(D)\,u$ is also a solution of equation $P(D)u = 0$, then from Corollary 4.10, we obtain

$$\begin{aligned}
&\varepsilon^{2\sigma(l+1)+2m}\sum_{0\neq\alpha}\varepsilon^{-2|\alpha|}\left\|\left(Q_{\mathcal{H}}^{l+1}(D)\,P^{(\alpha)}(D)\,u\right)\right\|^2_{L^2\left(\Omega_{\varepsilon(l+1)}\right)}\\
&\leq C\varepsilon^{2\sigma l+2m}\sum_{0\neq\alpha}\varepsilon^{-2|\alpha|}\left\|P^{(\alpha)}(D)\,Q_{\mathcal{H}}^{l}(D)\,u\right\|^2_{L^2(\Omega_{\varepsilon l})}
\end{aligned} \tag{4.7}$$

By the induction hypothesis, we have

$$\varepsilon^{2\sigma l+2m}\sum_{0\neq\alpha}\varepsilon^{-2|\alpha|}\left\|P^{(\alpha)}(D)\,Q_{\mathcal{H}}^{l}(D)\,u\right\|^2_{L^2(\Omega_{\varepsilon l})}\leq C_1^{2(l+1)},$$

consequently, we obtain

$$\varepsilon^{2\sigma(l+1)+2m}\sum_{0\neq\alpha}\varepsilon^{-2|\alpha|}\left\|\left(Q_{\mathcal{H}}^{l+1}(D)\,P^{(\alpha)}(D)\,u\right)\right\|^2_{L^2\left(\Omega_{\varepsilon(l+1)}\right)}\leq C_2^{2(l+2)},$$

hence $\forall j\in\mathbb{N}$, we have

$$\varepsilon^{2\sigma j+2m}\sum_{0\neq\alpha}\varepsilon^{-2|\alpha|}\left\|P^{(\alpha)}(D)\,Q_{\mathcal{H}}^{j}(D)\,u\right\|^2_{L^2(\Omega_{\varepsilon j})}\leq C_2^{2(j+1)} \tag{4.8}$$

The estimate (4.8) with $|\alpha| = m$ gives $\forall j\in\mathbb{N}$,

$$\left\|Q_{\mathcal{H}}^{j}(D)\,u\right\|^2_{L^2(\Omega_{\varepsilon j})}\leq\varepsilon^{-2\sigma j}C_2^{2(j+1)},$$

as $\varepsilon = \dfrac{\delta}{j}$, then

$$\left\|Q_{\mathcal{H}}^{j}(D)\,u\right\|^2_{L^2(\Omega_{\varepsilon j})}\leq\left(\frac{j}{\delta}\right)^{2\sigma j}C_2^{2(j+1)}\leq C^{2(j+1)}j^{2\sigma j},$$

hence

$$\left\|Q_{\mathcal{H}}^{j}(D)\,u\right\|_{L^2(\Omega_{j\varepsilon})}\leq C^{(j+1)}j^{\sigma j}$$

□

We denote $G^{s,\,\mathcal{H}}(\Omega)$ the multi-anisotropic Gevrey space associated with $\mathcal{H}$ and by $\mu_{\mathcal{H}}$ and μ_Q the respective formal orders of the Newton's polyhedrons $\mathcal{H}$ and $\Gamma(Q_{\mathcal{H}})$, then we have the following relations

$$\Gamma(Q_{\mathcal{H}}) = \sigma\mathcal{H} \quad\text{and}\quad \mu_Q = \sigma\mu_{\mathcal{H}}$$

The principal result of this paper is the following theorem.

Theorem 4.11. *Every solution $u\in\mathcal{D}'(\Omega)$ of the hypoelliptic equation $P(D)u = 0$ is a function of $G^{\frac{\sigma}{\mu_{\mathcal{H}}},\,\mathcal{H}}(\Omega)$.*

Proof. Theorem 4.11 says that every u solution of the hypoelliptic equation $Pu = 0$ is a Gevrey vector of the operator $Q_{\mathcal{H}}$, i.e., we have $u\in G^{\frac{\sigma}{\mu_Q}}(\Omega, Q_{\mathcal{H}})$. From Theorem 3.4 and as the operator $Q_{\mathcal{H}}$ is multi-quasielliptic, then we have $u\in G^{\frac{\sigma}{\mu_Q},\,\Gamma(Q_{\mathcal{H}})}(\Omega)$, and consequently $u\in G^{\frac{1}{\mu_{\mathcal{H}}},\,\sigma\mathcal{H}}(\Omega)$. A simple computation shows that in general $G^{s,\,\sigma\mathcal{H}}(\Omega) = G^{s\sigma,\,\mathcal{H}}\Omega$, hence $u\in G^{\frac{\sigma}{\mu_{\mathcal{H}}},\,\mathcal{H}}(\Omega)$. □

Remark 4.12. It is interesting to compare the result of the theorem with the microlocal Gevrey regularity result obtained in [2].

References

[1] C. Bouzar and R. Chaïli, Gevrey vectors of multi-quasielliptic systems, *Proc. Amer. Math. Soc.* **131**(5) (2003), 1565–1572.

[2] C. Bouzar and R. Chaïli, A Gevrey microlocal analysis of multi-anisotropic differential operators, *Rend. Sem. Mat. Univ. Pol. Torino,* **64**(3) (2006), 305–318.

[3] J. Friberg, Multi-quasielliptic polynomials, *Ann. Sc. Norm. Sup. Pisa. Cl. di. Sc.* **21** (1967), 239–260.

[4] S.G. Gindikin and L.R. Volevich, *The Method of Newton Polyhedron in the Theory of Partial Differential Equations*, Kluwer, 1992.

[5] G.H. Hakobyan, Estimates of the higher order derivatives of the solution of hypoelliptic equations, *Rend. Sem. Mat. Univ. Pol. Torino,* **61**(4) (2003), 443–459.

[6] L. Hörmander, *Linear Partial Differential Operators,* Springer-Verlag, 1969.

[7] L. Hörmander, *The Analysis of Linear Partial Differential Operators I. Distribution Theory and Fourier Analysis,* Second Edition, Springer-Verlag, 1990.

[8] V.P. Mikhaïlov, The behavior at infinity of a class of polynomials, *Proc. Steklov. Inst. Mat.* **91** (1967), 59–80.

[9] L. Rodino, *Linear Partial Differential Operators in Gevrey Spaces,* World Scientific, 1993.

[10] L. Zanghirati, Iterati di una classe di operatori ipoellíptici e classi generalizzate di Gevrey, *Suppl. Boll. U.M.I.* (1980), 177–195.

C. Bouzar
Department of Mathematics
Oran-Essenia University
Oran, Algeria
e-mail: `bouzar@univ-oran.dz`

A. Dali
Institut des Sciences exactes
Centre Universitaire de Bechar, Algeria
e-mail: `Ahmedalimat@yahoo.fr`

Operator Theory:
Advances and Applications, Vol. 189, 275–285

Modified Stockwell Transforms and Time-Frequency Analysis

Qiang Guo, Shahla Molahajloo and M.W. Wong

Abstract. We give results complementary to those in the paper [4] from the perspective of time-frequency analysis [1] to the effect that high frequencies can be amplified and low frequencies diminished. Time-frequency spectra for the chirp, the sum of the cosine functions and the Gaussian-modulated sinusoidal pulse are presented for comparisons. The property of the absolutely referenced phase information of modified Stockwell transforms is given in terms of Riesz transforms.

Mathematics Subject Classification (2000). Primary 65R10, 92A55, 94A12; Secondary 47G10, 47G30.

Keywords. Stockwell transforms, Stockwell spectra, signals, resolution of the identity formula, absolutely referenced phase information, Riesz transforms.

1. Introduction

Let $\varphi \in L^1(\mathbb{R}) \cap L^2(\mathbb{R})$ be such that $\int_{-\infty}^{\infty} \varphi(x)\, dx = 1$. Then, as a hybrid of the Gabor transform and the wavelet transform, the Stockwell transform $S_\varphi f$ of a signal f in $L^2(\mathbb{R})$ with respect to the window φ is defined by

$$(S_\varphi f)(b,\xi) = (2\pi)^{-1/2}|\xi| \int_{-\infty}^{\infty} e^{-ix\xi} f(x)\overline{\varphi(\xi(x-b))}dx$$

for all b in $\mathbb{R}$ and ξ in $\mathbb{R}\setminus\{0\}$. Putting the Stockwell transform in perspective, we note that for all $f \in L^2(\mathbb{R})$, $b \in \mathbb{R}$ and $\xi \in \mathbb{R}\setminus\{0\}$,

$$(S_\varphi f)(b,\xi) = (f, \varphi^{b,\xi})_{L^2(\mathbb{R})},$$

where $(\, ,\,)_{L^2(\mathbb{R})}$ is the inner product in $L^2(\mathbb{R})$ and

$$\varphi^{b,\xi} = (2\pi)^{-1/2} M_\xi T_{-b} D^1_\xi \varphi.$$

This research has been supported by the Natural Sciences and Engineering Research Council of Canada.

Here, M_ξ, T_{-b} and D^1_ξ are the modulation operator, the translation operator and the dilation operator defined, respectively, by

$$(M_\xi h)(x) = e^{ix\xi}h(x),$$
$$(T_{-b}h)(x) = h(x-b)$$

and

$$(D^1_\xi h)(x) = |\xi| h(\xi x)$$

for all x in $\mathbb{R}$ and all measurable functions h on $\mathbb{R}$. Applications of Stockwell transforms abound and can be found in [6, 7].

In the paper [4] are first introduced the modified Stockwell transforms S^s_φ parametrized by s, where $1 \le s \le \infty$. These modified Stockwell transforms include the classical Stockwell transforms when $s = 1$ and are reminiscent of the wavelet transforms as explained in Remark 1.2 when $s = 2$.

To recall, let $\varphi \in L^1(\mathbb{R}) \cap L^2(\mathbb{R})$ be such that $\int_{-\infty}^{\infty} \varphi(x)\,dx = 1$. Let $f \in L^2(\mathbb{R})$. Then for $1 \le s \le \infty$, we define the modified Stockwell transform $S^s_\varphi f$ of f by

$$(S^s_\varphi f)(b,\xi) = (f, \varphi^{b,\xi}_s)_{L^2(\mathbb{R})}, \quad b \in \mathbb{R},\ \xi \in \mathbb{R}\setminus\{0\},$$

where

$$\varphi^{b,\xi}_s = (2\pi)^{-1/2} M_\xi T_{-b} D^s_\xi \varphi$$

and the dilation operator D^s_ξ here is defined by

$$(D^s_\xi h)(x) = |\xi|^{1/s} h(\xi x) \tag{1.1}$$

for all x in $\mathbb{R}$ and all measurable functions h on $\mathbb{R}$. More explicitly,

$$(S^s_\varphi f)(b,\xi) = |\xi|^{-1/s'} (S_\varphi f)(b,\xi), \quad b \in \mathbb{R},\ \xi \in \mathbb{R}\setminus\{0\}, \tag{1.2}$$

where s' is the conjugate index of s, *i.e.*, $\frac{1}{s} + \frac{1}{s'} = 1$, and

$$(S^s_\varphi f)(b,\xi) = (2\pi)^{-1/2} |\xi|^{1/s} \int_{-\infty}^{\infty} e^{-ix\xi} f(x) \overline{\varphi(\xi(x-b))}\,dx$$

for all b in $\mathbb{R}$ and ξ in $\mathbb{R}\setminus\{0\}$.

In view of (1.2), we see that modified Stockwell transforms modulate frequencies in such a way that low frequencies are amplified and high frequencies reduced. This is illustrated in Figure 1 for $s = 1$, $s = 2$ and $s = 8$ when we use the Gaussian window φ given by

$$\varphi(x) = (2\pi)^{-1/2} e^{-x^2/2}, \quad x \in \mathbb{R}.$$

We note in particular that very low frequencies, which are almost undetected by the Stockwell transform, can be seen using the modified Stockwell transform with $s = 2$ and clearly manifested if the modified Stockwell transform with $s = 8$ is used.

With the new normalizing factor for the dilation operator in (1.1), we have the following result.

Theorem 1.1. *For all f in $L^2(\mathbb{R})$,*

$$(S^s_\varphi f)(b,\xi) = (2\pi)^{-1/2} e^{-ib\xi}|\xi|^{(1/2)-(1/s')}(\Omega_\psi f)(b, 1/\xi), \quad b \in \mathbb{R},\ \xi \in \mathbb{R}\setminus\{0\},$$

where

$$\psi(x) = e^{ix}\varphi(x), \quad x \in \mathbb{R},$$

and Ω_ψ is the wavelet transform corresponding to the mother wavelet ψ given by

$$(\Omega_\psi f)(b,a) = \frac{1}{\sqrt{a}} \int_{-\infty}^{\infty} f(x)\overline{\psi\left(\frac{x-b}{a}\right)}\, dx, \quad b \in \mathbb{R},\ a \in \mathbb{R}\setminus\{0\},$$

for all functions f in $L^2(\mathbb{R})$.

Details on wavelet transforms can be found in [2, 8] among others.

Remark 1.2. If we let $s = 2$, then we get the modified Stockwell transform S^2_φ, which comes closest to the wavelet transform. This is particularly so for applications for which the amplitude spectra alone are of interest, but the applicability of the computational techniques available for wavelet transforms is limited and complicated by converting the scaling variable to the frequency variable via the equation $a = 1/\xi$. More precisely, the numerical implementation to carry out $a = 1/\xi$ may introduce disturbing oscillations.

After the presentation of the results in [4] by M.W. Wong at the meeting of the National Centre for Medical Device Developments (NCMDD)[1] held at IBM Hawthorne in New York in February 2008, it was pointed out by the scientists in the audience that a variation of the modified Stockwell transforms to diminish low frequencies and amplify high frequencies should also be of interest. That this is possible can be done easily by looking at the modified Stockwell transforms for $0 < s < 1$ instead of $1 \le s \le \infty$. This is done in Section 2 by comparing the modified Stockwell transform of the chirp for $0 < s < 1$ with that for $1 \le s \le \infty$ given in [4]. This comparison is further illuminated in Section 3 with signals given by the sum of cosines and the Gaussian-modulated sinusoidal pulse. The property of the absolutely referenced phase information of the modified Stockwell transforms for $0 < s \le \infty$ and its connection with Riesz transforms are given in Section 4. At the NCMDD meeting, it was also suggested that dilation of the window φ by some function w of the frequency ξ should have good applications in time-frequency analysis. The study of variants of the Stockwell transforms in this direction led by M.W. Wong and his group has led to new research projects in mathematical analysis and applications.

[1]NCMDD is a new initiative and joint venture in which York University and IBM Markham in the Greater Toronto Area in Canada are key players. It is expected that the Southlake Regional Medical Centre, located in Newmarket, Ontario, Canada, will be an industrial partner.

2. A variation

Let $\varphi \in L^1(\mathbb{R}) \cap L^2(\mathbb{R})$ be such that $\int_{-\infty}^{\infty} \varphi(x)\,dx = 1$. Then for $0 < s < 1$, we define the modified Stockwell transform $S_\varphi^s f$ of a signal f by

$$(S_\varphi^s f)(b,\xi) = (2\pi)^{-1/2}|\xi|^{1/s} \int_{-\infty}^{\infty} e^{-ix\xi} f(x)\overline{\varphi(\xi(x-b))}dx$$

for all b in $\mathbb{R}$ and ξ in $\mathbb{R}\setminus\{0\}$. Equivalently,

$$(S_\varphi^s f)(b,\xi) = (f, \varphi_s^{b,\xi})_{L^2(\mathbb{R})},$$

where

$$\varphi_s^{b,\xi} = (2\pi)^{-1/2} M_\xi T_{-b} D_\xi^s \varphi.$$

Here, the modulation operator M_ξ and the translation operator T_{-b} are the same as before, but D_ξ^s is the dilation operator with s now in $(0,1)$ instead of $[1,\infty]$. Alternatively,

$$(S_\varphi^s f)(b,\xi) = |\xi|^{-1/s'}(S_\varphi f)(b,\xi), \quad b \in \mathbb{R},\ \xi \in \mathbb{R}\setminus\{0\}.$$

It is clear that the modified Stockwell transforms S_φ^s, $0 < s < 1$, amplify the high frequencies and diminish the low frequencies. That this is in fact the case can be illustrated by Figure 2 for the Gaussian window and for $s = 1/2$, $s = 1/4$ and $s = 1/8$, which should be looked at side by side with Figure 1.

Remark 2.1. Theorem 1.1 remains valid for the modified Stockwell transforms S_φ^s with $0 < s < 1$, but now the exponent in $|\xi|^{(1/2)-(1/s')}$ is always greater than $1/2$. Thus, the numerical implementations described in Remark 1.2 can be carried out without disturbing oscillations.

The resolution of the identity formula for the modified Stockwell transforms S_φ^s in Theorem 2.4 of [4] for $1 \le s \le \infty$ can now be extended to the full range $0 < s \le \infty$ in exactly the same way as in [4]. More precisely, we have the following result, which is an analog of Theorem 2.3 in [3] for modified Stockwell transforms.

Theorem 2.2. *Let $\varphi \in L^1(\mathbb{R}) \cap L^2(\mathbb{R})$ be such that $\int_{-\infty}^{\infty} \varphi(x)\,dx = 1$ and*

$$\int_{-\infty}^{\infty} \frac{|\hat{\varphi}(\xi-1)|^2}{|\xi|} d\xi < \infty,$$

where

$$\hat{\varphi}(\xi) = (\mathcal{F}\varphi)(\xi) = (2\pi)^{-1/2} \int_{-\infty}^{\infty} e^{-ix\xi}\varphi(x)\,dx, \quad \xi \in \mathbb{R}.$$

Then for $0 < s \le \infty$ and for all f and g in $L^2(\mathbb{R})$, we get

$$(f,g)_{L^2(\mathbb{R})} = \frac{1}{c_\varphi} \int_{-\infty}^{\infty}\int_{-\infty}^{\infty} (f, \varphi_s^{b,\xi})_{L^2(\mathbb{R})} (\varphi_s^{b,\xi}, g)_{L^2(\mathbb{R})} \frac{db\,d\xi}{|\xi|^{1-(2/s')}},$$

where

$$c_\varphi = \int_{-\infty}^{\infty} \frac{|\hat{\varphi}(\xi-1)|^2}{|\xi|} d\xi.$$

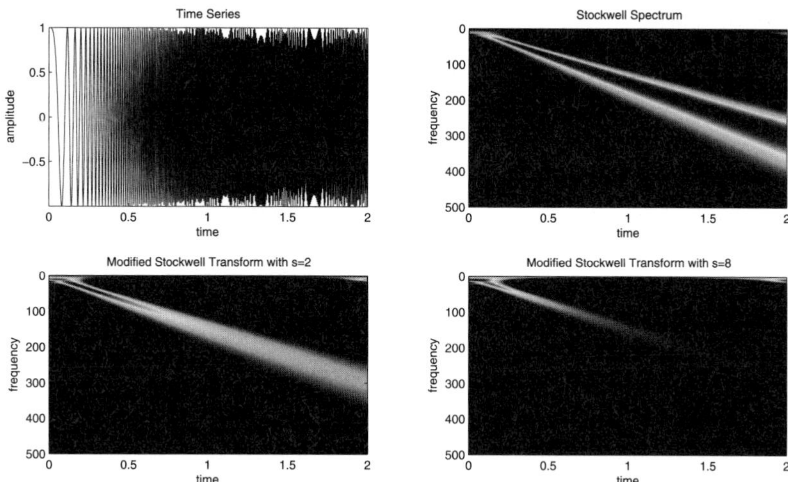

FIGURE 1. Modified Stockwell Spectra of the Chirp, $1 \leq s \leq \infty$

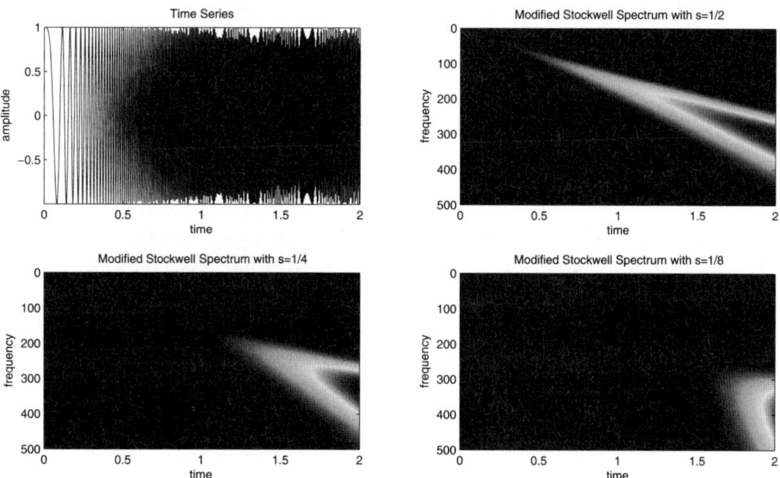

FIGURE 2. Modified Stockwell Spectra of the Chirp, $0 < s < 1$

An immediate corollary is that for $0 < s \leq \infty$, every signal f in $L^2(\mathbb{R})$ can be reconstructed from its time-frequency spectrum $S_\varphi^s f$ via the formula

$$f = \frac{1}{c_\varphi} \int_{-\infty}^{\infty} \int_{-\infty}^{\infty} (S_\varphi^s f)(b, \xi) \varphi_s^{b,\xi} \frac{db\, d\xi}{|\xi|^{1-(2/s')}}.$$

3. Two signals

We first give in this section the modified Stockwell transforms of the signal given by the sum of cosines for selected values of s in $1 \leq s \leq \infty$ and in $0 < s < 1$. Then we do the same for the Gaussian-modulated sinusoidal pulse.

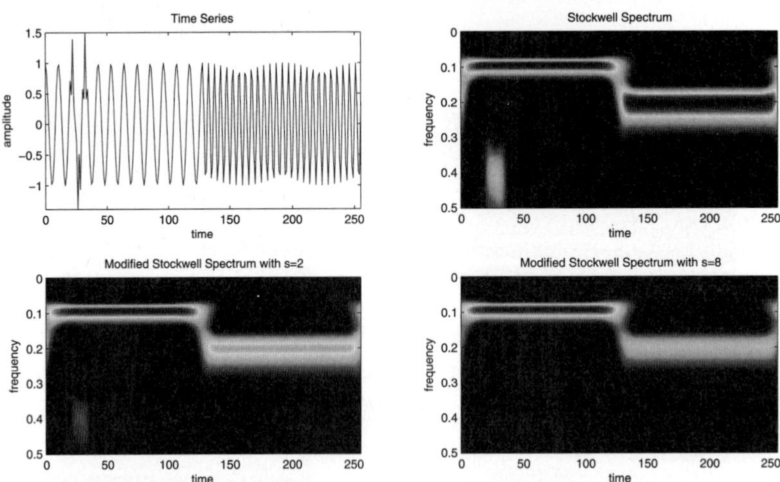

FIGURE 3. Modified Stockwell Spectra of Cosines, $1 \leq s \leq \infty$

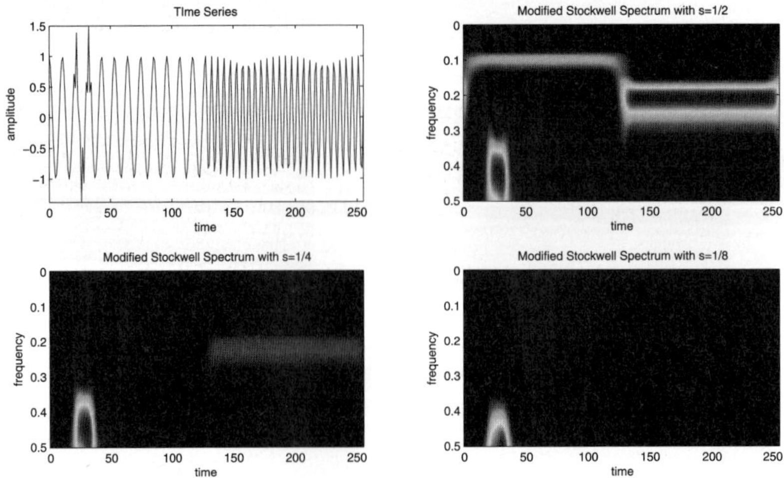

FIGURE 4. Modified Stockwell Spectra of Cosines, $0 < s < 1$

It is again manifested that frequencies are amplified and diminished as in the case of the chirp. The same is true for the Gaussian-modulated sinusoidal pulse.

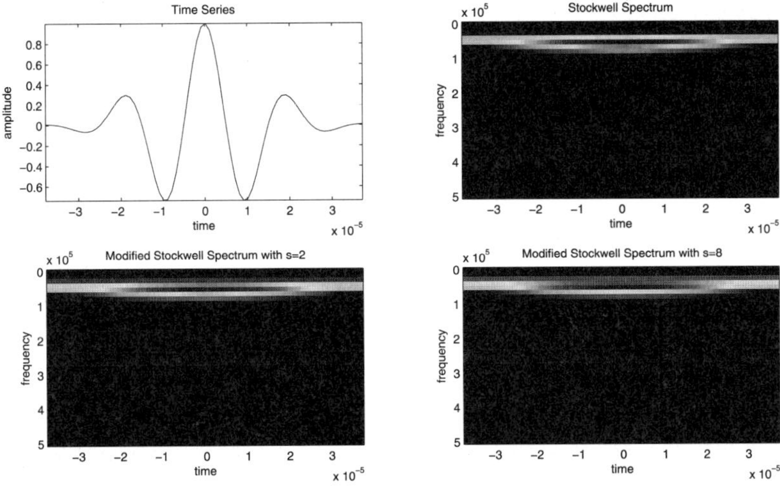

FIGURE 5. Modified Stockwell Spectra of the Pulse, $1 \leq s \leq \infty$

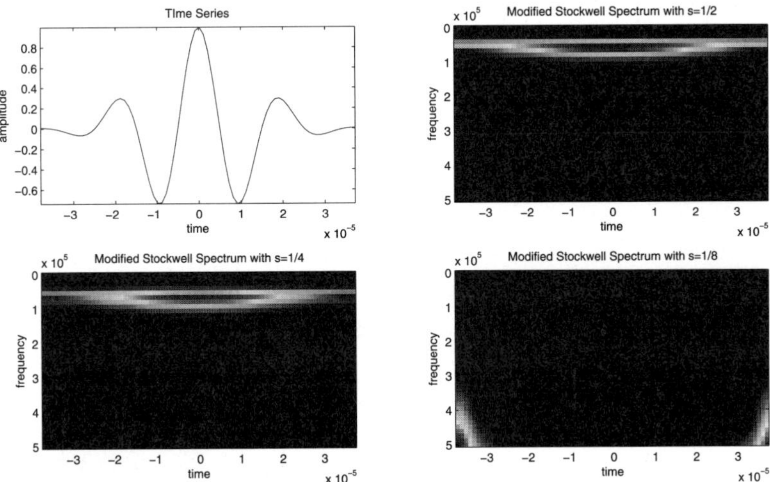

FIGURE 6. Modified Stockwell Spectra of the Pulse, $0 < s < 1$

4. Absolutely referenced phase information

The property of absolutely referenced phase information for the Stockwell transform can be formulated mathematically in the following theorem.

Theorem 4.1. *Let $\varphi \in L^1(\mathbb{R}) \cap L^2(\mathbb{R})$ be such that $\int_{-\infty}^{\infty} \varphi(x)\,dx = 1$. Then for all functions f in $L^1(\mathbb{R}) \cap L^2(\mathbb{R})$,*

$$\int_{-\infty}^{\infty} (S_\varphi f)(b,\cdot)\,db = \hat{f}.$$

Theorem 4.1 can be found in the original paper [7] and a rigorous proof can be found in [9]. By Theorem 4.1 and (1.2), we have the following version of the property of absolutely referenced phase information for modified Stockwell transforms, which is Theorem 8 in [4].

Theorem 4.2. *Let $\varphi \in L^1(\mathbb{R}) \cap L^2(\mathbb{R})$ be such that $\int_{-\infty}^{\infty} \varphi(x)\,dx = 1$. Then for all functions f in $L^1(\mathbb{R}) \cap L^2(\mathbb{R})$,*

$$\mathcal{F}^{-1} \int_{-\infty}^{\infty} (S_\varphi^s f)(b,\cdot)\,db = \mathcal{F}^{-1} |\cdot|^{-1/s'} \mathcal{F} f,$$

where $0 < s \leq \infty$ and $\frac{1}{s} + \frac{1}{s'} = 1$.

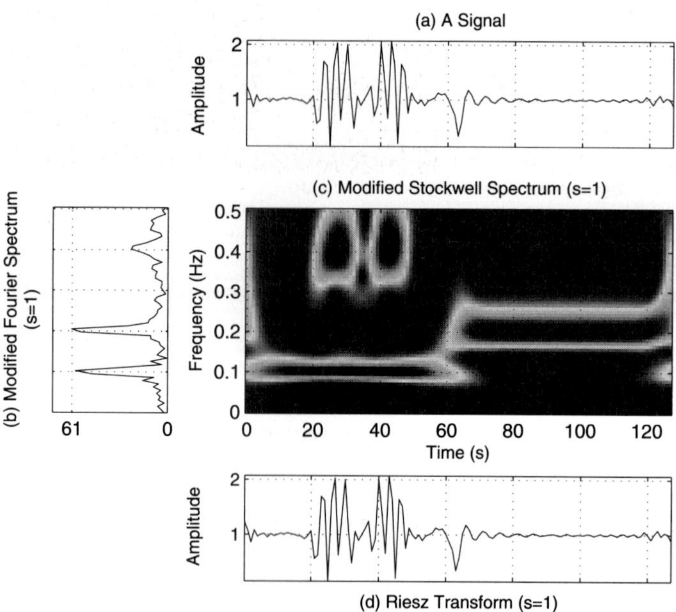

FIGURE 7. Images for $s = 1$

Remark 4.3. For $0 < s \leq \infty$, the function $\mathcal{F}^{-1}|\cdot|^{-1/s'}\mathcal{F}f$ is known as the Riesz transform, also known as the Riesz potential, of the signal f. Chapter V of the book [5] by Stein contains the mathematical analysis of Riesz transforms. Thus, Theorem 4.2 offers a new dimension for an understanding of Riesz transforms in terms of modified Stockwell transforms.

We display in Figure 7 the modified Stockwell transform, the modified Fourier transform and the Riesz transform of a signal for $s = 1$. In this case when the modified Stockwell transform coincides with the Stockwell transform, the modified Fourier transform and the Riesz transform coincide with, respectively, the Fourier transform and the original signal. The corresponding images for $s = 2$ and $s = 1/2$ are given in, respectively, Figure 8 and Figure 9.

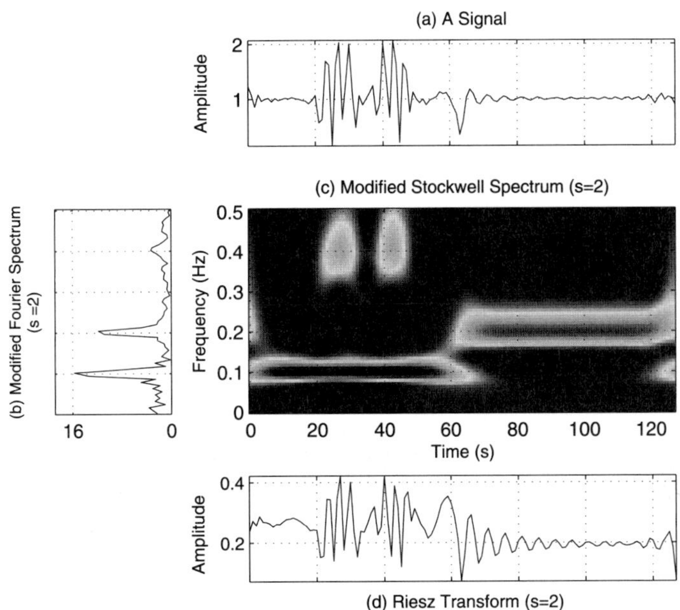

FIGURE 8. Images for $s = 2$

We see from the modified Stockwell spectrum in Figure 8 that indeed the low frequencies are amplified and high frequencies diminished. We also note that the Fourier transform is modified and the Riesz transform of the signal is now different from the original signal.

Since $0 < s < 1$ in Figure 9, the high frequencies are amplified and low frequencies diminished in the modified Stockwell spectrum, as confirmed by the figure.

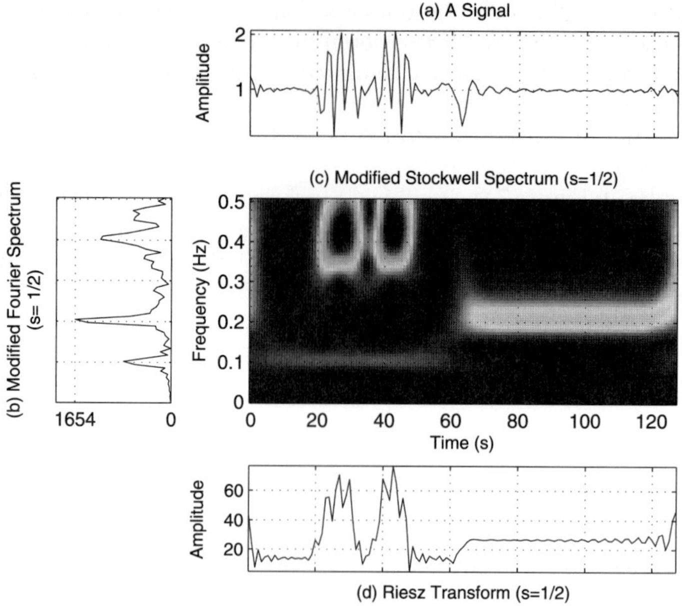

FIGURE 9. Images for $s = 1/2$

5. Conclusions and future research

The results in this paper complement those in [4] and are intended to reflect on the feedback on the presentation of the results in [4] by M.W. Wong at the NCMDD meeting held at IBM Hawthorne in New York in February 2008. More importantly, one of the goals of [4] and this paper is to document the Stockwell transform and its variants as new hybrids of the Gabor transform and the wavelet transform in time-frequency analysis. Most importantly, it is envisaged that new areas of mathematical analysis with applications can be cultivated.

References

[1] L. Cohen, *Time-Frequency Analysis,* Prentice-Hall, 1995.

[2] I. Daubechies, *Ten Lectures on Wavelets*, SIAM, 1992.

[3] J. Du, M.W. Wong and H. Zhu, Continuous and discrete inversion formulas for the Stockwell transform, *Integral Transforms Spec. Funct.* **18** (2007), 537–543.

[4] Q. Guo and M.W. Wong, Modified Stockwell transforms, *Mem. Accademia Sci. Torino*, to appear.

[5] E.M. Stein, *Singular Integrals and Differentiability Properties of Functions*, Princeton University Press, 1971.

[6] R.G. Stockwell, Why use the S transform?, in *Pseudo-Differential Operators: Partial Differential Equations and Time-Frequency Analysis*, Editors: L. Rodino, B.-W. Schulze and M.W. Wong, Fields Institute Communications Series **52**, American Mathematical Society, 2007, 279–309.

[7] R.G. Stockwell, L. Mansinha and R.P. Lowe, Localization of the complex spectrum: the S transform, *IEEE Trans. Signal Processing* **44** (1996), 998–1001.

[8] M.W. Wong, *Wavelet Transforms and Localization Operators*, Birkhäuser, 2002.

[9] M.W. Wong and H. Zhu, A characterization of the Stockwell spectrum, in *Modern Trends in Pseudo-Differential Operators*, Editors: J. Toft, M.W. Wong and H. Zhu, Operator Theory: Advances and Applications **172**, Birkhäuser, 2007, 251–257.

Qiang Guo, Shahla Molahajloo and M.W. Wong
Department of Mathematics and Statistics
York University
4700 Keele Street
Toronto, Ontario M3J 1P3
Canada
e-mail: ppguo@mathstat.yorku.ca
smollaha@mathstat.yorku.ca
mwwong@mathstat.yorku.ca

Operator Theory:
Advances and Applications, Vol. 189, 287–296

Localization Operators for Two-Dimensional Stockwell Transforms

Yu Liu

Abstract. We start with an introduction about the Stockwell transforms and the resolution of the identity formula. Localization operators corresponding to the Stockwell transforms are then defined. It is proved that under suitable conditions on the symbols, the localization operators are, in descending order of complexity, paracommutators, paraproducts and Fourier multipliers.

Mathematics Subject Classification (2000). Primary 47G10, 47G30.

Keywords. Stockwell transform, localization Operator, paracommutators, paraproducts, Fourier multiplier.

1. Introduction

1.1. The 1D Stockwell transforms

The Stockwell transform is a hybrid of the Gabor transform and the wavelet transform. For a signal f in $L^2(\mathbb{R})$, the one-dimensional Stockwell transform $S_\varphi f$ with respect to the window φ in $L^1(\mathbb{R}) \cap L^2(\mathbb{R})$ is the time-frequency representation of f given by

$$(S_\varphi f)(b,\xi) = (2\pi)^{-1/2}|\xi| \int_{-\infty}^{\infty} e^{-ix\xi} f(x)\overline{\varphi(\xi(x-b))}\, dx, \quad b,\xi \in \mathbb{R}.$$

More transparently,

$$(S_\varphi f)(b,\xi) = (f, \varphi^{b,\xi})_{L^2(\mathbb{R})}, \quad \text{where} \quad \varphi^{b,\xi} = (2\pi)^{-1/2} M_\xi T_{-b} D_\xi \varphi$$

and $(\ ,\ )_{L^2(\mathbb{R})}$ is the inner product in $L^2(\mathbb{R})$. Here, the modulation operator M_ξ, the translation operator T_{-b} and the dilation operator D_ξ are, respectively, defined by

$$(M_\xi h)(x) = e^{ix\xi}h(x), \quad (T_{-b}h)(x) = h(x-b) \quad \text{and} \quad (D_\xi h)(x) = |\xi|h(\xi x),$$

This research has been supported by the Natural Sciences and Engineering Research Council of Canada.

for all x in $\mathbb{R}$ and all measurable functions h on $\mathbb{R}$. The Stockwell transform $S_\varphi f$ can be looked at as a modified Gabor transform of f with a window, which is designed in such a way that the window is wide for low frequency and narrow for high frequency. Besides the modulation with respect to frequency ξ, a distinguishing feature of the Stockwell transform from the well-known wavelet transform is the normalizing factor in the dilation operator, which is $|\cdot|$ instead of $|\cdot|^{1/2}$. This feature is responsible for the property of the absolutely referenced phase information, which accounts for the fact that the Stockwell transform is very much akin to the Fourier transform. See Theorem 1.3 in this connection.

Applications of the Stockwell transform abound in the literature. See, for instance, [5, 13] and [6, 17] for applications in, respectively, geophysics and medical imaging.

In an attempt to reconstruct a signal f from its Stockwell spectrum $\{(S_\varphi f)(b,\xi) : b, \xi \in \mathbb{R}\}$, we have the following result in [4].

Theorem 1.1. *Let $\varphi \in L^2(\mathbb{R})$ be such that $\|\varphi\|_{L^2(\mathbb{R})} = 1$ and*

$$\int_{-\infty}^{\infty} \frac{|\hat{\varphi}(\xi-1)|^2}{|\xi|} d\xi < \infty. \tag{1.1}$$

Then for all signals f and g in $L^2(\mathbb{R})$,

$$(f,g)_{L^2(\mathbb{R})} = \frac{1}{c_\varphi} \int_{-\infty}^{\infty} \int_{-\infty}^{\infty} (f, \varphi^{b,\xi})_{L^2(\mathbb{R})} (\varphi^{b,\xi}, g)_{L^2(\mathbb{R})} \frac{db\, d\xi}{|\xi|}, \tag{1.2}$$

where

$$c_\varphi = \int_{-\infty}^{\infty} \frac{|\hat{\varphi}(\xi-1)|^2}{|\xi|} d\xi,$$

and $\wedge$ denotes the Fourier transform defined by

$$\hat{F}(\zeta) = (2\pi)^{-N/2} \int_{\mathbb{R}^N} e^{-ix\cdot\zeta} F(x)\, dx$$

for all F in $L^1(\mathbb{R}^N)$.

Remark 1.2. Theorem 1.1 is known as the Plancherel formula or the resolution of the identity formula for the one-dimensional Stockwell transform. The integrability condition (1.1) is the admissibility condition for a function φ in $L^2(\mathbb{R})$ to be a window. An important corollary of Theorem 1.1 is that every signal f can be reconstructed from its Stockwell spectrum by means of the inversion formula

$$f = \frac{1}{c_\varphi} \int_{-\infty}^{\infty} \int_{-\infty}^{\infty} (f, \varphi^{b,\xi})_{L^2(\mathbb{R})} \varphi^{b,\xi} \frac{db\, d\xi}{|\xi|}.$$

That the admissibility condition (1.1) is a necessary condition for the inversion formula for the Stockwell transform can be seen by letting $f = g = \varphi$ in (1.2). Details can be found in [3] and [14].

There is another inversion formula for the one-dimensional Stockwell transform, which is a contribution in [13] and is therein referred to as the property of absolutely referenced phase information. It is formulated in the following theorem.

Theorem 1.3. *Let $\varphi \in L^1(\mathbb{R}) \cap L^2(\mathbb{R})$ be such that*

$$\int_{-\infty}^{\infty} \varphi(x)\, dx = 1.$$

Then for all signals f in $L^2(\mathbb{R})$,

$$\hat{f} = \int_{-\infty}^{\infty} (S_\varphi f)(b, \cdot)\, db.$$

1.2. The 2D Stockwell transforms

The two-dimensional Stockwell transform is defined in the following way. For a signal f in $L^2(\mathbb{R}^2)$, the one-dimensional Stockwell transform $S_\varphi f$ with respect to the window φ in $L^1(\mathbb{R}^2) \cap L^2(\mathbb{R}^2)$ is the time-frequency representation of f given by

$$(S_\varphi f)(b,\xi) = (2\pi)^{-1}|\xi_1||\xi_2| \int_{\mathbb{R}^2} e^{-ix\xi} f(x)\overline{\varphi(\xi(x-b))}\, dx, \quad b, \xi = (\xi_1, \xi_2) \in \mathbb{R}^2.$$

More transparently,

$$(S_\varphi f)(b,\xi) = (f, \varphi^{b,\xi})_{L^2(\mathbb{R}^2)},$$

where

$$\varphi^{b,\xi} = (2\pi)^{-1} M_\xi T_{-b} D_\xi \varphi$$

and $(\,,\,)_{L^2(\mathbb{R}^2)}$ is the inner product in $L^2(\mathbb{R}^2)$. Here, the modulation operator M_ξ, the translation operator T_{-b} and the dilation operator D_ξ are, respectively, defined by

$$(M_\xi h)(x) = e^{ix\xi} h(x), \quad (T_{-b}h)(x) = h(x-b) \quad \text{and} \quad (D_\xi h)(x) = |\xi_1||\xi_2| h(\xi x)$$

for all x in $\mathbb{R}^2$ and all measurable functions h on $\mathbb{R}^2$.

Applications of two-dimensional Stockwell transforms can be found in [5], [9] and others.

In an attempt to reconstruct a signal f from its Stockwell spectrum $\{(S_\varphi f)(b,\xi) : b, \xi \in \mathbb{R}^2\}$, we have the following result in [8].

Theorem 1.4. *Let $\varphi \in L^2(\mathbb{R}^2)$ be such that $\|\varphi\|_{L^2(\mathbb{R}^2)} = 1$ and*

$$\int_{\mathbb{R}^2} \frac{|\hat{\varphi}(\xi - (1,1))|^2}{|\xi_1||\xi_2|}\, d\xi < \infty. \tag{1.3}$$

Then for all signals f and g in $L^2(\mathbb{R}^2)$,

$$(f,g)_{L^2(\mathbb{R}^2)} = \frac{1}{c_\varphi} \int_{\mathbb{R}^2}\int_{\mathbb{R}^2} (f, \varphi^{b,\xi})_{L^2(\mathbb{R}^2)} (\varphi^{b,\xi}, g)_{L^2(\mathbb{R}^2)} \frac{db\, d\xi}{|\xi_1||\xi_2|}, \tag{1.4}$$

where $c_\varphi = \int_{\mathbb{R}^2} \frac{|\hat{\varphi}(\xi-(1,1))|^2}{|\xi_1||\xi_2|} d\xi$, and $\wedge$ denotes the Fourier transform defined by

$$\hat{F}(\zeta) = (2\pi)^{-N/2} \int_{\mathbb{R}^N} e^{-ix\cdot\zeta} F(x)\, dx$$

for all F in $L^1(\mathbb{R}^N)$.

Remark 1.5. An corollary of Theorem 1.4 is that every signal f can be reconstructed from its Stockwell spectrum by means of the inversion formula

$$f = \frac{1}{c_\varphi} \int_{\mathbb{R}^2} \int_{\mathbb{R}^2} (f, \varphi^{b,\xi})_{L^2(\mathbb{R}^2)} \varphi^{b,\xi} \frac{db\, d\xi}{|\xi_1||\xi_2|}.$$

That the admissibility condition (1.3) is a necessary condition for the inversion formula for the Stockwell transform can be seen by letting $f = g = \varphi$ in (1.4). Details can be found in [8] .

There is another inversion formula for the two-dimensional Stockwell transform, which is a contribution in [13]. It is formulated in the following theorem.

Theorem 1.6. *Let $\varphi \in L^1(\mathbb{R}^2) \cap L^2(\mathbb{R}^2)$ be such that*

$$\int_{\mathbb{R}^2} \varphi(x)\, dx = 1.$$

Then for all signals f in $L^2(\mathbb{R}^2)$,

$$\hat{f} = \int_{\mathbb{R}^2} (S_\varphi f)(b, \cdot)\, db.$$

1.3. The localization operators

Let U be the upper half-plane given by

$$U = \{(b, a) : b \in \mathbb{R},\ a > 0\}.$$

Then with respect to the binary operation $\cdot$ on U defined by

$$(b_1, a_1) \cdot (b_2, a_2) = (b_1 + a_1 b_2, a_1 a_2), \quad (b_1, a_1), (b_2, a_2) \in U,$$

U becomes a non-abelian group in which (0,1) is the identity element and the inverse element of (b, a) is $\left(-\frac{b}{a}, \frac{1}{a}\right)$ for all (b, a) in U. In fact, it is a non-unimodular Lie group on which the left Haar measure and the right Haar measure are given, respectively, by

$$d\mu = \frac{db\, da}{a^2} \quad \text{and} \quad d\nu = \frac{db\, da}{a}.$$

It is commonly known as the affine group in the literature.

Let $H^2_+(\mathbb{R})$ be the subspace of $L^2(\mathbb{R})$ defined by

$$H^2_+(\mathbb{R}) = \{f \in L^2(\mathbb{R}) : \operatorname{supp}(\hat{f}) \subseteq [0, \infty)\},$$

where $\operatorname{supp}(\hat{f})$ is the support of the Fourier transform $\hat{f}$. The Fourier transform $\widehat{F}$ of a function F in $L^2(\mathbb{R}^n)$ to be used in this paper is the one defined by

$$\widehat{F}(\xi) = \lim_{R \to \infty} (2\pi)^{-n/2} \int_{\mathbb{R}^n} e^{-ix \cdot \xi} \chi_R(x) F(x)\, dx,$$

where $x \cdot \xi$ is the inner product of x and ξ in $\mathbb{R}^n$, χ_R is the characteristic function on the ball $\{x \in \mathbb{R}^n : |x| \leq R\}$, and the convergence is understood to occur in $L^2(\mathbb{R}^n)$. We define $H^2_-(\mathbb{R})$ to be the subspace of $L^2(\mathbb{R})$ by

$$H^2_-(\mathbb{R}) = \{f \in L^2(\mathbb{R}) : \operatorname{supp}(\hat{f}) \subseteq (-\infty, 0]\}.$$

$H^2_+(\mathbb{R})$ and $H^2_-(\mathbb{R})$ are known as the Hardy space and the conjugate Hardy space respectively.

It is convenient at this point to mention that the results in this section are stated in the context of the left Haar measure $d\mu$ and the Hardy space $H^2_+(\mathbb{R})$ only.

Now, we let $U(H^2_+(\mathbb{R}))$ be the set of all unitary operators on $H^2_+(\mathbb{R})$. It is a group with respect to the usual composition of mappings. Let $\pi : U \to U(H^2_+(\mathbb{R}))$ be the mapping defined by

$$(\pi(b,a)f)(x) = \frac{1}{\sqrt{a}} f\left(\frac{x-b}{a}\right), \quad x \in \mathbb{R},$$

for all (b,a) in U and all f in $H^2_+(\mathbb{R})$. Then it can be proved that π is an irreducible and unitary representation of U on $H^2_+(\mathbb{R})$. The *raison d'être* for $H^2_\pm(\mathbb{R})$ is that the unitary representation $\pi : U \to U(L^2(\mathbb{R}))$ is not irreducible.

In fact, π is a square-integrable representation of U on $H^2_+(\mathbb{R})$ in the sense that there exists a nonzero function φ in $H^2_+(\mathbb{R})$ such that

$$\int_0^\infty \int_{-\infty}^\infty |(\varphi, \pi(b,a)\varphi)_{L^2(\mathbb{R})}|^2 \frac{db\, da}{a^2} < \infty. \tag{1.5}$$

We call any function φ in $H^2_+(\mathbb{R})$ for which $\|\varphi\|_{L^2(\mathbb{R})} = 1$ and (1.1) is valid an admissible wavelet for the square-integrable representation π of U on $H^2_+(\mathbb{R})$. We have the following characterization of admissible wavelets.

Theorem 1.7. *The set $AW(\pi)$ of all admissible wavelets for the square-integrable representation π of U on $H^2_+(\mathbb{R})$ is given by*

$$AW(\pi) = \left\{ \varphi \in H^2_+(\mathbb{R}) : \|\varphi\|_{L^2(\mathbb{R})} = 1, \int_0^\infty \frac{|\hat{\varphi}(\xi)|^2}{|\xi|} d\xi < \infty \right\}.$$

The proof of Theorem 1.1 follows from the fact that the integral in (1.1) is equal to $2\pi \int_0^\infty \frac{|\hat{\varphi}(\xi)|^2}{|\xi|} d\xi$. A proof can be found in Daubechies [3], Wong [14] and others.

Let $\varphi \in AW(\pi)$. Then we define the wavelet transform $W_\varphi f$ of a function f in $H^2_+(\mathbb{R})$ by

$$(W_\varphi f)(b,a) = (f, \pi(b,a)\varphi)_{L^2(\mathbb{R})}, \quad (b,a) \in U.$$

Then we have the following resolution of the identity formula, which can also be found in Daubechies [3] and Wong [14].

Theorem 1.8. *let $\varphi \in AW(\pi)$ and let c_φ be the constant defined by*

$$c_\varphi = 2\pi \int_0^\infty \frac{|\hat{\varphi}(\xi)|^2}{|\xi|} d\xi.$$

Then for all f and g in $H^2_+(\mathbb{R})$,

$$(f,g)_{L^2(\mathbb{R})} = \frac{1}{c_\varphi} \int_0^\infty \int_{-\infty}^\infty (f, \pi(b,a)\varphi)_{L^2(\mathbb{R})} (\pi(b,a)\varphi, g)_{L^2(\mathbb{R})} \frac{db\, da}{a^2}.$$

Let $\varphi \in AW(\pi)$ and let F be a measurable function on U. Then we define the localization operator $L_{F,\varphi} : H^2_+(\mathbb{R}) \to H^2_+(\mathbb{R})$ associated to the symbol F and the admissible wavelet φ by

$$(L_{F,\varphi}f, g)_{L^2(\mathbb{R})} = \frac{1}{c_\varphi}\int_0^\infty \int_{-\infty}^\infty F(b,a)(f, \pi(b,a)\varphi)_{L^2(\mathbb{R})}(\pi(b,a)\varphi, g)_{L^2(\mathbb{R})}\frac{db\, da}{a^2}$$

for all f and g in $L^2(\mathbb{R})$. It is shown in Wong [15] that if

$$F(b,a) = \beta(b)\alpha(a), \quad (b,a) \in U,$$

where α and β are suitable functions on $(0,\infty)$ and $(-\infty,\infty)$ respectively, then the localization operator $L_{F,\varphi}$ is in fact a paracommutator studied in Janson and Peetre [7], Peng [10, 11] and Peng and Wong [12]. Furthermore, in Wong [15], it is proved that if F is a suitable function of b only, then $L_{F,\varphi}$ is given in terms of a paraproduct in the sense of Coifman and Meyer [2], and if F is a function of a only, then $L_{F,\varphi}$ is a Fourier multiplier.

The aim of this paper is to extend the results hitherto described to the two-dimensional Stockwell transform. Under conditions similar to those in Wong [15], we show that the localization operators are given by paracommutators, paraproducts and Fourier multipliers .

2. Localization operators for two-dimensional Stockwell transforms

Let $F \in L^p(\mathbb{R}^2 \times \mathbb{R}^2)$, $1 \le p \le \infty$, then for all $f \in L^2(\mathbb{R}^2)$, we define $L_{F,\varphi}f$ by

$$(L_{F,\varphi}f, g)_{L^2(\mathbb{R}^2)} = \frac{1}{c_\varphi}\int_{\mathbb{R}^2}\int_{\mathbb{R}^2} F(b,\xi)(S_\varphi f)(b,\xi)\overline{(S_\varphi g)(b,\xi)}\,\frac{db\, d\xi}{|\xi_1||\xi_2|}$$

for g in $L^2(\mathbb{R}^2)$.

The following two results are the Stockwell transform analog of Proposition 12.1 and Theorem 14.5 in the book [14] by Wong.

Theorem 2.1. *Let $\varphi \in L^2(\mathbb{R}^2)$ be an admissible wavelet for the Stockwell transform. Let $F \in L^p(\mathbb{R}^2 \times \mathbb{R}^2)$, $1 \le p \le \infty$, and $L_{F,\varphi} : L^2(\mathbb{R}^2) \to L^2(\mathbb{R}^2)$ be the localization operator associated to the symbol F and the admissible wavelet φ ,then there exists a unique bounded linear operator $L_{F,\varphi} : L^2(\mathbb{R}^2) \to L^2(\mathbb{R}^2)$ such that*

$$\|L_{F,\varphi}\|_* \le (1/c_\varphi)^{\frac{1}{p}}\, \|F\|_p,$$

where $\| \|_*$ is the norm of bounded operators over $L^2(\mathbb{R}^2)$.

Theorem 2.2. *Let $\varphi \in L^2(\mathbb{R}^2)$ be an admissible wavelet for the Stockwell transform. Let $F \in L^p(\mathbb{R}^2 \times \mathbb{R}^2)$, $1 \le p \le \infty$, and $L_{F,\varphi} : L^2(\mathbb{R}^2) \to L^2(\mathbb{R}^2)$ be the localization operator associated to the symbol F and the admissible wavelet φ then is $L_{F,\varphi}$ in the Schatten–von Neumann class S_p, especially $L_{F,\varphi}$ is compact, and*

$$\|L_{F,\varphi}\|_{S_p} \le \left(\frac{1}{c_\varphi}\right)^{\frac{1}{p}} \|F\|_p,$$

where $\| \|_{S_p}$ is the norm in S_p.

We give in the following sections explicit formulas for the localization operators $L_{F,\varphi} : L^2(\mathbb{R}^2) \to L^2(\mathbb{R}^2)$ under different conditions imposed on the symbol F. These results are extensions of results obtained in the paper [15] for the standard wavelet transforms on the affine group $\mathbb{U}$.

3. Paracommutators

Let F be a function on $\mathbb{R}^2 \times \mathbb{R}^2$ given by

$$F(b,\xi) = \beta(b)\alpha(\xi), \quad b,\xi \in \mathbb{R}^2,$$

where α and β are suitable functions on $\mathbb{R}^2$. Then using Plancherel's theorem and Fubini's theorem, we get for all f and g in $L^2(\mathbb{R}^2)$,

$$\begin{aligned}(L_{F,\varphi}f,g)_{L^2(\mathbb{R}^2)} &= \frac{1}{c_\varphi}\int_{\mathbb{R}^2}\int_{\mathbb{R}^2} F(b,\xi)(S_\varphi f)(b,\xi)\overline{(S_\varphi g)(b,\xi)}\frac{db\,d\xi}{|\xi_1||\xi_2|}\\ &= \frac{1}{c_\varphi}\int_{\mathbb{R}^2}\alpha(\xi)\left(\int_{\mathbb{R}^2}\beta(b)(S_\varphi f)(b,\xi)\overline{(S_\varphi g)(b,\xi)}\,db\right)\frac{d\xi}{|\xi_1||\xi_2|}\\ &= \frac{2\pi}{c_\varphi}\int_{\mathbb{R}^2}\int_{\mathbb{R}^2} A(\zeta,\eta)\hat{\beta}(\eta-\zeta)\hat{f}(\zeta)\overline{\hat{g}(\eta)}\,d\zeta\,d\eta, \qquad (3.1)\end{aligned}$$

where

$$A(\zeta,\eta) = \int_{\mathbb{R}^2}\alpha(\xi)\overline{\hat{\varphi}(\xi^{-1}(\zeta-\xi))}\hat{\varphi}(\xi^{-1}(\eta-\xi))\frac{d\xi}{|\xi_1||\xi_2|},$$

for all ξ and η in $\mathbb{R}^2$, where $\xi^{-1} = (\xi_1^{-1},\xi_2^{-1})$, for all $\xi = (\xi_1,\xi_2) \in \mathbb{R}^2$. Thus, the localization operator $L_{F,\varphi}$ is a paracommutator with Fourier kernel A and symbol β.

4. A Paraproduct connection

If the symbol F is independent of the second variable, i.e., it is given by

$$F(b,\xi) = \beta(b), b,\xi \in \mathbb{R}^2.$$

In order to simplify the computations that follow, let us introduce a notation. Let ξ is some point in $\mathbb{R}^2$. Then we define the mollifier u_ξ of a function u in $L^2(\mathbb{R}^2)$ by

$$u_\xi(x) = |\xi_1||\xi_2|e^{i(\xi_1x_1+\xi_2x_2)}u(\xi x), \quad x \in \mathbb{R}^2.$$

An easy computation gives

$$\widehat{u_\xi}(\zeta) = \hat{u}(\xi^{-1}(\zeta-\xi)), \quad \xi \in \mathbb{R}^2,$$

Where $\xi^{-1} = (\xi_1^{-1},\xi_2^{-2})$. Let ψ be the function on $\mathbb{R}^2$ defined by

$$\psi(x) = \overline{\varphi(-x)}, \quad x \in \mathbb{R}^2.$$

Then, we get

$$(L_{F,\varphi}f,g)_{L^2(\mathbb{R}^2)} = \frac{2\pi}{c_\varphi}\int_{\mathbb{R}^2}\int_{\mathbb{R}^2}\left(\int_{\mathbb{R}^2}\widehat{\psi_\xi}(\zeta)\widehat{\varphi_\xi}(\eta)\hat{f}(\zeta)\hat{\beta}(\eta-\zeta)\,\frac{d\xi}{|\xi_1||\xi_2|}\,d\zeta\right)\overline{\hat{g}(\eta)}\,d\eta.$$

Since

$$(u*v)^\wedge = 2\pi\hat{u}\hat{v}$$

for sufficiently nice functions u and v on $\mathbb{R}^2$, it follows from Plancherel's formula that for all f and g in $L^2(\mathbb{R}^2)$,

$$(L_{F,\varphi}f,g)_{L^2(\mathbb{R}^2)} = \frac{1}{c_\varphi}\int_{\mathbb{R}^2}\left(\int_{\mathbb{R}^2}((\beta(\psi_\xi * f))*\varphi_\xi)(x)\frac{d\xi}{|\xi_1||\xi_2|}\right)\overline{g(x)}\,dx \tag{4.1}$$

and hence

$$(L_{F,\varphi}f)(x) = \frac{1}{c_\varphi}\int_{\mathbb{R}^2}((\beta(\psi_\xi * f))*\varphi_\xi)(x)\frac{d\xi}{|\xi_1||\xi_2|},\quad x\in\mathbb{R}^2.$$

Further manipulations of (8.1) using Fubini's theorem give

$$(L_{F,\varphi}f,g)_{L^2(\mathbb{R}^2)} = \frac{1}{c_\varphi}\int_{\mathbb{R}^2}\beta(y)p_\psi(f,g)(y)\,dy$$

for all f and g in $L^2(\mathbb{R}^2)$, where $p_\psi(f,g)$ is the paraproduct of f and g with respect to ψ given by

$$p_\psi(f,g)(y) = \int_{\mathbb{R}^2}(\psi_\xi * f)(y)\overline{(\psi_\xi * g)(y)}\,\frac{d\xi}{|\xi_1||\xi_2|},\quad y\in\mathbb{R}^2.$$

5. Fourier multipliers

Let F be a function on $\mathbb{R}^2\times\mathbb{R}^2$ given by

$$F(b,\xi) = \alpha(\xi),\quad b,\xi\in\mathbb{R}^2.$$

Then, by (3.1), we obtain for all f and g in $L^2(\mathbb{R}^2)$,

$$\begin{aligned}&(L_{F,\varphi}f,g)_{L^2(\mathbb{R}^2)}\\ =\ &\frac{2\pi}{c_\varphi}\int_{\mathbb{R}^2}\int_{\mathbb{R}^2}\int_{\mathbb{R}^2}\alpha(\xi)\overline{\widehat{\varphi_\xi}(\zeta)}\widehat{\varphi_\xi}(\eta)\hat{f}(\xi)\overline{\hat{g}(\eta)}\frac{d\xi}{|\xi_1||\xi_2|}\,d\zeta\,d\eta.\end{aligned}$$

Let $\phi(x) = e^{-|x|^2/2}$, $x\in\mathbb{R}^2$. For all positive numbers ε, let I_ε be the number defined by

$$I_\varepsilon = \frac{2\pi}{c_\varphi}\int_{\mathbb{R}^2}\int_{\mathbb{R}^2}\int_{\mathbb{R}^2}\alpha(\xi)\overline{\widehat{\varphi_\xi}(\zeta)}\widehat{\varphi_\xi}(\eta)\hat{f}(\zeta)\overline{\hat{g}(\eta)}\phi_\varepsilon(\zeta-\eta)\frac{d\xi}{|\xi_1||\xi_2|}\,d\zeta\,d\eta, \tag{5.1}$$

where ϕ_ε is the Friedrich mollifier of ϕ. Then

$$I_\varepsilon = \frac{2\pi}{c_\varphi}\int_{\mathbb{R}^2}\left(\int_{\mathbb{R}^2}\alpha(\xi)\overline{\widehat{\varphi_\xi}(\zeta)}\hat{f}(\zeta)((\widehat{\varphi_\xi}\overline{\hat{g}})*\phi_\varepsilon)(\zeta)\frac{d\xi}{|\xi_1||\xi_2|}\right)d\zeta.$$

Now, there exists a sequence $\{\varepsilon_j\}_{j=1}^\infty$ of positive numbers such that $\varepsilon_j\to 0$, as $j\to\infty$ and

$$(\widehat{\varphi_\xi}\overline{\hat{g}})*\phi_{\varepsilon_j}\to 2\pi\widehat{\varphi_\xi}\overline{\hat{g}}$$

in $L^2(\mathbb{R}^2)$ and almost everywhere on $\mathbb{R}^2$ as $j \to \infty$. Thus,

$$I_{\varepsilon_j} \to \frac{4\pi^2}{c_\varphi} \int_{\mathbb{R}^2} \left(\int_{\mathbb{R}^2} \alpha(\xi) \overline{\widehat{\varphi_\xi}} \hat{f}(\zeta) \widehat{\varphi_\xi}(\zeta) \overline{\hat{g}(\zeta)} \frac{d\xi}{|\xi_1||\xi_2|} \right) d\zeta.$$

On the other hand, since the Fourier transform of the function $e^{-\frac{|\cdot|^2\epsilon^2}{2}}$ is $\hat{\phi}_\epsilon$, compare (3.1) and (5.1), we obtain

$$I_\epsilon = \frac{1}{c_\varphi} \int_{\mathbb{R}^2} \int_{\mathbb{R}^2} \alpha(\xi) e^{-\frac{|b|^2\epsilon^2}{2}} (S_\varphi f)(b,\xi) \overline{(S_\varphi g)(b,\xi)} \frac{db\, d\xi}{|\xi_1||\xi_2|}.$$

Thus, using Lebesgue's dominated convergence theorem, we see that

$$I_{\varepsilon_j} \to (L_{F,\varphi} f, g)_{L^2(\mathbb{R}^2)}$$

as $j \to \infty$. Therefore for all f and g in $L^2(\mathbb{R}^2)$,

$$(L_{F,\varphi} f, g)_{L^2(\mathbb{R}^2)} = (T_m f, g)_{L^2(\mathbb{R}^2)},$$

where T_m is the Fourier multiplier with symbol m given by

$$m(\zeta) = \frac{4\pi^2}{c_\varphi} \int_{\mathbb{R}^2} \alpha(\xi) |\widehat{\varphi_\xi}(\zeta)|^2 \frac{d\xi}{|\xi_1||\xi_2|}, \quad \xi \in \mathbb{R}^2.$$

References

[1] J.-P. Antoine, R. Murenzi, P. Vandergheynst and S.T. Ali, *Two-Dimensional Wavelets and their Relatives*, Cambridge University Press, 2004.

[2] R.R. Coifman and Y. Meyer, Au Delà des Opérateurs Pseudo-Différentiels, *Astérisque* **57**, 1978.

[3] I. Daubechies, *Ten Lectures on Wavelets*, SIAM, 1992.

[4] J. Du, M.W. Wong and H. Zhu, Continuous and Discrete Inversion Formulas for the Stockwell transform, *Integral Transforms Spec. Funct.*, **18** (2007), 537–543.

[5] M.G. Eramian, R.A. Schincariol, L. Mansinha and R.G. Stockwell, Generation of aquifer heterogeneity maps using two dimensional special texture segmentation techniques, *Math. Geology* **31** (1999), 327–348.

[6] B.G. Goodyear, H. Zhu, R.A. Brown and J.R. Mitchell, Removal of phase artifacts from fMRI data using a Stockwell transform filter improves brain activity detection, *Magn. Reson. Med.* **51** (2004), 16–21.

[7] S. Janson and J. Peetre, Paracommutators-boundedness and Schatten-von Neumann properties, *Trans. Amer. Math. Soc.* **305** (1988), 467–504.

[8] Y. Liu and M.W. Wong , Inversion Formulas for Two-Dimensional Stockwell Transforms, in *Pseudo-Differential Operators: Partial Differential Equations and Time-Frequency Analysis*, Editors: L. Rodino, B.-W. Schulze and M.W. Wong, Fields Institute Communications Series, American Mathematical Society, 2007, 323–330.

[9] L. Mansinha, R.G. Stockwell and R.P. Lowe, Pattern analysis with two-dimensional spectral localization: applications of 2-dimensional S-transforms, *Physica A* **239** (1997), 286–295.

[10] L. Peng, On the compactness of commutators, *Ark. Mat.* **26** (1988), 315–325.

[11] L. Peng, Wavelets and paracommutators, *Ark. Mat.* **31** (1993), 83–99.

[12] L. Peng and M.W. Wong, Compensated compactness and paracommutators, *J. London Math. Soc.* **62**(2) (2000), 505–520.

[13] R.G. Stockwell, L. Mansinha and R.P. Lowe, Localization of the complex spectrum: the S-transform, *IEEE Trans. Signal Processing* **44** (1996), 998–1001.

[14] M.W. Wong, *Wavelet Transforms and Localization Operators,* Birkhäuser, 2002.

[15] M.W. Wong, Localization operators on the affine group and paracommutators, in *Progress in Analysis*, Editors: H.G.W. Begehr, R.P. Gilbert and M.W. Wong, World Scientific, 2003, 663–669.

[16] M.W. Wong and H. Zhu, A characterization of Stockwell spectra, in *Modern Trends in Pseudo-Differential Operators*, Editors: J. Toft, M.W. Wong and H. Zhu, Birkhäuser, 2007, 251–257.

[17] H. Zhu, B.G. Goodyear, M.L. Lauzon, R.A. Brown, G.S. Mayer, L. Mansinha, A.G. Law and J.R. Mitchell, A new multiscale Fourier analysis for MRI, *Med. Phys.* **30** (2003), 1134–1141.

Yu Liu
Department of Mathematics and Statistics
York University
4700 Keele Street
Toronto
Ontario M3J 1P3, Canada
e-mail: hanliu@yorku.ca

Operator Theory:
Advances and Applications, Vol. 189, 297–306

Pseudo-Differential Operators on $\mathbb{S}^1$

Shahla Molahajloo and M.W. Wong

Abstract. Pseudo-differential operators on the unit circle $\mathbb{S}^1$ with center at the origin are defined in terms of symbols on $\mathbb{S}^1 \times \mathbb{Z}$. Results on the boundedness and compactness of these pseudo-differential operators on $L^2(\mathbb{S}^1)$ are presented. In addition, we prove a result on the L^p-boundedness for $1 < p < \infty$.

Mathematics Subject Classification (2000). Primary 47G30; Secondary 42A45.

Keywords. Pseudo-differential operators, L^2-boundedness, L^p-boundedness, L^2-compactness, Hilbert–Schmidt operators, Fourier multipliers, Littlewood–Paley theory.

1. Introduction

For $m \in (-\infty, \infty)$, let S^m be the set of all functions σ in $C^\infty(\mathbb{R}^n \times \mathbb{R}^n)$ such that for all multi-indices α and β, there exists a positive constant $C_{\alpha,\beta}$ for which

$$|(D_x^\alpha D_\xi^\beta \sigma)(x, \xi)| \leq C_{\alpha,\beta}(1 + |\xi|)^{m-|\beta|}, \quad x, \xi \in \mathbb{R}^n.$$

A function σ in S^m is called a symbol. Let σ be a symbol. Then we define the pseudo-differential operator T_σ on the Schwartz space $\mathcal{S}$ by

$$(T_\sigma \varphi)(x) = (2\pi)^{-n/2} \int_{\mathbb{R}^n} e^{ix\cdot\xi} \sigma(x, \xi) \hat{\varphi}(\xi)\, d\xi, \quad x \in \mathbb{R}^n,$$

for all functions φ in $\mathcal{S}$, where

$$\hat{\varphi}(\xi) = (2\pi)^{-n/2} \int_{\mathbb{R}^n} e^{-ix\cdot\xi} \varphi(x)\, dx, \quad \xi \in \mathbb{R}^n.$$

It is easy to prove that T_σ maps $\mathcal{S}$ into $\mathcal{S}$ continuously. A much deeper fact is that if $\sigma \in S^0$, then T_σ can be extended to a bounded linear operator from $L^p(\mathbb{R}^n)$ into $L^p(\mathbb{R}^n)$ for $1 < p < \infty$. See, for instance, Theorem 10.7 in the book [13] by Wong.

Although the phase space $\mathbb{R}^n \times \mathbb{R}^n$ is used in defining symbols of pseudo-differential operators on $\mathbb{R}^n$, a more appropriate setting is, loosely speaking, the

This research has been supported by the Natural Sciences and Engineering Research Council of Canada.

phase space $\mathbb{G} \times \widehat{\mathbb{G}}$ when we are interested in pseudo-differential operators on $\mathbb{G}$, where $\mathbb{G}$ is a Lie group and $\widehat{\mathbb{G}}$ is the set of all irreducible and unitary representations of $\mathbb{G}$. In the case of $\mathbb{R}^n$, $\widehat{\mathbb{R}^n}$ can be identified with $\mathbb{R}^n$ and hence the phase space $\mathbb{R}^n \times \mathbb{R}^n$ used for pseudo-differential operators on $\mathbb{R}^n$ is consistent with this big picture. In this paper the focus is on pseudo-differential operators on $\mathbb{S}^1$ with symbols defined on the phase space $\mathbb{S}^1 \times \mathbb{Z}$, where $\mathbb{S}^1$ is the unit circle centered at the origin and $\mathbb{Z}$ is the dual group $\widehat{\mathbb{S}^1}$. To be more precise, let σ be a measurable function on $\mathbb{S}^1 \times \mathbb{Z}$. Then for every function f in $L^1(\mathbb{S}^1)$, we define the function $T_\sigma f$ on $\mathbb{S}^1$ by

$$(T_\sigma f)(\theta) = \sum_{n\in\mathbb{Z}} \sigma(\theta, n)\hat{f}(n)e^{in\theta}, \quad \theta \in [-\pi, \pi],$$

where

$$\hat{f}(n) = \frac{1}{2\pi}\int_{-\pi}^{\pi} e^{-in\theta} f(\theta)\, d\theta, \quad n \in \mathbb{Z}.$$

Among the most interesting problems are the ones on finding good conditions on σ to produce bounded and compact pseudo-differential operators $T_\sigma : L^p(\mathbb{S}^1) \to L^p(\mathbb{S}^1)$, $1 < p < \infty$. Some results in this direction are given in this paper. Related works can be found in [1, 2, 3, 5, 6, 7] among others.

We begin in Section 2 with an L^p formula for $1 \le p < \infty$. This is then used in Section 3 to prove that a pseudo-differential operator $T_\sigma : L^2(\mathbb{S}^1) \to L^2(\mathbb{S}^1)$ is Hilbert–Schmidt if and only if $\sigma \in L^2(\mathbb{S}^1 \times \mathbb{Z})$. In Section 4, necessary conditions and sufficient conditions on symbols σ are given for the corresponding pseudo-differential operators $T_\sigma : L^2(\mathbb{S}^1) \to L^2(\mathbb{S}^1)$ to be bounded linear operators. Sufficient conditions on symbols σ are given in Section 5 to guarantee that the corresponding pseudo-differential operators $T_\sigma : L^2(\mathbb{S}^1) \to L^2(\mathbb{S}^1)$ are compact. We give in Section 6 sufficient conditions on symbols σ to give bounded linear operators $T_\sigma : L^p(\mathbb{S}^1) \to L^p(\mathbb{S}^1)$, $1 < p < \infty$.

The results in this paper are valid for the multi-dimensional torus using multiple Fourier series in, *e.g.*, the book [10] by Stein and Weiss. For the sake of simpler notation and greater transparency, we have chosen to work only with the unit circle $\mathbb{S}^1$ with center at the origin.

2. An L^p formula, $1 \le p < \infty$

For $j \in \mathbb{Z}$, we let e_j be the function defined by

$$e_j(\theta) = e^{ij\theta}, \quad \theta \in [-\pi, \pi].$$

Then we have the following formula.

Theorem 2.1. *Let $\sigma \in L^p(\mathbb{S}^1 \times \mathbb{Z})$, $1 \le p < \infty$. Then*

$$\sum_{j\in\mathbb{Z}} \|T_\sigma e_j\|^p_{L^p(\mathbb{S}^1)} = \|\sigma\|^p_{L^p(\mathbb{S}^1\times\mathbb{Z})}.$$

Proof. For $j \in \mathbb{Z}$, we get

$$(T_\sigma e_j)(\theta) = \sum_{n\in\mathbb{Z}} \sigma(\theta, n)\widehat{e_j}(n)e^{in\theta}, \quad \theta \in [-\pi, \pi]. \tag{2.1}$$

But

$$\widehat{e_j}(n) = \frac{1}{2\pi}\int_{-\pi}^{\pi} e^{-in\theta}e^{ij\theta}\,d\theta = \frac{1}{2\pi}\int_{-\pi}^{\pi} e^{i(j-n)\theta}\,d\theta = \begin{cases} 1, & n = j, \\ 0, & n \neq j. \end{cases} \tag{2.2}$$

So, by (2.1) and (2.2),

$$(T_\sigma e_j)(\theta) = \sigma(\theta, j)e^{ij\theta} = \sigma(\theta, j)e_j(\theta), \quad \theta \in [-\pi, \pi]. \tag{2.3}$$

Hence

$$\sum_{j\in\mathbb{Z}} \|T_\sigma e_j\|^p_{L^p(\mathbb{S}^1)} = \sum_{j\in\mathbb{Z}} \int_{-\pi}^{\pi} |\sigma(\theta, j)|^p\,d\theta = \|\sigma\|^p_{L^p(\mathbb{S}^1\times\mathbb{Z})},$$

as asserted. □

3. Hilbert–Schmidt operators

Let us begin with the fact that a bounded linear operator A on a complex and separable Hilbert space X is a Hilbert–Schmidt operator $\Leftrightarrow$ there exists an orthonormal basis $\{\varphi_k\}_{k=1}^{\infty}$ for X such that

$$\sum_{k=1}^{\infty} \|A\varphi_k\|_X^2 < \infty,$$

where $\| \, \|_X$ denotes the norm in X. Moreover, if A is a Hilbert–Schmidt operator on X, then the Hilbert–Schmidt norm $\|A\|_{HS}$ of A is given by

$$\|A\|_{HS}^2 = \sum_{k=1}^{\infty} \|A\varphi_k\|_X^2,$$

where $\{\varphi_k\}_{k=1}^{\infty}$ is any orthonormal basis for X. So, using Theorem 2.1 and the fact that $\{(2\pi)^{-1/2}e_k\}_{k=-\infty}^{\infty}$ is an orthonormal basis for $L^2(\mathbb{S}^1)$, we have the following result.

Theorem 3.1. *The pseudo-differential operator $T_\sigma : L^2(\mathbb{S}^1) \to L^2(\mathbb{S}^1)$ is a Hilbert–Schmidt operator $\Leftrightarrow$ $\sigma \in L^2(\mathbb{S}^1 \times \mathbb{Z})$. Moreover, if $T_\sigma : L^2(\mathbb{S}^1) \to L^2(\mathbb{S}^1)$ is a Hilbert–Schmidt operator, then*

$$\|T_\sigma\|_{HS} = (2\pi)^{-1/2}\|\sigma\|_{L^2(\mathbb{S}^1\times\mathbb{Z})}.$$

4. L^2-boundedness

The proof of Theorem 2.1 gives a necessary condition on a measurable function σ on $\mathbb{S}^1 \times \mathbb{Z}$ for $T_\sigma : L^2(\mathbb{S}^1) \to L^2(\mathbb{S}^1)$ to be a bounded linear operator. To wit, let σ be a measurable function on $\mathbb{S}^1 \times \mathbb{Z}$ and let $f \in L^2(\mathbb{S}^1)$. Then for $n \in \mathbb{Z}$, we can use (2.3) to obtain $(T_\sigma e_n)(\theta) = \sigma(\theta, n)e_n(\theta)$, $\theta \in [-\pi, \pi]$, and hence

$$\|T_\sigma e_n\|^2_{L^2(\mathbb{S}^1)} = \int_{-\pi}^{\pi} |\sigma(\theta, n)|^2 d\theta.$$

So, if $T_\sigma : L^2(\mathbb{S}^1) \to L^2(\mathbb{S}^1)$ is a bounded linear operator, we can get a positive constant C such that

$$\|T_\sigma e_n\|^2_{L^2(\mathbb{S}^1)} \leq C\|e_n\|^2_{L^2(\mathbb{S}^1)}, \quad n \in \mathbb{Z},$$

which is then the same as

$$\int_{-\pi}^{\pi} |\sigma(\theta, n)|^2 d\theta \leq 2\pi C, \quad n \in \mathbb{Z}.$$

Therefore a necessary condition for $T_\sigma : L^2(\mathbb{S}^1) \to L^2(\mathbb{S}^1)$ to be a bounded linear operator is

$$\sup_{n\in\mathbb{Z}} \int_{-\pi}^{\pi} |\sigma(\theta, n)|^2 d\theta < \infty. \tag{4.1}$$

The following example shows that (4.1) is not sufficient for $T_\sigma : L^2(\mathbb{R}^n) \to L^2(\mathbb{R}^n)$ to be a bounded linear operator.

Example 4.1. Let σ be the function on $\mathbb{S}^1 \times \mathbb{Z}$ defined by

$$\sigma(\theta, n) = e^{-in\theta}, \quad \theta \in [-\pi, \pi],\, n \in \mathbb{Z},$$

and let f be the function on $\mathbb{S}^1$ defined by

$$f(\theta) = \sum_{n=1}^{\infty} \frac{1}{n}\, e^{in\theta}, \quad \theta \in [-\pi, \pi].$$

Then it is easy to see that $f \in L^2(\mathbb{S}^1)$, but $T_\sigma f \notin L^2(\mathbb{S}^1)$.

The following theorem gives sufficient conditions for a pseudo-differential operator $T_\sigma : L^2(\mathbb{S}^1) \to L^2(\mathbb{S}^1)$ to be a bounded linear operator.

Theorem 4.2. *Let σ be a measurable function on $\mathbb{S}^1 \times \mathbb{Z}$. Suppose that we can find a positive constant C and a function w in $L^1(\mathbb{Z})$ such that*

$$|\hat{\sigma}(m, n)| \leq C|w(m)|, \quad m, n \in \mathbb{Z}, \tag{4.2}$$

where

$$\hat{\sigma}(m, n) = \frac{1}{2\pi} \int_{-\pi}^{\pi} e^{-im\theta} \sigma(\theta, n)\, d\theta.$$

Then $T_\sigma : L^2(\mathbb{S}^1) \to L^2(\mathbb{S}^1)$ is a bounded linear operator and

$$\|T_\sigma\|_* \leq C\|w\|_{L^1(\mathbb{Z})},$$

where $\| \, \|_*$ is the norm in the C^*-algebra of all bounded linear operators on $L^2(\mathbb{S}^1)$.

Proof. Let $f \in C^2(\mathbb{S}^1)$. Then we have

$$\begin{aligned}\|T_\sigma f\|^2_{L^2(\mathbb{S}^1)} &= \int_{-\pi}^{\pi} \left| \sum_{n\in\mathbb{Z}} \sigma(\theta, n)\hat{f}(n)e^{in\theta}\right|^2 d\theta \\ &= \int_{-\pi}^{\pi} \left| \sum_{n\in\mathbb{Z}}\sum_{m\in\mathbb{Z}} \hat{\sigma}(m, n)\hat{f}(n)e^{i(m+n)\theta}\right|^2 d\theta.\end{aligned}$$

So,

$$\begin{aligned}\|T_\sigma f\|^2_{L^2(\mathbb{S}^1)} &= \int_{-\pi}^{\pi} \left| \sum_{n\in\mathbb{Z}}\sum_{k\in\mathbb{Z}} \hat{\sigma}(k-n, n)\hat{f}(n)e^{ik\theta}\right|^2 d\theta \\ &= \int_{-\pi}^{\pi} \left| \sum_{k\in\mathbb{Z}}\left\{\sum_{n\in\mathbb{Z}} \hat{\sigma}(k-n, n)\hat{f}(n)\right\}e^{ik\theta}\right|^2 d\theta. \qquad (4.3)\end{aligned}$$

Using (4.3) and the orthogonality of the functions $\{e_j\}_{j=-\infty}^{\infty}$ in $L^2(\mathbb{S}^1)$,

$$\begin{aligned}\|T_\sigma f\|^2_{L^2(\mathbb{S}^1)} &= 2\pi \sum_{k\in\mathbb{Z}}\left|\sum_{n\in\mathbb{Z}} \hat{\sigma}(k-n, n)\hat{f}(n)\right|^2 \\ &\le 2\pi \sum_{k\in\mathbb{Z}}\left(\sum_{n\in\mathbb{Z}} |\hat{\sigma}(k-n, n)||\hat{f}(n)|\right)^2. \qquad (4.4)\end{aligned}$$

Using (4.2), we get

$$\|T_\sigma f\|^2_{L^2(\mathbb{S}^1)} \le 2\pi C^2 \sum_{k\in\mathbb{Z}}\left(\sum_{n\in\mathbb{Z}} |w(k-n)|\,|\hat{f}(n)|\right)^2 \le 2\pi C^2 \sum_{k\in\mathbb{Z}} |(|w| * |\hat{f}|)(k)|^2,$$

where $|w| * |\hat{f}|$ is the convolution of $|w|$ and $|\hat{f}|$. Finally, using Young's inequality, we have

$$\|T_\sigma f\|^2_{L^2(\mathbb{S}^1)} \le 2\pi C^2 \|w\|^2_{L^1(\mathbb{Z})}\|\hat{f}\|^2_{L^2(\mathbb{Z})} = C^2\|w\|^2_{L^1(\mathbb{Z})}\|f\|^2_{L^2(\mathbb{S}^1)}. \qquad (4.5)$$

Since $C^2(\mathbb{S}^1)$ is dense in $L^2(\mathbb{S}^1)$, it follows that (4.5) holds for all functions f in $L^2(\mathbb{S}^1)$. □

Remark 4.3. In order to justify the interchange of the two sums in (4.3), we note that by Fubini's theorem, it is sufficient to prove that

$$\sum_{k\in\mathbb{Z}}\sum_{n\in\mathbb{Z}} |\hat{\sigma}(k-n, n)||\hat{f}(n)| < \infty.$$

Since $f \in C^2(\mathbb{S}^1)$, it follows that

$$|\hat{f}(n)| \le O(n^{-2})$$

as $|n| \to \infty$. Hence $\hat{f} \in L^1(\mathbb{Z})$. Using an argument in the proof of Theorem 4.2, we have

$$\sum_{k\in\mathbb{Z}}\sum_{n\in\mathbb{Z}} |\hat{\sigma}(k-n, n)||\hat{f}(n)| \le C\|w\|_{L^1(\mathbb{Z})}\|\hat{f}\|_{L^1(\mathbb{Z})} < \infty.$$

Remark 4.4. The sufficient conditions on the symbol σ to ensure the boundedness of $T_\sigma : L^2(\mathbb{S}^1) \to L^2(\mathbb{S}^1)$ in [5, 6] require a certain number of derivatives of σ with respect to θ. By means of Theorem 4.2 in this paper and Bernstein's theorem, we see that all we need is that the symbol σ is in the Lipschitz class Λ^α, $\alpha > \frac{1}{2}$. See Section 3 in Chapter VI of the book [14] by Zygmund for results in this direction.

5. L^2-compactness

Let σ be a measurable function on $\mathbb{S}^1 \times \mathbb{Z}$ such that $T_\sigma : L^2(\mathbb{S}^1) \to L^2(\mathbb{S}^1)$ is compact. Since $e_n \to 0$ weakly in $L^2(\mathbb{S}^1)$ as $|n| \to \infty$, it follows from the compactness of $T_\sigma : L^2(\mathbb{S}^1) \to L^2(\mathbb{S}^1)$ that $\|T_\sigma e_n\|_{L^2(\mathbb{S}^1)} \to 0$ as $|n| \to \infty$. By (2.3), we see that

$$\int_{-\pi}^{\pi} |\sigma(\theta, n)|^2 d\theta \to 0 \tag{5.1}$$

as $|n| \to \infty$. That the condition (5.1) is not enough for $T_\sigma : L^2(\mathbb{S}^1) \to L^2(\mathbb{S}^1)$ to be compact can be illustrated by the following example.

Example 5.1. Let σ be the function on $\mathbb{S}^1 \times \mathbb{Z}$ defined by

$$\sigma(\theta, n) = \begin{cases} \frac{1}{\ln n} e^{-in\theta}, & n > 1, \\ 0, & n \leq 1, \end{cases}$$

for all θ in $[-\pi, \pi]$. If we let f be the function on $\mathbb{S}^1$ defined by

$$f(\theta) = \sum_{n=1}^{\infty} \frac{1}{n} e^{in\theta}, \quad \theta \in [-\pi, \pi],$$

then $f \in L^2(\mathbb{S}^1)$. But

$$(T_\sigma f)(\theta) = \sum_{n=2}^{\infty} \frac{1}{n \ln n}, \quad \theta \in [-\pi, \pi].$$

So, $T_\sigma f$ is not even in $L^2(\mathbb{S}^1)$.

The following theorem gives a sufficient condition for the L^2-compactness of pseudo-differential operators on $L^2(\mathbb{S}^1)$.

Theorem 5.2. *Let σ be a measurable function on $\mathbb{S}^1 \times \mathbb{Z}$. Suppose that we can find a function w in $L^1(\mathbb{Z})$ and a function C on $\mathbb{Z}$ such that*

$$\lim_{|n| \to \infty} C(n) = 0$$

and

$$|\hat{\sigma}(m, n)| \leq C(n)|w(m)|, \quad m, n \in \mathbb{Z}. \tag{5.2}$$

Then $T_\sigma : L^2(\mathbb{S}^1) \to L^2(\mathbb{S}^1)$ is compact.

Proof. For all positive integers N, we define the function σ_N on $\mathbb{S}^1 \times \mathbb{Z}$ by

$$\sigma_N(\theta, n) = \begin{cases} \sigma(\theta, n), & |n| \leq N, \\ 0, & |n| > N, \end{cases}$$

for all θ in $[-\pi, \pi]$ and n in $\mathbb{Z}$. Then for $N = 1, 2, \ldots$,

$$\sum_{n\in\mathbb{Z}} \int_{-\pi}^{\pi} |\sigma_N(\theta, n)|^2 d\theta = \sum_{n=-N}^{N} \int_{-\pi}^{\pi} |\sigma(\theta, n)|^2 d\theta < \infty$$

and hence by Theorem 3.1, $T_{\sigma_N} : L^2(\mathbb{S}^1) \to L^2(\mathbb{S}^1)$ is a Hilbert–Schmidt operator. Let $\tau_N = \sigma - \sigma_N$. Then by the definition of σ_N, we have

$$\tau_N(\theta, n) = \begin{cases} 0, & |n| \leq N, \\ \sigma(\theta, n), & |n| > N, \end{cases}$$

for all θ in $[-\pi, \pi]$. Let ε be a given positive number. Then there exists a positive integer N_0 such that

$$|C(n)| < \varepsilon$$

whenever $|n| > N_0$. So, for $N \geq N_0$,

$$\|(T_{\sigma_N} - T_\sigma)f\|^2_{L^2(\mathbb{S}^1)} = \int_{-\pi}^{\pi} |(T_{\tau_N} f)(\theta)|^2 d\theta = \int_{-\pi}^{\pi} \left| \sum_{n\in\mathbb{Z}} \tau_N(n, \theta) \hat{f}(n) e^{in\theta} \right|^2 d\theta.$$

Now, we use the same argument in the derivation of (4.4) to get

$$\begin{aligned} \|(T_{\sigma_N} - T_\sigma)f\|^2_{L^2(\mathbb{S}^1)} &\leq 2\pi \sum_{k\in\mathbb{Z}} \left(\sum_{n\in\mathbb{Z}} |\hat{\tau}(k-n, n)||\hat{f}(n)| \right)^2 \\ &= 2\pi \sum_{k\in\mathbb{Z}} \left(\sum_{|n|>N} |\hat{\sigma}(k-n, n)||\hat{f}(n)| \right)^2. \end{aligned}$$

Using (5.2), we get

$$\begin{aligned} \|(T_{\sigma_N} - T_\sigma)f\|^2_{L^2(\mathbb{S}^1)} &\leq 2\pi \sum_{k\in\mathbb{Z}} \left(\sum_{|n|>N} C(n)|w(k-n)|\,|\hat{f}(n)| \right)^2 \\ &\leq 2\pi\varepsilon^2 \sum_{k\in\mathbb{Z}} \left(\sum_{n\in\mathbb{Z}} |w(k-n)|\,|\hat{f}(n)| \right)^2. \end{aligned}$$

Using the same argument in the the derivation of (4.5), we have

$$\|(T_{\sigma_N} - T_\sigma)f\|^2_{L^2(\mathbb{S}^1)} \leq \varepsilon^2 B \|f\|^2_{L^2(\mathbb{S}^1)},$$

where $B = \|w\|^2_{L^1(\mathbb{Z})}$. Thus, for $N \geq N_0$,

$$\|T_\sigma - T_{\sigma_N}\|_* \leq \sqrt{B}\varepsilon.$$

In other words, T_σ is the limit in norm of the sequence $\{T_{\sigma_N}\}_{N=1}^{\infty}$ and so must be compact. □

6. L^p-boundedness

The following lemma gives sufficient conditions on a function σ on $\mathbb{Z}$ to ensure that the corresponding pseudo-differential operator $T_\sigma : L^p(\mathbb{S}^1) \to L^p(\mathbb{S}^1)$, $1 < p < \infty$, is a bounded linear operator. Pseudo-differential operators T_σ, where σ is a function on $\mathbb{Z}$ only, are often known as Fourier multipliers.

Lemma 6.1. *Let σ be a measurable function on $\mathbb{Z}$ and let k be the smallest integer greater than $\frac{1}{2}$. Suppose that there exists a positive constant C such that*

$$|(\partial^j \sigma)(n)| \leq C\langle n\rangle^{-j}, \quad n \in \mathbb{Z},$$

for $0 \leq j \leq k$, where ∂^j is the difference operator given by

$$(\partial^j \sigma)(n) = \sum_{l=0}^{j} (-1)^{j-l} \binom{j}{l} \sigma(n+l), \quad n \in \mathbb{Z},$$

and

$$\langle n\rangle = (1 + |n|^2)^{1/2}, \quad n \in \mathbb{Z}.$$

Then for $1 < p < \infty$, $T_\sigma : L^p(\mathbb{S}^1) \to L^p(\mathbb{S}^1)$ is a bounded linear operator and there exists a positive constant B, depending on p only, such that

$$\|T_\sigma f\|_{L^p(\mathbb{S}^1)} \leq BC\|f\|_{L^p(\mathbb{S}^1)}, \quad f \in L^p(\mathbb{S}^1). \tag{6.1}$$

Remark 6.2. The condition on the number k of "derivatives" in Lemma 6.1 can best be understood in the context of the multi-dimensional torus in which k should be the smallest integer greater than $d/2$, where d is the dimension of the torus.

The proof of Lemma 6.1 entails the use of the Littlewood–Paley theory in Fourier series in, *e.g.*, Chapter XV of the book [14] by Zygmund. See, in particular, Theorem 4.14 in Chapter XV of [14] in this connection. Extensions of Lemma 6.1 to the context of compact Lie groups are attributed to Weiss [11, 12] and the Littlewood–Paley theory for compact Lie groups can be found in Stein [9]. Analogs of Lemma 6.1 for $\mathbb{R}^n$ can be found in the works of Hörmander [4] and Stein [8].

The following theorem gives sufficient conditions for the L^p-boundedness of pseudo-differential operators on $\mathbb{S}^1$. The ideas for the result and its proof come from Theorem 10.7 in the book [13] by Wong.

Theorem 6.3. *Let σ be a measurable function on $\mathbb{S}^1 \times \mathbb{Z}$ and let k be the smallest integer greater than $\frac{1}{2}$. Suppose that we can find a positive constant C and a function w in $L^1(\mathbb{Z})$ such that*

$$|(\partial_n^j \hat{\sigma})(m,n)| \leq C|w(m)|\langle n\rangle^{-j}, \quad m,n \in \mathbb{Z}, \tag{6.2}$$

for $0 \leq j \leq k$, where ∂_n^j is the partial difference operator with respect to the variable n in $\mathbb{Z}$. Then for $1 < p < \infty$, $T_\sigma : L^p(\mathbb{S}^1) \to L^p(\mathbb{S}^1)$ is a bounded linear operator. Moreover, there exists a positive constant B depending only on p such that

$$\|T_\sigma\|_* \leq BC\|w\|_{L^1(\mathbb{Z})}\|f\|_{L^p(\mathbb{S}^1)},$$

where $\| \, \|_*$ here denotes the norm in the Banach algebra of all bounded linear operators on $L^p(\mathbb{S}^1)$.

Proof. Let $f \in L^p(\mathbb{S}^1)$. Then by Fubini's theorem,

$$\begin{aligned}(T_\sigma f)(\theta) &= \sum_{n\in\mathbb{Z}} \hat{f}(n)\sigma(\theta,n)e^{in\theta} \\ &= \sum_{n\in\mathbb{Z}} \hat{f}(n)\left\{\sum_{m\in\mathbb{Z}} \hat{\sigma}(m,n)e^{im\theta}\right\}e^{in\theta} \\ &= \sum_{m\in\mathbb{Z}} e^{im\theta}\left\{\sum_{n\in\mathbb{Z}} \hat{\sigma}(m,n)\hat{f}(n)e^{in\theta}\right\} \\ &= \sum_{m\in\mathbb{Z}} e^{im\theta}(T_{\sigma_m} f)(\theta)\end{aligned}$$

for all θ in $[-\pi,\pi]$, where

$$\sigma_m(n) = \hat{\sigma}(m,n) = \frac{1}{2\pi}\int_{-\pi}^{\pi} e^{-im\theta}\sigma(\theta,n)\,d\theta, \quad m,n\in\mathbb{Z}.$$

By (6.2),

$$|(\partial^j\sigma_m)(n)| = |(\partial_n^j\hat{\sigma})(m,n)| \le C|w(m)|\langle n\rangle^{-j}, \quad m,n\in\mathbb{Z},$$

for $0 \le j \le k$. Therefore by Lemma 6.1, there exists a positive constant B depending only on p such that

$$\|T_{\sigma_m} f\|_{L^p(\mathbb{S}^1)} \le BC|w(m)|\,\|f\|_{L^p(\mathbb{S}^1)}, \quad m\in\mathbb{Z}. \tag{6.3}$$

Then by (6.3) and Minkowski's inequality, we get

$$\begin{aligned}\|T_\sigma f\|_{L^p(\mathbb{S}^1)} &= \left\{\int_{-\pi}^{\pi}\left|\sum_{m\in\mathbb{Z}} e^{im\theta}(T_{\sigma_m} f)(\theta)\right|^p d\theta\right\}^{\frac{1}{p}} \\ &\le \sum_{m\in\mathbb{Z}}\left\{\int_{-\pi}^{\pi}|(T_{\sigma_m} f)(\theta)|^p d\theta\right\}^{\frac{1}{p}} \\ &= \sum_{m\in\mathbb{Z}} \|T_{\sigma_m} f\|_{L^p(\mathbb{S}^1)} \\ &\le BC\|w\|_{L^1(\mathbb{Z})}\|f\|_{L^p(\mathbb{S}^1)},\end{aligned}$$

and this completes the proof. □

Acknowledgment

We are grateful to Professor M.S. Agranovich for very useful comments on the results and the presentation of this paper, and to Professor Ville Turunen for pointing out the preprint [6] to M.W. Wong at the conference on pseudo-differential operators held at Växjö University in Sweden on June 23–27, 2008.

References

[1] M.S. Agranovich, Spectral properties of elliptic pseudodifferential operators on a closed curve, (Russian) *Funktsional. Anal. i Prilozhen.* **13** (1979), 54–56.

[2] M.S. Agranovich, Elliptic pseudodifferential operators on a closed curve, (Russian) *Trudy Moskov. Mat. Obshch.* **47** (1984), 22–67, 246.

[3] B.A. Amosov, On the theory of pseudodifferential operators on the circle, (Russian) *Uspekhi Mat. Nauk.* **43** (1988), 169–170; Translation in *Russian Math. Surveys* **43** (1988), 197–198.

[4] L. Hörmander, Estimates for translation invariant linear operators in L^p spaces, *Acta Math.* **104** (1960), 93–139.

[5] M. Ruzhansky and V. Turunen, On the Fourier analysis of operators on the torus, in *Modern Trends in Pseudo-Differential Operators*, Editors: J. Toft, M.W. Wong and H. Zhu, Birkhäuser, 2007, 87–105.

[6] M. Ruzhansky and V. Turunen, Quantization of pseudo-differential operators on the torus, preprint, arXiv:0805.2892.

[7] J. Saranen and G. Vainikko, *Periodic Integral and Pseudodifferential Equations with Numerical Approximation*, Springer-Verlag, 2002.

[8] E.M. Stein, *Singular Integrals and Differentiability Properties of Functions*, Princeton University Press, 1970.

[9] E.M. Stein, *Topics in Harmonic Analysis Related to Littlewood–Paley Theory*, Third Printing with Corrections, Princeton University Press, 1985.

[10] E.M. Stein and G. Weiss, *Introduction to Fourier Analysis on Euclidean Spaces*, Princeton University Press, 1971.

[11] N.J. Weiss, Multipliers on compact Lie groups, *Proc. Nat. Acad. Sci. U.S.A.* **68** (1971), 930–931.

[12] N.J. Weiss, L^p estimates for bi-invariant operators on compact Lie groups, *Amer. J. Math.* **94** (1972), 103–118.

[13] M.W. Wong, *An Introduction to Pseudo-Differential Operators*, Second Edition, World Scientific, 1999.

[14] A. Zygmund, *Trigonometric Series,* Third Edition, Volumes I & II Combined, Cambridge University Press, 2002.

Shahla Molahajloo and M.W. Wong
Department of Mathematics and Statistics
York University
4700 Keele Street
Toronto
Ontario M3J 1P3, Canada
e-mail: smollaha@mathstat.yorku.ca
mwwong@mathstat.yorku.ca

Operator Theory:
Advances and Applications, Vol. 189, 307–322

On Pseudo-Differential Operators on Group SU(2)

Michael Ruzhansky and Ville Turunen

Abstract. In this paper we will outline elements of the global calculus of pseudo-differential operators on the group SU(2). This is a part of a more general approach to pseudo-differential operators on compact Lie groups that will appear in [17].

Mathematics Subject Classification (2000). Primary 35S05; Secondary 22E30.

Keywords. Pseudo-differential operator, Lie groups, SU(2).

1. Introduction

The paper is devoted to outline a global approach to pseudo-differential operators on matrix group SU(2), without resorting to local charts. This can be done by presenting functions on the group by Fourier series obtained from the representations of the group. Due to non-commutativity, the Fourier coefficients become matrices of varying dimension. A pseudo-differential operator can be presented as a convolution operator-valued mapping on the group. The corresponding Fourier coefficient matrices provide a natural global symbol of the pseudo-differential operator. Consequently, a global calculus and full symbols of pseudo-differential operators can be obtained on the 3-dimensional sphere, with further geometric implications.

Let us first recall and fix the notation for the standard notions of the Fourier analysis. In view of our applications, in the formulae below we will emphasize the fact that the dual group $\widehat{\mathbb{R}^n}$ of $\mathbb{R}^n$ is isomorphic to $\mathbb{R}^n$, thus keeping the notation of $\widehat{\mathbb{R}^n}$. The Fourier transform

$$(f \mapsto \widehat{f}) : \mathcal{S}(\mathbb{R}^n) \to \mathcal{S}(\widehat{\mathbb{R}^n})$$

is defined by

$$\widehat{f}(\xi) = \int_{\mathbb{R}^n} f(x)\ \mathrm{e}^{-\mathrm{i}2\pi x\cdot\xi}\ \mathrm{d}x,$$

The first author was supported by a Royal Society grant and by the EPSRC Grant EP/E062873/01.

and the Fourier inversion formula is

$$f(x) = \int_{\widehat{\mathbb{R}^n}} \widehat{f}(\xi)\, \mathrm{e}^{\mathrm{i}2\pi x\cdot\xi}\, \mathrm{d}\xi.$$

Operator $A : \mathcal{S}(\mathbb{R}^n) \to \mathcal{S}(\mathbb{R}^n)$ is a pseudo-differential operator of order $m \in \mathbb{R}$ if

$$(Af)(x) = \int_{\widehat{\mathbb{R}^n}} \sigma_A(x,\xi)\, \widehat{f}(\xi)\, \mathrm{e}^{\mathrm{i}2\pi x\cdot\xi}\, \mathrm{d}\xi,$$

where the symbol $\sigma_A \in C^\infty(\mathbb{R}^n \times \widehat{\mathbb{R}^n})$ of A satisfies symbolic inequalities

$$\left|\partial_\xi^\alpha \partial_x^\beta \sigma_A(x,\xi)\right| \leq C_{A\alpha\beta m}\, \langle\xi\rangle^{m-|\alpha|},$$

for all multi-indices α and β and all $x \in \mathbb{R}^n$ and $\xi \in \widehat{\mathbb{R}^n}$ (see, e.g., Kohn and Nirenberg [11], Hörmander [9]). In this case we write $A \in \Psi^m(\mathbb{R}^n)$ and $\sigma_A \in S^m(\mathbb{R}^n)$.

On a compact manifold M without boundary the space $\Psi^m(M)$ of pseudo-differential operators can be defined via localisations. One disadvantage of this definition is that it often destroys the geometric and other global structures of M. The main problem is that there is no canonical way to write the phase of a pseudo-differential operator when working in local coordinates. However, if one fixes a connection on the manifold, one can define the phase in a global way in terms of the connection (see, e.g., Widom [32], Safarov [18], Sharafutdinov [21]).

If G is a Lie group, we may construct another global representation of pseudo-differential operators on G and their calculus by relying on the globally defined group structure. Moreover, such analysis can be extended to symmetric spaces, where we have a transitive action $G \times M \to M$ of a Lie group G on a manifold M. In this way we obtain the calculus on M as a "shadow" from the calculus on G. For example, for $M = \mathbb{S}^{n-1}$ we can take $G = \mathrm{SO}(n)$, and for the complex sphere $M = \{x \in \mathbb{C}^n : \|x\|_{\mathbb{C}^n} = 1\}$ we can take $G = \mathrm{SU}(n)$. A version of pseudo-differential operators on SU(2) for functions with finite Fourier series (i.e., for trigonometric polynomials) was analysed in [8].

The important starting point for our analysis is the global Fourier analysis on the Lie group G. Two important ingredients that are needed on G are

(a_G) the Fourier transform;

(b_G) multiplication and its relation to the Fourier transform.

Analogies of these operations on $\mathbb{R}^n$ are

($\mathrm{a}_{\mathbb{R}^n}$) $x \mapsto \mathrm{e}^{\mathrm{i}x\cdot\xi}$ $(\xi \in \mathbb{R}^n)$ and the Fourier integral;

($\mathrm{b}_{\mathbb{R}^n}$) multiplication and the Fourier transform related by

$$\widehat{(-\mathrm{i}x)^\alpha f} = \partial_\xi^\alpha \widehat{f}.$$

On torus $\mathbb{T}^n = (\mathbb{R}/2\pi\mathbb{Z})^n$ we have the following analogies:

($\mathrm{a}_{\mathbb{T}^n}$) $x \mapsto \mathrm{e}^{\mathrm{i}x\cdot\xi}$ $(\xi \in \mathbb{Z}^n)$ and the Fourier series;

($\mathrm{b}_{\mathbb{T}^n}$) multiplication and the Fourier transform are related by

$$\widehat{x^{(\alpha)} f} = \triangle_\xi^\alpha \widehat{f},$$

where for $n = 1$, we define $x^{(\alpha)} = \left(\mathrm{e}^{-\mathrm{i}x} - 1\right)^\alpha$, and similarly in higher dimensions.

The relation between the toroidal and Euclidean representations of pseudo-differential operators in one dimension (i.e., on the circle $\mathbb{T}^1$) was given by Agranovich ([1, 2, 3]), who showed that $A \in \Psi^m(\mathbb{T}^1)$ if and only if it can be written as

$$(Af)(x) = \sum_{\xi \in \mathbb{Z}^1} \sigma_A(x,\xi)\ \widehat{f}(\xi)\ \mathrm{e}^{\mathrm{i}x\cdot\xi},$$

where the toroidal symbol $\sigma_A \in C^\infty(\mathbb{T}^1 \times \mathbb{Z}^1)$ of A satisfies

$$\left|\triangle_\xi^\alpha \partial_x^\beta \sigma_A(x,\xi)\right| \leq C_{A\alpha\beta m}\ \langle\xi\rangle^{m-|\alpha|},$$

and where the difference operator $\triangle_\xi$ is defined by

$$(\triangle_\xi \sigma)(\xi) = \sigma(\xi+1) - \sigma(\xi).$$

For $\Psi^m(\mathbb{T}^n)$ the natural analogy holds with partial difference operators $\triangle_\xi^\alpha$. This result has been extended to a more comprehensive analysis of pseudo-differential operators and the corresponding toroidal representations of Fourier integral operators in all dimensions in [14] and [15]. We also mention the related paper by Amosov [4] on pseudo-differential operators on the circle, and more general characterisations of pseudo-differential operators on the torus by McLean [13]. We note that such discrete analysis has applications for numerical mathematics (e.g., [19], [20], [30]).

2. Calculus on compact Lie groups G

A global description and calculus of pseudo-differential operators is possible on general compact Lie groups G and homogeneous spaces of the type G/K with $K \cong \mathbb{T}^n$. The main ingredients of this analysis are

(a_G) the Fourier transform based on the usual Haar measure μ_G on G together with irreducible unitary representations and corresponding Fourier series;

(b_G) multiplication and the Fourier series related by the exponential mapping, using Taylor polynomials in deformed exponential coordinates;

(c_G) global pseudo-differential operator calculus exploiting the Fourier series.

Let us now outline these elements in more detail.

(a_G) Fourier transform

Let $\mathbb{C}^{c\times d}$ denote the space of complex matrices with c rows and d columns, and let $\mathrm{U}(d) \subset \mathbb{C}^{d\times d}$ be the set of d-dimensional unitary matrices. Let $\mathcal{D}(G) = C^\infty(G)$ denote the test function space. The unitary dual $\widehat{G}$ consists of equivalence classes $[\xi]$ of irreducible unitary representations ξ of G. In the sequel, from each class $[\xi] \in \widehat{G}$, we choose one representative $\xi : G \to \mathrm{U}(\dim(\xi))$, so that the unitary matrices $\xi(x) = \left(\xi_{mn}(x)\right)_{m,n}$ define $\dim(\xi)^2$ functions $\xi_{mn} \in C^\infty(G)$. Fourier coefficient $\widehat{f}(\xi) \in \mathbb{C}^{\dim(\xi)\times\dim(\xi)}$ of $f \in \mathcal{D}(G)$ is defined by

$$\widehat{f}(\xi) = \int_G f(x)\ \xi(x)\ \mathrm{d}\mu_G(x) = \left(\int_G f(x)\ \xi_{mn}(x)\ \mathrm{d}\mu_G(x)\right)_{m,n},$$

so that the Fourier series becomes

$$f(x) = \sum_{[\xi]\in\widehat{G}} \dim(\xi)\ \mathrm{Tr}\left(\widehat{f}(\xi)\ \xi^*(x)\right),$$

where $\xi^*(x) = \xi(x)^*$, and Tr is the trace functional (i.e., the sum of the diagonal elements in a matrix). We can also notice the useful relation $\widehat{f*g}(\xi) = \widehat{f}(\xi)\ \widehat{g}(\xi)$. If $A : \mathcal{D}(G) \to \mathcal{D}(G)$ is a linear continuous map, then it can be written as

$$Af(x) = \sum_{[\xi]\in\widehat{G}} \dim(\xi)\ \mathrm{Tr}\left(\sigma_A(x,\xi)\ \widehat{f}(\xi)\ \xi^*(x)\right),$$

where the (matrix) symbol of A is defined as the mapping $(x,\xi) \mapsto \sigma_A(x,\xi) \in \mathbb{C}^{\dim(\xi)\times\dim(\xi)}$, where

$$\sigma_A(x,\xi) = \xi(x)\ (A(\xi^*))(x) = \left(\sum_k \xi_{mk}(x)\ (A\overline{\xi_{nk}})(x)\right)_{m,n}.$$

(b$_G$) Multiplication and the Fourier series

The Taylor expansion of a smooth function $f : G \to \mathbb{C}$ nearby the neutral element $I \in G$ becomes

$$f(x) \approx \sum_{|\alpha|\le N} \frac{1}{\alpha!}\ x^{(\alpha)}\ (\partial_x^{(\alpha)} f)(I),$$

with $x^{(\alpha)} \approx x^\alpha$ in, e.g., the exponential coordinates. We can then define the difference operators by the relation

$$\widehat{x^{(\alpha)} f}(\xi) =: \triangle_\xi^\alpha \widehat{f}(\xi).$$

(c$_G$) Pseudo-differential operators

The following theorem relates pseudo-differential operators with their globally defined symbols:

Theorem 2.1 ([27]). *$A \in \mathcal{L}(\mathcal{D}(G))$ belongs to $\Psi^m(G)$ if and only if*

$$\sigma_A \in S^m(G) = \bigcap_{k=0}^{\infty} S_k^m(G).$$

In this theorem, the symbol classes $S_k^m(G)$ are defined in the following recursive way:

Definition 2.2. Symbol $\sigma_A \in S_0^m(G)$ if and only if

$$\left\|\triangle_\xi^\alpha \partial_x^\beta \sigma_A(x,\xi)\right\| \le C_{A\alpha\beta m}\ \langle\xi\rangle^{m-|\alpha|},$$

where the norm is the usual operator norm, see Section 5 for details. Here, the weight is

$$\langle\xi\rangle = (1-\lambda_\xi)^{1/2},$$

where $\lambda_\xi \leq 0$ is the eigenvalue of the bi-invariant Laplacian $\Delta : C^\infty(G) \to C^\infty(G)$ on the eigenspace spanned by the matrix element functions $\xi_{mn} \in C^\infty(G)$. Furthermore, $\sigma_A \in S^m_{k+1}(G)$ if and only if

$$\begin{aligned}\sigma_A &\in S^m_k(G),\\ \sigma_{\partial_j}\sigma_A - \sigma_A\sigma_{\partial_j} &\in S^m_k(G),\\ \left(\triangle^\gamma_\xi \sigma_{\partial_j}\right)\sigma_A &\in S^{m+1-|\gamma|}_k(G),\\ \left(\triangle^\gamma_\xi \sigma_A\right)\sigma_{\partial_j} &\in S^{m+1-|\gamma|}_k(G),\end{aligned}$$

where $|\gamma| > 0$ and $1 \leq j \leq \dim(G)$. Then we define

$$S^m(G) = \bigcap_{k=0}^{\infty} S^m_k(G).$$

We note that the proof of Theorem 2.1 is based on the commutator characterisation of pseudo-differential operators (see, e.g., [5], [7], [6], [26]; see also [25] for a different characterisation on the sphere, studying smoothness of operator-valued mappings). It is well known that if $A \in \Psi^{m_A}$ and $B \in \Psi^{m_B}$ then $AB, BA \in \Psi^{m_A+m_B}$, but the commutator satisfies

$$[A, B] = AB - BA \in \Psi^{m_A+m_B-1}.$$

The commutator characterisation uses this property to characterise pseudo-differential operators:

Theorem 2.3. *$A \in \Psi^m(M)$ if and only if*

$$[D_{k+1}, [D_k, \ldots [D_1, A] \cdots]] \in \mathcal{L}(H^m(M), H^0(M))$$

for every sequence of smooth vector fields D_k on M, where $H^m(M)$ is the Sobolev space of order $m \in \mathbb{R}$ on M.

Details of these constructions with applications to the analysis on Lie groups will appear in [17]. We also note that this applies also to non-invariant pseudo-differential operators (compared to, e.g., [22], [23]).

3. Calculus on $G = \mathrm{SU}(2)$

Let us now explain the constructions of the previous section in more detail in the case of the matrix group SU(2). In this case the special unitary group of complex rotations

$$G = \mathrm{SU}(2) = \{x \in \mathbb{C}^{2\times 2} : \ x^* = x^{-1}, \ \det(x) = 1\}$$

acts transitively on the unit sphere in $\mathbb{C}^2$, so the analysis also applies for pseudo-differential operator on spheres in $\mathbb{C}^2$. The group SU(2) is especially important since $G \cong \mathbb{S}^3$, the 3-sphere with the quaternionic structure. In particular, we obtain the global calculus of pseudo-differential operators on $\mathbb{S}^3$, which allows further extensions to general closed simply connected 3-manifolds by the pullback by the Ricci flow (details of this will appear in [16]).

Let us define basic rotations $\omega_j(t) \in G$ $(j = 1, 2, 3)$ by

$$\begin{pmatrix} \cos\frac{t}{2} & \mathrm{i}\sin\frac{t}{2} \\ \mathrm{i}\sin\frac{t}{2} & \cos\frac{t}{2} \end{pmatrix}, \quad \begin{pmatrix} \cos\frac{t}{2} & -\sin\frac{t}{2} \\ \sin\frac{t}{2} & \cos\frac{t}{2} \end{pmatrix}, \quad \begin{pmatrix} \mathrm{e}^{\mathrm{i}t/2} & 0 \\ 0 & \mathrm{e}^{-\mathrm{i}t/2} \end{pmatrix}.$$

The Euler angles ϕ, θ, ψ of $x \in G$ satisfy

$$x = x(\phi, \theta, \psi) := \omega_3(\phi)\ \omega_2(\theta)\ \omega_3(\psi),$$

where $-\pi < \phi \leq \pi$, $0 \leq \theta \leq \pi$, $-2\pi < \psi \leq 2\pi$. The Euler angles define local coordinates for G whenever $0 < \theta < \pi$ (with natural interpretation when $\phi = \pi$ or $\psi = 2\pi$). Let us now define the principal ingredients for the analysis.

($\mathbf{a}_G$) Fourier transform

The Haar integral on $G = \mathrm{SU}(2)$ in the Euler angles is given by

$$\begin{aligned} f \mapsto & \int_G f\ \mathrm{d}\mu_G \\ &= \frac{1}{16\pi^2} \int_{-2\pi}^{2\pi} \int_0^{\pi} \int_{-\pi}^{\pi} f(x)\ \sin(\theta)\ \mathrm{d}\phi\ \mathrm{d}\theta\ \mathrm{d}\psi. \end{aligned}$$

The unitary dual of $G = \mathrm{SU}(2)$ is

$$\widehat{G} \cong \left\{ t^l \in \mathrm{Hom}(G, \mathrm{U}(2l+1)) :\ l \in \frac{1}{2}\mathbb{N} \right\},$$

where $\mathrm{U}(d) \subset \mathbb{C}^{d\times d}$ is the unitary matrix group of dimension d, and functions $t^l_{mn} \in C^\infty(G)$ are products of exponentials and Legendre functions.

The Fourier transform $f \mapsto \widehat{f}$ on $G = \mathrm{SU}(2)$ becomes

$$\widehat{f}(l) = \widehat{f}(t^l) := \int_G f(x)\ t^l(x)\ \mathrm{d}\mu_G(x),$$

with the inverse Fourier transform given by

$$f = \sum_{l \in \mathbb{N}/2} (2l+1)\ \mathrm{Tr}\left(\widehat{f}(l)\ (t^l)^* \right).$$

For details of these formulae see, e.g., Vilenkin [31] and Zelobenko [33]. We also note that the Fourier transform $\widehat{f}(\xi)$ of $f \in \mathcal{D}(G)$ is defined for every $\xi \in [\xi] \in \widehat{G}$, i.e., $\widehat{f}(\xi) : V_\xi \to V_\xi$ is linear on the representation vector space V_ξ of the irreducible representation $\xi : G \to \mathrm{Aut}(V_\xi)$ of G. If $\xi \sim \eta$ (i.e., $[\xi] = [\eta] \in \widehat{G}$) then $\widehat{f}(\xi)$ and $\widehat{f}(\eta)$ are related by intertwining.

($\mathbf{b}_G$) Taylor polynomials

Lie group $G = \mathrm{SU}(2)$ has Lie algebra $\mathfrak{g} = \mathrm{su}(2) = \{X(z) :\ z \in \mathbb{R}^3\}$, where

$$X(z) = \frac{1}{2} \begin{pmatrix} \mathrm{i}z_3 & \mathrm{i}z_1 - z_2 \\ \mathrm{i}z_1 + z_2 & -\mathrm{i}z_3 \end{pmatrix}.$$

If $\{e_j\}_{j=1}^3$ is the standard basis of $\mathbb{R}^3$ and $X_j = X(e_j)$ (Pauli matrices) then we have commutator relations:

$$[X_1, X_2] = X_3, \quad [X_2, X_3] = X_1, \quad [X_3, X_1] = X_2.$$

Now, let us describe the vector fields in the basis of the Lie algebra. For $Y \in \mathrm{su}(2)$, let

$$A_Y : C^\infty(\mathrm{SU}(2)) \to C^\infty(\mathrm{SU}(2))$$

be the left-invariant differential operator defined by

$$A_Y f(x) = \frac{\mathrm{d}}{\mathrm{d}t} f(x \exp(tY))\Big|_{t=0}.$$

Let us denote $A_j := A_{X_j}$. For a multi-index $\beta \in \mathbb{N}^3$, define the left-invariant differential operator

$$\partial^\beta := A_1^{\beta_1} A_2^{\beta_2} A_3^{\beta_3}.$$

Let us also define creation, annihilation, and neutral operators by the following formulae:

$$\begin{aligned}\partial_+ &:= \mathrm{i}A_1 - A_2,\\ \partial_- &:= \mathrm{i}A_1 + A_2,\\ \partial_0 &:= \mathrm{i}A_3.\end{aligned}$$

In Euler angles these operators can be expressed as

$$\begin{aligned}\partial_+ &= \mathrm{e}^{-\mathrm{i}\psi}\left(\mathrm{i}\,\frac{\partial}{\partial\theta} - \frac{1}{\sin(\theta)}\,\frac{\partial}{\partial\phi} + \frac{\cos(\theta)}{\sin(\theta)}\,\frac{\partial}{\partial\psi}\right),\\ \partial_- &= \mathrm{e}^{\mathrm{i}\psi}\left(\mathrm{i}\,\frac{\partial}{\partial\theta} + \frac{1}{\sin(\theta)}\,\frac{\partial}{\partial\phi} - \frac{\cos(\theta)}{\sin(\theta)}\,\frac{\partial}{\partial\psi}\right),\\ \partial_0 &= \mathrm{i}\frac{\partial}{\partial\psi}.\end{aligned}$$

Note that the invariant Laplace operator on G is given by

$$\Delta = A_1^2 + A_2^2 + A_3^2.$$

Let $t := \|z\|_{\mathbb{R}^3}/2$. The Rodrigues formula states that we have

$$\exp(X(z)) = \begin{pmatrix}1 & 0\\ 0 & 1\end{pmatrix}\cos(t) + X(z)\frac{\sin(t)}{t}.$$

Define now functions $x_{ij} \in C^\infty(G)$ by

$$x = \begin{pmatrix}x_{11}(x) & x_{12}(x)\\ x_{21}(x) & x_{22}(x)\end{pmatrix} \in G.$$

These functions (or their linear combinations) will play the role of coordinates. We can now define the Taylor polynomial $x^{(\alpha)} \in C^\infty(G)$ ($\alpha \in \mathbb{N}^3$) by setting

$$x^{(\alpha)} = (x_{12})^{\alpha_1}\ (x_{21})^{\alpha_2}\ (x_{11} - x_{22})^{\alpha_3}.$$

The corresponding difference operators $\triangle_\xi^\alpha$ are then defined by

$$\triangle_\xi^\alpha \widehat{f}(\xi) := \widehat{x^{(\alpha)} f}(\xi).$$

In the Taylor expansion formula, Taylor polynomials are multiplied by the partial differential operators of corresponding orders. For this purpose, we define

$$\partial^{(\alpha)} = \sum_{|\beta| \leq |\alpha|} c_{\alpha\beta}\, \partial^\beta,$$

so that we have

$$\partial^{(\alpha)} x^{(\beta)}|_{x=I} = \alpha!\, \delta_{\alpha,\beta}.$$

Now the Taylor expansion for a function $f \in \mathcal{D}(G)$ at the neutral element $I \in G$ becomes

$$f(x) \sim \sum_{\alpha \geq 0} \frac{1}{\alpha!}\, x^{(\alpha)}\, (\partial^{(\alpha)} f)(I).$$

Let us finally briefly recall the irreducible unitary representations for the group SU(2). Group SU(2) acts naturally on space $\mathbb{C}[z_1, z_2]$ by the formula

$$T : \mathrm{SU}(2) \to \mathrm{GL}(\mathbb{C}[z_1, z_2]), \quad (T(x)f)(z) := f(zx).$$

Invariant subspaces $V_l \subset \mathbb{C}[z_1, z_2]$ of homogeneous $2l$-degree polynomials, where $l \in \mathbb{N}/2$, give the restriction

$$T_l : \mathrm{SU}(2) \to \mathrm{GL}(V_l), \quad (T_l(x)f)(z) = f(zx).$$

If T_∞ is an irreducible representation of SU(2) on a vector space V_∞ then T_∞ is equivalent to T_l, where $\dim(V_\infty) = 2l+1$. Operators $T_l(x)$ can be viewed as matrices and now we will recall the matrix elements of T_l, which can be written as

$$t^l_{mn}(\phi, \theta, \psi) = \mathrm{e}^{-\mathrm{i}(m\phi + n\phi)}\, P^l_{mn}(\cos(\theta)),$$

where

$$P^l_{mn}(x) = c^l_{mn} \frac{(1-x)^{(n-m)/2}}{(1+x)^{(m+n)/2}} \left(\frac{\mathrm{d}}{\mathrm{d}x}\right)^{l-m} \left[(1-x)^{l-n}(1+x)^{l+n}\right],$$

with constants

$$c^l_{mn} = 2^{-l} \frac{(-1)^{l-n}\, \mathrm{i}^{n-m}}{\sqrt{(l-n)!\,(l+n)!}} \sqrt{\frac{(l+m)!}{(l-m)!}}.$$

Multiplication of two elements $t^{l'}_{m'n'}$ and t^l_{mn} can be expressed as

$$t^{l'}_{m'n'}\, t^l_{mn} = \sum_{k=|l-l'|}^{l+l'} C^{ll'(l+k)}_{m'm(m'+m)}\, C^{ll'(l+k)}_{n'n(n'+n)}\, t^{l+k}_{(m'+m)(n'+n)},$$

with the Clebsch–Gordan coefficients $C^{ll'(l+k)}_{m'm(m'+m)}$. In particular, let us denote

$$\begin{pmatrix} t_{--} & t_{-+} \\ t_{+-} & t_{++} \end{pmatrix} = t^{1/2} = \begin{pmatrix} t^{1/2}_{-1/2,-1/2} & t^{1/2}_{-1/2,+1/2} \\ t^{1/2}_{+1/2,-1/2} & t^{1/2}_{+1/2,+1/2} \end{pmatrix}$$

and $x^{\pm} := x \pm 1/2$ for $x \in \mathbb{R}$. Then we have the following multiplication formulae

$$\begin{aligned}
(2l+1)t^l_{mn}t_{--} &= \sqrt{(l-m+1)(l-n+1)}\; t^{l^+}_{m^-n^-} - \sqrt{(l+m)(l+n)}\; t^{l^-}_{m^-n^-},\\
(2l+1)t^l_{mn}t_{++} &= \sqrt{(l+m+1)(l+n+1)}\; t^{l^+}_{m^+n^+} - \sqrt{(l-m)(l-n)}\; t^{l^-}_{m^+n^+},\\
(2l+1)t^l_{mn}t_{-+} &= \sqrt{(l-m+1)(l+n+1)}\; t^{l^+}_{m^-n^+} - \sqrt{(l+m)(l-n)}\; t^{l^-}_{m^-n^+},\\
(2l+1)t^l_{mn}t_{+-} &= \sqrt{(l+m+1)(l-n+1)}\; t^{l^+}_{m^+n^-} - \sqrt{(l-m)(l+n)}\; t^{l^-}_{m^+n^-}.
\end{aligned}$$

4. Differences for symbols on SU(2)

Let us define functions

$$q_+ := t_{-+}, \tag{4.1}$$

$$q_- := t_{+-}, \tag{4.2}$$

$$q_0 := t_{++} - t_{--}. \tag{4.3}$$

The key property of these functions is that each $q_+, q_-, q_0 \in \mathcal{D}(G)$ vanishes at the neutral element $I \in G$.

Let $\sigma = \widehat{s}$, which means that $\sigma : \mathcal{D}(G) \to \mathcal{D}(G)$ is the left convolution operator with convolution kernel $s \in \mathcal{D}'(G)$:

$$\sigma f = s * f.$$

Let us define "difference operators" $\triangle_+, \triangle_-, \triangle_0$ acting on symbols σ by

$$\triangle_+\sigma := \widehat{q_+\, s}, \tag{4.4}$$

$$\triangle_-\sigma := \widehat{q_-\, s}, \tag{4.5}$$

$$\triangle_0\sigma := \widehat{q_0\, s}. \tag{4.6}$$

These difference operators will play the role of derivatives in the definition and calculus of pseudo-differential operators. In general, they depend on the structure of the dual space $\widehat{G}$ of the group G. For example, in the Euclidean space we have $\widehat{\mathbb{R}^n} \cong \mathbb{R}^n$, so their analogues are simply the differential operators ∂^α_ξ. For the torus, we have $\widehat{\mathbb{T}^n} \cong \mathbb{Z}^n$, and the corresponding difference operators $\triangle^\alpha_\xi$ have been analysed, e.g., in [14] and [26].

We can now define the conditions on globally defined pseudo-differential symbols on SU(2) in terms of these difference operators. First let us show that a pseudo-differential operator can be globally parametrised by a matrix symbol $\sigma_A(x,\xi)$, where $x \in \mathrm{SU}(2)$, ξ is a discrete variable corresponding to the dual of SU(2), and the size of the matrix $\sigma_A(x,\xi)$ increases with ξ. In fact, using the structure of the dual of SU(2), we can identify ξ with a half-integer in $\mathbb{N}/2$, so that $\sigma_A(x,\xi)$ becomes a matrix of the size $(2\xi+1)\times(2\xi+1)$. In general, the matrix symbol of a continuous linear operator $A : \mathcal{D}(G) \to \mathcal{D}(G)$ will be written as $\sigma_A(x,\xi)$, and it is defined for every $x \in G$ and $\xi \in [\xi] \in \widehat{G}$. If $[\xi] = [\eta]$ then $\sigma_A(x,\xi)$ and $\sigma_A(x,\eta)$ are related and allow a geometric interpretation. We refer to [16] for further details, and will only give here the description based on the identification of the dual of SU(2) with the set $\mathbb{N}/2$ of half-integers.

By using the Fourier series, we can conclude that a continuous linear operator $A : \mathcal{D}(G) \to \mathcal{D}(G)$ belongs to $\Psi^m(G)$ if and only if it can be written as

$$\begin{aligned}(Af)(x) &= \sum_{\xi\in\mathbb{N}/2} (2\xi+1)\ \mathrm{Tr}\left(\sigma_A(x,\xi)\ \widehat{f}(\xi)\ t^\xi(x)^*\right) \\ &= \sum_{\xi\in\mathbb{N}/2} (2\xi+1) \sum_{m=-\xi}^{\xi}\sum_{n=-\xi}^{\xi} \overline{t^\xi_{mn}(x)} \left(\sum_{k=-\xi}^{\xi} \sigma_A(x,\xi)_{mk}\ \widehat{f}(\xi)_{kn}\right),\end{aligned}$$

where the symbol

$$\sigma_A(x,\xi) = t^\xi(x)\ (A(t^\xi)^*)(x) = \left(\sum_{k=-\xi}^{\xi} t^\xi_{mk}(x)\ (A\overline{t^\xi_{nk}})(x)\right)_{m,n} = \left(\sigma_A(x,\xi)_{mn}\right)_{m,n}$$

satisfies symbol inequalities that will be described below. With a natural embedding $\mathcal{D}(\mathbb{S}^2) \hookrightarrow \mathcal{D}(G)$, we also note that every pseudo-differential operator $B \in \Psi^m(\mathbb{S}^2)$ has an extension $A \in \Psi^m(G)$ such that $A|_{\mathcal{D}(\mathbb{S}^2)} = B$, see [27].

Let us consider the following example: for $\xi \in \mathbb{N}/2$ and a $(2\xi+1)\times(2\xi+1)$ matrix A, let A_{mn} denote the matrix element on the mth row and nth column, where $|m|,|n| \le \xi$ such that $m,n \in \{\pm\xi, \pm(\xi-1), \pm(\xi-2), \cdots\}$. The symbols of the first-order partial differential operators $\partial_+, \partial_-, \partial_0$ are then given by

$$\sigma_{\partial_+}(x,\xi)_{mn} = \begin{cases} -\sqrt{(\xi-n)(\xi+n+1)}, & \text{if } m = n+1, \\ 0, & \text{otherwise.}\end{cases}$$

$$\sigma_{\partial_-}(x,\xi)_{mn} = \begin{cases} -\sqrt{(\xi+n)(\xi-n+1)}, & \text{if } m = n-1, \\ 0, & \text{otherwise.}\end{cases}$$

$$\sigma_{\partial_0}(x,\xi)_{mn} = \begin{cases} n, & \text{if } m = n, \\ 0, & \text{otherwise,}\end{cases}$$

$$\sigma_I(x,\xi)_{mn} = \begin{cases} 1, & \text{if } m = n, \\ 0, & \text{otherwise,}\end{cases}$$

where $I = (f \mapsto f) : \mathcal{D}(G) \to \mathcal{D}(G)$ is the identity operator. Moreover, we have the following properties:

$$\triangle_+\sigma_\Delta = -\sigma_{\partial_-}, \quad \triangle_-\sigma_\Delta = -\sigma_{\partial_+}, \quad \triangle_0\sigma_\Delta = -\sigma_{\partial_0},$$

where Δ is the invariant Laplacian of G,

$$\sigma_I = \triangle_+\sigma_{\partial_+} = \triangle_-\sigma_{\partial_-} = \triangle_0\sigma_{\partial_0},$$

and

$$0 = \triangle_+\sigma_I = \triangle_-\sigma_I = \triangle_0\sigma_I.$$

Furthermore, $\triangle_\alpha\sigma_{\partial_\eta}(x,\xi) = 0$, if $\alpha,\eta \in \{+,-,0\}$ are such that $\alpha \neq \eta$.

5. Symbol inequalities

In this section we study inequalities describing symbols of pseudo-differential operators on $G = \mathrm{SU}(2)$. For a vector $v = (v_j)_{j=1}^n \in \mathbb{C}^d$ we use the Euclidean norm $\|v\|_{\mathbb{C}^d}$ given by

$$\|v\|_{\mathbb{C}^d}^2 := \sum_{j=1}^{d} |v_j|^2,$$

and for a matrix $M \in \mathbb{C}^{d\times d}$ the corresponding operator norm

$$\|M\| = \|M\|_{\mathbb{C}^{d\times d}} := \sup\left\{\|Mv\|_{\mathbb{C}^d} :\ v \in \mathbb{C}^d,\ \|v\|_{\mathbb{C}^d} \leq 1\right\}.$$

We notice that we have the operator norm equality

$$\|f \mapsto f * a\|_{\mathcal{L}(L^2(G))} = \sup\left\{\|\widehat{a}(\xi)\| :\ \xi \in \widehat{G}\right\}.$$

In order to describe matrix symbols of standard pseudo-differential operators on SU(2), we need to check conditions of Theorem 2.1. First, $\sigma_A \in S_0^m(G)$, i.e., it satisfies

$$\left\|\triangle_\xi^\alpha \partial_x^\beta \sigma_A(x,\xi)\right\| \leq C_{\alpha\beta}\ \langle\xi\rangle^{m-|\alpha|} \tag{5.1}$$

for all multi-indices α, β, all $x \in \mathrm{SU}(2)$, and all $\xi \in \mathbb{N}/2$. On $G = \mathrm{SU}(2)$, the weight $\langle\xi\rangle$ becomes

$$\langle\xi\rangle = \left(1 + \xi + \xi^2\right)^{1/2}.$$

We note that condition $(\triangle_\xi^\gamma \sigma_{\partial_j})\sigma_A \in S_k^{m+1-|\gamma|}(G)$ is automatically satisfied due to the example presented at the end of the previous section.

We will now show that in order to satisfy conditions of Theorem 2.1, matrices $\sigma_A(x,\xi)$ must have a certain "rapid off-diagonal decay" property. By the example in the previous section, $\sigma_{\partial_0}(x,\xi)_{ij} = i\ \delta_{ij}$, so that

$$\begin{aligned}
&[\sigma_{\partial_0}(x,\xi), \sigma_A(x,\xi)]_{ij} \\
&= \sum_k \left(\sigma_{\partial_0}(x,\xi)_{ik}\ \sigma_A(x,\xi)_{kj} - \sigma_A(x,\xi)_{ik}\ \sigma_{\partial_0}(x,\xi)_{kj}\right) \\
&= \sum_k \left(i\ \delta_{ik}\ \sigma_A(x,\xi)_{kj} - \sigma_A(x,\xi)_{ik}\ k\ \delta_{kj}\right) \\
&= (i-j)\ \sigma_A(x,\xi)_{ij}.
\end{aligned}$$

Let us iterate such a commutator $p \in \mathbb{Z}^+$ times, to obtain that the symbol

$$(x,\xi) \mapsto \left((i-j)^p \sigma_A(x,\xi)_{ij}\right)_{i,j}$$

must belong to $S^m(G) \subset S_0^m(G)$, regardless of p. Thus the pseudo-differential symbol σ_A must have "*rapid off-diagonal decay*":

$$\sup_{\xi,i,j}\ \langle\xi\rangle^{-m} \langle i-j\rangle^p\ |\sigma_A(x,\xi)_{ij}| < \infty$$

– notice that in a matrix $\left(a_{ij}\right)_{i,j}$, the distance from the location of the matrix element a_{ij} to the diagonal is $|i-j|$.

What about the commutators $[\sigma_{\partial_+}, \sigma_A]$ and $[\sigma_{\partial_-}, \sigma_A]$? Due to the symmetries of the symbols of $\partial_\pm$, studying the case of σ_{∂_+} will be enough. By the example in the previous section,

$$\sigma_{\partial_+}(x,\xi)_{ij} = -\sqrt{(\xi-i)(\xi+i+1)}\;\delta_{i+1,j},$$

so that

$$\begin{aligned}
&[\sigma_{\partial_+}(x,\xi), \sigma_A(x,\xi)]_{ij}\\
&= \sum_k \left(\sigma_{\partial_+}(x,\xi)_{ik}\; \sigma_A(x,\xi)_{kj} - \sigma_A(x,\xi)_{ik}\; \sigma_{\partial_+}(x,\xi)_{kj}\right)\\
&= \sum_k \Big(-\sqrt{(\xi-i)(\xi+i+1)}\;\delta_{i+1,k}\;\sigma_A(x,\xi)_{kj}\\
&\qquad\qquad +\sigma_A(x,\xi)_{ik}\;\sqrt{(\xi-k)(\xi+k+1)}\;\delta_{k+1,j}\Big)\\
&= -\sqrt{(\xi-m)(\xi+m+1)}\;\sigma_A(x,\xi)_{i+1,j}\\
&\quad + \sqrt{(\xi-n+1)(\xi+n)}\;\sigma_A(x,\xi)_{i,j-1}.
\end{aligned}$$

At first sight, this commutator may look baffling: is this some sort of weighted difference operator acting on σ_A "along the diagonal"? We have to understand this first-order commutator condition properly before we may consider higher-order iterated commutators.

So how to understand the behaviour of $[\sigma_{\partial_+}(x,\xi), \sigma_A(x,\xi)]$? We claim that this is a variant of the "rapid off-diagonal decay" property. Let us explain what we mean by this. Briefly: $\partial_0 = \mathrm{i}A_3$, so the A_3-symbol commutator condition means "rapid off-diagonal decay". "Badly behaving" operators ∂_+, ∂_- are linear combinations of left-invariant vector fields A_1, A_2, which are conjugates to A_3 (and hence essentially similar in behaviour); then the idea is to exploit the diffeomorphism invariance of $\Psi^m(G)$.

We know from the local theory that the pseudo-differential operator class $\Psi^m(G)$ is diffeomorphism-invariant: if $\phi : G \to G$ is a diffeomorphism and $A \in \Psi^m(G)$, then also $A_\phi \in \Psi^m(G)$, where

$$A_\phi f = A(f\circ\phi)\circ\phi^{-1}.$$

Especially, let us consider the inner automorphisms

$$\phi_u = (x \mapsto u^{-1}xu) : G \to G,$$

where $u \in G$. Such $\phi_u : G \to G$ is a diffeomorphism that maps one-parametric subgroups to one-parametric subgroups. If $\xi \in [\xi] \in \widehat{G}$ then

$$\begin{aligned}
\widehat{f\circ\phi_u}(\xi) &= \int_G f(u^{-1}xu)\;\xi(x)\;\mathrm{d}\mu_G(x) = \int_G f(x)\;\xi(uxu^{-1})\;\mathrm{d}\mu_G(x)\\
&= \xi(u)\int_G f(x)\;\xi(x)\;\mathrm{d}\mu_G(x)\;\xi(u)^* = \xi(u)\;\widehat{f}(\xi)\;\xi(u)^*,
\end{aligned}$$

which is similar to $\widehat{f}(\xi)$ by transform $\xi(u)$. Recall the left-invariant vector fields A_1, A_2, A_3 out of which operators $\partial_+, \partial_-, \partial_0$ were defined as linear combinations. Thus we are interested in commutators $[\sigma_{A_j}(x,\xi), \sigma_A(x,\xi)]$, and we already know the good behaviour of the case A_3, since $\partial_0 = \mathrm{i}A_3$. Now,

$$\sigma_{A_1}(x,\xi) = t^\xi(v_1)^* \ \sigma_{A_3}(x,\xi) \ t^\xi(v_1),$$
$$\sigma_{A_2}(x,\xi) = t^\xi(v_2)^* \ \sigma_{A_3}(x,\xi) \ t^\xi(v_2),$$

where $v_j \in G$ and $t^\xi : \mathrm{SU}(2) \to \mathrm{U}(2\xi+1)$ is the usual irreducible unitary matrix representation. Hence

$$\begin{aligned}[\sigma_{A_1}(x,\xi), \sigma_A(x,\xi)] &= \left[t^\xi(v_1)^* \sigma_{A_3}(x,\xi) t^\xi(v_1), \sigma_A(x,\xi)\right] \\ &= t^\xi(v_1)^* \left[\sigma_{A_3}(x,\xi), t^\xi(v_1)\sigma_A(x,\xi)t^\xi(v_1)^*\right] t^\xi(v_1).\end{aligned}$$

We may assume that here $A \in \Psi^m(G)$ is left-invariant, i.e., $A = (f \mapsto f*a)$, so that $\sigma_A(x,\xi) = \widehat{a}(\xi)$ for every $\xi \in \mathbb{N}/2$ (in these symbol commutator properties, we may always fix $x \in G$, even if A is not left-invariant). Now $((x,\xi) \mapsto \widehat{a}(\xi)) \in S^m(G)$, and due to the diffeomorphism-invariance of $\Psi^m(G)$, also

$$(x,\xi) \mapsto t^\xi(v_1) \ \widehat{a}(\xi) \ t^\xi(v_1)^*$$

must belong to $S^m(G)$, thus "decaying rapidly off-diagonal". Thereby we already understand the behaviour of the commutator

$$\left[\sigma_{A_3}(x,\xi), t^\xi(v_1)\sigma_A(x,\xi)t^\xi(v_1)^*\right],$$

and this is similar (by transform $t^\xi(v_1)^*$, see above) to

$$[\sigma_{A_1}(x,\xi), \sigma_A(x,\xi)].$$

The same kind of reasoning applies to the commutator

$$[\sigma_{A_2}(x,\xi), \sigma_A(x,\xi)].$$

Thus the symbol commutator condition is satisfied if we require that

$$(x,\xi) \mapsto t^\xi(u) \ \sigma_A(x,\xi) \ t^\xi(u)^*$$

"decays rapidly off-diagonal" for every $u \in G$.

6. Global calculus

Properties $(\mathrm{a}_{\mathbb{R}^n})$ and $(\mathrm{b}_{\mathbb{R}^n})$ can be used to construct the global calculus of the introduced matrix symbols. For example, we have the well-known composition formula for pseudo-differential operators A and B in $\mathbb{R}^n$:

$$\sigma_{AB}(x,\xi) \sim \sum_{\alpha \geq 0} \frac{1}{\alpha!} \ (\partial_\xi^\alpha \sigma_A)(x,\xi) \ (\partial_x^\alpha \sigma_B)(x,\xi),$$

see, e.g., [11], [10], [12]. On the torus $\mathbb{T}^n$, properties ($\mathrm{a}_{\mathbb{T}^n}$) and ($\mathrm{b}_{\mathbb{T}^n}$) allow to describe the composition of pseudo-differential operators in terms of toroidal symbols in an analogous way:

$$\sigma_{AB}(x,\xi) \sim \sum_{\alpha\geq 0} \frac{1}{\alpha!} \, (\triangle_\xi^\alpha \sigma_A)(x,\xi) \, (\partial_x^{(\alpha)} \sigma_B)(x,\xi).$$

This formula was obtained in [29] and was further extended to compositions with toroidal Fourier integral operators in [14]. In the case of SU(2), we can use properties (a_G) and (b_G) to show that it also has the form

$$\sigma_{AB}(x,\xi) \sim \sum_{\alpha\geq 0} \frac{1}{\alpha!} \, (\triangle_\xi^\alpha \sigma_A)(x,\xi) \, (\partial_x^{(\alpha)} \sigma_B)(x,\xi).$$

In fact, this formula can be also viewed in the following way. Let

$$Q^\alpha \sigma_A(x) = \pi(y \mapsto \check{q}_\alpha(y) \; s_A(x)(y))$$

with $\check{q}_\alpha \in C^\infty(G)$ satisfying $\check{q}_\alpha(\exp(z)) = z^\alpha$ for z in a small neighbourhood of the origin on the Lie algebra $\mathfrak{g}$ of G. Then we have the following elements of the calculus:

Theorem 6.1 (M.E. Taylor [24]**).**

$$\sigma_{A^*}(x) \sim \sum_{\alpha\geq 0} \frac{1}{\alpha!} \, Q^\alpha \, \partial_x^\alpha \, \sigma_A(x)^*,$$

$$\sigma_{AB}(x) \sim \sum_{\alpha\geq 0} \frac{1}{\alpha!} \, (Q^\alpha \sigma_A(x)) \, \partial_x^\alpha \sigma_B(x).$$

Theorem 6.2. *Let $A \in \Psi^m(G)$ with expansion $A \sim \sum_{j=0}^\infty A_j$, $A_j \in \Psi^{m-j}(G)$, and assume that $x \mapsto \sigma_{A_0}(x)^{-1}$ is a symbol of $B_0 \in \Psi^{-m+1-\varepsilon}(G)$ (for some $\varepsilon > 0$). Then A is elliptic with a parametrix B, $\sigma_B \sim \sum_{k=0}^\infty \sigma_{B_k}$, where*

$$\sigma_{B_0}(x) = \sigma_{A_0}(x)^{-1},$$

$$\sigma_{B_N}(x) = -\sigma_{B_0}(x) \sum_{k=0}^{N-1} \sum_{j=0}^{N-k} \sum_{\gamma: j+k+|\gamma|=N} (Q^\gamma \sigma_{A_j}(x)) \, \partial_x^\gamma \sigma_{B_k}(x).$$

For further details of these results we refer to [24], [27], [28], and for the relation to introduced symbol classes we refer to [17].

Acknowledgements

The authors wish to express their gratitude to Matania Ben-Artzi for valuable comments leading to an improvement of the manuscript.

References

[1] M.S. Agranovich, Spectral properties of elliptic pseudodifferential operators on a closed curve, *Funktsional. Anal. i Prilozhen* **13** (1979), 54–56 (in Russian).

[2] M.S. Agranovich, On elliptic pseudodifferential operators on a closed curve, *Trans. Moscow Math. Soc.* **47** (1985), 23–74.

[3] M.S. Agranovich, Elliptic operators on closed manifolds, *Itogi Nauki i Tehniki, Ser. Sovrem. Probl. Mat. Fund. Napravl,* **63** (1990), 5–129 (in Russian). (English translation in Encyclopaedia Math. Sci. **63** (1994), 1–130.)

[4] B.A. Amosov, *On the theory of pseudodifferential operators on the circle, Uspekhi Mat. Nauk* **43** (1988), 169–170 (in Russian). (English translation in Russian Math. Surveys **43** (1988), 197–198.)

[5] R. Beals, Characterization of pseudo-differential operators and applications, *Duke Math. J.* **44** (1977), 45–57.

[6] H.O. Cordes, On pseudodifferential operators and smoothness of special Lie group representations, *Manuscripta Math.* **28** (1979), 51–69.

[7] J. Dunau, Fonctions d'un opérateur elliptique sur une variété compacte, *J. Math. Pures et Appl.* **56** (1977), 367–391.

[8] F. Geshwind and N.H. Katz, Pseudodifferential operators on SU(2), *J. Fourier Anal. Appl.* **3** (1997), 193–205.

[9] L. Hörmander, Pseudo-differential operators, *Comm. Pure Appl. Math.* **18** (1965), 501–517.

[10] L. Hörmander, *The Analysis of Linear Partial Differential Operators III-IV,* Berlin: Springer-Verlag, 1985.

[11] J.J. Kohn and L. Nirenberg, On the algebra of pseudo-differential operators, *Comm. Pure Appl. Math.* **18** (1965), 269–305.

[12] H. Kumano-go, *Pseudo-Differential Operators,* MIT Press, Cambridge, Mass.-London, 1981.

[13] W. McLean, Local and global description of periodic pseudodifferential operators, *Math. Nachr.* **150** (1991), 151–161.

[14] M. Ruzhansky and V. Turunen, On the Fourier analysis of operators on the torus, in *Modern Trends in Pseudo-Differential Operators,* Editors: J. Toft, M.W. Wong and H. Zhu, Operator Theory: Advances and Applications **172**, Birkhäuser, 2007, 87–105,

[15] M. Ruzhansky and V. Turunen, On pseudo-differential and Fourier integral operators on the torus, in preparation.

[16] M. Ruzhansky and V. Turunen, Pseudo-differential operators on spheres, in preparation.

[17] M. Ruzhansky and V. Turunen, *Pseudo-Differential Operators and Symmetries,* Birkhäuser, to appear.

[18] Yu. Safarov, Pseudodifferential operators and linear connections, *Proc. London Math. Soc.* **74** (1997), 379–416.

[19] J. Saranen and G. Vainikko, *Periodic Integral and Pseudodifferential Equations with Numerical Approximation,* Springer Monographs in Mathematics, Springer-Verlag, Berlin, 2002.

[20] J. Saranen and W.L. Wendland, The Fourier series representation of pseudodifferential operators on closed curves, *Complex Variables, Theory Appl.* **8** (1987), 55–64.

[21] V.A. Sharafutdinov, Geometric symbol calculus for pseudodifferential operators I *Siberian Adv. Math.* **15** (2005), 81–125. [Translation of Mat. Tr. **7** (2004), 159–206].

[22] H. Stetkær, Invariant pseudo-differential operators, *Math. Scand.* **28** (1971), 105–123.

[23] R. Strichartz, Invariant pseudo-differential operators on a Lie group, *Ann. Scuola Norm. Sup. Pisa* **26** (1972), 587–611.

[24] M.E. Taylor, *Noncommutative Microlocal Analysis.* Mem. Amer. Math. Soc. **52** (1984).

[25] M.E. Taylor, Beals–Cordes-type characterizations of pseudodifferential operators, *Proc. Amer. Math. Soc.* **125** (1997), 1711–1716.

[26] V. Turunen, Commutator characterization of periodic pseudodifferential operators, *Z. Anal. Anw.* **19** (2000), 95–108.

[27] V. Turunen, *Pseudodifferential Calculus on Compact Lie Groups and Homogeneous Spaces,* Ph.D. Thesis, Helsinki University of Technology, 2001.

[28] V. Turunen, Pseudodifferential calculus on the 2-sphere, *Proc. Estonian Acad. Sci. Phys. Math.* **53**(3) (2004), 156–164.

[29] V. Turunen and G. Vainikko, On symbol analysis of periodic pseudodifferential operators, *Z. Anal. Anw.* **17** (1998), 9–22.

[30] G. Vainikko, An integral operator representation of classical periodic pseudodifferential operators, *Z. Anal. Anw.* **18** (1999), 687–699.

[31] N. Vilenkin, *Special Functions and the Theory of Group Representations,* Trans. Math. Monographs **22**, Amer. Math. Soc., 1968.

[32] H. Widom, A complete symbolic calculus for pseudodifferential operators, *Bull. Sci. Math.* **104** (1980), 19–63.

[33] D. Zelobenko, *Compact Lie groups and their Representations.* Trans. Math. Monographs **40**, Amer. Math. Soc., 1973.

Michael Ruzhansky
Department of Mathematics
Imperial College London
United Kingdom
e-mail: `m.ruzhansky@imperial.ac.uk`

Ville Turunen
Institute of Mathematics
Helsinki University of Technology
Finland
e-mail: `ville.turunen@tkk.fi`

Operator Theory:
Advances and Applications, Vol. 189, 323–332

Sampling and Pseudo-Differential Operators

Alip Mohammed and M.W. Wong

Abstract. Discrete formulas for pseudo-differential operators based on the Shannon–Whittaker sampling formula and the Poisson summation formula are given.

Mathematics Subject Classification (2000). Primary 42B10, 47G30.

Keywords. Fourier transforms, pseudo-differential operators, Shannon–Whittaker sampling formula, Poisson summation formula.

1. Introduction

Pseudo-differential operators, first introduced by Kohn and Nirenberg [18] and immediately modified by Hörmander [15], have been used with great success in partial differential equations. The usefulness in quantum physics has been manifested via the representation of pseudo-differential operators envisaged by Weyl [26], and studied in detail in [16] and [27] among others. Fuelled by wavelets, pseudo-differential operators have been studied extensively for the past two decades as localization operators [5, 6, 9, 10, 11, 12, 29] and wavelet multipliers [17, 19, 30, 31, 32].

Applications of pseudo-differential operators to sciences and engineering entail computations. Prompted by the success stories of the fast Fourier transform [7, 8, 25], it is reasonable to seek computational techniques based on the Fourier transform. Among the various manifestations of the pseudo-differential operators in the literature hitherto described, the original representation in [18], studied extensively in the book [28], is most intimately and explicitly tied to the Fourier transform. The aim of this paper is to give discrete formulas for these pseudo-differential operators based on sampling related to the Fourier transform. They

This research has been supported by the Natural Sciences and Engineering Research Council of Canada.

are the Shannon–Whittaker sampling formula and the Poisson summation formula. Accounts of these sampling formulas abound and can be found in, *e.g.*, Chapter 8 of [13] and Part C of Chapter 7 in [24].

We first recall in Section 2 the pseudo-differential operators studied exclusively in the paper [18] and the book [28]. Its relations with the Rihaczek distribution in time-frequency analysis [21] recently studied in [3, 4] and [20] are described. The Shannon–Whittaker sampling formula is recalled in Section 3 and used to give a sampling formula for pseudo-differential operators. The Poisson summation formula is recalled in Section 4 and a sampling formula based on it is derived. This is an amazingly simple and beautiful formula and its connections with recent developments in magnetic resonance imaging (MRI) are its *raison d'être*.

2. Pseudo-differential operators

Let $m \in (-\infty, \infty)$. Then we denote by S^m the set of all functions σ in $C^\infty(\mathbb{R}^n \times \mathbb{R}^n)$ such that for all multi-indices α and β, there exists a positive constant $C_{\alpha,\beta}$ for which

$$|(D_x^\alpha D_\xi^\beta \sigma)(x,\xi)| \le C_{\alpha,\beta}(1+|\xi|)^{m-|\beta|}, \quad x,\xi \in \mathbb{R}^n.$$

We call a function in S^m a symbol of order m. Let $\sigma \in S^m$. Then we define the pseudo-differential operator T_σ corresponding to the symbol σ by

$$(T_\sigma \varphi)(x) = (2\pi)^{-n/2} \int_{\mathbb{R}^n} e^{ix\cdot\xi} \sigma(x,\xi) \hat{\varphi}(\xi)\, d\xi, \quad x \in \mathbb{R}^n,$$

for all φ in the Schwartz space $\mathcal{S}(\mathbb{R}^n)$, where the Fourier transform $\hat{\varphi}$ is defined by

$$\hat{\varphi}(\xi) = (2\pi)^{-n/2} \int_{\mathbb{R}^n} e^{-ix\cdot\xi} \varphi(x)\, dx, \quad \xi \in \mathbb{R}^n.$$

It can be proved easily that T_σ maps $\mathcal{S}(\mathbb{R}^n)$ into $\mathcal{S}(\mathbb{R}^n)$ continuously.

Let $\sigma \in \mathcal{S}(\mathbb{R}^n \times \mathbb{R}^n)$. Then for all functions φ and ψ in $\mathcal{S}(\mathbb{R}^n)$,

$$\begin{aligned}
(T_\sigma \varphi, \psi)_{L^2(\mathbb{R}^n)} &= \int_{\mathbb{R}^n} (T_\sigma \varphi)(x) \overline{\psi(x)}\, dx \\
&= (2\pi)^{-n/2} \int_{\mathbb{R}^n} \int_{\mathbb{R}^n} e^{ix\cdot\xi} \sigma(x,\xi) \hat{\varphi}(\xi) \overline{\psi(x)}\, dx\, d\xi \\
&= (2\pi)^{-n/2} \int_{\mathbb{R}^n} \int_{\mathbb{R}^n} \sigma(x,\xi) e^{ix\cdot\xi} \hat{\varphi}(\xi) \overline{\psi(x)}\, dx\, d\xi \\
&= (2\pi)^{-n/2} \int_{\mathbb{R}^n} \int_{\mathbb{R}^n} \sigma(x,\xi) R(\varphi,\psi)(x,\xi)\, dx\, d\xi, \qquad (2.1)
\end{aligned}$$

where $R(\varphi,\psi)$ is the Rihaczek transform of φ and ψ introduced in [20] and is defined by

$$R(\varphi,\psi)(x,\xi) = e^{ix\cdot\xi} \hat{\varphi}(\xi) \overline{\psi(x)}, \quad x,\xi \in \mathbb{R}^n.$$

It is obvious that $R(\varphi,\psi)$ is a function in $\mathcal{S}(\mathbb{R}^n \times \mathbb{R}^n)$ and the formula (2.1) can be used to define pseudo-differential operators with symbols in the space $\mathcal{S}'(\mathbb{R}^n \times \mathbb{R}^n)$

of all tempered distributions on $\mathbb{R}^n \times \mathbb{R}^n$. To wit, let $\sigma \in \mathcal{S}'(\mathbb{R}^n \times \mathbb{R}^n)$. Then the pseudo-differential operator T_σ with symbol σ is defined by

$$(T_\sigma \varphi)(\psi) = (2\pi)^{-n/2}\sigma(R(\varphi, \overline{\psi})), \quad \varphi, \psi \in \mathcal{S}(\mathbb{R}^n).$$

3. The Shannon–Whittaker sampling formula

We begin with the classical Shannon–Whittaker sampling theorem.

Theorem 3.1. *Let $f \in L^2(\mathbb{R})$ be such that*

$$\hat{f}(\xi) = 0, \quad |\xi| > \pi, \qquad \text{and} \qquad \sum_{k=-\infty}^{\infty} |f(k)| < \infty.$$

Then

$$f(x) = \sum_{k=-\infty}^{\infty} f(k) \operatorname{sinc} \pi(x-k), \quad x \in \mathbb{R},$$

where

$$\operatorname{sinc} x = \frac{\sin x}{x}, \quad x \in \mathbb{R},$$

and the convergence of the series is uniform on every compact subset of $\mathbb{R}$.

In signal analysis, a function f in $L^2(\mathbb{R})$ for which the support of $\hat{f}$ is contained in $[-L, L]$ is known as a *bandlimited signal* or, more precisely, an *L-bandlimited signal.* By Theorem 3.1, we have the following immediate result by suitable scaling.

Theorem 3.2. *Let f be an L-bandlimited signal such that*

$$\sum_{k=-\infty}^{\infty} \left| f\left(\frac{k\pi}{L}\right) \right| < \infty.$$

Then

$$f(x) = \sum_{k=-\infty}^{\infty} f\left(\frac{k\pi}{L}\right) \operatorname{sinc}(Lx - k\pi), \quad x \in \mathbb{R},$$

where the convergence of the series is uniform on every compact subset of $\mathbb{R}$.

Remark 3.3. Theorem 3.2 says that an L-bandlimited signal f is determined by its values on the uniformly spaced sample $\left\{\frac{k\pi}{L} : k \in \mathbb{Z}\right\}$. We call the number $\frac{\pi}{L}$ the sample spacing and the number $\frac{L}{\pi}$ the sample rate or the *Nyquist rate.* Sampling at a rate lower than $\frac{L}{\pi}$ is called undersampling and sampling at a rate higher than $\frac{L}{\pi}$ is called oversampling. With oversampling, we can obtain a more rapidly convergent Shannon–Whittaker sampling formula in which the sinc function is replaced by another suitable function. Details are given on pages 279–281 of [13].

Theorem 3.4. *Let $\sigma \in L^2(\mathbb{R} \times \mathbb{R})$. Then for all functions f in $L^2(\mathbb{R})$ such that $\hat{f}(\xi) = 0$ whenever $|\xi| > L$, and $\sum_{k=-\infty}^{\infty} \left|f\left(\frac{k\pi}{L}\right)\right| < \infty$, we get*

$$(T_\sigma f)(x) = (2\pi)^{-1/2} \sum_{k=-\infty}^{\infty} f\left(\frac{k\pi}{L}\right) (\check{\sigma}(x, \cdot) * T_{-k\pi/L} D_L \mathrm{sinc})(x)$$

for almost all x in $\mathbb{R}$, where $\check{\sigma}$ is the inverse Fourier transform of σ with respect to the second variable, $T_{-k\pi/L}$ and D_L are the translation operator and dilation operator defined by

$$(T_{-k\pi/L} h)(x) = h\left(x - \frac{k\pi}{L}\right), \quad x \in \mathbb{R},$$

and

$$(D_L h)(x) = h(Lx), \quad x \in \mathbb{R},$$

for all measurable functions h on $\mathbb{R}$.

Proof. For a fixed x in $\mathbb{R}$, if we denote by M_x the modulation operator defined by

$$(M_x h)(y) = e^{ixy} h(y), \quad y \in \mathbb{R},$$

and by $\hat{\sigma}$ the Fourier transform of σ with respect to the second variable, then we get by Theorem 3.2

$$\begin{aligned}
(T_\sigma f)(x) &= (2\pi)^{-1/2} \int_{-\infty}^{\infty} e^{ix\xi} \sigma(x, \xi) \hat{f}(\xi)\, d\xi \\
&= (2\pi)^{-1/2} \int_{-\infty}^{\infty} (M_x \sigma(x, \cdot))(\xi) \hat{f}(\xi)\, d\xi \\
&= (2\pi)^{-1/2} \int_{-\infty}^{\infty} (T_{-x} \hat{\sigma}(x, \cdot))(y) f(y)\, dy \\
&= (2\pi)^{-1/2} \int_{-\infty}^{\infty} \hat{\sigma}(x, y - x) f(y)\, dy \\
&= (2\pi)^{-1/2} \int_{-\infty}^{\infty} \check{\sigma}(x, x - y) f(y)\, dy \\
&= (2\pi)^{-1/2} \int_{-\infty}^{\infty} \check{\sigma}(x, x - y) \sum_{k=-\infty}^{\infty} f\left(\frac{k\pi}{L}\right) \mathrm{sinc}\,(Ly - k\pi)\, dy \\
&= (2\pi)^{-1/2} \sum_{k=-\infty}^{\infty} f\left(\frac{k\pi}{L}\right) \int_{-\infty}^{\infty} \check{\sigma}(x, x - y)\, \mathrm{sinc}\,(Ly - k\pi)\, dy \\
&= (2\pi)^{-1/2} \sum_{k=-\infty}^{\infty} f\left(\frac{k\pi}{L}\right) (\check{\sigma}(x, \cdot) * T_{-k\pi/L} D_L \mathrm{sinc})(x)
\end{aligned}$$

for almost all x in $\mathbb{R}$, as required. □

Remark 3.5. The interchange of the order of integration and summation in the proof of Theorem 3.4 can be justified. Indeed, using the assumptions that $\sigma \in L^2(\mathbb{R} \times \mathbb{R})$ and

$$\sum_{k=-\infty}^{\infty} \left| f\left(\frac{k\pi}{L}\right) \right| < \infty,$$

we can apply Schwarz' inequality, Plancherel's theorem and the fact that the sinc function is in $L^2(\mathbb{R})$ to obtain

$$\begin{aligned} &\sum_{k=-\infty}^{\infty} \left| f\left(\frac{k\pi}{L}\right) \right| \int_{-\infty}^{\infty} |\check{\sigma}(x, x-y))| \, |\mathrm{sinc}\,(Ly - k\pi)| \, dy \\ &\leq \sum_{k=-\infty}^{\infty} \left| f\left(\frac{k\pi}{L}\right) \right| \left\{ \int_{-\infty}^{\infty} |\check{\sigma}(x,y)|^2 dy \right\}^{1/2} \left\{ \int_{-\infty}^{\infty} |\mathrm{sinc}\, y|^2 dy \right\}^{1/2} L^{-1/2} \\ &< \infty \end{aligned}$$

for almost all x in $\mathbb{R}$. Thus, by Fubini's theorem, the order of integration and summation can be interchanged.

For applications in imaging, we need to extend the Shannon–Whittaker sampling formula and the corresponding sampling formula in Theorem 3.4 to $\mathbb{R}^n$. First, we have the following n-dimensional Shannon–Whittaker sampling formula.

Theorem 3.6. *Let f be an L-bandlimited signal in $\mathbb{R}^n$, i.e.,*

$$\hat{f}(\xi) = 0, \quad \xi \notin \prod_{j=1}^{n} [-L_j, L_j].$$

Suppose that

$$\sum_{\alpha_1, \alpha_2, \ldots, \alpha_n = -\infty}^{\infty} \left| f\left(\frac{\alpha\pi}{L}\right) \right| < \infty,$$

where

$$\frac{\alpha\pi}{L} = \left(\frac{\alpha_1 \pi}{L_1}, \frac{\alpha_2 \pi}{L_2}, \ldots, \frac{\alpha_n \pi}{L_n} \right).$$

Then

$$f(x) = \sum_{\alpha_1, \alpha_2, \ldots, \alpha_n = -\infty}^{\infty} f\left(\frac{\alpha\pi}{L}\right) \prod_{j=1}^{n} \mathrm{sinc}\,(L_j x - \alpha_j \pi), \quad x \in \mathbb{R}^n.$$

An n-dimensional analogue of Theorem 3.4 is then given in the following theorem.

Theorem 3.7. *Let $\sigma \in L^2(\mathbb{R}^n \times \mathbb{R}^n)$. Then for all functions f in $L^2(\mathbb{R}^n)$ such that*

$$\hat{f}(\xi) = 0, \quad \xi \notin \prod_{j=1}^{n} [-L_j, L_j],$$

and

$$\sum_{\alpha_1,\alpha_2,\dots,\alpha_n=-\infty}^{\infty} \left| f\left(\frac{\alpha\pi}{L}\right)\right| < \infty,$$

we get for almost all x in $\mathbb{R}^n$,

$$(T_\sigma f)(x) = (2\pi)^{-n/2} \sum_{\alpha_1,\alpha_2,\dots,\alpha_n=-\infty}^{\infty} f\left(\frac{\alpha\pi}{L}\right) \left(\check{\sigma}(x,\cdot) * \prod_{j=1}^{n} T_{-\alpha_j\pi/L_j} D_{L_j}\mathrm{sinc}\right)(x).$$

4. The Poisson summation formula

In magnetic resonance imaging (MRI) [14], the data of the images are collected in the Fourier domain. In other words, the Fourier transforms of the images are sampled. Thus, the Poisson summation formula is useful in this connection.

Let $f \in L^1(\mathbb{R})$. Then for every positive number L, we define the function f_L on $\mathbb{R}$ by

$$f_L(x) = \sum_{k=-\infty}^{\infty} f(x + 2kL), \quad x \in \mathbb{R}.$$

It is easy to see that f_L is a periodic function and the period is $2L$. Moreover, $f_L \in L^1[-L, L]$.

We can now recall the Poisson summation formula.

Theorem 4.1. *Let $f \in L^1(\mathbb{R})$ be such that*

$$\sum_{k=-\infty}^{\infty} \left|\hat{f}\left(\frac{k\pi}{L}\right)\right| < \infty.$$

Then at each point x of continuity of f_L,

$$f_L(x) = \sum_{k=-\infty}^{\infty} f(x + 2kL) = \sqrt{\frac{\pi}{2}}\frac{1}{L} \sum_{k=-\infty}^{\infty} \hat{f}\left(\frac{k\pi}{L}\right) e^{ik\pi x/L}.$$

It follows from the Poisson summation formula that if f is a function in $L^2(\mathbb{R})$ such that

$$f(x) = 0, \quad |x| > L,$$

then

$$f(x) = \sqrt{\frac{\pi}{2}}\frac{1}{L} \sum_{k=-\infty}^{\infty} \hat{f}\left(\frac{k\pi}{L}\right) e^{ik\pi x/L}, \quad x \in [-L, L],$$

where the convergence of the series is uniform on $[-L, L]$.

Theorem 4.2. *Let $\sigma \in L^2(\mathbb{R} \times \mathbb{R})$ be such that*

$$\check{\sigma}(x,\cdot) \in L^1(\mathbb{R})$$

for almost all x in $\mathbb{R}$. Then for all functions f in $L^2(\mathbb{R})$ such that

$$f(x) = 0, \quad |x| > L,$$

we get

$$(T_\sigma f)(x) = \sqrt{\frac{\pi}{2}} \frac{1}{L} \sum_{k=-\infty}^{\infty} \hat{f}\left(\frac{k\pi}{L}\right) e^{ik\pi x/L} \sigma\left(x, \frac{k\pi}{L}\right)$$

for almost all x in $\mathbb{R}$.

Proof. From the proof of Theorem 3.4, we have

$$(T_\sigma f)(x) = (2\pi)^{-1/2} \int_{-\infty}^{\infty} \check{\sigma}(x, x-y) f(y)\, dy, \quad x \in \mathbb{R}.$$

So, using the Poisson summation formula in Theorem 4.2, we get

$$\begin{aligned}
(T_\sigma f)(x) &= \frac{1}{2L} \sum_{k=-\infty}^{\infty} \hat{f}\left(\frac{k\pi}{L}\right) \int_{-\infty}^{\infty} \check{\sigma}(x, x-y) e^{ik\pi y/L} dy \\
&= \frac{1}{2L} \sum_{k=-\infty}^{\infty} \hat{f}\left(\frac{k\pi}{L}\right) \int_{-\infty}^{\infty} \check{\sigma}(x, z) e^{ik\pi(x-z)/L} dz \\
&= \sqrt{\frac{\pi}{2}} \frac{1}{L} \sum_{k=-\infty}^{\infty} \hat{f}\left(\frac{k\pi}{L}\right) e^{ik\pi x/L} \sigma\left(x, \frac{k\pi}{L}\right)
\end{aligned}$$

for almost all x in $\mathbb{R}$, as asserted. □

For applications in imaging, n-dimensional analogues of the Poisson summation formula and the corresponding sampling formula for pseudo-differential operators can be formulated. First, we have the following n-dimensional Poisson summation formula.

Theorem 4.3. *For positive numbers $L_1, L_2, \ldots, L_n$, let $f \in L^1(\mathbb{R}^n)$ be such that*

$$\sum_{\alpha_1, \alpha_2, \ldots, \alpha_n = -\infty}^{\infty} \left| \hat{f}\left(\frac{\alpha\pi}{L}\right) \right| < \infty.$$

Then at each point x of continuity of f_L, where

$$f_L(y) = \sum_{\alpha_1, \alpha_2, \ldots, \alpha_n = -\infty}^{\infty} f(y_1 + 2\alpha_1 L_1, y_2 + 2\alpha_2 L_2, \ldots, y_n + 2\alpha_n L_n), \quad y \in \mathbb{R}^n,$$

we get

$$f_L(x) = \left(\frac{\pi}{2}\right)^{n/2} \prod_{j=1}^{n} \frac{1}{L_j} \sum_{\alpha_1, \alpha_2, \ldots, \alpha_n = -\infty}^{\infty} \hat{f}\left(\frac{\alpha\pi}{L}\right) \prod_{j=1}^{n} e^{i\alpha_j \pi x/L_j}.$$

The corresponding sampling formula for pseudo-differential operators is given in the following theorem.

Theorem 4.4. *Let $\sigma \in L^2(\mathbb{R}^n \times \mathbb{R}^n)$ be such that*

$$\check{\sigma}(x, \cdot) \in L^1(\mathbb{R}^n)$$

for almost all x in $\mathbb{R}^n$. Then for all functions f in $L^2(\mathbb{R}^n)$ such that

$$f(x) = 0, \quad x \notin \prod_{j=1}^{n} [-L_j, L_j],$$

we get

$$(T_\sigma f)(x) = \left(\frac{\pi}{2}\right)^{n/2} \prod_{j=1}^{n} \frac{1}{L_j} \sum_{\alpha_1,\alpha_2,\ldots,\alpha_n=-\infty}^{\infty} \hat{f}\left(\frac{\alpha\pi}{L}\right) \prod_{j=1}^{n} e^{i\alpha_j \pi x/L_j} \sigma\left(x, \frac{\alpha\pi}{L}\right)$$

for almost all x in $\mathbb{R}^n$.

5. Conclusions

Two sampling formulas for pseudo-differential operators are given in this paper. Fast computation is the motivation for these formulas. The first is based on the Shannon–Whittaker sampling formula and works for bandlimited signals. The second formula entails the use of the Poisson summation formula and is tailored for time-limited signals in, *e.g.*, MRI. The results are presented in the framework of modern mathematical analysis with precise conditions to guarantee convergence and interchange of order of integration in different circumstances. No attempt, however, is made to present the results under more general hypotheses. In the case of the discrete formula for pseudo-differential operators built on the Poisson summation formula, related results can be found in the work [22] on the Fourier analysis of operators on the torus, which in turn can be traced to the original works of Agranovich [1, 2]. The numerical analysis of pseudo-differential operators can be found in [23] among others.

References

[1] M.S. Agranovich, Spectral properties of elliptic pseudodifferential operators on a closed curve, (Russian) *Funktsional. Anal. i Prilozhen.* **13** (1979), 54–56.

[2] M.S. Agranovich, Elliptic pseudodifferential operators on a closed curve, (Russian) *Trudy Moskov. Mat. Obshch.* **47** (1984), 22–67, 246.

[3] P. Boggiatto, G. De Donno and A. Oliaro, Uncertainty principle, positivity and L^p-boundedness for generalized spectrograms, *J. Math. Anal. Appl.* **335** (2007), 93–112.

[4] P. Boggiatto, G. De Donno and A. Oliaro, A unified point of view on time-frequency representations and pseudo-differential operators, in *Pseudo-Differential Operators: Partial Differential Equations and Time-Frequency Analysis*, Fields Institute Communications Series, American Mathematical Society, 2007, 383–399.

[5] P. Boggiatto, A. Oliaro and M.W. Wong, L^p boundedness and compactness of localization operators, *J. Math. Anal. Appl.* **322** (2006), 193–206.

[6] P. Boggiatto and M.W. Wong, Two-wavelet localization operators on $L^p(\mathbb{R}^n)$ for the Weyl–Heisenberg group, *Integral Equations Operator Theory* **49** (2004), 1–10.

[7] R.N. Bracewell, *The Fourier Transform and Its Applications*, Third Edition, McGraw-Hill, 2000.

[8] E.O. Brigham, *The Fast Fourier Transform*, Prentice-Hall, 1974.

[9] E. Cordero and K. Gröchenig, Time-frequency analysis of localization operators, *J. Funct. Anal.* **205** (2003), 107–131.

[10] E. Cordero and L. Rodino, Wick calculus: a time-frequency approach, *Osaka J. Math.* **42** (2005), 43–63.

[11] S. Dang and L. Peng, Reproducing spaces and time-frequency localization operators, *Appl. Anal.* **80** (2001), 431–447.

[12] S. Dang and L. Peng, Reproducing spaces and localization operators, *Acta Math. Sin. (Engl. Ser.)* **20** (2004), 255–260.

[13] C.L. Epstein, *Introduction to the Mathematics of Medical Imaging*, Second Edition, SIAM, 2008.

[14] C.L. Epstein, Introduction to magnetic resonance imaging for mathematicians, *Ann. Inst. Fourier (Grenoble)* **54** (2004), 1697–1716.

[15] L. Hörmander, Pseudo-differential operators and hypoelliptic equations, in *Singular Integrals*, Proc. Symp. Pure Math. Vol. X, American Mathematical Society, 1967, 138–183.

[16] L. Hörmander, The Weyl calculus of pseudo-differential operators, *Comm. Pure Appl. Math.* **32** (1979), 359–433.

[17] Z. He and M.W. Wong, Wavelet multipliers and signals, *J. Austal. Math. Soc. Ser B* **40** (1999), 437–446.

[18] J.J. Kohn and L. Nirenberg, On the algebra of pseudo-differential operators, *Comm. Pure Appl. Math.* **18** (1965), 269–305.

[19] Y. Liu, A. Mohammed and M.W. Wong, Wavelet multipliers on $L^p(\mathbb{R}^n)$, *Proc. Amer. Math. Soc.* **136** (2008), 1009–1018.

[20] A. Mohammed and M.W. Wong, Rihaczek transforms and pseudo-differential operators, in *Pseudo-Differential Operators: Partial Differential Equations and Time-Frequency Analysis*, Fields Institute Communications Series, American Mathematical Society, 2007, 375–382.

[21] A.W. Rihaczek, Signal energy distribution in time and frequency, *IEEE Trans. Inform. Theory* **14** (1968), 369–374.

[22] M. Ruzhansky and V. Turunen, On the Fourier analysis of operators on the torus, in *Modern Trends in Pseudo-Differential Operators*, Birkhäuser, 2007, 87–105.

[23] J. Saranen and G. Vainikko, *Periodic Integral and Pseudodifferential Equations with Numerical Approximation*, Springer-Verlag, 2002.

[24] J.S. Walker, *Fourier Analysis*, Oxford University Press, 1988.

[25] J.S. Walker, *Fast Fourier Transforms*, Second Edition, CRC Press, 1996.

[26] H. Weyl, *The Theory of Groups and Quantum Mechanics*, Dover, 1950.

[27] M.W. Wong, *Weyl Transforms*, Springer-Verlag, 1998.

[28] M.W. Wong, *An Introduction to Pseudo-Differential Operators*, Second Edition, World Scientific, 1999.

[29] M.W. Wong, *Wavelet Transforms and Localization Operators*, Birkhäuser, 2002.

[30] M.W. Wong and Z. Zhang, Trace class norm inequalities for wavelet multipliers, *Bull. London Math. Soc.* **34** (2002), 739–744.

[31] M.W. Wong and H. Zhu, Matrix representations and numerical computations of wavelet multipliers, in *Wavelets, Multiscale Systems and Hypercomplex Analysis*, Birkhäuser, 2006, 173–182.

[32] Z. Zhang, *Localization Operators and Wavelet Multipliers*, Ph.D. Dissertation, York University, 2003.

Alip Mohammed and M.W. Wong
Department of Mathematics and Statistics
York University
4700 Keele Street
Toronto
Ontario M3J 1P3, Canada
e-mail: alipm@mathstat.yorku.ca
mwwong@mathstat.yorku.ca